Springer Water

The book series Springer Water comprises a broad portfolio of multi- and interdisciplinary scientific books, aiming at researchers, students, and everyone interested in water-related science. The series includes peer-reviewed monographs, edited volumes, textbooks, and conference proceedings. Its volumes combine all kinds of water-related research areas, such as: the movement, distribution and quality of freshwater; water resources; the quality and pollution of water and its influence on health; the water industry including drinking water, wastewater, and desalination services and technologies; water history; as well as water management and the governmental, political, developmental, and ethical aspects of water.

Amrit Kumar Jha · Nitish Kumar
Editors

Cadmium Toxicity in Water

Challenges and Solutions

Springer

Editors
Amrit Kumar Jha
Krishi Vigyan Kendra Sahibganj
Birsa Agricultural University
Sahibganj, Jharkhand, India

Nitish Kumar
Department of Biotechnology
Central University of South Bihar
Gaya, Bihar, India

ISSN 2364-6934 ISSN 2364-8198 (electronic)
Springer Water
ISBN 978-3-031-54004-2 ISBN 978-3-031-54005-9 (eBook)
https://doi.org/10.1007/978-3-031-54005-9

This Springer imprint is published by the registered company Springer Nature Switzerland AG
The registered company address is: Gewerbestrasse 11, 6330 Cham, Switzerland

Preface

Cadmium (Cd) is a highly toxic, non-essential, and non-biodegradable heavy metal with half-life of 20 years. It poses a health risk for both humans and animals. It is naturally occurring in the environment as a pollutant that is derived from agricultural and industrial sources. Exposure to cadmium primarily occurs through the ingestion of contaminated food and water. Its tracer amount in the environment, soil, water, and biological systems can pose major issues for all living things, and its bioaccumulation in the food chain is particularly risky for the well-being of people and animals. The existing literature demonstrates that non-biodegradable character and continuous use results in accumulation of cadmium concentration in the environment and causes various ill effects such as neurotoxicity, change in psychological and behavioral development of different organisms. In soil, speciation of cadmium greatly affects its bioavailability and thus its toxicity on plants and microbes. Many plants and bacteria have evolved to develop detoxification mechanisms to counter the toxic effect of cadmium.

The book sheds light on this global environmental issue and proposes solutions to contamination through multi-disciplinary approaches. This book contains three sections. First section describes the different sources and distribution of cadmium in environment. Second section explains the health risks linked to cadmium toxicity. Third section addresses sustainable mitigation strategies and the potential applications of recent biological technology in providing solutions. We provide an overview of the bioremediation treatments promoted by plants (phytoremediation), fungi, or bacteria that could be applied to areas polluted by cadmium. These restoration processes have the advantage of being environmentally friendly and cost-effective solutions that exploit plants to immobilize and extract contaminants from soil and water and fungi and bacteria to degrade them. Phytoremediation is an extensively studied and mature practice, with many in the field applications where numerous plant species have been employed. This book is a valuable resource to students, academics, researchers, and environmental professionals doing field work on cadmium contamination throughout the world.

Gaya, Bihar, India Nitish Kumar
Sahibganj, Jharkhand, India Amrit Kumar Jha

Acknowledgments

First and foremost, we would like to praise and thank God, the almighty, who has granted countless blessing, knowledge, and opportunity to accomplish the work. Thanks to all the authors of the various chapters for their contributions. It had been a bit of a long process from the initial outlines to developing the full chapters and then revising them in light of reviewer's comments. We sincerely acknowledge the author's willingness to go through this process. We also acknowledge the work and knowledge of the members of our review panels, many of which had to be done at short notice. Thanks to all the people at Springer Nature especially Dr. Margaret Deignan and Ms. Saranya Kalidoss with whom we corresponded for their advice and facilitation in the production of this book.

Gaya, Bihar, India — Nitish Kumar
Sahibganj, Jharkhand, India — Amrit Kumar Jha

Contents

About the Editors

Dr. Amrit Kumar Jha (Editor) is Scientist (Soil Science) and Head at Krishi Vigyan Kendra, Sahibganj, Birsa Agricultural University, Ranchi, Jharkhand, India. Dr. Jha completed his doctoral research in subject Soil science and Agricultural Chemistry from Birsa Agricultural University, Ranchi, Jharkhand, India. He has more than 18 years of research experience in the field of soil chemistry, soil fertility, and soil pollution. He has published more than 40 research articles in cadmiuming international and national journals, more than 11 technical bulletins and two books. He has received many awards such as Young Scientist Award conferred by Society of Krishi Vigyan in 2018, Best KVK Scientist Award conferred by Indian Society of Extension Education in 2018, and Best KVK Scientist Award conferred by Society of Krishi Vigyan in 2020.

Dr. Nitish Kumar (Co-editor) is Senior Assistant Professor at the Department of Biotechnology, Central University of South Bihar, Gaya, Bihar, India. Dr. Kumar completed his doctoral research at the Council of Scientific & Industrial Research—Central Salt & Marine Chemicals Research Institute, Bhavnagar, Gujarat, India. He has published more than 70 research articles in cadmiuming international and national journals, more than 20 book chapters and 10 books with Springer and Taylor & Francis. He has a wide area of research experience in the field of Agriculture & Crop Improvement and Microbial & Environmental Biotechnology. Dr. Kumar is a recipient of the Young Scientist Award from the Science and Engineering Research Board (SERB) in 2014. He has received many awards/fellowships/projects from various prestigious government organizations like CSIR, DBT, ICAR and SERB-DST, BRNS-BARC, among others. He is Reviewer for various International Journals and serves as Associate Editor of the journal *Gene* (Elsevier).

Source and Distribution of Cadmium in Environment

Cadmium in Environment—An Overview

Ankush, Ritambhara, Shubham Lamba, Deepika, and Ram Prakash

Abstract Cadmium (Cd), a naturally occurring heavy metal, has garnered significant attention due to its widespread presence in the environment and potential adverse effects on ecosystems and human health. This chapter provides a comprehensive overview of the sources, distribution, fate, and toxicological aspects of cadmium in the environment. Industrial processes, mining activities, and agricultural practices are significant contributors to cadmium contamination in the environment, impacting soil, water, and air quality. Additionally, atmospheric deposition of combustion emissions and urbanization also play crucial roles in the dissemination of cadmium pollutants. These diverse anthropogenic activities can lead to the release of cadmium into the environment through various pathways, posing risks to ecosystems and human health. Cadmium's high mobility is influenced by various factors, including pH, redox state, and ionic strength. Despite this, Cd can endure in solution via water-soluble complexes like $CdCl^{-}$ and $Cd(SO_4)_2{}^{2-}$, and interacts with dissolved organic matter. Overall, cadmium toxicity is a serious health and environmental concern, necessitating vigilant monitoring and precautionary measures to safeguard human health and ecological integrity. This overview aims to aid researchers, policymakers, and environmental practitioners in tackling cadmium contamination and fostering sustainable environmental management.

Keywords Cadmium · Environment · Health · Heavy metal · Remediation · Soil · Toxicity

1 Introduction

Cadmium is a soft, silvery-white metal belonging to Group 12 of the periodic table. With a symbol "Cd" and atomic number 48, it has a relatively low melting point and boiling point, making it suitable for various industrial applications. Cadmium is often considered a transition metal, and its chemical properties resemble those of zinc and

Ankush (✉) · Ritambhara · S. Lamba · Deepika · R. Prakash
Department of Soil Science, CCS HAU, Hisar, Haryana 125004, India
e-mail: ankushdhanda@hau.ac.in

A. K. Jha and N. Kumar (eds.), *Cadmium Toxicity in Water*, Springer Water,
https://doi.org/10.1007/978-3-031-54005-9_1

mercury. In its elemental form, cadmium is relatively rare in nature. However, it is commonly found as an impurity in zinc, lead, and copper ores. This association with other metals has led to the unintentional release of cadmium into the environment during mining, smelting, and refining processes. Cadmium possesses unique toxicological properties, which make it a significant environmental concern. It is highly persistent in the environment and can accumulate in soils, sediments, and organisms, leading to potential biomagnification along the food chain. Cadmium, being non-essential to the body, poses a toxic threat with detrimental effects, particularly on bones. Chronic exposure to cadmium is associated with adverse health effects, especially on the kidneys, respiratory system, and bones, making it a significant public health issue. Moreover, toxic nature of Cd extends to micro-organisms, plants, and animals in the environment, as it persists and resists breakdown into less harmful compounds.

2 Background and Discovery of Cadmium

In 1817, the German chemist ***Friedrich Stromeyer*** isolated cadmium as a byproduct during the analysis of a zinc carbonate ore sample. Stromeyer named the newly discovered element after the Latin word "*Cadmia*" which referred to zinc oxide. Throughout the nineteenth and twentieth centuries, researchers investigated cadmium's properties and applications. It was utilized in various industrial processes, including electroplating, battery production, and pigments.

3 Sources of Cadmium

Cadmium enters the environment through a combination of natural processes and human activities, contributing to its widespread occurrence and potential environmental and health impacts. Different sources of Cd are expressed in Fig. 1.

3.1 Natural Sources

1. *Geological Weathering*: Cadmium is a naturally occurring element found in the Earth's crust, primarily associated with zinc, lead, and copper ores. The process of weathering and erosion of these rocks and minerals gradually releases cadmium into the soil and water, contributing to its presence in the environment. In general, Cd concentrations in sedimentary rocks (0.01–2.6 mg/ kg) are higher than those in igneous rocks (0.07–0.25 mg/kg) or metamorphic rocks (0.11–1.0 mg/kg).
2. *Volcanic Activity*: Volcanic eruptions can emit cadmium-containing volcanic ash and gases into the atmosphere. While the amount of cadmium released

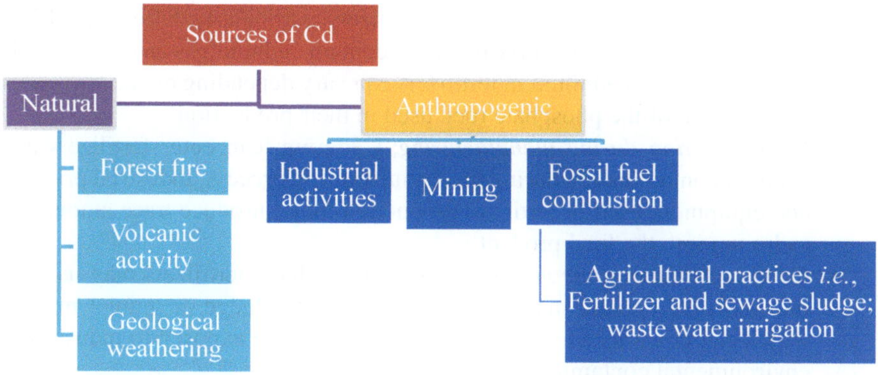

Fig. 1 Various sources of Cd in environment

through volcanic activity is relatively small compared to other sources, it can still contribute to localized contamination of soil and water following deposition.

3. *Forest Fires*: Wildfires release cadmium stored in vegetation and organic matter into the atmosphere and soil. Although the immediate impact of wildfires on cadmium mobilization may be limited, it can become a concern in the aftermath, as the released cadmium may gradually find its way into the ecosystem.

3.2 Anthropogenic Sources

1. *Industrial Activities*: Industrial processes, such as metal smelting, refining, and battery manufacturing, generate cadmium as a byproduct. Effluents and emissions from these industries can discharge substantial amounts of cadmium into the environment, affecting nearby soil and water bodies.
2. *Mining and Metallurgical Operations*: Mining activities targeting zinc, lead, and copper ores can liberate significant amounts of cadmium into the environment. As mines often contain cadmium along with other metals, the extraction process can lead to its release into the surrounding environment, impacting soil and water quality.
3. *Fossil Fuel Combustion*: Combustion of fossil fuels in power plants, vehicles, and industrial facilities releases cadmium as part of particulate matter and gaseous emissions. These emissions can be transported over long distances and subsequently deposited on land and water surfaces.
4. *Fertilizers and Sewage Sludge*: Some fertilizers, especially phosphate fertilizers, contain cadmium as an impurity. When applied to agricultural fields, cadmium can accumulate in the soil. Similarly, sewage sludge, used as a soil amendment, can also contribute to cadmium contamination, particularly if it originates from industrial wastewater. Phosphatic fertilizers can introduce cadmium into the soil through several mechanisms:

- *Cd content in fertilizers*: Phosphatic fertilizers, particularly those derived from phosphate rock deposits, may naturally contain cadmium as an impurity. The concentration of cadmium in fertilizers can vary depending on the source and composition of the phosphate rock used in their production.
- *Contamination during manufacturing*: Cadmium can enter fertilizers as a result of contamination during the manufacturing process. Industrial practices and equipment used in fertilizer production may introduce trace amounts of cadmium into the final product.
- *Cd in recycled organic materials*: Some phosphatic fertilizers may incorporate recycled organic materials, such as sewage sludge or animal manure, which can contain cadmium if these materials were exposed to industrial or environmental contamination.

5. *Electronic Waste*: Improper disposal and recycling of electronic waste release cadmium and other toxic substances into the environment. Electronic devices, such as batteries and circuit boards, can contain cadmium, and their improper handling can lead to its dispersion in soil and water.
6. *Contaminated Water*: Industrial discharges and improper waste disposal can lead to cadmium contamination of water bodies. Cadmium can be transported through waterways, affecting aquatic ecosystems and potentially entering the food chain.
7. *Urban Runoff*: Urbanization and storm water runoff from impervious surfaces can transport cadmium from urban areas to nearby water bodies and soil. Urban runoff can carry cadmium from various sources, including vehicle emissions, industrial sites, and paved surfaces.
8. *Other Products*: Certain consumer products, such as some paints, plastics, and batteries, contain cadmium. While the overall contribution of consumer products to cadmium contamination may be relatively small, their improper disposal can lead to localized hotspots of cadmium pollution.

4 Factors Affecting Cd Availability

Enhanced Cd concentrations in soils and sediments are commonly associated with the presence of specific mineral components, such as clay minerals, carbonates, organic matter, and hydrous oxides. Additionally, certain physicochemical conditions, like increased pH levels and/or anoxic (low oxygen) conditions, can also contribute to the accumulation of Cd in these environmental matrices. While Cd tends to bind to the less stable fractions, such as exchangeable, carbonate, and hydrous oxide fractions, other heavy metals like lead (Pb) and copper (Cu) are more strongly bound to the organic and sulfidic fractions. This disparity in binding behavior provides a plausible explanation for the distinctive geochemical behavior of Cd, making it more prone to mobilization when compared to other heavy metals. The preferential association of Cd with less stable fractions renders it susceptible to changes in environmental conditions, allowing it to readily enter the soil solution and become mobile. Consequently, certain physicochemical events, such as fluctuations in pH or the development of anoxic conditions, can lead to increased Cd mobility and potential migration to surrounding environments. The availability of cadmium in the environment can be influenced by several factors, which play crucial roles in determining its mobility, bioavailability, and potential impact on ecosystems and human health. These factors can be broadly categorized as follows:

1. *Soil Properties*: Soil plays a significant role in regulating cadmium availability. Soil pH is a critical factor, as acidic conditions can enhance cadmium mobility and increase its leaching potential. Organic matter in soil can also impact cadmium availability. Organic matter can form complexes with cadmium, influencing its solubility and potential for plant uptake. Additionally, clay minerals in soil have a high affinity for cadmium, leading to its sorption and retention in the soil matrix.
2. *Redox Conditions*: Redox conditions, representing the availability of oxygen in the soil environment, can affect cadmium speciation and availability. Under reducing conditions, such as waterlogged or anaerobic soils, cadmium may form more soluble complexes, increasing its potential for transport through soil water and groundwater. Conversely, under oxidizing conditions, cadmium may precipitate as less soluble forms, reducing its mobility.
3. *Presence of Ligands*: Cadmium can form various complexes with ligands present in the soil solution. Organic matter, inorganic anions (e.g., chloride, sulfate), and dissolved organic matter can form water-soluble complexes with cadmium, affecting its solubility and bioavailability. These complexes can enhance or hinder cadmium mobility and uptake by plants and other organisms.
4. *Anthropogenic Activities*: Human activities play a significant role in cadmium availability in the environment. Industrial processes, mining, and the use of fertilizers and sewage sludge can introduce cadmium into the soil. Anthropogenic sources contribute to elevated cadmium levels in soil and water, increasing its availability for uptake by plants and subsequent transfer through the food chain.
5. *Soil Amendments and Fertilizers*: The use of certain soil amendments and fertilizers can influence cadmium availability. For example, phosphate fertilizers can increase cadmium mobility and uptake by forming water-soluble complexes with cadmium in the soil. In acidic soils, this can lead to enhanced cadmium uptake by plants, posing risks to food safety.
6. *Plant Uptake and Root Exudates*: Plants can influence cadmium availability in the soil through their uptake and root exudation. Some plant species have a higher affinity for cadmium uptake, leading to its accumulation in plant tissues. Root exudates can also impact cadmium solubility and mobility in the rhizosphere, affecting its availability to plants.
7. *Climate and Hydrological Conditions*: Climate and hydrological factors, such as rainfall patterns and drainage, can affect cadmium transport in the environment. High rainfall and poor drainage can lead to cadmium leaching from soil and its transport to groundwater and surface water bodies. This can result in contamination of water resources and potential exposure to aquatic organisms and humans.

8. *Adsorption and Desorption*: Cadmium's interaction with soil particles, including adsorption and desorption processes, plays a crucial role in its availability. Adsorption onto soil surfaces can reduce cadmium availability in the soil solution, while desorption can release previously sorbed cadmium back into the soil solution, influencing its mobility.

5 Environmental Distribution and Transport

5.1 Fate and Transport of Cadmium in Air, Water, and Soil Compartments

Cadmium is a highly toxic heavy metal released into the environment through industrial processes, agricultural practices, and natural sources [18]. Once released, cadmium undergoes intricate processes of fate and transport, leading to its distribution in the air, water, and soil compartments of the environment. Cadmium can exist in the atmosphere as particulate matter, aerosols, or gaseous compounds. It is susceptible to long-range transportation via air currents and subsequent deposition on land or water surfaces through dry or wet deposition [14]. Inhaling airborne cadmium particles poses health risks to humans and animals [18].

Various pathways introduce cadmium into water bodies, including industrial discharges, agricultural runoff, and atmospheric deposition [18]. Upon entering aquatic environments, cadmium can undergo adsorption onto suspended particles, sedimentation, and bioaccumulation in aquatic organisms. The mobility of cadmium in water is influenced by factors such as pH, temperature, and the presence of organic matter [2]. Its binding to sediment particles can result in long-term impacts on aquatic ecosystems [2], necessitating careful monitoring and assessment.

Cadmium is deposited in soil either through direct atmospheric deposition or the application of fertilizers and sewage sludge [18]. Due to its low mobility, cadmium mainly binds to clay minerals and organic matter in the soil [2]. However, under specific conditions, such as acidic soils or intense agricultural practices, cadmium may leach into groundwater, posing a threat to drinking water sources and aquatic systems [1, 2].

5.2 Atmospheric Deposition and Atmospheric Transport of Cadmium

Atmospheric deposition and atmospheric transport play crucial roles in the distribution and fate of cadmium (Cd) in the environment. Cadmium can be released into the atmosphere through various anthropogenic sources, such as industrial emissions, fossil fuel combustion, and waste incineration, as well as natural sources like volcanic eruptions and forest fires [15]. Once in the atmosphere, cadmium can undergo several processes that influence its deposition and transport to different environmental compartments. Cadmium can be directly deposited onto surfaces through dry particles and aerosols settling out of the air influenced by factors such as particle size, wind speed, and surface characteristics [12]. Once cadmium particles settle on land or water surfaces, they can be washed into the soil and water bodies during rain events or other weather conditions. The atmospheric transport of cadmium involves the movement of cadmium-laden air masses over long distances before eventual deposition. Winds and weather patterns can carry cadmium particles and aerosols across regional and even continental scales [14]. This process is particularly significant for cadmium emissions originating from distant sources, which can impact regions far away from the original emission site. The fate of cadmium during atmospheric transport is influenced by factors such as the chemical form of cadmium (e.g., particulate, gaseous, or bound to aerosols), atmospheric stability, and meteorological conditions [14]. Some forms of cadmium may have a longer atmospheric residence time and can be transported farther before deposition.

5.3 Bioaccumulation of Cadmium in Food Chains

Bioaccumulation is the process by which certain substances, such as heavy metals like cadmium (Cd), accumulate in living organisms at higher concentrations than what is present in their surrounding environment [7]. In the context of food chains, bioaccumulation of cadmium can have significant implications for both human health and ecosystem integrity. Cadmium enters food chains primarily through its uptake by plants from contaminated soil or water [20]. Plants can absorb cadmium from the soil, and it is subsequently transferred to herbivores when they consume these contaminated plants. From herbivores, cadmium can then move up the food chain to higher trophic levels when carnivores or omnivores feed on contaminated herbivores. The process continues until cadmium reaches the highest trophic level, often apex predators. As cadmium moves up the food chain, it undergoes biomagnification, a process where its concentration increases at each trophic level [17]. This means that organisms higher in the food chain have higher cadmium levels than those lower in the chain, even if they are exposed to the same environmental concentrations. Biomagnification is a result of the progressive inability of organisms to metabolize and excrete cadmium efficiently, leading to its accumulation in tissues. Bioaccumulation of cadmium in food chains can pose health risks to humans, especially those who consume cadmium-contaminated foods, such as fish, shellfish, and certain vegetables [7]. Chronic exposure to cadmium through contaminated food has been linked to various adverse health effects, including kidney damage, bone mineral density loss, and increased risks of cardiovascular diseases and certain cancers. Pregnant women, infants, and children are particularly vulnerable to cadmium's toxic effects. In addition to human health concerns, the bioaccumulation of cadmium in food chains can have detrimental effects on ecosystems [20]. High cadmium levels in predatory species may lead to population declines or alter community dynamics. Additionally, cadmium can affect reproductive success and growth rates of organisms, potentially disrupting entire ecosystems.

6 Toxicity and Health Effects

Cadmium (Cd) is a toxic heavy metal that poses significant health risks to humans and animals when present at elevated levels in the environment. It is released into the air, water, and soil through various industrial processes, mining activities, and the use of cadmium-containing products [17]. Cadmium can accumulate in living organisms, leading to both acute and chronic toxicity.

6.1 Mechanisms of Cadmium Toxicity in Humans and Animals

1. Cellular and Molecular Interference: Cadmium toxicity primarily occurs through its ability to interfere with essential cellular processes. Once absorbed into the body, cadmium ions can replace essential metals, such as zinc and calcium, in various enzymes and proteins. This disruption can impair critical cellular functions, including DNA repair, cell signaling, and antioxidant defense systems, leading to cellular damage and dysfunction [13].
2. Oxidative Stress: Cadmium exposure triggers the generation of reactive oxygen species (ROS) in cells. ROS are highly reactive molecules that can cause damage to lipids, proteins, and DNA, leading to oxidative stress. Prolonged oxidative stress can contribute to tissue damage and increase the risk of various diseases, including cancer, cardiovascular disorders, and neurodegenerative conditions [21].
3. Kidney and Bone Damage: Cadmium has a strong affinity for the kidneys and bones, where it tends to accumulate over time. Chronic exposure to cadmium can cause severe damage to the kidneys, leading to nephrotoxicity and impaired renal function. In bones, cadmium disrupts the balance of bone resorption and formation, contributing to reduced bone density and increased fracture risk [17].
4. Carcinogenicity: Cadmium is classified as a Group 1 human carcinogen by the International Agency for Research on Cancer (IARC). Prolonged exposure to cadmium has been linked to an increased risk of developing lung cancer and prostate cancer in humans. The exact mechanisms of cadmium-induced carcinogenesis are complex and involve multiple cellular pathways [21].
5. Reproductive and Developmental Effects: Cadmium exposure can also impact reproductive health and development. It has been associated with adverse effects on male and female fertility, including reduced sperm quality and disrupted

hormone regulation. In pregnant women, exposure to cadmium can lead to developmental issues in the fetus, affecting the nervous and immune systems [13].

6. Endocrine Disruption: Cadmium can interfere with the normal functioning of the endocrine system, which regulates various physiological processes through hormone signaling. Disruption of hormone balance can lead to reproductive, metabolic, and developmental disorders [13].

6.2 Acute and Chronic Health Effects of Cadmium Exposure

Cadmium (Cd) exposure can lead to both acute and chronic health effects, depending on the level and duration of exposure. The toxic effects of cadmium can affect various organ systems in the body:

1. Acute Health Effects: Acute exposure to high levels of cadmium can result in immediate health effects, particularly on the respiratory and gastrointestinal systems. Inhalation of cadmium fumes or dust can cause acute respiratory distress, including coughing, chest pain, and difficulty breathing. Ingestion of large amounts of cadmium can lead to acute gastrointestinal symptoms such as nausea, vomiting, and abdominal pain [19].
2. Chronic Health Effects: Chronic exposure to lower levels of cadmium over extended periods is of greater concern due to its cumulative nature and potential long-term health effects. The kidneys are the primary target organ for chronic cadmium toxicity. Cadmium accumulates in the kidneys, causing renal dysfunction and an increased risk of chronic kidney disease. Long-term exposure to cadmium has also been associated with reduced bone mineral density, leading to osteoporosis and an increased risk of fractures [7].
3. Cardiovascular Effects: Chronic cadmium exposure has been linked to adverse cardiovascular effects, including hypertension, atherosclerosis, and an increased risk of heart disease. Cadmium may contribute to cardiovascular damage through its ability to promote oxidative stress and inflammation [14].
4. Reproductive and Developmental Effects: Cadmium can adversely affect the reproductive system, leading to reduced fertility in both males and females. In pregnant women, cadmium exposure may lead to adverse developmental effects in the fetus, affecting growth and neurological development [21].

6.3 Carcinogenic Potential of Cadmium

Cadmium is classified as a Group 1 human carcinogen by the International Agency for Research on Cancer (IARC). Long-term exposure to cadmium has been associated with an increased risk of several types of cancer:

1. Lung Cancer: Occupational exposure to cadmium, particularly in industries involving metal smelting, has been linked to an elevated risk of lung cancer. Inhalation of cadmium-containing fumes and dust is the primary route of exposure for lung cancer development [6].
2. Prostate Cancer: Studies have suggested a potential association between cadmium exposure and an increased risk of prostate cancer. The mechanism by which cadmium may promote prostate cancer is still under investigation [6].
3. Kidney Cancer: Chronic exposure to cadmium is strongly linked to an increased risk of kidney cancer [6].
4. Other Cancers: Some evidence suggests a possible link to other cancers, such as breast and pancreatic cancer. However, further research is needed in this domain [6].

7 Remediation of Cd Toxicity

The innumerable possible cadmium toxicity effects (as discussed in the previous sections) on terrestrial and biological ecosystems necessitate its remediation and mitigation from these systems with help of certain strategies and mechanisms (Fig. 2).

1. **Defense strategies of plants**

Plants can tolerate Cd toxicity via two major mechanisms: (i) tolerance and (ii) Avoidance. Plant limit uptake of Cd under avoidance mechanism and accumulation of Cd occur inside plants in tolerance strategy. Some of the plants are hyperaccumulating and show no phytotoxicity. Plants cope with Cd exposure with the help of

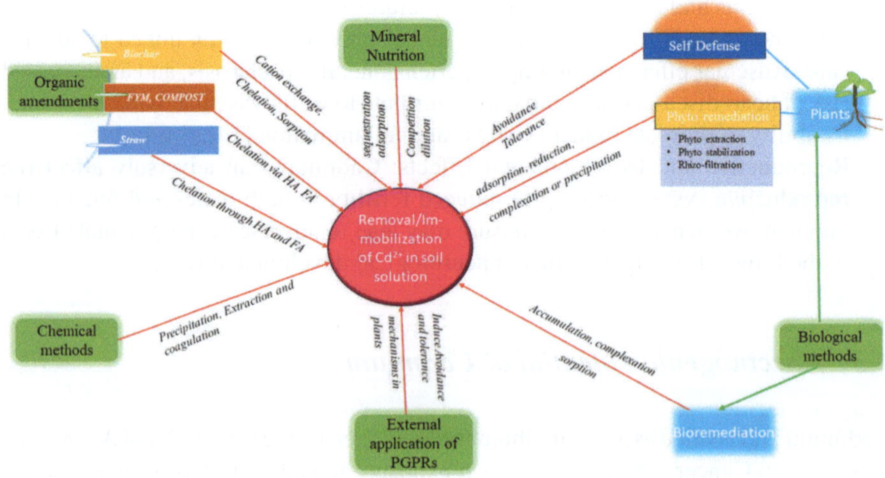

Fig. 2 Major strategies and their mechanisms for remediation of Cd pollution in soil

certain stress signaling molecules viz, salicylic acid, ethylene, jasmonic acid, humic acid, nitric oxide, and ethylene. Hyperaccumulating plants utilize sequestration, detoxification, and storage of trace metal ions, including Cd, in plant shoots. These processes occur in cuticles, epidermis, and trichomes, where photosynthetic activity is least affected. Metals are excluded from guard cells and subordinate stomatal cells, protecting functionary cells from Cd ions' phytotoxic effects.

2. **Mineral nutrition in minimizing cadmium toxicity**

Plant nutrients can effectively prevent the toxicity of Cd in plants, but understanding their interaction with soil is crucial. Essential plant nutrients impact the availability of Cd and its toxicity through direct and indirect effects. Direct effects involve minimizing Cd solubility through mechanisms like sequestration, adsorption, and competition for membrane transporters. Indirect effects involve dilution of Cd concentration, alleviating physiological stress, and improving productivity [5]. High doses of N and other nutrients can cause soil acidification, enhancing the solubility and bioavailability of toxic metals like Cd.

3. **Organic Amendments**

   (a) ***Biochar***: Biochar which is formed through pyrolysis of substances of organic nature, has been quite effective in Cd remediation as it reduces the Cd availability thereby preventing accumulation and toxicity of Cd in plants [4]. Biochar parameters, such as feedstock type, pyrolysis temperature, and retention time, play a crucial role in regulating Cd adsorption in contaminated agricultural soils. Biochar helps in remediation via its adsorption ability of the trace metals driven mainly by surface complexation and ion exchange. Urban sewage sludge-derived biochar significantly increases Cd removal, while legume-derived biochar has been found to have a higher Cd pollution reduction ability than that produced from non-legume.
   (b) ***Organic manure and compost***: Besides improving soil structure and fertility, Farmyard manure (FYM) has the capability of altering the soil's physicochemical properties and leading to the transformation of Cd in contaminated soils. Fulvic acid and humic acid in FYM possess functional groups viz. carboxyl and phenolic-OH groups which form metal complexes affecting metals' bioavailability, transport, and solubility in soils. However, anomalies are found in the reviewed literature w.r.t the effects of FYM application on Cd mobility. The ratio of humic acid to fulvic acid is determining factor in regulating Cd solubility. Fresh manures have a low humic acid/fulvic acid ratio while composted residues and aged manures have high humic/fulvic acid content.
   (c) ***Plant growth regulators***: Exogenous application of plant growth regulators, such as salicylic acid and abscisic acid, has been shown to create plant resistance against various stresses. These compounds help plants reduce oxidative damage caused by Cd, minimize seedling transpiration,

and improve tolerance. Other phytochromes, such as gibberellins, nitric oxide, and daminozide, are widely used to minimize the adverse effects of Cd on cultivable plants.

4. **Phytoremediation**

Phytoremediation is an evolving strategy that utilizes chemical, biochemical, and physical interactions to mineralize contaminants in polluted areas. It involves various pathways and depends on the form of pollutants. Phytoremediation plants have attributes like rapid growth, high biomass production, a deep root system, and a high bioaccumulation coefficient. Cd remediation by plants depends on soil toxicity, pH, nutrient concentrations, and temperature. Phytoremediation plants, from various families (for e.g., Poaceae, Violaceae, Lamiaceae, Fabaceae, Cyperaceae, Euphorbioceae), use mechanisms like phytoextraction, rhizo-filtration, stabilization, and volatilization out of which phytoextraction and rhizo-filtration are chief methods used by plants for alleviating Cd contamination. Phyto-extraction via rapeseed extracts can reduce Cd levels by 60% in soil than the control [9]. Rhizo-filtration involves plants with hairy, longer roots that filter pollutants, like rapeseed and sunflower. Phyto-stabilization involves plants growing in contaminated sites, facilitating metal immobilization in soils through adsorption, reduction, complexation, or precipitation. This method is effective for fine-textured soil with high organic content.

5. **Bioremediation**

Microbial flora, including fungi, bacteria, and algae, can effectively eliminate organic and metal contaminants from agricultural and industrial effluents. Bacillus and Pseudomonas species effectively remove pollutants like Cd [8]. Mycorrhizae association with plants can reduce pollutants transfer by acting as a barrier, binding metals to fungi's hyphae. Fungi's vesicles, spores, and extraradical mycelia play crucial roles in chelating pollutants and metal accumulation. Inoculated plants may generate Cd complex chelating molecules, such as phytochelatins, glutathione, and metallothioneins. Glomalin produced by mycorrhizal mycelia binds more metals, promoting trace metal immobilization and host plant adaptation. Certain species of fungi like *Piriformospora indica* and *Trichoderma* spp. are versatile plant growth promoters and biological control agents, adapting to high pollutant concentrations. Investigating their volatilization, biosorption, bioaccumulation, and phytobial remediation capacities is crucial for reducing environmental pollution.

6. **Chemical methods**

Under chemical methods, coagulation via barium acetate in the electroplating industry, precipitation by the addition of hydroxides of sodium, calcium and magnesium extraction by cementation has been found effective for Cd removal. Various other phosphorous-based and aqueous nitrogen-donor extracts can also be used for Cd extraction. However, the solvent is needed in quite large amounts during the process, leading to its failure. Therefore, these methods are not recommended for removing low cadmium.

8 Recent Advancements in the Cd Remediation Strategies

Most of the conventional approaches of remediation are not able to provide a promising and acceptable solution and many of them are expensive and invasive. So, advancement in this domain is going on for more effective remediation of Cd pollution in the environment.

1. **Nano-materials for Removal**: Nanotechnology-based materials are being developed to remove cadmium from water and soil. These materials, such as nanoparticles and nanocomposites, offer high surface area and reactivity for efficient cadmium adsorption.
2. **Risk Assessment and Modeling**: Advances in computational modeling and risk assessment tools allow researchers to better predict the fate and transport of cadmium in the environment. This helps in identifying areas of concern and developing targeted mitigation strategies.
3. **Microbial fermentation**: Recent research has shown that microbial fermentation is a viable technique for eliminating contaminants from food. Scientists have been studying optimal conditions for reducing Cd in rice noodles using mixed Lactobacillus fermentation, demonstrating efficient reduction, improved physicochemical properties, and quality improvement in rice flour.
4. **Regulation and Policy**: Governments and international organizations have been working to establish regulations and guidelines for Cd exposure limits in different contexts (Fig. 3).

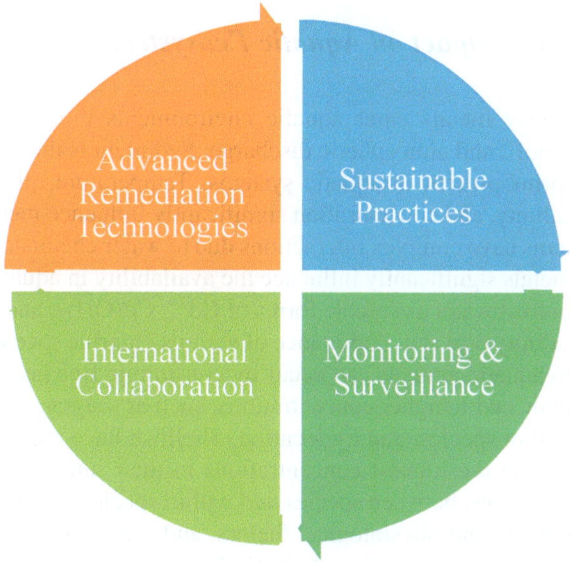

Fig. 3 Strategies for environmental cadmium management

(a) *Advanced Remediation Technologies*: Developing innovative and cost-effective techniques for removing Cd from the environment will remain a priority. This might involve nanotechnology, bioremediation, and soil management practices.
(b) *Sustainable Practices*: Encouraging sustainable agricultural practices and reducing the use of Cd-containing products can prevent further contamination and exposure.
(c) *Monitoring and Surveillance*: Developing efficient monitoring systems to track Cd levels in different environmental compartments will be crucial for early detection and response to pollution events.
(d) *International Collaboration*: As Cd pollution is a global issue, international collaboration on research and sharing best practices will be important for addressing the problem effectively.

9 Ecological Impacts of Cadmium Toxicity

Onset of Industries led to significant increase in pollution by Cd in ecosystem of earth viz. terrestrial and aquatic ecosystems, leading to the deterioration of soil-plant-human health and aquatic lives as well by entering the food chain and undergoing biomagnification to toxic concentrations. Here in this we will briefly discuss impact of cadmium toxicity on aquatic ecosystems, soil health, plant and human health.

9.1 Impact on Aquatic Ecosystem

Heavy metals enter aquatic environments through industrial effluent, terrestrial runoff, and atmospheric discharge. Sediment is the primary repository, with soluble forms present in aquatic systems. Temperature, pH, dissolved oxygen, hardness, salinity, and concentration significantly influence metal availability. Marine ecosystems have complex interactions due to water chemistry and hydrodynamics of heavy metals significantly influence the availability in aquatic environments. Cd is present in freshwater as soluble forms of Cd^{2+}, $Cd(OH)_2$, and $CdCO_3$; while, with increased salinity of water, dominance of various chlorine species is increased. Cd bio-transfer to aquatic animals can occur by consuming free heavy metal ions but primarily via plants with higher concentrations, such as seagrass. Cd levels vary between species, within species, and by location. Shellfish have higher Cd levels than fish, with fish having the highest concentrations in their intestine, liver, gill, and muscle tissue. Variations between species and within species may be due to environmental factors, genetic and physiological factors, and seasonal variations [10]. Higher Cd concentrations were found in shellfish during rainy seasons than dry seasons, likely due to more runoff bringing more mobile Cd into the ocean. Aquatic organisms are largely affected by Cd accumulation, with low exposure causing growth impairment in fishes. Invertebrates can be more susceptible to Cd exposure, with even low concentrations causing mortality in them.

9.2 Impact on Plant Health and Their Response to Cd Toxicity

Cd toxicity causes various phytotoxic indications, including chlorophyll loss, growth inhibition, H_2O imbalance, phosphorus and nitrogen deficits, and oxidative stress. It can replace Zn, Ca, and Fe in proteins, react with sulfur-containing proteins, interfere with mitochondrial electron transfer chains, and inhibit water transport, leading to proline accumulation. Heavy metal pollution affects water, air, and soil, causing crop loss and health hazards. Plants respond to stress through antioxidative systems, affecting their physiological and biochemical responses. Heavy metal toxicity in plants causes chromosomal aberrations and decreased cell division. Genotoxic responses vary among species, with concentration, oxidation state, and exposure affecting genotoxic responses. Plants have developed an anti-oxidative defense system, including enzymes like ascorbate peroxidase, catalase, superoxide dismutase, glutathione, peroxidase, and guaiacol peroxidase. Oxidative stress and reactive oxygen species (ROS) are crucial in aerobic organisms, causing cell structure alteration and mutagenesis. Heavy metals increase ROS production, disrupting redox status and causing membrane damage. Plants face physiological challenges due to imbalances in ROS formation and removal through antioxidative defense mechanisms. Metal-induced genotoxicity involves DNA damage and heavy metal production, with complex mechanisms affecting genotoxic responses. Plants have developed an anti-oxidative defense system, including enzymes like ascorbate peroxidase, catalase, superoxide dismutase, glutathione, peroxidase, and guaiacol peroxidase.

9.3 Impact on Soil Health

Soil health management is crucial for sustainable agriculture, and soil biology plays a crucial role. Soil microbes are an essential component of the ecosystem. Soil microbes maintain soil fertility and soil health through organic matter disintegration and nutrient cycling. However, they can be negatively impacted by stress factors and heavy metal contamination from anthropogenic activities like mining, sewage sludge, and pesticides. Microbial viability decreases with increasing levels of heavy metal contamination. Heavy metal toxicity can lead to reduced litter decomposition, decreased organic carbon mineralization, and negative impacts on nitrogen transformation. Soil respiration rate is a key indicator of metal toxicity. Nitrogen transformation is negatively impacted by heavy metal application, as seen in blue green algae grown on sludge-amended soil with low extractable heavy metal concentrations metals (Zn = 30 μg g^{-1} soil, Cu = 15 μg g^{-1}, Ni and Cd = 2 μg g^{-1}) [3] with nitrification being more sensitive to pollution. Soil enzyme activities, such as arylsulfatase, alkaline phosphatase, b glucosidase, cellulase, dehydrogenase, invertase, protease, and urease, are sensitive to Cd toxicity [16]. Pollution caused by Cd can significantly affect soil ecosystems and biological activity, lowering the soil microbial population.

10 Policies and Implications

1. *Environmental Regulations*: Many countries have established regulations to control the release of cadmium into the environment. These regulations often limit cadmium emissions from industrial processes, agricultural activities, and waste disposal. The aim is to minimize exposure to cadmium for both humans and ecosystems. The Joint Food and Agriculture Organization of the United Nations (FAO)/WHO Expert Committee on Food Additives (JECFA) established a provisional tolerable monthly intake for cadmium in 2010 of 25 μg/kg body weight, based on meta-analysis of epidemiological studies on the relationship between urinary cadmium and beta-2-microglobulin (a marker of renal tubular effects). In light of the long half-life of cadmium in humans, the Committee decided to express the tolerable intake as a monthly value. The guideline for cadmium in drinking-water is 3 μg/L, and in the air is 5 ng/m^3 (annual average).
2. *E-waste Management*: Cadmium is commonly found in electronic waste (e-waste). Effective management of e-waste is crucial to prevent cadmium and other hazardous substances from leaching into soil and water. Recycling and proper disposal of electronic products are essential components of these policies.
3. *Food Safety*: Cadmium can accumulate in crops through contaminated soil and water [11]. Many countries have established maximum allowable levels of cadmium in food products to ensure consumer safety. Agricultural practices that reduce cadmium uptake by plants, such as using clean water sources and soil management techniques are encouraged.
4. *Consumer Products*: Regulations on cadmium content in consumer products, such as batteries, paints and plastics help limit exposure to the general population. These regulations aim to prevent the release of cadmium during product use and disposal.
5. *Mining and Industry*: Policies related to mining and industrial processes involving cadmium focus on minimizing environmental contamination through proper waste management and treatment. Efforts to reduce cadmium emissions include using pollution control technologies and adopting cleaner production practices.

11 Conclusion

The persistence of cadmium in the environment is a concern because it resists degradation into less toxic forms. Cadmium can accumulate in soil, sediments, and biota, posing risks to ecosystems and human health. Cadmium availability in the environment is a complex interplay of various factors, including soil properties, redox conditions, the presence of ligands, human activities, and biotic interactions. Understanding the sources and pathways of cadmium contamination is vital for implementing effective environmental management strategies and regulatory measures to mitigate its impact and safeguard environmental and human well-being.

References

1. Agency for Toxic Substances and Disease Registry (ATSDR). (2012). *Toxicological profile for cadmium*. Department of Health and Human Services, Public Health Service.
2. Alloway, B. J. (2013). *Heavy metals in soils: Trace metals and metalloids in soils and their bioavailability*. Springer Science & Business Media.
3. Brookes, P. C., McGrath, S. P., & Heijnen, C. (1986). Metal residues in soils previously treated with sewage-sludge and their effects on growth and nitrogen fixation by blue-green algae. *Soil Biology and Biochemistry, 18*(4), 345–353.
4. Farooq, M., Ullah, A., Usman, M., & Siddique, K. H. (2020). Application of zinc and biochar help to mitigate cadmium stress in bread wheat raised from seeds with high intrinsic zinc. *Chemosphere, 260*, 127652.
5. Forster, S. M., Rickertsen, J. R., Mehring, G. H., & Ransom, J. K. (2018). Type and placement of zinc fertilizer impacts cadmium content of harvested durum wheat grain. *Journal of plant nutrition, 41*(11), 1471–1481.
6. International Agency for Research on Cancer (IARC). (2012). Cadmium and cadmium compounds. *IARC Monographs on the Evaluation of Carcinogenic Risks to Humans, 100C*, 121–145.
7. Jarup, L., & Akesson, A. (2009). Current status of cadmium as an environmental health problem. *Toxicology and Applied Pharmacology, 238*(3), 201–208.
8. Karimpour, M., Ashrafi, S. D., Taghavi, K., Mojtahedi, A., Roohbakhsh, E., & Naghipour, D. (2018). Adsorption of cadmium and lead onto live and dead cell mass of Pseudomonas aeruginosa: A dataset. *Data in brief, 18*, 1185–1192.
9. Kathal, R., Malhotra, P., & Chaudhary, V. (2016). Phytoremediation of cadmium from polluted soil. *Journal of Bioremediation and Biodegradation, 7*, 376.
10. Mok, J. S., Yoo, H. D., Kim, P. H., Yoon, H. D., Park, Y. C., Lee, T. S., & Kim, J. H. (2015). Bioaccumulation of heavy metals in oysters from the southern coast of Korea: Assessment of potential risk to human health. *Bulletin of Environmental Contamination and Toxicology, 94*, 749–755.
11. Mubeen, S., Ni, W., He, C., & Yang, Z. (2023). Agricultural strategies to reduce cadmium accumulation in crops for food safety. *Agriculture, 13*(2), 471.
12. Munthe, J., Hultberg, H., Bahlmann, E., Dellfeldt, J., Wangberg, I., Lofgren, S., & Kocman, D. (2015). Atmospheric mercury and trace elements at rural sites in Sweden: Long-term ambient measurements and source identification. *Science of the Total Environment, 538*, 105–115.
13. Nordberg, G. F., Jin, T., Wu, X., Lu, J., Chen, L., & Lei, L. (2004). Prevalence of kidney dysfunction in humans–relationship to cadmium dose, metallothionein, immunological and metabolic factors. *Biochimie, 89*(12), 1229–1237.
14. Nriagu, J. O. (Ed.). (1998). *Cadmium in the environment* (2nd edn.). Wiley.

15. Pacyna, J. M., & Pacyna, E. G. (2001). An assessment of global and regional emissions of trace metals to the atmosphere from anthropogenic sources worldwide. *Environmental Reviews, 9*(4), 269–298.
16. Pan, J., & Yu, L. (2011). Effects of Cd or/and Pb on soil enzyme activities and microbial community structure. *Ecological Engineering, 37*(11), 1889–1894.
17. Satarug, S., Baker, J. R., Urbenjapol, S., Haswell-Elkins, M., Reilly, P. E., Williams, D. J., & Moore, M. R. (2003). A global perspective on cadmium pollution and toxicity in non-occupationally exposed population. *Toxicology Letters, 137*(1–2), 65–83.
18. Satarug, S., Garrett, S. H., Sens, M. A., & Sens, D. A. (2010). Cadmium, environmental exposure, and health outcomes. *Environmental Health Perspectives, 118*(2), 182–190.
19. Satarug, S., & Moore, M. R. (2004). Adverse health effects of chronic exposure to low-level cadmium in foodstuffs and cigarette smoke. *Environmental Health Perspectives, 112*(10), 1099–1103.
20. Storelli, M. M., & Barone, G. (2011). Bioaccumulation of trace metals in fish from the Mediterranean Sea and Atlantic Ocean. *Environmental Monitoring and Assessment, 178*(1–4), 361–368.
21. Waalkes, M. P. (2000). Cadmium carcinogenesis. *Mutation Research/Fundamental and Molecular Mechanisms of Mutagenesis, 533*(1–2), 107–120.

Cadmium in Soil and Plants: A Review

Emmanuely Zephaline Nungula, Muhammad Ali Raza, Jamal Nasar, Sagar Maitra, Mahmoud F. Seleiman, Shivani Ranjan, Smruti Ranjan Padhan, Sumit Sow, Dinkar Jagannath Gaikwad, and Harun H. Gitari

Abstract Heavy metals are serious pollutants and their accumulation in the environment imposes great effects on environmental ecology. Cadmium contamination is one of the heavy metals that is currently becoming a serious concern worldwide due to its high toxicity effects leading to worse results on both environmental and human health. This is particularly due to the consumption of plants with cadmium deposition. This chapter makes a critical review by highlighting sources of cadmium

E. Z. Nungula · H. H. Gitari (✉)
Department of Agricultural Science and Technology, School of Agriculture and Environmental Sciences, Kenyatta University, Nairobi, Kenya
e-mail: harun.gitari@ku.ac.ke

E. Z. Nungula
Department of Environmental and Sustainable Development, Mzumbe University, Morogoro, Tanzania

M. A. Raza
College of Agronomy, Sichuan Agricultural University, Chengdu, China

J. Nasar
Institute of Rice Industry Technology Research, Key Laboratory of Functional Agriculture, College of Agricultural Sciences, Guizhou University, Guiyang 550025, China

S. Maitra
Centurion University of Technology and Management, Gajapati, Odisha, India

M. F. Seleiman
Department of Plant Production, College of Food and Agriculture Sciences, King Saud University, Riyadh 11451, Saudi Arabia

S. Ranjan · S. Sow
Department of Agronomy, Dr. Rajendra Prasad Central Agricultural University, Pusa, Samastipur 848125, India

S. R. Padhan
Division of Agronomy, ICAR-Indian Agricultural Research Institute, Pusa Campus, New Delhi 110012, India

D. J. Gaikwad
Department of Biochemistry and Crop Physiology, M.S. Swaminathan School of Agriculture, Centurion University of Technology and Management, Gajapati, Odisha, India

A. K. Jha and N. Kumar (eds.), *Cadmium Toxicity in Water*, Springer Water,
https://doi.org/10.1007/978-3-031-54005-9_2

in soil, its accumulation, permissible levels, uptake mechanism and translocation in plant tissue, and its toxicity effects on the plant and humans. Also, the scope of this paper describes various remediation strategies currently available for reducing cadmium accumulation in the soil as well as in plants.

Keywords Heavy metal contamination · Cadmium toxicity · Cadmium uptake mechanism · Cadmium accumulation · Translocation mechanism

1 Introduction

Heavy metals are serious pollutants and their accumulation in the soil and plant is currently becoming a popular cause of environmental pollution [31, 43, 78]. Availability of heavy metals in soil and plants depends on naturally occurring factors such as pedogenic and lithogenic processes, human activities such as mining, waste disposal (municipal solid waste disposal sites), soil runoff, metal industries, combustion of fossil fuels, and fertilizer application such as phosphate fertilizer [4]. The increasing heavy metal concentration in soils is highly influenced by different anthropogenic factors including agricultural processes such as the application of sludge, pesticides, and inorganic fertilizer to agricultural land [7, 58, 79]. Municipal, industrial, and domestic waste disposal sites are also key sources of heavy metals [7, 78, 81]. Above all, heavy metal ions pollutions in soil are mainly due to industrialization, which represents a great environmental threat due to their toxicity [3, 31].

Environmental pollution with metals including lead, arsenic, cadmium, and mercury imposes a big health risk to human beings, plants, and animals [48, 60, 81]. Globally, pollution due to heavy metals contamination accounts for two out of ten major environmental events [81, 87]. In many African and Asian countries farming using untreated wastewater for irrigation contribute to over 50% of the vegetables supplied to cities. Nonetheless, most of these waters contain heavy metals such as arsenic, lead, cadmium, and mercury, which are not essential for plant growth and lack any known physiological function [61, 81]. Consumption of vegetables that have accumulated heavy metals having been grown on polluted lands is disastrous to human health [38].

Accumulation of heavy metals in the soil is a serious threat to the soil's ecosystem as well as plants. For instance, cadmium accumulation in the soil triggers the deceleration of soil microbial diversity (bacteria, fungi, and actinomycetes) [53, 54]. Such metals also affect microbial cell metabolism and enzyme activities of catalase, sucrase, urease, and acid phosphatase [21, 54]. Cadmium accumulation in soil usually is linked with the uptake, translocation, and deposition of cadmium in plants which has some direct and indirect effects on plant physiological process that reduces growth, development, and yield [36, 57]. In addition, the metals can alter some of the physiological functions in plants by creating complexes with oxygen, nitrogen, and sulphur ligands [12, 36]. Specifically, cadmium accumulation interferes with the mineral uptake mechanism by affecting roots, membrane functioning of the plant cell, protein metabolism, seed germination, and water relations [12, 66]. The element inhibits net photosynthesis by reducing oxygen evolution in photosystem II in the chloroplasts of plants, which may cause metabolic disturbance by interfering with essential biochemical processes [33, 36].

2 Heavy Metal Contamination

2.1 Heavy Metal Contamination on the Soil and Plants

Heavy metals refer to materials with a specific gravity of greater than 4 g/cm^3 [6]. The metals form part of the earth's crust and include materials like mercury, lead, zinc, copper, and cadmium among others [28, 60, 92]. Heavy metal contamination in the soil refers to overrated deposition and accumulation of these toxic materials either from natural sources or anthropogenic activities like industrialization and urbanization [28]. Such contamination on the soils is mainly derived from sewage systems, poor deposition of industrial effluents, quarry activities like mining, and the use of agrochemicals. High levels of metal ions accumulation pose a great environmental risk in newly industrialized countries such as China, due to the detrimental effect on plants, animals as well as human health [98]. In most developing countries, heavy metal contamination in the cities is a popular threat due to improper sewage disposal systems and open dumpsites. For instance, in Kenya, contamination of soils with heavy metals in the cities possess a great risk to the agricultural lands [63, 98].

Heavy metals pollution in agricultural land is mainly due to their accumulation from different human activities sources including; industries, mines, leaded gasoline and paints, inorganic fertilizer use, poorly decomposed animal manure, human waste sludge, wastewater irrigation, fossils utilization, and atmospheric deposition [81] (Fig. 1). Wastewater effluents have relatively low concentrations, but long-term irrigation results in heavy metal accumulation [81]. Soil acts as a sequester for metal ions interrupting, the soil-plant cycle in which they exert their toxicity [81]. Additionally, they have adverse effects on soil ecology and soil enzyme activities of urease, catalase, invertase, and acid phosphatase that play key roles in the mineralization process occurring in the soil [17, 72]. Metal ions alter the composition and characteristics of soil microorganisms affecting soil biological and biochemical properties [72, 81]. The effects on biological and biochemical properties are determined by pH, clay content, and organic content in the soil [81]. Pollution of soils with heavy metals poses a great risk to agricultural soils, that's difficult to remediate [17].

Contamination of soils contributes to the great risk worldwide due to its toxicity, multiple sources, non-biodegradable, and its accumulation characteristics [34], with a significant proportion entering humans through contaminated food and drinking water [6]. According to Wuana and Okieimen [105], the use of wastewater for irrigation accounting for 50% of the garden products such as vegetables supplied to the urban and peri-urban areas in most developing counties given that irrigation

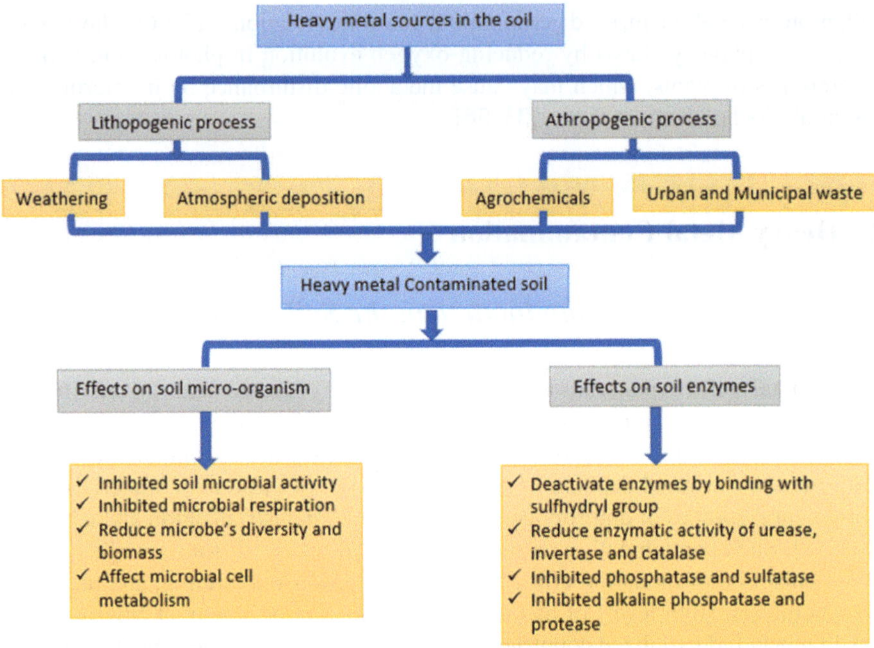

Fig. 1 Summary of the effects of heavy metals on Soil ecology

farming is considered instrumental in achieving vegetable demand [60]. This process increases heavy metal contamination in agricultural land, which significantly reduces their capacity to support crops [48, 72, 87]. Heavy metals impose a high toxicity effect once they accumulate in the plant system [81]. The metals have phytotoxic effects in the plants including the formation of reactive oxygen species, enhancing lipid peroxidation, imposing a change in intercellular glutathione levels, and also they inhibit enzymes activity (Fig. 2). This finally damages the cellular components such as lipids, protein, and nucleic acid resulting in the death of plants [72, 99].

2.2 *Transfer and Mobility of Heavy Metal in Plant Tissue*

Plants have differential uptake of metal ions from the soil, with some being hyperaccumulators, and can be effectively used for phytoremediation in polluted agricultural lands. The plant uptake and translocate metals from the soil through the use of proton pumps, co and anti-transporters, and channels [94]. Lead uptake is non-selective and depends on low-affinity cation transporters and is independent of H^+/ATPase pumps, commonly competing for calcium channels for uptake [46]. Transfer Factor (TF) is the ratio of the concentration in plant tissues to the respective concentration of the metal ions in the growth medium. Soil characteristics influence metals' mobility,

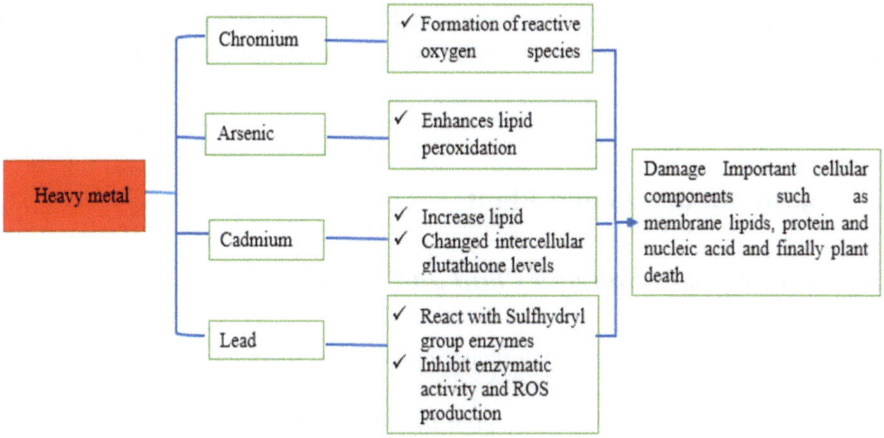

Fig. 2 Heavy metal physiological damage to plants

bioavailability, and ecotoxicological risk [1, 94]. Soil properties that influence the uptake include the soil pH, organic matter, the amount of phosphorous, carbonates, and cation exchange capacity. Soil microbial activities mobilize the metal ions hence, raising their bioavailability for plants [94].

The bioavailability of heavy metals has a negative correlation with soil pH where they dissolve easily with a reduction in soil redox potential and hence usually accumulate in poorly drained soils [23, 46]. For instance, lead exists in the aqueous form $Pb(H_2O_6)^{2+}$ in acidic conditions, while in basic soil it exists as hydroxyl ions (OH^-). Also, high amounts of clay in the soil promote metal adsorption through ion exchange capacity and specific adsorption mechanisms. Soil organic carbon influences microbial activities and solubility of metals and bioavailability of the ions to plants [46].

The translocation index or the mobility index refers to the ratio of the shoot concentration to the respective concentration on the roots. The amount of metals translocated varies depending on the crop species, with some transporting less to the leaves but much to the stems [51]. The translocation of metal ions from roots to the shoots is regulated and hence most crops do not accumulate higher than the metabolic needs [94]. The transport of metals to the shoot is associated with evapotranspiration [46]. A significant amount of heavy metals absorbed by plants accumulated in the roots with only a small portion usually transported to the above-ground plant tissue [46, 57, 87]. The apoplast transport of heavy metals is limited due to the nature of most of them binding with galacturonic, carboxyl, and glucuronic acid, hence providing a partial barrier on translocation from roots to the shoots [23, 108].

Plants growing in contaminated and polluted environments accumulate heavy metals in tissues [6, 46]. The consumption of such plants for food presents a great health risk to both animals and humans [44]. Heavy metals cause a lot of problems in human health even in small quantities such as carcinogenic, toxicity to the nervous, renal, and pulmonary systems, and skin [44]. Production of vegetable crops in polluted soils with metal ions results in higher concentrations in edible plant tissues, hence the need to reduce the transfer of the pollutants and improve crop productivity [108].

3 Cadmium in Soil and Plant

3.1 Cadmium in Soil and Plant and Its Sources

Cadmium is a naturally occurring heavy metal with, silvery-white with an atomic number of 48, in period 5, group 12 in the periodic table [33]. The concentration of cadmium is approximately 0.15 ppm in the earth's crust and it usually occurs in the form of greenockite (CdS), as a divalent cation [12, 107]. Cadmium is an element that is in most plant-available forms in agricultural soils, which imposes great physiological function effects due to its accumulation on the plants' tissues [40]. In assessing the senescence effects of cadmium contaminated in the soil quantifying the amount of plant available Cd is more effective, rather than the total content [12, 49]. In achieving this goal of assessing the deleterious, effect of cadmium different techniques has been established, including chemical analysis like sequential and single extraction procedures, which have been applied to assess the available reliable amount of Cd in a cost-efficient manner and the method has been used in a range of soils type and sediments to identify various source and amount [12, 49].

Cadmium has different sources that include natural and anthropogenic sources triggered by human activities. Cadmium atmospheric emission includes; suspended particles of oxide, sulphide, sulphate, and chloride of cadmium, which are chemically stable and can stay for a prolonged period in the troposphere of about one to four weeks. Such suspension depends on their size and climatic conditions after which the particles get deposited on the soil and plants by wet and dry deposition [40]. Naturally, of the atmospheric cadmium emission, soil particles contribute 24×10^3 tonnes worldwide compared with forest and bush fires, sea salt, volcanoes, and meteoric with 13×10^3, 2×10^3, 1.6×10^3, and 0.017×10^3 tonnes, respectively [45, 65]. For the anthropogenic sources of atmospheric cadmium emission, non-ferrous metal production leads by emitting 2.171×10^3 tonnes followed by fossil fuel combustion, production of iron and steel, urban and municipal waste disposal, and cement production with 0.691×103, 0.064×10^3, 0.04×10^3 and 0.017×10^3 tonnes, respectively [45, 74] (Table 1).

Industry activities produce a huge amount of cadmium in both forms of particulate and soluble, though its effect is likely to be less in affecting soils directly. Its indirect effects through draining of waste into water sources and groundwater in soil impose high pollution and increased amount of cadmium in the soil [8]. As pointed out by Kabir et al. [42] and Andreas et al., the mining and metal industries are the highest producers of cadmium with the maximum production of 289 mg kg^{-1} followed by the textile industry (83.6 mg kg^{-1}), nonmetallic mineral production industries

Table 1 Global cadmium emissions source to the atmosphere

Nature of source	Specific source	Cadmium (tonnes)
Natural sources	Meteoric dust	0.017×10^3
	Volcanoes	1.6×10^3
	Sea salt	2×10^3
	Forest and bushfires	13×10^3
	Soil particles	24×10^3
Anthropogenic sources	Non-ferrous metal production	2.171×10^3
	Combustion of fossil fuel	0.691×10^3
	Production process (iron and steel)	0.064×10^3
	Urban and municipal waste	0.04×10^3
	Cement production	0.017×10^3

Source Modified from Kubier et al. [45]

(72 mg kg^{-1}), fertilizers and chemical industries (2.13 mg kg^{-1}) and leather industries (1.26 mg kg^{-1}). For these industries, the mean cadmium production is such that the textile industry (42 mg kg^{-1}) > mining and metal industries (37.6 mg kg^{-1}) > nonmetallic mineral production industries (25.8 mg kg^{-1}) > leather industries (0.63 mg kg^{-1}) > fertilizers and chemical industries (0.51 mg kg^{-1}) (Fig. 3) [42].

Rocks and minerals also play a role in the provision of cadmium in the soil through the weathering process. Based on the study by Tabelin et al. [93], and Smolders and Mertens [89], concentration of cadmium is between 1×10^{-2} and 26×10^{-1} mg kg^{-1} in sedimentary rocks, which is higher compared to the concentration in igneous rocks and metamorphic rocks with respective ranges of 2×10^{-2} to 25×10^{-2} mg kg^{-1} and

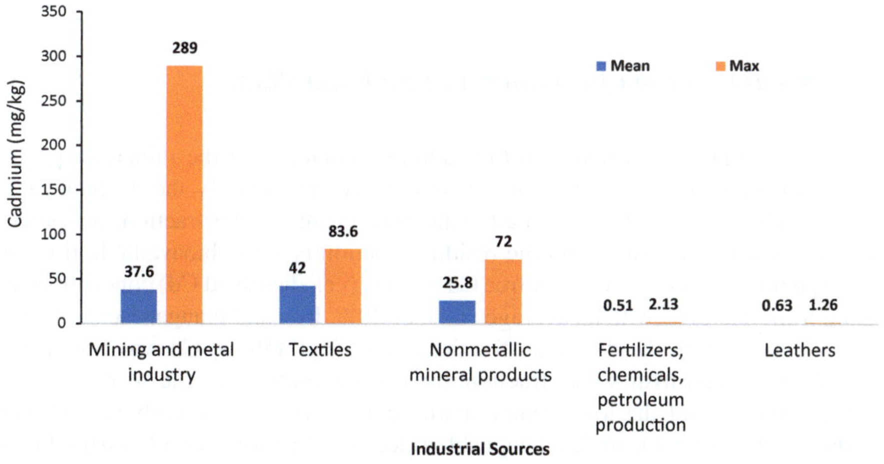

Fig. 3 Cadmium content in the soil as affected by different Industry production

Table 2 Rock materials and their cadmium concentration

Rock type	Rock materials	Cd (mg kg^{-1})
Sedimentary	Carbonate stone	12×10^{-3}
	Sandstone	28×10^{-3}
	Red shales	3×10^{-2}
	Limestone	1×10^{-2}
	Organic sediment	5×10^{-1}
	Red clay	56×10^{-2}
	Bituminous shale	8×10^{-1}
	Bentonite	14×10^{-1}
	Marlstone	26×10^{-1}
Igneous	Ultramafic rocks	2×10^{-2}
	Mafic rocks	11×10^{-2}
	Granitic rocks	12×10^{-2}
	Basalt	22×10^{-2}
	Obsidian	25×10^{-2}
Metamorphic	Metamorphic Gneisses	4×10^{-2}
	Schists	2×10^{-2}

Source Smolders and Mertens [89]

2×10^{-2} to 4×10^{-2} mg kg^{-1}, respectively (Table 2). Similarly, Greenockite (CdS) has a higher percentage of cadmium at 77.8%, whereas Sphalerite (Zn, Cd)S has the lowest percentage of < 2%. Nonetheless, in terms of concentration, the highest and lowest Cd amount are in Galena (< 3000 mg kg^{-1}) of cadmium and Apatite (0.14–0.15 mg kg^{-1}), respectively [14, 45, 96] (Table 3).

3.2 *Accumulation of Cadmium in Soil and Plant*

Cadmium availability depends on the cadmium content and the minerals species present in the soil [37]. Cadmium available fraction is mainly the exchangeable fraction, while the Fe–Mn oxide, carbonate, and organic matter fractions are potentially bioavailable fractions and the residual fraction is a non-bioavailable fraction [77]. Speciation studies show that free Cd species (i.e., dissolved Cd) concentrations varied from 0 to 60% with an average value of 20%. Iron and manganese bound Cd fractions were often the dominant fractions and about 45% of Cd was found in a Fe–Mn oxides and hydroxides fraction and organic matter was the second largest (39%), regardless of the low organic matter content of the soil, carbonate (13%), residue fraction (2.5%), and exchangeable fractions (1%) were largely insignificant [32]. A slight change in Cd fractions was observed six-weeks after plant seedings,

Table 3 Mineral composition and its cadmium concentration

Mineral	Composition	Cd (mg kg^{-1})
Greenockite	CdS	77.80%
Otavite	$CdCO_3$	65.18%
Smithonite	$ZnCO_3$	< 2.35%
Sphalerite	(Zn,Cd)S	< 2%
Galena	PbS	< 3000
Tetrahedrite	(Cu, Fe, Zn, Ag)12SbAs$_4$S1$_3$	80 to 2000
Anglesite	$PbSO_4$	120 to > 1000
Bindheimite	$Pb_2Sb_2O6(O,OH)$	100–1000
Limonite	Hydrous iron oxides	< 1000
Mn-Oxides	Hydrous manganese oxides	< 1000
Chalcopyrite	$CuFeS_2$	< 110
Marcasite	FeS_2	< 50
Calcite	$CaCO_3$	< 1–23
Pyromorphite	$Pb_5Cl(PO_4)_3$	< 1–8
Scorodite	$FeAsO_4 \cdot 2H_2O$	< 1–5.8
Silicates		0.03–5.8
Arsenopyrite	FeAsS	< 5
Magnetite	Fe_3O_4	< 0.31
Apatite	Ca5(F,Cl)$(PO_4)_3$	0.14–0.15

Source Thornton [96]

where the fraction of metal bound to oxides and hydroxides decreased from 45 to 29%, and the carbonate bound and exchangeable Cd fractions increased from 13 to 19% and 1 to 8%, respectively [101].

Cadmium accumulation in the soil is affected by soil factors, such as organic matter, pH, redox potential (Eh), and microbial community proliferation, with soil pH being the most important factor followed by redox potential. Soil redox potential affects Cd accumulation by varying electron acceptors and redox status. Cadmium usually exists as a $CdCO_3$ precipitate and CdS under reduced conditions and in a Cd^{2+} or as soluble salts under oxidizing conditions [32, 83]. Due to the reasons that reducing conditions in the soil affects organic matter mineralization and soil mineral content, this implies similar effects on Cd availability. Redox potential is related to Fe redox cycling due to varying dry and wet conditions, this scenario can cause many changes in soils including immobilization of Cd [26]. Cadmium is usually adsorbed onto Fe–Mn oxy-hydroxides surface, but reducing conditions result in the dissolution of Fe–Mn oxy-hydroxides and thereby increased cadmium concentration in the soil. However, reducing conditions coupled with pH increase due to lime application reduces soluble fractions of Cd in soils this is attributed to the precipitation of metal sulfide and relatively high pH. On the other side, the oxidizing conditions support decreased cadmium availability in the soil. Decrease of cadmium to soil solution is enhanced under an oxidizing environment (up to + 676 mV) as compared to a reducing environment (270 mV) [75].

Cd bioavailability is also affected by soil pH, by which the acidic pH usually results in greater Cd bioavailability implying an increase of Cd with decreasing soil pH. Acidic pH also increases Cd transformation from a non-available like Fe and manganese oxide bound, and carbonate bound forms to simpler more available form [100]. A recent study on cadmium uptake by rice and availability in the soil, reveals that the concentrations of cadmium in the rice grain were negative and correlated significantly with the pH of the soil [30]. Also, field experiments showed an increase in pH from 5.0 to 7.0 giving a significant Cd concentrations reduction in rice grain. Meanwhile, long-term flooding implies low redox potential, and high soil pH reduces Cd availability since long-term flooding increased pH and negative charges in the soils and thus increased Cd adsorption onto colloids, which significantly decreased exchangeable Cd in soils [27].

Cadmium availability also depends on the amount and sources of organic matter present in the soil. The commonly known human-made organic source that contains high levels of Cd is sewage sludge [30, 32, 83]. Cadmium availability in the soil is affected by organic matter due to induced physicochemical properties change in the soil, such as alteration in the amount pH, EC, and micro- and macronutrients [37]. Organic amendments to soils initially increase Cd immobilization through an increase in the surface charge which influence greatly the Cd retention and immobilization [32]. However, this reduction in Cd availability influenced by Organic Matter does not impose an effect on the reduction of the total Cd soil concentrations. Materials with high content in organic matter result in decreased Cd availability in soil, due to the chemistry fact that the organic matter could not bind Cd due to the high cation exchange capacity of the material [30, 32].

For example, 75% of the cadmium concentration in wheat grain is reduced by the addition of chicken manure [83]. Compost and biogas slurry amendments also significantly decrease Cd bioavailability. The application of biochar from rice husk and municipal solid waste compost significantly affected the adsorption and availability of Cd by converting Cd fractions into more unavailable forms. Biochar decreased Cd uptake by 19% for jack beans and 42% for Mucunaaterrima. Wheat straw biochar application also lower Cd by up to 43% in soil and up to 61% in a rice crop [83, 100]. Organic matter mineralization results in a reduction in oxygen levels and the amount of active Cd in soils can increase as Cd retention decreases [30, 100]. For example, biochar amendment decreased the bioavailable Cd fraction by 56% in soil. Rice straw and bamboo biochar were used to examine changes in Cd speciation in soil. Biochar of rice straw reduced the extractable acid Cd by 11% and increased the organically bound Cd by 37% [77].

Microorganisms also influence Cd availability in soils via the mobilization processes such as protonation, chelation, and chemical transformation where sorption or precipitation process by microorganisms hinders Cd availability. Soluble Cd is usually reduced chemically by microbes and fixed through microbial adsorption. Arbuscular mycorrhizal fungi showed a considerable influence on Cd availability and uptake [77, 100]. Rhizosphere bacteria have a significant effect on phytoremediation by altering the availability of metals through oxidation/reduction reactions, releasing siderophores and organic acids, by biosorption, and by changing soil pH. Microbially produced organic acids with low molecular weight are important factors determining the solubility of cadmium in soil because organic acids bind cadmium ions strongly by complexation processes in the soil solution [27, 30]. The production of extracellular polymeric substances, proteins, and mucopolysaccharides by plant-associated microbes significantly contributes to the complexing of poisonous cadmium and decreases their availability and mobility in soils [37]. Some microorganisms can change Cd mobility using oxidation or reduction reactions. For example, Iron/Sulfur oxidizing bacteria possess the capability to increase Cd availability by acidification reaction in the soils while iron-reducing and oxidizing bacteria synergistically improved the availability of Cd [100].

3.3 Permissible Levels of Cadmium in Soil and Plant Tissue

The permissible levels of cadmium in soil and plant tissues are usually examined by the ecological environment effects in various countries by which the methods of determining Maximum Allowable Concentration (MAC) also vary between countries. The permissible amounts change depending on the country or the regulating body. China uses the MAC to establish quality standards and develop maximum allowable levels of cadmium in agricultural soils [76]. Germany and France developing MAC in agricultural soil by not allowing any contamination of cadmium in the food chain with regards to plant products [2]. Cadmium allowable levels in Japan are usually examined in the food standard, although the United States Environment Protection Agency (USA–EPA) soil screening guidance for cadmium is based on the human health risk assessment approach. Also, Britain sets soil guideline values to assess the risks posed by exposure of humans to cadmium contamination to health [91].

The permissible levels of cadmium in soils for agriculture are 0.03 mg kg^{-1} and in plant tissues are 0.002 mg kg^{-1} as per WHO guidelines. Agricultural production of crops beyond these 21 permissible levels of lead and cadmium poses a great human health risk [24]. The USA–EPA set allowable limits of 0.01 mg kg^{-1} in agricultural soils and 0.2 mg kg^{-1} in vegetable dry weight also USA–EPA limits cadmium concentrations in irrigation water at 5 mg L^{-1} and 0.01 mg L^{-1} respectively. Other countries like Holland allow levels between 8 and 5 mg kg^{-1}, and China allows cadmium levels of 3–6 mg kg^{-1}. Slovakian sales network also allows cadmium levels from 0.03 to 0.10 mg kg^{-1} in vegetables [91]. Also, according to Kenya National Environment Management Authority (NEMA) allowable limits for cadmium concentrations in plant tissues is 0.05–0.2 mg kg^{-1} with the authority remaining silent on concentrations in agricultural soils.

4 Mechanism of Cadmium Uptake and Translocation in Plant Tissue

Plants' mechanism response to the increased levels of cadmium varies depending greatly on the ability of plant species to take up and translocate cadmium. Transportation of cadmium within plants usually is in the form of metallo-organic compounds, although, the mechanism is quite complex [39]. The availability of cadmium is associated with various parameters including,cadmium concentration, temperature, soil pH, and redox potential [23]. Therefore cadmium uptake mechanism by the plant roots is usually linked up with the competition in the exchange complexes between the cadmium and other nutrient minerals with similar valence charges [15, 39]. This competition can come up with a reduction of exchangeable bases in the plant tissues, and antagonism effects of cadmium with zinc though their active absorption occurs in some plants like lettuce [19, 39]. But also mineral nutrients with dissimilar chemical characteristics to cadmium can be affected just like nitrate [80].

An electrochemical potential difference of Cd^{2+} concentration between cytosol and the root apoplasts regulates cadmium uptake across the plasma membrane. Higher negative potential supplies enough energy that drives Cd^{2+} uptake by the plant roots at the low level of Cd^{2+} [19, 55]. Cd^{2+} uptake by plant roots shows its biphasic nature with saturable components correlated by membrane transport systems namely as; Saturable cation movement and diffusive movement [50, 55]. Also, cadmium absorption occurs as complexes form organic (phytometallophore) and inorganic cadmium ($CdCl^{+}$, $CdCl_2$, and $CdSO_4$). For instance, the Zn (II) phytometallophore complexes are absorbed by plant roots because of high specificity in the binding sites to the plasma membrane of the root, which allows the transport of cadmium [47, 97].

In the xylem pathway, the process needs some conditions including a pH of between 5.0 and 6.0, higher oxidizing redox potential, and low organic compounds (sugars, peptides, and proteins) concentration. Cadmium forms bonds with sulphydryl, N, and O ligand groups. Therefore in the xylem pathway phytochelatins, glutathione, cysteine, different organic citrate and amino acids are very important in transporting cadmium from roots to shoots because they stimulate Cd uptake in the plant roots [56, 86]. At the Phloem pathway, the alkaline pH condition between pH 7.0–8.0 and plenty of amino acids, organic acids, sugars, peptides, and proteins are required to ensure Cd^{2+} transport in the phloem pathway, which favors the stability of sulphydryl-containing ligands as carriers of Cadmium [56]. However, it is reported that phytochelatins and phytometallophores play a vital role in cadmium transport across the phloem into the seeds and grains [95].

Cadium penetrates and enters into the root system using the xylem via apoplastic and symplastic way and phloem pathway thereby reaching the plant tissues of aerial parts (Fig. 4) [103, 106]. Due to the variation in metal ions mobility in the plant's tissues, the cadmium concentration is greater in the root than in the aerial plant tissues [61, 70, 79]. The logic implies that the high amount of cadmium retained in the plant roots and small portions are translocated and stored to plant shoots, the flowing order of cadmium in the plant tissue is higher in the root followed by leaves parts, then

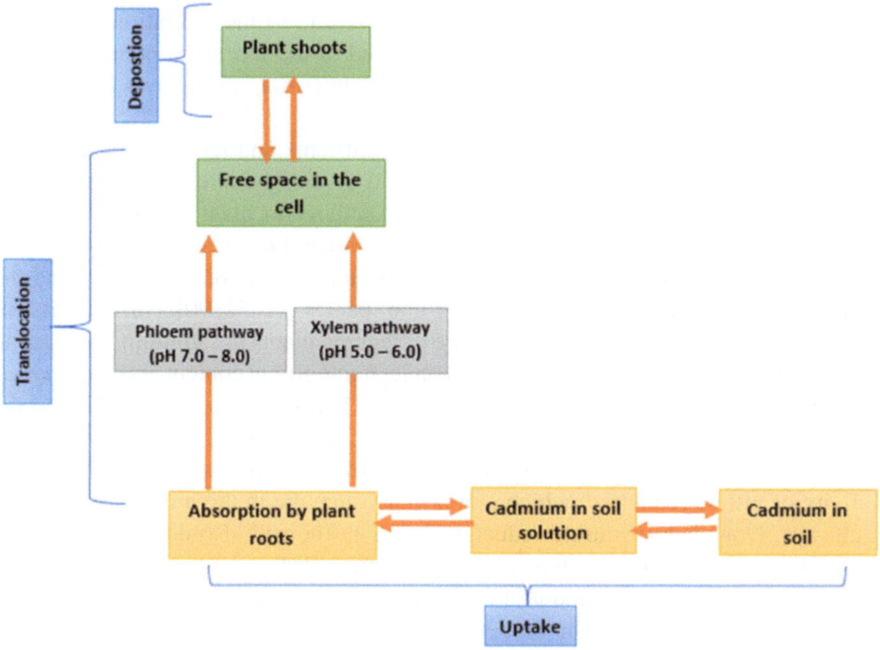

Fig. 4 Schematic flow indicating cadmium update, translocation, and deposition

fruits parts and the last parts with a low concentration of cadmium is a seed [13, 16, 88].

5 Mechanism of Cadmium Toxicity in Plant Tissue and Its Impacts

Cadmium imposes great toxicity in plant tissues once taken up through plant roots. Plant metabolism processes are negatively affected by cadmium accumulation, which reduces physiological function and processes within the plants [10]. Cadmium inhibits the activity of nitrate reductase enzyme and hence reduces the nitrate uptake and its transportation from plant roots to shoots. This results in a reduction in leaf area, plant height, root length, and fresh and dry mass accumulation [20, 36]. At low concentrations, cadmium is less toxic to plants as compared to high doses where the phytotoxic symptoms include stunted growth and chlorosis. Cadmium chlorosis effects are due to changes in iron to zinc (Fe:Zn) ratio and inhibition of iron, calcium, magnesium, manganese, potassium, and phosphorous uptake by plants [36]. When in plants, cadmium increases non-photochemical and decreases chlorophyll and carotenoids amount. Similarly, in some plant species, cadmium influences a decrease in the synthesis and level of chlorophyll [59]. Stunted growth of plants associated with the accumulation of cadmium is due to the inhibition of protein synthesis and its phytotoxicity in plants observed in the form of protein loss. This can be well proven in grains grown under cadmium stress, which gives low protein content [36, 67].

Cadmium inhibits the availability of plant nutrients and reduces the microbial population, transport of Ca, Mg, P, and K, and diversity due to its toxicity effect [10, 22]. It inhibits nitrate reductase enzyme activity in plant shoots, absorption of nitrate by plants, and nitrogen fixation in legumes [41]. Cadmium presence in the soil lower the biomass of nitrogen-fixing bacteria (*Phaseolus vulgaris)* and the number of nodules formed per plant in legume species [22, 52, 90]. Nitrogen assimilation is severely affected by the exposure of legume plants to cadmium. There is an occurrence of a positive correlation between leghemoglobin content and nitrogenase enzyme activities with cadmium accumulation. There is also a decrease due to the oxidation stress posted by cadmium that triggers deterioration in the formation of nodules [9, 18]. Cadmium in the soil affects nitrate assimilation by inhibiting the activity of nitrate reductase, which is a primary enzyme in the nitrate assimilation pathways. Nonetheless, nitrate reductase levels in the soil also depend on various environmental conditions such as pH, and moisture [25].

Plants possess a quite amount of antioxidant systems as a protection mechanism against oxidative damage. The first enzyme in the detoxifying process is Superoxidase (SOD) which converts oxygen (O_2) radicals to peroxide (H_2O_2) at an extremely high rate [68]. Cadmium has oxidative stress that mainly occurred in two main ways, first by bringing on oxygen-free radical production and second by lowering concentrations of enzymatic and non-enzymatic antioxidants [21]. Defense systems of plants consist of several metabolites (glutathione, tocopherol, and superoxide dismutases ascorbate) and activated oxygen enzymatic scavengers (catalases and peroxidases). In the higher plant, the uptake of heavy metal toxic quantity can lead to peroxidase induction [21, 25]. Cadmium accumulation in plants can hinder or energize the performance of several antioxidative enzymes [54]. For example in the leaves of *Halianthus annuus*, cadmium accumulation reduces the activity of superoxidase, catalase, glutathione reductase, glutathione reductase, ascorbate peroxidase and dehydroascorbate reductase, enhanced lipid peroxidation, and increased lipoxygenase activity. Also, the leaves of soybean and *Oryza sativa* induce the activity of peroxidase [54].

Cadmium reduces cell division in meristematic cells hence reducing plant growth and biomass yield [36]. It reduces the root length but increases the root diameter due to an increase in parenchyma cells and cortex tissue cells, which alleviates salt and water stress in plants as a response to Cadmium stress [64, 85]. It reduces chlorophyll synthesis and alters chloroplast structure hence reducing the rate of photosynthesis. Inhibition in photosynthesis is also due to the inhibition of enzymatic activities involved in carbon (iv) oxide fixation, promoting lipid peroxidation, and altering nitrogen and sulphur metabolism [36, 66, 71]. Cadmium reduces the viability and germination rate of seeds. It also inhibits pollen germination by altering endomembrane organelles, inhibiting endocytosis and exocytosis, and the formation of acidic vacuoles [12, 33].

Cadmium is hazardous to humans and the environment even at very small concentrations with effects on pregnancy, lactation, kidney, liver, pancreas, heart, and testis [33]. Human exposure occurs through inhaling or ingestion of contaminated food especially leafy vegetables [12], and significantly through cigarette smoking [66]. It causes chronic toxicity in humans [66]. Cd has carcinogenic effects and causes osteotoxicity, cardiovascular abnormalities [12], renal disorders, hypertension, and diabetes [33]. Additionally, it disrupts the endocrine system, male infertility, insulin resistance, and suppresses the immune system [12]. Its toxicity in the body is through causing epigenetic alteration in DNA expression, reduced tubule transport pathways inhibit heme-synthesis, and competitive interference with the physiology of zinc, and magnesium exchangeable sites and extracellular precipitation deposition on cell walls [12]. Metals such as lead at higher concentrations damage the cells, and the semi-permeability function of cell membranes and tonoplast, and hence some metals may be transported via symplast [87].

6 Cadmium Toxicity Remediation Strategies

The remediation strategies of cadmium may involve the use of plants, a process referred to as phytoremediation. This is the specific type of bioremediation that strictly involve the use of plants for removing or remediating pollutants. It is a feasible method simply because the plants are naturally increasing the level of bioavailability of some pollutants by storing them in their shoots and leaves [104]. The accumulated pollutants in the plants, which are mainly stored in an inactive form in the cells get removed through harvesting of the affected plants to remove the stored pollutant from the ecosystem or it can be emitted to the atmosphere through the volatilization process [73]. Phytoremediation with hyper-accumulator plants is most attractive for the removal of heavy metal pollutants from soil because of the larger number of heavy metals that such hyper-accumulators can tolerate and sequester.

The choice of plant is significant in phytoremediation because a high root surface area to soil contact is important for enhanced metal deposition [37]. High biomass yield and rapid growth rate are the main characteristics considered in selecting plants suitable for phytoextraction. This can include naturally growing weeds and grasses such as *Ludwigia parviflora*, *Cyperus rotundus*, *Marselia quadrifolia* and *Cyperus kylinga*, which are all reported to grow well in heavy metal contaminated areas and are thus good choices for the remediation [5, 29, 31]. Coupling of phytoremediation together with metal-chelating compounds or symbiotic relationships with mycorrhiza is a significant technique in phytoremediation. More so cadmium can be cleaned up by mycorrhiza, which involves the chelation of the metal in the fungus body where it gets adsorbed in the cell walls chitin, and precipitate as polyphosphate granules [102].

Bioremediation refers to the involvement of biological materials like plants and microbes in achieving pollutant cleaning or removal. These biological materials may deteriorate the pollutant by complete removal (mineralization) or by transforming the pollutant into less toxic forms (biotransformation) and this is done by the majority of bacteria and fungi available in soil [38, 62, 73, 102]. In a given environment, the microbial remediation of heavy metals is affected and influenced by several factors like ionic strength, pH, calcium concentration, methylation of cadmium, binding of cadmium to cell walls, and heavy metal concentration [11]. In most bacteria, the cell wall contains different functional groups that are hydroxyl, sulfhydryl, phosphate, and carboxyl work as a Cd adsorption complex, which means that a significant number of bacterial biomass can decrease Cd bioavailability in soil. Thus they are more effective in sandy than in clay soils due to the favorable environmental conditions that support bacteria survival [82]. This can be the best and most efficient solution because Cd immobilized on G9 *Pseudomonas fluorescens* and Tp8 *Bacillus subtilis* successfully remove Cd from agricultural soil [84]. Bacteria reduce sulfate by precipitating it into metal sulfide and thus lowering the amount of cadmium available in the plant. The application of alginate for microbial soil inoculation is also capable of reducing Cd [35, 82].

In addition, various soil amendments have been successfully used for Cd remediation. These soil amendments include humus and lime, Fe/Mn oxide, paper mill sludge, alkaline organic treatment, poultry manure, bentonite, leaf litter, beringite, sewage sludge, charcoal, silica, mineral rock phosphate and phosphate, clay minerals, limestone, silica slag, vermicompost, cattle manure, and compost, peat [37, 69, 73, 82].

7 Conclusion

Heavy metal accumulation in soil like cadmium is a serious threat to soil ecosystems, plants, and food systems. Cadmium deposition in the soil results in not only a reduction of soil microbial activities and microbe biomass of bacteria, fungi, and actinomycetes but also enzymatic activities of invertase, urease, catalase, sulfatase, and acidic and alkaline phosphatase. Cadmium accumulation also highly affects plant metabolism by the formation of reactive oxygen species, enhancing lipid peroxidation, imposing changes in intercellular glutathione levels, and also inhibiting enzyme activity, which finally damages the cellular components such as lipids, protein, and nucleic acid. Adoption of various remediation strategies aimed at reducing or removing cadmium accumulation in the soil and plant such as bioremediation and phytoremediation is one of the best environmentally friendly ways that can help in achieving these goals while protecting the environment as well as human health for better and sustainable economic development.

References

1. A-El-Amier, Y., Alghanem, S., & Alzuaibr, F. (2017). Bioaccumulation and translocation of heavy metals from coastal soil by wild halophytes. *American Journal of Environmental Protection, 5*(2), 52–60. https://doi.org/10.12691/env-5-2-4
2. Adams, M., Zhao, F., McGrath, S., Nicholson, F., & Chambers, B. (2004). Predicting cadmium concentrations in wheat and barley grain using soil properties. *Journal of Environment Quality, 33*, 532–541.
3. Afzal, M., Shabir, G., Iqbal, S., Mustafa, T., Khan, Q., & Khalid, Z. (2013). Assessment of heavy metal contamination in soil and groundwater at leather industrial area of Kasur, Pakistan. *CLEAN—Soil, Air, Water, 42*(8), 1133–1139. https://doi.org/10.1002/clen.201100715
4. Akinola, M., & Ekiyoyo, T. A. (2006). Accumulation of lead, cadmium and chromium in some plants cultivated along the bank of river Ribila at Odo–nla area of Ikorodu, Lagos state, Nigeria. *Journal of Environmental Biology, 27*(3), 597–599.
5. Alhammad, B. A., Roy, D. K., Ranjan, S., Padhan, S. R., Sow, S., Nath, D., Seleiman, M. F., & Gitari, H. (2023). Conservation tillage and weed management influencing weed dynamics, crop performance, soil properties, and profitability in rice-wheat-greengram system in Eastern Indo-Gangetic Plains. *Agronomy, 13*, 1953. https://doi.org/10.3390/agronomy13071953
6. Ali, M., & Nas, F. S. (2018). The effect of lead on plants in terms of growing and biochemical parameters: A review. *MOJ Ecology & Environmental Sciences, 3*(4), 2. https://doi.org/10.15406/mojes.2018.03.00098
7. Alloway, B. J. (1995). *Heavy metal in soils* (p. 368). Blackie Academic and Professional.
8. Babula, P., Adam, V., Opatrilova, R., Zehnalek, J., Havel, L., & Kizek, R. (2009). Uncommon heavy metals, metalloids and their plant toxicity: A review. *Organic Farming, Pest Control and Remediation of Soil Pollutants*, 275–317.
9. Balestrasse, K. B., Gardey, L., Gallego, S. M., & Tomaro, M. L. (2001). Response of antioxidant defence system in soybean nodules and roots subjected to cadmium stress. *Functional Plant Biology, 28*(6), 497–504.
10. Benavides, M. P., Gallego, S. M., & Tomaro, M. L. (2005). Cadmium toxicity in plants, Brazilian. *Journal of Plant Physiology, 17*(1), 21–34. https://doi.org/10.1590/s1677-04202005000100003
11. Beolchini, F., Dell Anno, A., De Propris, L., Ubaldini, S., Cerrone, F., & Danovaro, R. (2009). Auto- and heterotrophic acidophilic bacteria enhance the bioremediation efficiency of sediments contaminated by heavy metals. *Chemosphere, 74*(10), 1321–1326.
12. Bernhoft, R. A. (2013). Cadmium toxicity and treatment. *The Scientific World Journal, 2013*, 1–7. https://doi.org/10.1155/2013/394652
13. Blum, W. H. (1997). Cadmium uptake by the higher plants. *Proceeding of Extended Abstracts from the fourth International Conference on the Biogeochemistry of Trace Elements* (pp. 109–110). University of California.
14. Bostick, B. C., Fendorf, S., & Fendorf, M. (2000). Disulfide disproportionation and CdS formation upon cadmium sorption on FeS_2. *Geochimica et Cosmochimica Acta, 64*, 247–255.
15. Burzynski, M. (1988). The uptake and accumulation of phosphorus and nitrates and the activity of nitrate reductse in cucumber seedlings treated with Pb and Cd. *Acta Societatis Botanicorum Poloniae, 57*, 349–359.
16. Cataldo, C. D., Garland, T. R., & Widung, R. E. (1983). Cadmium uptake, kinetics in intact soybean plants. *Plant Physiology, 73*, 844–848.
17. Chibuike, G. U., & Obiora, S. C. (2014). Heavy metal polluted soils: Effect on plants and bioremediation methods. *Applied and Environmental Soil Science, 2014*, 1–12. https://doi.org/10.1155/2014/752708
18. Comba, M. E., Benavides, M. P., Gallego, S. M., & Tomaro, M. L. (1997). Relationship between nitrogen fixation and oxidative stress induction in nodules of salt–treated soybean plants. *Phyton, 48*.
19. Costa, G., & Morel, J. L. (1994). Water relations gas exchange and amino acid content in Cd treated lettuce. *Plant Physiology and Biochemistry, 32*, 561–570.

20. Das, P., Samantaray, S., & Rout, G. R. (1997). Studies on cadmium toxicity in plants: A review. *Environmental Pollution, 98*(1), 29–36. https://doi.org/10.1016/s0269-7491(97)00110-3
21. Demirevska-Kepova, K., Simova-Stoilova, L., Stoyanova, Z. P., & Feller, U. (2006). Cadmium stress in barley: Growth, leaf pigment, and protein composition and detoxification of reactive oxygen species. *Journal of Plant Nutrition, 29*(3), 451–468. 61.
22. Dhingra, H. R., & Priefer, U. B. (2006). Impact of cadmium on structural and functional aspects of pea (*Pisum sativum* L.) root nodules. *Journal of Plant Biology–New Delhi, 33*(3), 201.
23. Epstein, E., & Bloom, A. J. (2005). *Mineral nutrition of plant principles and perspective* (2nd ed.). Sinauer Associates, Inc.
24. FAO/WHO. (1995). *Report of the 27th session of the codex committee on food additives and contaminants.* Codex Alımentarıus Commıssıon.
25. Fernández-Pascual, M., De Lorenzo, C., De Felipe, M. R., Rajalakshmi, S., Gordon, A. J., Thomas, B. J., & Minchin, F. R. (1996). Possible reasons for relative salt stress tolerance in nodules of white lupin cv. Multolupa. *Journal of Experimental Botany, 47*(11), 1709–1716. 49.
26. Gallego, S. M., Pena, L. B., Barcia, R. A., Azpilicueta, C. E., Iannone, M. F., & Rosales, E. P. (2012). Unravelling cadmium toxicity and tolerance in plants: Insight into regulatory mechanisms. *Environmental Experimental Botany, 83*, 33–46. https://doi.org/10.1016/j.envexpbot.2012.04.006
27. Gigliotti, G., & Massaccesi, L. (2013). Cadmium chemical speciation and absorption in plant in a polluted soil.
28. Gweyi-Onyango, J. P., & Osei-Kwarteng, M. (2011). Safe vegetable production with wastewater in developing countries: Demystifying negative notions. *African Journal of Horticultural Science, 5*, 70–83.
29. Hamid, Y., Tang, L., Sohail, M. I., Cao, X., Hussain, B., Aziz, M. Z., Usman, M., He, Z. L., & Yang, X. (2019). An explanation of soil amendments to reduce cadmium phytoavailability and transfer to food chain. *Science Total Environment, 660*, 80–96.
30. Han, X. Q., Xiao, X. Y., Guo, Z. H., Xie, Y. H., Zhu, H. W., Peng, C., & Liang, Y. Q. (2018). Release of cadmium in contaminated paddy soil amended with NPK fertilizer and lime under water management. *Ecotoxicology of Environment Safety, 159*, 38–45.
31. Hassan, M. J., Raza, M. A., Rehman, S. U., Ansar, M., Gitari, H., Khan, I., Wajid, M., Ahmed, M., Shah, G. A., Peng, Y., & Li, Z. (2020). Effect of cadmium toxicity on growth, oxidative damage, antioxidant defense system and cadmium accumulation in two sorghum cultivars. *Plants, 9*, 1575. https://doi.org/10.3390/plants9111575
32. Healey, J. R., Nason, M. A., Williamson, J. C., & Jones, D. L. (2009). Heavy metal fractionation during the co–composting of biosolids, deinking paper fibre and green waste. *Bioresource Technology, 100*(18), 4220–4226. https://doi.org/10.1016/j.biortech.2009.02.046. PMID: 19386494.
33. Honey S, Bless B, Neetu R. (2015). The Characteristics, toxicity and effects of cadmium. *International Journal of Nanotechnology and Nanoscience*, 3 (August 2016), 1–9.
34. Hu, B., Jia, X., Hu, J., Xu, D., Xia, F., & Li, Y. (2017). Assessment of Heavy Metal Pollution and Health Risks in the Soil–Plant–Human System in the Yangtze River Delta, China. *International Journal of Environmental Research and Public Health, 14*(9), 1042. https://doi.org/10.3390/ijerph14091042
35. Iravani S, Varma RS. (2020). Bacteria in Heavy Metal Remediation and Nanoparticle Biosynthesis. *ACS Sustainable Chemical and Engineering*, 2020.
36. Ismael, M. A., Elyamine, A. M., Moussa, M. G., Cai, M., Zhao, X., & Hu, C. (2019). Cadmium in plants: Uptake, toxicity, and its interactions with selenium fertilizers. *Metallomics, 11*(2), 255–277. https://doi.org/10.1039/c8mt00247a
37. Jackel, U., Russo, S., & Schnell, S. (2005). Enhanced iron reduction by iron supplement: A strategy to reduce methane emission from paddies. *Soil Biology and Biochemistry, 37*, 2150–2154.

38. Jaishankar, M., Tseten, T., Anbalagan, N., Mathew, B., & Beeregowda, K. N. (2014). Toxicity, mechanism and health effects of some heavy metals. *Interdisciplinary Toxicology, 7*(2), 60–72. https://doi.org/10.2478/intox-2014-0009
39. Jarvis, C., Jones, L. H. P., & Hopper, M. J. (1976). Cadmium uptake from solution by plants and its transport from roots to shoots. *Plant and Soil, 44*, 179–191.
40. Javed, M. T., Tanwir, K., Akram, M. S., Shahid, M., Niazi, N. K., Lindberg, S. (2019). Phytoremediation of cadmium–polluted water/sediment by aquatic macrophytes: Role of plant–induced pH changes. In *Cadmium toxicity and tolerance in plants* (pp. 495–529). Academic Press.
41. Jena, J., Maitra, S., Hossain, A., Pramanick, B., Gitari, H. I., Praharaj, S., Shankar, T., Palai, J. B., Rathore, A., Mandal, T. K., & Jatav, H. S. (2022). Role of legumes in cropping system for soil ecosystem improvement. In H. S. Jatav & V. D. Rajput (Eds.), *Ecosystem services: Types, management and benefits*. Nova Science Publishers, Inc. 415 Oser Avenue, Suite N Hauppauge, NY, 11788 USA. https://doi.org/10.52305/PFZA6988
42. Kabir, E., Ray, S., Kim, K. H., Yoon, H. O., Jeon, E. C., Kim, Y. S., Cho, Y. S., Yun, S. T., & Brown, R. J. C. (2012). Current status of trace metal pollution in soils affected by industrial activities. *Science World Journal*. https://doi.org/10.1100/2012/916705
43. Kevrešan, S., Krstić, B., Popović, M., Kovačev, L., Pajević, S., & Kandrač, J. (1997). Biochemical changes in sugar beet lines in dependence on soil moisture. *Biologia Plantarum, 40*(2), 245–250.
44. Kiende, J. (2012). Assessment of heavy metal concentrations in urban grown vegetables in Thika Town, Kenya. *African Journal of Food Science, 6*(3), 1. https://doi.org/10.5897/ajfs10.072
45. Kubier, A., Wilkin, R. T., & Pichler, T. (2019). Cadmium in soils and groundwater: A review. *Applied Geochemistry, 108*, 104388.
46. Kumar, A., Kumal, A., Cabral-Pinto, M., Chaturvedi, A., Shabnam, A., Subrahmanyam, G., Raju, G., Dipak, K., Malyan, S., Kumar, S., Khan, S., & Yadav, K. (2020). Lead toxicity: Health hazards, influence on food chain, and sustainable remediation approaches. *International Journal of Environmental Research and Public Health, 17*(7), 2179.https://doi.org/10.3390/ijerph17072179
47. Kuppelwieser, H., & Feller, U. (1991). Transport of Rb and Sr to the ear in mature excised shoots of wheat effects of temperature and stem length on Rb removal from the xylem. *Plant and Soil, 132*, 281–299.
48. Lans, C., Boer, F., & Flings, A. (2012). Vegetable chains in Kenya: Production and consumption of vegetables in the Nairobi metropolis (1130). Wageningen UR Greenhouse Horticulture. Retrieved on June 13, 2023 from https://edepot.wur.nl/216710
49. Lu, Y., Ma, J., Teng, Y., He, J., Christie, P., Zhu, L., Ren, W., Zhang, M., & Deng, S. (2018). Effect of silicon on growth, physiology, and cadmium translocation of tobacco (*Nicotiana tabacum* L.) in cadmium–contaminated soil. *Pedosphere, 28*(4), 680–689.
50. McLaughlin, M. J., Tiller, K. G., Naidu, R., & Stevens, D. P. (1996). Reviewd: The behaviour and impact of contaminants in fertilizers. *Austrian Journal of Soil Resourcs, 34*, 1–54.
51. Mehes-Smith. (2013). Mobility of heavy metals in plants and soil: A case study from a mining region in Canada. *American Journal of Environmental Sciences, 9*(6), 483–493. https://doi.org/10.3844/ajessp.2013.483.493
52. Mirriam, A., Mugwe, J., Raza, M. F., Seleiman, M. A., Maitra, S., & Gitari, H. I. (2022). Aggrandizing soybean yield, phosphorus use efficiency and economic returns under phosphatic fertilizer application and inoculation with *Bradyrhizobium. Journal of Soil Science and Plant Nutrition, 22*, 5086–5098. https://doi.org/10.1007/s42729-022-00985-8
53. Mirzapour, M., Valizadeh, N., Heydarzadeh, S., Rahimi, A., & Gitari, H. (2022). Factors affecting the production of secondary metabolites in medicinal plants. In: G. Ozyazici (Ed.), *New development on medicinal and aromatic plants-II*. Iksad Publishing House (pp. 3–32). ISBN: 978-625-8213-33-1. https://iksadyayinevi.com/home/new-development-on-medicinal-and-aromatic-plants-ii

54. Mohan, B. S., & Hosetti, B. B. (2006). Phytotoxicity of cadmium on the physiological dynamics of *Salvinia natans* L. grown in macrophyte ponds. *Journal of Environmental Biology, 27*(4), 701–704. 62.
55. Morel, R. G., Palacios, I., Gomez, J., & Navarro-Pedreno, M. J. (1994). Distribution and accumulation of heavy metal (Cd, Ni and Cr) in tomato plants. *Fresenius Environmental Bulletin, 3*, 395–399.
56. Mori, S., Nishizawa, N., Hayashi, H., Chino, M., Yoshimura, E., & Ishihara, J. (1991). Why are young rice plants highly susceptible to iron deficiency. In Y. Chen & Y. Hadav (Eds.), *Iron nutrition and interactions in plants* (pp. 175–188). Kluwer Academic Publishing.
57. Muui, C., Ngugi, M. M., Gweyi-Onyango, J. P., & Gitari, H. I. (2022). Influence of silicon on translocation, compartmentation and uptake of lead in leafy vegetables. *International Journal of Agriculture, Environment, and Biotechnology, 15*, 33–50. https://doi.org/10.30954/0974-1712.01.2022.5
58. Mwakidoshi, E. R., Gitari, H. I., Muindi, E. M., Wamukota, A. W., Seleiman, M. F., & Maitra, S. (2023). Smallholder farmers' knowledge of the use of bioslurry as a soil fertility amendment for potato production in Kenya. *Land Degradation & Development, 34*, 2214–2227. https://doi.org/10.1002/ldr.4601
59. Nasar, J., Khan, W., Khan, M. Z., Gitari, H. I., Gbolayori, J. F., Moussa, A. A., Mandozai, A., Rizwan, N., Anwari, G., & Maroof, S. M. (2021). Photosynthetic activities and photosynthetic nitrogen use efficiency of maize crop under different planting patterns and nitrogen fertilization. *Journal of Soil Science and Plant Nutrition., 21*, 2274–2284. https://doi.org/10.1007/s42729-021-00520-1
60. Ngugi, M. M., Gitari, H. I., Muui, C., & Gweyi-Onyango, J. P. (2021). Cadmium mobility, uptake and accumulation in spinach, kale and amaranths vegetables as influenced by Silicon fertilization. *Bioremediation Journal., 26*, 113–127. https://doi.org/10.1080/10889868.2021.1924111
61. Ngugi, M. M., Gitari, H. I., Muui, C. W., & Gweyi-Onyango, J. P. (2022). Growth tolerance, concentration, and uptake of heavy metals as ameliorated by silicon application in vegetables. *International Journal of Phytoremediation., 24*, 1543–1556. https://doi.org/10.1080/15226514.2022.2045251
62. Nyawade, S. O., Karanja, N. N., Gachene, C. K. K., Gitari, H. I., Schulte-Geldermann, E., & Parker, M. L. (2019). Short-term dynamics of soil organic matter fractions and microbial activity in smallholder legume intercropping systems. *Applied Soil Ecology., 142*, 123–135. https://doi.org/10.1016/j.apsoil.2019.04.015
63. Nzeve, J. K. (2015). Assessment of heavy metal contamination in Masinga Reservoir, Kenya. M.Sc. Thesis, Kenyatta University, Kenya.
64. Ochieng' I. O., Ranjani, S., Seleiman, M. F., Padhan, S. R., Psiwa, R., Sow, S., Wasonga, D. O., & Gitari, H. I. (2023). Increasing rainwater use efficiency, gross return, and grain protein of rain-fed maize under nitrate and urea nitrogen forms. *Notulae Botanicae Horti Agrobotanici Cluj-Napoca, 51*, 13293. https://doi.org/10.15835/nbha51313293
65. Pacyna, J. M., & Pacyna, E. G. (2001). An assessment of global and regional emissions of trace metals to the atmosphere from anthropogenic sources worldwide. *Environmental Review., 9*, 269–298.
66. Parada, A., Pardiñas, U., Salazar-Bravo, J., D'Elía, G., & Palma, R. (2013). Dating an impressive neotropical radiation: Molecular time estimates for the sigmodontinae (rodentia) provide insights into its historical biogeography. *Molecular Phylogenetics and Evolution, 66*(3), 960–968.
67. Phetsombat, S., Kruatrachue, M., Pokethitiyook, P., & Upatham, S. (2006). Toxicity and bioaccumulation of cadmium and lead in *Salvinia cucullata. Journal of Environmental Biology, 27*(4), 645–652.
68. Pál, M., Horváth, E., Janda, T., Páldi, E., & Szalai, G. (2006). Physiological changes and defense mechanisms induced by cadmium stress in maize. *Journal of Plant Nutrition and Soil Science, 169*(2), 239–246. 58.

69. Rahimi, A., Gitari, H., Lyons, G., Heydarzadeh, S., Tuncturk, M., & Tuncturk, R. (2023). Effects of vermicompost, compost and animal manure on vegetative growth, physiological and antioxidant activity characteristics of *Thymus vulgaris* L. under water stress (2022). *Yuzuncu Yil University Journal of Agricultural Sciences, 32*, 40–53. https://doi.org/10.29133/yyutbd.1124458
70. Ramas, I., Esteban, E., Lucena, J. J., & Gorate, A. (2002). Cadmium uptake and subcellular distribution in plants of Lactuca sp. Cd–Mn interaction. *Plant Science, 162*, 761–767.
71. Raza, M. A., Yasin, H. S., Gul, H., Qin, R., Din, A. M. U., Khalid, M. H. B., Hussain, S., Gitari, H., Saeed, A., Wang, J., Rezaei-Chiyaneh, E., Sabagh, A. E., Manzoor, A., Fatima, A., Ahmad, S., Yang, F., Sikalicky, M., & Yang, W. (2022). Maize/soybean strip intercropping produces higher crop yields and saves water under semi-arid conditions. *Frontiers in Plant Science, 13*, 1006720. https://doi.org/10.3389/fpls.2022.1006720
72. Rehman, S., De Castro, F., Marini, P., Aprile, A., Benedetti, M., & Fanizzi, F. P. (2023). Vermibiochar: A novel approach for reducing the environmental impact of heavy metals contamination in agricultural land. *Sustainability, 15*(12), 9380. https://doi.org/10.3390/su15129380
73. Riaz, U., Murtaza, G., & Farooq, M. (2018). Influence of different sewage sludges and composts on growth, yield, and trace elements accumulation in rice and wheat. *Land Degradation Development, 29*(5), 1343–1352.
74. Richardson, M. G., Garrett, R., Mitchell, I. A., Mah-Paulson, M., & Hackbarth, T. (2001). Critical review on natural global and regional emissions of six trace metals to the atmosphere international lead zinc research organisation, the international copper association, and the nickel producers. *Environmental Research Association* (p. 52).
75. Rinklebe, J., Shaheen, S. M., & Yu, K. (2016). Release of As, Ba, Cd, Cu, Pb, and Sr under pre–definite redox conditions in different rice paddy soils originating from the U.S.A. and Asia. *Geoderma, 270*, 21–32. https://doi.org/10.1016/j.geoderma.2015.10.011
76. Rodrigues, S., Pereira, E., Duarte, A., & Römkens, P. (2012). Derivation of soil to plant transfer functions for metals and metalloids: Impact of contaminant's availability. *Plant and Soil, 361*, 329–341.
77. Rodriguez, L., Ruiz, E., & Alonso-Azcarate, J. (2009). Heavy metal distribution and chemical speciation in tailings and soils around a Pb Zn mine in Spain. *Journal Environmental Management, 2009*(90), 1106–1116. https://doi.org/10.1016/j.jenvman.2008.04.007
78. Sairaam, M., Maitra, S., Praharaj, S., Nath, S., Shankar, T., Sahoo, U., Santosh, D. T., Sagar, L., Panda, M., Priya, G. S., Ashwini, T. R., Gaikwad, D. J., Hossain, A., Pramanick, B., Jatav, H. S., Gitari, H. I., & Aftab, T. (2023). An insight into the consequences of emerging contaminants in soil and water and plant responses. In T. Aftab (Ed.), *Emerging contaminants and plants*. Springer. https://doi.org/10.1007/978-3-031-22269-6_1
79. Salt, D. E., & Rauser, W. E. (1995). MgATP–dependent transport of phytochelatins across the tonoplast of oat roots. *Plant Physiology, 107*(4), 1293–1301.
80. Sanita di Toppi, L., Lambardi, M., Pazzagli, L., Cappugi, G., Durante, M., & Gabbrielli, R. (1998). Response to cadmium in carrot in vitro plants and cell suspension cultures. *Plant Science, 137*, 119–129.
81. Sankhla, M., Kumari, M., Nandan, M., Kumar, R., & Agrawal, P. (2016). Heavy metals contamination in water and their hazardous effect on human health—A review. *SSRN Electronic Journal*. https://doi.org/10.2139/ssrn.3428216
82. Sarin, C., & Sarin, S. (2010). Removal of cadmium and zinc from soil using immobilized cell of biosurfactant producing bacteria. *Environment Asian, 3*(2), 49–53. https://doi.org/10.14456/ea.2010.23
83. Sarwar, N., Malhi, S. S., Zia, M. H., Naeem, A., Bibi, S., & Farid, G. (2010). Role of mineral nutrition in minimizing cadmium accumulation by plants. *Journal of Science and Food Agriculture, 90*(6), 925–937. https://doi.org/10.1002/jsfa.3916. PMID: 20355131.
84. Sebastian, A., & Prasad, M. N. V. (2013). Cadmium minimization in rice. A review. *Agronomy Sustainable Development, 34*(1), 155–173. https://doi.org/10.1007/s13593-013-0152-y

85. Seleiman, M. F., Aslam, M. T., Alhammad, B. A., Hassan, M. U., Maqbool, R., Chattha, M. U., Khan, I., Gitari, H. I., Uslu, O. S., Roy, R., & Battaglia, M. L. (2021). Salinity stress in wheat: Effects, mechanisms and management strategies. *Phyton-International Journal of Experimental Botany, 91*, 667–694. https://doi.org/10.32604/phyton.2022.017365
86. Senden, M. H., Van der, A. J., Meer, G. M., Verburg, T. G., & Wolterbeek, H. T. (1994). Effect of cadmium on the behaviour of citric acid in isolated tomato xylem cell walls. *Journal of Experimental. Botany, 45*, 597–606.
87. Sharma, P., & Dubey, R. S. (2005). Lead toxicity in plants. *Brazilian Journal of Plant Physiology, 17*(1), 35–52. https://doi.org/10.1590/s1677-04202005000100004
88. Sharma, M., Gautam, K. H., & Handique, A. K. (2006). Toxic heavy metal stress in paddy: Metal accumulation profile and development of a novel stress protein in seed. *Indian Journal of Plant Physiology, 11*, 227–233.
89. Smolders, E., & Mertens, J. (2013). Cadmium. In J. B. Alloway (Ed.), *Heavy metals in soils—Trace metals and metalloids in soils and their bioavailability* (3rd ed., pp. 283–299). Springer.
90. Sousa, W. S., Soratto, R. P., Peixoto, D. S., Campos, T. S., da Silva, M. B., Souza, A. G. V., Teixeira, I. R., & Gitari, H. I. (2022). Effects of *Rhizobium* inoculum compared with mineral nitrogen fertilizer on nodulation and seed yield of common bean. A meta-analysis. *Agronomy for Sustainable Development, 42*(3), 52. https://doi.org/10.1007/s13593-022-00784-6
91. Stanovič, R., Július, Á., Ľuboš, H., Michal, K., & Tomáš, T. (2014). Cadmium and nickel content in frozen spinach from Slovakia sales network. *JMBFS, 3*, 289–290.
92. Su, C., Jiang, L., & Zhang, W. (2014). A review on heavy metal contamination in the soil worldwide: Situation, impact and remediation techniques. *Environmental Skeptics & Critics, 3*(2), 24–38.
93. Tabelin, C. B., Igarashi, T., Villacorte-Tabelin, M., Park, I., Opiso, E. M., Ito, M., & Hiroyoshi, N. (2018). Arsenic, selenium, boron, lead, cadmium, copper, and zinc in naturally contaminated rocks: A review of their sources, modes of enrichment, mechanisms of release, and mitigation strategies. *Science Total Environment, 645*, 1522–1553.
94. Tangahu, B., Sheikh Abdullah, S., Basri, H., Idris, M., Anuar, N., & Mukhlisin, M. (2011). A review on heavy metals (As, Pb, and Hg) uptake by plants through phytoremediation. *International Journal of Chemical Engineering, 2011*, 1–31. https://doi.org/10.1155/2011/939161
95. Taylor, K., Albrigo, L. G., & Chase, C. D. (1988). Zinc complexation in the phloem of blight affected citrus. *Journal of American Society of Horticulture Science, 113*, 407–411.
96. Thornton, I. (1986). Geochemistry of cadmium. In H. Mislin & O. Ravera (Eds.), *Cadmium in the environment* (pp. 7–12). Birkhaeuser.
97. Von Wiren, N., Marschner, H., & Romheld V. (1996). Roots of iron-efficient maize also absorb phytosiderophore chelated zinc. *Plant Physiology. 111*, 1119 –1125.
98. Wambua, J. M., Korir, N., Ngene, S. M., & Gweyi-Onyango, J. (2018). Heavy metal concentration in amaranthus, cowpea, black nightshade and kale in response to wastewater irrigation in Kitui County, Kenya. *African Journal of Rural Development, 3*(4), 1025–1033.
99. Wambua, J. M., Ngene, S., Korir, N. K., Ntinyari, W., & Gweyi-Onyango, J. P. (2019). Differential biomass accumulation among African leafy vegetables as affected by wastewater irrigation in Kitui County, Kenya. *International Journal of Plant & Soil Science*. https://doi.org/10.9734/ijpss/2019/v27i630095
100. Wang, A. S., Angle, J. S., Chaney, R. L., Delorme, T. A., & Reeves, R. D. (2006). Soil pH effects on uptake of Cd and Zn by *Thlaspi caerulescens. Plant and Soil, 281*, 325–337. https://doi.org/10.1007/s11104-005-4642-9
101. Wang, Q. Y., Zhou, D. M., Cang, L., Li, L. Z., & Wang, P. (2009). Solid/solution Cu fractionations/speciation of a Cu contaminated soil after pilot–scale electrokinetic remediation and their relationships with soil microbial and enzyme activities. *Environmental Pollution, 157*(8–9), 2203–2208. https://doi.org/10.1016/j.envpol.2009.04.013
102. Wang, Y., Zhu, C., Yang, H., & Zhang, X. (2017). Phosphate fertilizer affected rhizospheric soils: Speciation of cadmium and phytoremediation by *Chlorophytum comosum. Environment Science and Pollution Resources International, 24*(4), 3934–3939.

103. Welch, R. M. (1995). Micronutrient nutrition of plants. Critical review. *Plant Science, 14*, 49–82.
104. Wenzel, W. W. (2009). Rhizosphere processes and management in plantassisted bioremediation (phytoremediation) of soils. *Plant and Soil, 321*, 385–408.
105. Wuana, R., & Okieimen, F. (2011). Heavy metals in contaminated soils: A review of sources, chemistry, risks and best available strategies for remediation. *ISRN Ecology, 2011*, 1–20. https://doi.org/10.5402/2011/402647
106. Yang, M. G., Lin, X., & Yang, X. E. (1998). Impact of Cd on growth and nutrient accumulation of different plant species. *China Journal of Applied Ecology, 19*, 89–94.
107. Yu, H. Y., Liu, C., Zhu, J., Li, F., Deng, D. M., Wang, Q., & Liu, C. (2016). Cadmium availability in rice paddy fields from a mining area: The effects of soil properties highlighting iron fractions and pH value. *Environmental Pollution, 209*, 38–45.
108. Zama, E., Reid, B., Sun, G., Yuan, H., Li, X., & Zhu, Y. (2018). Silicon (Si) biochar for the mitigation of arsenic (As) bioaccumulation in spinach (*Spinacia oleracean*) and improvement in the plant growth. *Journal of Cleaner Production, 189*, 386–395. https://doi.org/10.1016/j.jclepro.2018.04.056

Health Risks Linked to Cadmium Toxicity

Health Risks Associated with Cadmium Toxicity

Govindarao Kuna, Simhachalam Gullipalli, and Vivek Chintada

Abstract The chapter begins with an introduction that familiarizes readers with cadmium as a toxic heavy metal and its various sources, including industrial activities, soil and water contamination, and cigarette smoke. The objective of the chapter is then presented, which is to delve into the health risks associated with cadmium toxicity. The second section focuses on cadmium uptake and accumulation in the body. It explains the different routes through which cadmium enters the body, such as inhalation, ingestion, and dermal absorption. Furthermore, the chapter explores the bioavailability and absorption mechanisms of cadmium in different organs. The factors influencing the accumulation and distribution of cadmium within the body are also discussed in detail. Moving on to the core of the chapter, the third section examines the wide-ranging health effects caused by cadmium exposure. Both acute and chronic effects are explored, with an emphasis on the organs most heavily impacted by cadmium, including the kidneys, liver, lungs, and bones. Furthermore, the chapter delves into the mechanisms by which cadmium induces its toxic effects, such as oxidative stress, disruption of cellular processes, and interference with essential nutrients. The subsequent sections offer a deeper exploration of specific health risks associated with cadmium toxicity. The chapter discusses the link between cadmium exposure and kidney disease, detailing the mechanisms through which cadmium affects renal function and presenting epidemiological evidence supporting the association. Similarly, the respiratory effects of cadmium are examined, highlighting the risk of chronic obstructive pulmonary disease (COPD) and lung cancer due to cadmium-induced inflammation, fibrosis, and impaired lung function. Additionally, the detrimental impact of cadmium on bone health is explored, with an emphasis on the interference with bone metabolism, decreased bone mineral density, and increased risk of osteoporosis and fractures. The chapter also addresses other health risks related to cadmium toxicity, such as cardiovascular diseases, reproductive disorders, and cancer. The mechanisms by which cadmium contributes to the development of

G. Kuna · S. Gullipalli
Department of Zoology, Acharya Nagarjuna University, Guntur, Andhra Pradesh 522510, India

V. Chintada (✉)
Department of Zoology, Sri Venkateswara University, Tirupati, Andhra Pradesh 517502, India
e-mail: vivek.chintada@gmail.com

A. K. Jha and N. Kumar (eds.), *Cadmium Toxicity in Water*, Springer Water,
https://doi.org/10.1007/978-3-031-54005-9_3

these conditions are discussed in depth, underlining the urgent need for preventative measures and interventions to reduce cadmium exposure. In conclusion, this book chapter provides a comprehensive overview of the health risks associated with cadmium toxicity. By exploring the mechanisms through which cadmium induces toxicity and examining its impact on various organs and systems, the chapter highlights the importance of awareness, prevention, and regulations to mitigate cadmium exposure. It concludes by offering recommendations for future research and actions to minimize the adverse health effects of cadmium.

Keywords Cadmium toxicity · Heavy metal exposure · Bioavailability · Absorption mechanisms · Cadmium distribution · Acute health effects · Chronic health effects · Health risks

1 Introduction

Cadmium (Cd) is a toxic heavy metal that poses significant health risks to humans and ecosystems. It is widely distributed in the environment and can be found in varying concentrations in soil, water, and air [25]. Cadmium is primarily released into the environment through industrial activities such as mining, smelting, and battery manufacturing. It can also enter the environment through agricultural practices, including the use of phosphorous-based fertilizers, sewage sludge application, and irrigation with contaminated water [4, 57]. Once released, cadmium can persist in the environment for a long time and bioaccumulate in various organisms, including plants, animals, and humans. It enters the human body through multiple routes, including inhalation of air contaminated with cadmium particles, ingestion of food and water containing cadmium, and dermal absorption in certain occupational settings [55, 61].

Excessive cadmium exposure has been linked to a wide range of adverse health effects. It is classified as a human carcinogen by the International Agency for Research on Cancer (IARC) and has been associated with various types of cancers, including lung, prostate, and kidney cancer [2, 22, 23]. Moreover, cadmium is known to be nephrotoxic, leading to severe kidney damage and impairing kidney function. Chronic exposure to cadmium has been linked to the development of chronic kidney disease (CKD) and renal cancer [58]. In addition to its renal effects, cadmium also poses respiratory risks. Inhalation of cadmium fumes or dust can lead to respiratory symptoms such as coughing, wheezing, and pulmonary inflammation. It has been associated with an increased risk of chronic obstructive pulmonary disease (COPD) and lung cancer [13, 50, 51].

Furthermore, cadmium has deleterious effects on the cardiovascular system. It has been implicated in the development of hypertension, endothelial dysfunction, and atherosclerosis, thereby increasing the risk of cardiovascular diseases, including heart disease and stroke [60]. Cadmium toxicity also affects bone health. Chronic exposure to cadmium inhibits osteoblast activity, promotes osteoclast activity, and interferes with bone metabolism, leading to decreased bone mineral density and increased risk of osteoporosis and fractures [7].

In conclusion, cadmium is a toxic heavy metal with diverse health risks. Its presence in the environment and widespread exposure through various routes raise concerns about potential adverse effects on human health. Understanding the health risks associated with cadmium toxicity is crucial for implementing effective preventive measures and reducing exposure to this hazardous metal.

2 Cadmium Sources, Such as Industrial Activities, Contamination of Soil and Water, and Cigarette Smoke

Cadmium (Cd) is a toxic heavy metal that can be found in various environmental sources, including industrial activities, contamination of soil and water, and cigarette smoke. Industrial activities, such as mining, smelting of non-ferrous metals, and battery manufacturing, are major contributors to cadmium pollution [25]. These industries release cadmium into the atmosphere and surrounding environment through emissions, waste disposal, and effluents. Additionally, industrial wastewater containing cadmium can contaminate water bodies, further contributing to its presence in the soil and water [50, 51].

Contamination of soil and water is another significant source of cadmium exposure. Agricultural practices, including the use of phosphate-based fertilizers and the application of sewage sludge, can introduce cadmium into the soil [4]. Cadmium has a long half-life in soil, which increases the risk of uptake by plants and subsequent entry into the food chain. Irrigation with contaminated water can also lead to the accumulation of cadmium in crops [37]. Furthermore, cigarette smoke is a significant source of cadmium exposure, particularly for smokers. Tobacco plants readily accumulate cadmium from the soil, and during smoking, this cadmium is released in the form of tobacco smoke. It has been reported that tobacco smoke contains high levels of cadmium, which can result in significant exposure for smokers and passive smokers [6].

A study by Liu et al. [36] found that industrial emissions, agricultural practices, and tobacco smoke were the three major sources of cadmium exposure in rural areas in China. The authors highlighted the importance of reducing cadmium emissions from industrial activities, implementing proper waste disposal practices, and establishing regulations to limit cadmium content in agricultural products [36]. To address cadmium contamination in soil and water, measures such as soil management strategies, proper waste disposal, and water treatment techniques can be implemented [65]. In addition, public awareness campaigns on the hazards of cadmium exposure and the promotion of alternative, cleaner industrial practices can contribute to reducing cadmium pollution [50, 51, 53].

In summary, industrial activities, contamination of soil and water, and cigarette smoke are significant sources of cadmium exposure. The understanding of these sources and their contributions to cadmium pollution is crucial in developing effective strategies to mitigate cadmium contamination and reduce the associated health risks.

3 Health Risks Associated with Cadmium Toxicity

The objective of this chapter is to provide a comprehensive exploration of the health risks associated with cadmium toxicity. Cadmium is a toxic heavy metal that poses significant health hazards to humans and ecosystems. Understanding the potential adverse effects of cadmium exposure is essential for assessing and managing the risks to human health. Cadmium has been extensively studied for its toxicological properties and its association with various health conditions. Chronic exposure to cadmium has been linked to a wide range of adverse health effects, including renal dysfunction, respiratory diseases, cardiovascular disorders, impaired bone health, and increased cancer risk [22, 23, 27, 29, 61, 66].

One of the major health risks associated with cadmium toxicity is its nephrotoxicity. Chronic exposure to cadmium can lead to severe damage to the kidneys, impairing their function and potentially causing chronic kidney disease (CKD) [45]. Cadmium is known to accumulate in the kidneys, leading to renal tubular dysfunction, proteinuria, and increased risk of various renal disorders [54]. Cadmium exposure has also been linked to respiratory health risks. Inhalation of cadmium fumes or dust can lead to respiratory symptoms such as coughing, wheezing, and pulmonary inflammation. Long-term exposure to cadmium has been associated with an increased risk of chronic obstructive pulmonary disease (COPD) and lung cancer [6, 63].

Moreover, cadmium toxicity has detrimental effects on cardiovascular health. It has been implicated in the development of hypertension, endothelial dysfunction, and atherosclerosis, thereby increasing the risk of cardiovascular diseases, including heart disease and stroke [3, 60]. Furthermore, chronic cadmium exposure can adversely affect bone health. Cadmium interferes with bone metabolism, inhibiting osteoblast activity, promoting osteoclast activity, and leading to decreased bone mineral density. This increases the risk of osteoporosis and fractures. Additionally, cadmium has been classified as a human carcinogen and linked to the development of various types of cancers, including lung, prostate, and kidney cancer. Its carcinogenic potential is associated with promoting tumor growth, angiogenesis, and metastasis.

By exploring these health risks associated with cadmium toxicity, this chapter aims to enhance the understanding of the potential dangers posed by cadmium exposure. Such knowledge can contribute to the development of preventive measures, awareness campaigns, and regulations to minimize cadmium contamination and protect public health.

4 Cadmium Uptake and Accumulation in the Body

Cadmium is a toxic heavy metal that can accumulate in the human body. Understanding how cadmium enters the body through various routes, including inhalation, ingestion, and dermal absorption, is crucial for assessing the potential health risks associated with exposure to this element.

Cadmium Entry Routes

Cadmium can enter the body through inhalation of fine cadmium-containing particles or metal fumes present in occupational and environmental settings. Industries involved in metal refining, battery production, and mining are notable sources of airborne cadmium [1]. Inhalation is considered a significant route of exposure, particularly in occupational settings where high concentrations of cadmium can be present [25].

Ingestion

The consumption of cadmium-contaminated food and water is another major route of exposure. Cadmium enters the food chain through uptake by plants from contaminated soil or water sources. Shellfish, rice, cereals, and leafy vegetables are known to accumulate cadmium (EFSA). Tobacco smoking is also a significant source of cadmium exposure, with its absorption occurring mainly through inhalation and subsequent ingestion via tobacco residues [5].

Dermal Absorption

Although dermal absorption is not as significant as inhalation or ingestion, it can still contribute to the overall cadmium burden in the body. Occupational exposure to cadmium-containing compounds, such as cadmium salts and battery electrolytes, can lead to dermal absorption [3, 42]. Additionally, exposure to certain cosmetics, such as hair dyes, may introduce cadmium into the body through dermal absorption [12].

Cadmium can enter the human body through inhalation, ingestion, and dermal absorption, with different routes contributing to varying degrees of exposure. Inhalation and ingestion are the primary routes of concern, particularly in occupational settings and through the consumption of contaminated food and water. Understanding these routes of entry is crucial for implementing preventive strategies to minimize cadmium exposure and protect human health.

5 Bioavailability and Absorption Mechanisms of Cadmium in Different Organs

Understanding the bioavailability and absorption mechanisms of cadmium (Cd) in different organs is crucial for comprehending its toxicological effects and developing effective preventive strategies. This discussion aims to explain the bioavailability of cadmium and the mechanisms governing its absorption in various organs within the human body, supported by relevant references.

1. **Bioavailability of Cadmium**

Bioavailability refers to the extent to which a substance is absorbed and becomes available for biological processes. Cadmium poses unique challenges due to its complex chemistry and varying forms in different exposure sources. Environmental factors and solubility influence cadmium's bioavailability. Studies have shown that cadmium binds to proteins and forms complexes, affecting its solubility and bioavailability [46]. Furthermore, pH, redox potential, presence of ligands, and interactions with other metals in the environment can alter its behavior and absorption [2].

2. **Absorption Mechanisms in Different Organs**

Cadmium absorption occurs via various mechanisms in different organs, including the gastrointestinal tract, lungs, kidneys, and liver.

(a) **Gastrointestinal Tract**

Ingested cadmium is primarily absorbed in the gastrointestinal (GI) tract. The absorption of cadmium in the small intestine is influenced by several factors, such as its solubility, chemical form, and interactions with transport mechanisms. Studies have indicated that dietary factors, particularly essential nutrients like iron and calcium, can compete with cadmium for absorption, reducing its uptake in the GI tract [14]. Additionally, the metal can cross intestinal epithelial cells via passive diffusion, specific transporters, or endocytosis [55] (Klaassen et al., 2013).

(b) **Lungs**

Inhalation of airborne cadmium particles can lead to its absorption in the lungs. Particles can be deposited in the respiratory tract, mainly in the alveolar region, where they can undergo dissolution or interaction with lung lining fluids. Cadmium can cross the alveolar epithelium through a combination of passive diffusion, active transporters, and endocytosis [27].

(c) **Kidneys**

Cadmium has a strong affinity for the kidneys, where it accumulates and causes significant damage. Renal absorption of cadmium occurs through the proximal tubules. Metallothioneins (MTs) play a crucial role in the uptake, intracellular transport, and detoxification of cadmium within renal cells [28]. Specific transporters, such as divalent metal transporter-1 (DMT1) and zinc transporter-2 (ZnT2), are involved in cadmium absorption in the kidney [21].

(d) **Liver**

In the liver, cadmium can be taken up by hepatocytes through various mechanisms. The metal can bind to metal transporters, such as divalent metal transporter-1 (DMT1), organic ion transporters (OATs), and copper transporters (Ctr1), for its entry into hepatocytes [18]. Once inside the liver, cadmium can induce oxidative stress and disrupt cellular functions, contributing to liver toxicity [11].

Cadmium's bioavailability and absorption mechanisms in different organs involve complex processes influenced by multiple factors, such as solubility, chemical form, competing nutrients, and specific transporters. Understanding these mechanisms is essential for evaluating the toxicological effects of cadmium and developing preventive strategies to minimize human exposure to this toxic metal.

6 Factors Affecting the Accumulation and Distribution of Cadmium Within the Body

The accumulation and distribution of cadmium (Cd) within the human body are influenced by various factors that determine the extent and severity of its toxicological effects. Understanding these factors is crucial for assessing Cd-associated health risks and implementing effective preventive measures. This discussion aims to explore the primary factors affecting the accumulation and distribution of Cd in the body, citing relevant references.

1. **Absorption and Bioavailability**

The bioavailability and absorption of Cd play a crucial role in its accumulation within the body. Factors such as solubility, chemical form, and interaction with other substances can affect Cd's absorption through routes such as ingestion, inhalation, and dermal exposure. For instance, the gastrointestinal tract readily absorbs Cd, influenced by dietary factors like the presence of iron and calcium, which can compete with Cd for absorption [14]. Inhalation of airborne Cd particles and dermal absorption are other routes that contribute to its accumulation.

2. **Renal Reabsorption and Excretion**

Cd is known for its affinity to accumulate in the kidneys, with the proximal tubules being the primary site of Cd storage. The factors affecting renal accumulation and distribution include renal transporters and metallothionein (MT) proteins. Renal MTs, especially MT2A, play a crucial role in Cd retention and subsequent nephrotoxicity [49]. Additionally, renal transporters such as divalent metal transporter 1 (DMT1) and metal transporter protein 1 (MTP1) influence Cd reabsorption, leading to its prolonged presence in the kidneys [2, 34]. The balance between Cd uptake and efflux transporters is vital in determining renal Cd accumulation.

3. **Metallothionein Expression**

Metallothioneins (MTs) are a family of low-molecular-weight proteins involved in regulating essential and toxic metal homeostasis. Cd can bind to MTs, leading to the formation of Cd-MT complexes that protect cells from Cd-induced toxicity. The expression and availability of MTs in different tissues and organs play a significant role in Cd accumulation and distribution [10]. Genetic variations in MT genes can impact individual susceptibility to Cd-induced toxicity and influence Cd distribution patterns within the body.

4. **Binding to Intracellular Targets**

Cd accumulates in various intracellular compartments, including the liver, lungs, and bone. The binding of Cd to intracellular targets, such as metal-binding proteins, enzymes, and DNA, influences its distribution within cells and tissues. Cd can displace other essential metals like zinc and calcium, disrupting cellular functions. The affinity of Cd for binding proteins and other cellular components determines its intracellular localization and toxicological impact [33].

The accumulation and distribution of Cd within the body are influenced by factors such as absorption and bioavailability, renal reabsorption and excretion, expression of metallothioneins, and specific binding to intracellular targets. Understanding these factors is crucial for assessing and mitigating the health risks associated with Cd exposure. Further research in this area is needed to fully comprehend the complex mechanisms underlying the accumulation and distribution of Cd in the human body.

7 Cadmium-Related Health Effects

Acute and chronic health effects caused by cadmium exposure

Cadmium exposure has been linked to several acute health effects. Acute exposure to high levels of cadmium can result in symptoms such as nausea, vomiting, abdominal cramps, and diarrhea [1]. Ingestion of food or water contaminated with high cadmium levels has been known to cause acute poisoning, which can manifest as acute gastrointestinal distress [26]. However, it is important to note that chronic exposure to low levels of cadmium poses a greater health risk. Chronic cadmium exposure primarily occurs through occupational exposure, contaminated food, or tobacco smoke inhalation [15, 19]. The chronic exposure route is particularly concerning, as it has been associated with a wide range of adverse health effects.

Cadmium is known to have toxic effects on the kidneys. Prolonged exposure to cadmium can lead to renal dysfunction, including tubular damage, proteinuria, and impaired kidney function [52]. Chronic cadmium exposure has also been linked to the development of renal diseases, such as chronic kidney disease and kidney cancer

[8]. Another significant health effect of chronic cadmium exposure is its impact on the skeletal system. Cadmium accumulates in bone tissue, disrupting calcium metabolism, and impairing bone health [41]. This can lead to an increased risk of osteoporosis and fractures among individuals exposed to cadmium.

The respiratory system is also affected by chronic cadmium exposure. Inhalation of cadmium-containing dust or fumes can cause lung damage and respiratory disturbances, including a condition called cadmium-induced pneumonitis [35]. Chronic exposure to cadmium has also been associated with an increased risk of developing lung cancer [22, 23]. Other adverse health effects associated with chronic cadmium exposure include cardiovascular abnormalities, such as hypertension and increased risk of cardiovascular diseases [40], as well as potential effects on the reproductive system [64]. The toxic effects of cadmium are not limited to one specific organ but can affect multiple systems in the body.

In conclusion, cadmium exposure poses a significant risk to human health. Acute exposure can lead to gastrointestinal symptoms, while chronic exposure to low levels can cause severe damage to the kidneys, skeletal system, respiratory system, and other vital organs. Recognizing and minimizing exposure to cadmium is crucial for preserving health and preventing associated diseases.

8 Target Organs Affected by Cadmium, Such as the Kidneys, Liver, Lungs, and Bones

Cadmium is known to affect various organs in the human body. Let's discuss some of the primary target organs and their susceptibility to cadmium toxicity.

1. **Kidneys**

The kidneys are particularly vulnerable to cadmium toxicity. Chronic exposure to cadmium can lead to severe damage to the renal tubules and impairment of kidney function. This often manifests as tubular proteinuria, indicating protein leakage into the urine [24]. Cadmium-induced renal dysfunction can progress to chronic kidney disease, with an increased risk of renal failure [44]. The accumulation of cadmium in the kidneys is primarily due to the binding of cadmium to metallothionein, a protein that plays a role in cadmium detoxification [19, 32].

2. **Liver**

The liver is another target organ affected by cadmium toxicity, although to a lesser extent compared to the kidneys. Cadmium exposure has been associated with hepatotoxic effects, including hepatic dysfunction and damage to liver cells [15]. Cadmium accumulation in the liver can disrupt cellular processes and contribute to oxidative stress, which has been linked to liver injury [48].

3. **Lungs**

Inhalation of cadmium-containing dust or fumes represents a significant route of exposure to this toxic metal. Consequently, the lungs are highly susceptible to the harmful effects of cadmium. Chronic exposure to cadmium in the respiratory system can lead to various respiratory conditions, such as chronic obstructive pulmonary disease (COPD) [56]. Cadmium-induced pneumonitis, characterized by inflammation of lung tissue, can also occur [39]. Additionally, long-term exposure to cadmium is associated with an increased risk of lung cancer [22, 23].

4. **Bones**

Cadmium has a high affinity for binding to bone tissue, leading to its accumulation in bones over time. This accumulation disrupts calcium metabolism and impairs bone health. Chronic cadmium exposure is known to contribute to an increased risk of osteoporosis and fractures [41]. The mechanisms underlying cadmium-induced bone damage are complex and involve disturbances in osteoblast and osteoclast function, leading to impaired bone formation and increased bone resorption [9].

It is important to note that cadmium can also impact other organs, such as the cardiovascular system [47] and the reproductive system [59]. The toxic effects of cadmium are often systemic and can manifest in multiple organs simultaneously.

9 Mechanisms Through Which Cadmium Induces Toxicity, Including Oxidative Stress, Disruption of Cellular Processes, and Interference with Essential Nutrients

Cadmium (Cd) is a highly toxic heavy metal that poses considerable risks to human health and environmental well-being. Understanding the underlying toxic mechanisms is crucial for developing effective strategies to mitigate its harmful effects. This discussion aims to explore the mechanisms by which cadmium induces toxicity, focusing on oxidative stress, disruption of cellular processes, and interference with essential nutrients.

Oxidative Stress

Cadmium has been shown to generate reactive oxygen species (ROS) and disrupt the balance between the production and elimination of free radicals, leading to oxidative stress. ROS, such as superoxide anion ($O_2^{\bullet-}$) and hydroxyl radicals ($^{\bullet}OH$), can cause damage to lipids, proteins, and DNA within cells. Elevated levels of ROS result in the depletion of antioxidant defense systems, including glutathione and superoxide dismutase, further exacerbating oxidative stress [62]. Studies have indicated the involvement of Cd-induced oxidative stress in various organs, including the liver, kidneys, and lungs [20].

Disruption of Cellular Processes

Cadmium disrupts vital cellular processes by affecting multiple molecular targets. It can interfere with calcium homeostasis, disrupting calcium-dependent signaling pathways involved in cell proliferation, differentiation, and apoptosis [29]. Cd can also disrupt mitochondrial function, leading to impaired ATP synthesis, electron transport chain dysfunction, and increased production of ROS [30]. Furthermore, cadmium exposure has been linked to endoplasmic reticulum (ER) stress, which perturbs protein folding and triggers the unfolded protein response (UPR), ultimately affecting cellular viability and function [16].

Interference with Essential Nutrients

Cadmium can interfere with the metabolism and cellular uptake of essential nutrients, exacerbating its toxic effects. One of the primary mechanisms involves cadmium mimicking essential metals, such as zinc (Zn) and calcium (Ca), leading to their displacement and disruption of normal cellular functions [31]. Cd can competitively bind to metallothioneins (MTs), which are involved in the regulation of zinc and copper homeostasis, impairing their protective role against oxidative stress [43]. Additionally, Cd can interfere with the absorption, transport, and metabolism of essential minerals, such as iron (Fe) and selenium (Se), which are crucial for various biological processes, including antioxidant defense and DNA repair [17, 38].

10 Conclusions

Cadmium-induced toxicity is a complex process involving multiple mechanisms, including oxidative stress, disruption of cellular processes, and interference with essential nutrients. Oxidative stress mediated by ROS generation plays a central role in cadmium toxicity. Disruption of cellular processes, such as calcium homeostasis, mitochondrial function, and ER stress, further contributes to its detrimental effects. Furthermore, cadmium interferes with essential nutrients, compromising their bioavailability and disrupting their cellular metabolism. Investigating these mechanisms will enable the development of effective strategies to mitigate cadmium-induced toxicity.

References

1. Agency for Toxic Substances and Disease Registry. (2012). *Toxicological profile for cadmium*. U.S. Department of Health and Human Services.
2. Akaras, N., Ileriturk, M., Gur, C., Kucukler, S., Oz, M., & Kandemir, F. M. (2023). The protective effects of chrysin on cadmium-induced pulmonary toxicity; a multi-biomarker approach. *Environmental Science and Pollution Research*, 1–16.

3. Al Marshoudi, M., Al Reasi, H. A., Al-Habsi, A., & Barry, M. J. (2023). Additive effects of microplastics on accumulation and toxicity of cadmium in male zebrafish. *Chemosphere, 334*, 138969.
4. Alloway, B. J. (2013). *Heavy metals in soils: Trace metals and metalloids in soils and their bioavailability* (3rd edn.). Springer.
5. Bernhard, D., Rossmann, A., & Wick, G. (2005). Metals in cigarette smoke. *IUBMB Life, 57*(12), 805–809. https://doi.org/10.1080/15216540500371076
6. Bernhard, D., & Rossmann, A. (2013). Wick, smoke, and inflammation: New insights into the regulation of cadmium transport by the lung epithelial barrier and cigarette smoke exposure. *Environmental Health Perspectives, 121*(3), 287–295.
7. Brodziak-Dopierała, B., Kwapuliński, J., Sobczyk, K., & Kulczyńska, M. (2017). Effects of environmental and occupational exposure to cadmium on bone density. *International Journal of Environmental Research and Public Health, 14*(2), 172.
8. Chen, H., Lu, W., & Guo, Y. (2016). Chronic effects of cadmium on bone in mice: Mitochondrial oxidative damage and the activation of MAPK signaling pathways. *Environmental Toxicology and Pharmacology, 41*, 109–116.
9. Choudhury, H., et al. (2018). Cadmium induces the generation of reactive oxygen species (ROS), endoplasmic reticulum stress, and mitochondrial dysfunction in osteoblastic cells. *Journal of Biomedical Materials Research Part A, 106*(9), 2473–2480.
10. Cruz, N. G., Cruz, F. A. B., Jesus, G. C. L., & Crespo-Lopez, M. E. (2018). *Metallothionein role in cadmium bioaccumulation and repartition in target organs.*
11. Da Silva, A. O. F., Bezerra, V., Meletti, P. C., Simonato, J. D., & dos Reis Martinez, C. B. (2023). Cadmium effects on the freshwater teleost *Prochilodus lineatus*: Accumulation and biochemical, genotoxic, and behavioural biomarkers. *Environmental Toxicology and Pharmacology, 99*, 104121.
12. European Commission. Scientific Committee on Consumer Safety. (SCCS). (2013). Opinion on Cadmium in Cosmetics.
13. Ferraro, P. M., Costanzi, S., & Naticchia, A. (2010). Cadmium exposure and body mass index: A meta-analysis of prospective studies. *International Journal of Obesity, 34*(6), 893–898.
14. Flanagan, S. V., Johnston, R. B., & Zheng, Y. (2013). Arsenic in tube well water in Bangladesh: Health and economic impacts and implications for arsenic mitigation.
15. Fowler, B. A., & Nordberg, G. F. (Eds.). (2013). *Handbook on the toxicology of metals* (4th ed.). Elsevier.
16. Fu, J., et al. (2017). Cadmium-induced endoplasmic reticulum stress in chicken myocardial cells via the Ca^{2+}-, mitochondria- and apoptosis-dependent pathways. *Toxicology, 390*, 50–57.
17. Gailer, J., et al. (2004). Cadmium: Cellular effects, modifications of biomolecules, modulation of DNA repair and genotoxic consequences (A Review). *Bioinorganic Chemistry and Applications, 2*(1–2), 33–52.
18. Gailer, J., George, G. N., Pickering, I. J., Madden, S., & Avdeef, A. (2008). Cytosolic cadmium (II)-, copper(II)-, and zinc(II)-binding proteins in degenerative storage diseases.
19. Górska, A., Markiewicz-Gospodarek, A., Markiewicz, R., Chilimoniuk, Z., Borowski, B., Trubalski, M., & Czarnek, K. (2023). Distribution of iron, copper, zinc and cadmium in glia, their influence on glial cells and relationship with neurodegenerative diseases. *Brain Sciences, 13*(6), 911.
20. Hossain, E., et al. (2018). Cadmium-induced oxidative stress, apoptosis and genotoxicity in human renal proximal tubular cells. *Environmental Toxicology and Pharmacology, 57*, 184–194.
21. Huang, J., Li, H., Ren, X., Zhao, Y., Nie, G., & Wu, W. (2019). *Roles of zinc transporters in cadmium transport and accumulation in HepG2 cells.*
22. International Agency for Research on Cancer. (2012a). *Arsenic, metals, fibres, and dusts* (Vol. 100 C). IARC.
23. International Agency for Research on Cancer. (2012). Cadmium and cadmium compounds. *IARC Monographs on the Evaluation of Carcinogenic Risks to Humans, 100*(1), 121–145.

24. International Programme on Chemical Safety (IPCS). (1992). *Environmental health criteria 134: Cadmium*. World Health Organization.
25. Jarup, L. (2003). Hazards of heavy metal contamination. *British Medical Bulletin, 68*(1), 167–182. https://doi.org/10.1093/bmb/ldg032
26. Järup, L. (2002). Hazards of heavy metal contamination. *British Medical Bulletin, 68*(1), 167–182.
27. Jarup, L., & Akesson, A. (2009). *Current status of cadmium as an environmental health problem.*
28. Jin, T., Nordberg, M., Frech, W., Dumont, X., Bernard, A., Ye, T., et al. (2014). *Cadmium biomonitoring and renal dysfunction among a population environmentally exposed to cadmium from smelting in China (ChinaCad).*
29. Jomova, K., et al. (2012). Arsenic: Toxicity, oxidative stress, and human disease. *Journal of Applied Toxicology: JAT, 31*(2), 95–107.
30. Joseph, P. (2009). Mechanisms of cadmium carcinogenesis. *Toxicology and Applied Pharmacology, 238*(3), 272–279.
31. Klaassen, C., et al. (2009). Metallothionein: An intracellular protein to protect against cadmium toxicity. *Annual Review of Pharmacology and Toxicology, 49*(1), 243–264.
32. Klaassen, C. D., et al. (2001). Cadmium. In O. Ballatori (Ed.), *Heavy metal toxicity: From molecules to organ systems* (pp. 143–157). Wiley.
33. Liu, L., Li, Y., Liang, C., Wu, C., Hong, F., & Zhang, W. (2019). A review of cadmium deposition and toxicity in the fetal and neonatal environments.
34. Liu, L., Yu, L., & Lin, X. (2018). Metal transporters involved in cadmium accumulation and tolerance in *Arabidopsis thaliana.*
35. Liu, Y., et al. (2014). Lung injury induced by cadmium in rats and the protective effect of erythropoietin. *Inhalation Toxicology, 26*(13), 799–805.
36. Liu, Y., Xiao, Z., Liao, X., Xia, W., Zhang, B., & Zhang, X. (2008). Cadmium contamination in soil and rice and its health risk assessment among inhabitants in Lechang, China. *Journal of Environmental Sciences, 20*(6), 664–669.
37. Marino, D., de Bolós, C., & Sacco, D. (2012). Cadmium in soil and crops: A review. *Emir Journal of Food and Agriculture, 24*(6), 549–568.
38. Mohd, F. H., Idrus, S. N. S., Abdullah, M. R., & Othman, M. S. (2023). Effect of cadmium on the developmental instability of *Anabas testudineus* through evaluation of fluctuating asymmetry of eye diameter. *Journal of Environmental Protection, 14*(5), 392–403.
39. Navas-Acien, A., et al. (2004). Lung cancer in a U.S. population with low to moderate arsenic exposure. *Environmental Health Perspectives, 112*(14), 1464–1469.
40. Navas-Acien, A., et al. (2009). Low-level cadmium exposure and cardiovascular outcomes in elderly women: A population-based study. *Environmental Health Perspectives, 117*(10), 1576–1580.
41. Nawrot, T. S., et al. (2010). Cadmium exposure in the population: From health risks to strategies of prevention. *BioMetals, 23*(5), 769–782.
42. Nordberg, G. F., Nogawa, K., & Nordberg, M. (2014). Cadmium. In: *Handbook on the toxicology of metals* (4th edn, pp. 665–712). Academic Press.
43. Nordberg, M., et al. (2020). Metallothioneins: Roles in copper homeostasis and cadmium toxicity. *FASEB Journal, 34*(11), 14441–14457.
44. Nordberg, G. F., et al. (2007). *Handbook on the toxicology of metals* (3rd ed.). Elsevier.
45. Nordberg, G. F., Jin, T., Wu, X., Lu, J., Chen, L., Lei, L., & Hong, F. (2009). Prevalence of kidney dysfunction in humans—relationship to cadmium dose, metallothionein, immunological and biochemical indices. *Biochimie, 91*(10), 1282–1285.
46. Nordberg, G. F., Nogawa, K., & Nordberg, M. (2015). Cadmium. In *Handbook on the toxicology of metals* (4th edn., pp. 667–716). Academic Press.
47. Ouyang, L., & Zhang, J. (2020). Cadmium and its cardiovascular toxicities: Epidemiological, pathological, and mechanistic evidence. *Boletín Médico del Hospital Infantil de México, 77*(2), 98–109.

48. Patra, R. C., & Swarup, D. (2005). Oxidative stress in lead and cadmium toxicity and its amelioration. *Veterinary Medicine International, 1*(2), 1–9.
49. Qu, W., Xu, G., Tang, B., Chen, D., Assaad, H., & Hu, Z. (2018). Metallothionein gene isoforms, mRNA expression levels, and protein content in different organs of aging rats and rats with acute cadmium exposure.
50. Rajabian, F., Rajabian, A., & Tayarani-Najaran, Z. (2023a). Protective effect of betanin on cadmium-induced toxicity in MRC-5 cells.
51. Rajabian, F., Rajabian, A., & Tayarani-Najaran, Z. (2023b). The antioxidant activity of Betanin protects MRC-5 cells against cadmium induced toxicity. *Biological Trace Element Research*, 1–9.
52. Satarug, S., & Moore, M. R. (2004). Adverse health effects of chronic exposure to low-level cadmium in foodstuffs and cigarette smoke. *Environmental Health Perspectives, 112*(10), 1099–1103.
53. Satarug, S., Baker, J. R., Urbenjapol, S., Haswell-Elkins, M., Reilly, P. E. B., & Williams, D. J. (2003). A global perspective on cadmium pollution and toxicity in non-occupationally exposed population. *Toxicology Letters, 137*(1–2), 65–83.
54. Satarug, S., Moore, M. R., & Robson, M. (2004). Does cadmium play a role in the etiology of prostate cancer? *Toxicology and Applied Pharmacology, 198*(3), 250–261.
55. Shayan, M., Mehri, S., Razavi, B. M., & Hosseinzadeh, H. (2023). Minocycline protects PC12 cells against cadmium-induced neurotoxicity by modulating apoptosis. *Biological Trace Element Research, 201*(4), 1946–1954.
56. Sorensen, M. (2014). Cadmium, occupational, and environmental exposure, and chronic obstructive pulmonary disease. *COPD: Journal of Chronic Obstructive Pulmonary Disease, 11*(6), 605–609.
57. Srivastava, A., Kumari, A., Jagdale, P., Ayanur, A., Pant, A. B., & Khanna, V. K. (2023). Potential of quercetin to protect cadmium induced cognitive deficits in rats by modulating NMDA-R mediated downstream signaling and PI3K/AKT—Nrf2/ARE signaling pathways in hippocampus. *NeuroMolecular Medicine*, 1–15.
58. Staessen, J. A., Roels, H. A., Emelianov, D., Kuznetsova, T., Thijs, L., Vangronsveld, J., Fagard, R., & Bernard, A. (1999). Environmental exposure to cadmium, forearm bone density, and risk of fractures: Prospective population study. *The Lancet, 353*(9159), 1140–1144.
59. Telisman, S., et al. (2007). Reproductive toxicity of low-level lead exposure in men. *Environmental Research, 105*(2), 256–266.
60. Tinkov, A. A., Sinitskii, A. I., Gritsenko, V. A., Skalnaya, M. G., Demidov, V. A., Nikonorov, A. A., et al. (2018). Link between hypertension and cadmium metabolism: Evidence from experimental and human studies. *Journal of Trace Elements in Medicine and Biology, 49*, 252–261.
61. USEPA. (2012). *Cadmium fact sheet: Support document for the recommended maximum contaminant level and maximum residual disinfectant level goals.*
62. Valko, M., et al. (2006). Free radicals and antioxidants in normal physiological functions and human disease. *The International Journal of Biochemistry & Cell Biology, 39*(1), 44–84.
63. Waalkes, M. P., & Liu, J. (2007). Diabetic effects of cadmium exposure in a rat model. *Toxicological Sciences, 98*(2), 398–408.
64. Wirth, J. J., et al. (2010). Cadmium exposure and endometrial cancer risk: a large midwestern U.S. population-based case-control study. *PloS One, 5*(5), e0012081.
65. Yoon, J., Cao, X., Zhou, Q., & Ma, L. Q. (2006). Accumulation of Pb, Cu, and Zn in native plants growing on a contaminated Florida site. *Science of the Total Environment, 368*(2–3), 456–464.
66. Zhang, H., Chen, Y., Zhu, G., Liu, X., & Huang, T. (2016). Cadmium exposure and cardiovascular disease in the 2005 Korea national health and nutrition examination survey. *Biological Trace Element Research, 173*(2), 351–358.

Cadmium and Its Neurotoxic Effects

H. Dhilleswara Rao, Vivek Chintada, and K. Veeraiah

Abstract Cadmium, a toxic heavy metal, has gained significant attention due to its detrimental effects on the nervous system. This book chapter provides a comprehensive overview of cadmium's neurotoxic effects, highlighting the importance of studying its impact on neurohealth. The chapter begins with an introduction that outlines cadmium as a toxic heavy metal and emphasizes the need to understand its neurotoxic effects. It then delves into the sources and exposure routes of cadmium, including natural and anthropogenic sources, occupational exposure, and dietary intake. The implications of cadmium contamination in food are also discussed in this section. Next, the chapter explores the absorption and distribution of cadmium within the body, focusing on the mechanisms of absorption, tissue distribution, and factors influencing cadmium metabolism. This understanding sets the stage for a detailed exploration of the neurological effects of cadmium. It highlights cadmium-induced neurotoxicity, its manifestations, and the impacts on the central nervous system (CNS). The chapter also investigates the neurodevelopmental effects, cognitive impairments, behavioral changes, and psychological disorders associated with cadmium exposure. The mechanisms underlying cadmium neurotoxicity are then examined, including oxidative stress and reactive oxygen species (ROS) generation, mitochondrial dysfunction, neurotransmitter and synapse function impairment, inflammation, immune responses in the CNS, and epigenetic alterations. The chapter further explores the connection between cadmium exposure and neurodegenerative diseases, including Alzheimer's and Parkinson's disease, along with other conditions associated with cadmium exposure. The vulnerability of certain populations, such as children and the aging population, to cadmium neurotoxicity is addressed, as well as potential gender differences. The chapter also discusses various methods and techniques for detecting and assessing cadmium neurotoxicity, including biomarkers, measuring cadmium concentration in biological samples, and neurological assessment methods. Moreover, preventive measures and mitigation strategies are outlined

H. Dhilleswara Rao · K. Veeraiah
Department of Zoology, Acharya Nagarjuna University, Guntur, Andhra Pradesh 522510, India

V. Chintada (✉)
Department of Zoology, Sri Venkateswara University, Tirupati, Andhra Pradesh 517502, India
e-mail: vivek.chintada@gmail.com

A. K. Jha and N. Kumar (eds.), *Cadmium Toxicity in Water*, Springer Water,
https://doi.org/10.1007/978-3-031-54005-9_4

to address cadmium neurotoxicity. Occupational safety measures and environmental policies are discussed, along with dietary guidelines and pharmacological interventions such as chelating agents. In conclusion, the chapter summarizes the neurotoxic effects of cadmium, emphasizes the significance of further research, and provides strategies for reducing cadmium exposure and safeguarding neurohealth. This comprehensive examination of cadmium and its neurotoxic effects contributes to our understanding of the potential risks associated with cadmium exposure and informs future research endeavors to protect individuals from its deleterious effects on the nervous system.

Keywords Cadmium · Neurotoxic · Environmental pollutant · Neurological disorders

1 Introduction

Cadmium is a toxic heavy metal that has become a major concern for both environmental and human health due to its widespread use in various industrial processes and products. It is released into the environment through anthropogenic activities such as mining, smelting, and the production of batteries, pigments, plastics, and fertilizers [4, 24, 26].

Cadmium has a long half-life and can persist in the environment for decades. It accumulates in soils, plants, and aquatic systems, thereby entering the food chain. Humans are primarily exposed to cadmium through ingestion of contaminated food, particularly in areas where there is high cadmium content in the soil [24, 29]. Cadmium toxicity is a result of its ability to disrupt several important cellular processes. It binds to sulfhydryl groups of proteins, interfering with their structure and function. It can also induce oxidative stress and trigger the generation of reactive oxygen species that can damage cellular components such as DNA, lipids, and proteins [12].

Exposure to cadmium has been associated with a wide range of adverse health effects in humans. Chronic exposure to low levels of cadmium can lead to kidney damage, as the kidneys are the major target organ for cadmium accumulation and toxicity. Cadmium can also cause skeletal damage by replacing calcium in bones, leading to decreased bone density and increased risk of fractures [5, 33]. Furthermore, cadmium has been classified as a human carcinogen by various international agencies such as the International Agency for Research on Cancer (IARC) and the U.S. Environmental Protection Agency (EPA). It has been linked to the development of lung cancer, prostate cancer, and kidney cancer [4].

Importance of studying Cadmium's neurotoxic effects

While most research on cadmium toxicity has focused on its effects on the kidneys and other organs, there is growing evidence suggesting that cadmium may also have neurotoxic effects. The central nervous system (CNS) is particularly vulnerable to the toxic effects of heavy metals, including cadmium [37]. Studies have shown that cadmium can cross the blood–brain barrier and accumulate in the brain, where it can interfere with normal neuronal function. Research in animals has demonstrated that cadmium exposure can lead to neurobehavioral alterations, including impaired learning and memory, altered motor function, and increased anxiety-like behavior [1, 37].

Furthermore, cadmium has been found to induce neuroinflammation, oxidative stress, and mitochondrial dysfunction in the brain. These mechanisms can contribute to neuronal damage and cell death, resulting in long-term neurological impairments [7]. In recent years, there has been a growing interest in understanding the potential link between cadmium exposure and neurodevelopmental disorders such as autism spectrum disorders and attention deficit hyperactivity disorder (ADHD). Several epidemiological studies have reported associations between prenatal or early-life exposure to cadmium and an increased risk of neurodevelopmental abnormalities in children [37]. Given the potential neurotoxic effects of cadmium and its widespread environmental contamination, studying its impact on the central nervous system is crucial. Understanding the mechanisms underlying cadmium-induced neurotoxicity can help in the development of strategies to mitigate exposure and protect vulnerable populations [10, 37].

2 Neurological Effects of Cadmium

Cadmium is a toxic heavy metal widely encountered in various industrial and environmental settings, posing a significant threat to human health. Extensive research has been conducted to understand the neurological effects of cadmium exposure. This review aims to summarize and discuss the scientific literature on cadmium-induced neurotoxicity, its manifestation, impacts on the central nervous system (CNS), neurodevelopmental effects, cognitive impairments, as well as behavioral changes and psychological disorders.

Cadmium-Induced Neurotoxicity and its Manifestation

Cadmium has been shown to exert neurotoxic effects through oxidative stress, mitochondrial dysfunction, and disruption of calcium homeostasis [28]. Upon exposure, cadmium readily enters the brain, accumulating primarily in the hippocampus, cortex, and cerebellum [23]. The manifestation of cadmium-induced neurotoxicity includes neuronal death, reactive gliosis, and alterations in neurotransmitter systems [36]. Animal studies have demonstrated that cadmium exposure leads to morphological changes in neurons, such as dendritic damage and synaptic loss [34].

Impacts on the Central Nervous System (CNS)

The central nervous system is particularly vulnerable to cadmium toxicity due to its high metabolic activity and dependence on calcium signaling [19]. Studies conducted on both animals and human populations have shown that cadmium exposure impacts the CNS by disrupting neurotransmitter function, impairing synaptic transmission, and altering neuronal plasticity [20]. Additionally, cadmium has been linked to the disruption of blood–brain barrier integrity, facilitating the entry of other toxic substances into the brain [32].

Neurodevelopmental Effects and Cognitive Impairments

Cadmium exposure during critical periods of neurodevelopment has been associated with long-term neurobehavioral deficits. Animal studies have revealed that prenatal or early-life exposure to cadmium can lead to impaired neurogenesis, altered neuronal connectivity, and deficits in learning and memory [1, 43]. Human studies have reported associations between cadmium exposure and cognitive impairments, including decreased intelligence quotient (IQ), attention deficits, and lower scores in cognitive function tests [31]. These findings highlight the importance of reducing cadmium exposure, particularly in vulnerable populations, to safeguard neurodevelopment and cognitive health.

Behavioral Changes and Psychological Disorders

Mounting evidence suggests that cadmium exposure may contribute to behavioral changes and psychological disorders. Studies have shown that chronic exposure to cadmium is linked to increased anxiety-like behavior and depression-like symptoms in animals [40]. In humans, cadmium exposure has been associated with psychiatric disorders, including depression and attention-deficit hyperactivity disorder (ADHD) [11]. Mechanisms involved in these behavioral and psychological changes include alterations in neurotransmitter systems, inflammation, and oxidative stress [38].

The neurological effects of cadmium are broad and encompass various aspects of brain structure and function. Cadmium-induced neurotoxicity manifests as neuronal damage, reactive gliosis, and disruption of neurotransmitter systems. Its impacts on the CNS include impaired synaptic transmission, altered neuronal plasticity, and compromised blood–brain barrier integrity. Cadmium's neurodevelopmental effects are characterized by deficits in neurogenesis, neuronal connectivity, and cognitive function. Moreover, it is associated with behavioral changes, such as anxiety and depression, as well as psychological disorders. This comprehensive review emphasizes the importance of minimizing cadmium exposure to protect against these detrimental neurological effects and underscores the need for further research to elucidate the underlying mechanisms.

3 Mechanisms of Cadmium Neurotoxicity

Cadmium is a toxic heavy metal that exerts detrimental effects on the nervous system. Understanding the mechanisms underlying cadmium-induced neurotoxicity is crucial for developing effective strategies to mitigate its harmful consequences. This review aims to provide an in-depth analysis of the mechanisms involved in cadmium neurotoxicity, focusing on oxidative stress and reactive oxygen species (ROS) generation, mitochondrial dysfunction and energy metabolism disruption, impairment of neurotransmitters and synapse function, inflammation, immune responses in the central nervous system (CNS), as well as epigenetic alterations and gene expression dysregulation.

Oxidative Stress and Reactive Oxygen Species (ROS) Generation

Cadmium-induced neurotoxicity is closely associated with oxidative stress, which arises from an imbalance between the production of reactive oxygen species (ROS) and antioxidant defense systems. Cadmium readily generates ROS through the Fenton reaction and can deplete antioxidants by binding to sulfhydryl groups and interfering with antioxidant enzyme activities [25]. The resulting oxidative stress leads to various cellular damage, including lipid peroxidation, protein oxidation, and DNA damage [41].

Mitochondrial Dysfunction and Energy Metabolism Disruption

Cadmium has been shown to disrupt mitochondrial function, leading to energy metabolism impairment in the nervous system. By accumulating in mitochondria, cadmium interferes with mitochondrial respiration, reduces ATP production, and impairs the electron transport chain [16]. This disruption of cellular energy metabolism can further exacerbate oxidative stress and compromise neuronal viability and function [29, 39].

Impairment of Neurotransmitters and Synapse Function

Cadmium exposure has been reported to disrupt the balance of neurotransmitters and impair synapse function in the CNS. Studies have shown that cadmium affects the release, reuptake, and metabolism of various neurotransmitters, including glutamate, dopamine, and serotonin [17]. These alterations in neurotransmitter systems can disrupt synaptic transmission and neuronal communication, contributing to cognitive and behavioral abnormalities [8].

Inflammation and Immune Responses in the CNS

Cadmium induces neuroinflammation and activates immune responses within the CNS. Upon exposure, cadmium triggers the activation of glial cells, including microglia and astrocytes, leading to the release of pro-inflammatory cytokines, chemokines, and reactive oxygen species [5, 18]. This inflammatory response can cause damage to neurons, impair synaptic plasticity, and contribute to neurodegenerative processes [15].

Epigenetic Alterations and Gene Expression Dysregulation

Emerging evidence suggests that cadmium exposure can induce epigenetic alterations and dysregulate gene expression in the nervous system. Cadmium has been shown to modify DNA methylation patterns, histone modifications, and microRNA expression profiles [6]. These epigenetic changes can lead to abnormal gene expression, affecting neurodevelopment, synaptic plasticity, and oxidative stress response pathways [3]. Cadmium neurotoxicity involves a complex interplay of mechanisms, including oxidative stress and ROS generation, mitochondrial dysfunction and energy metabolism disruption, impairment of neurotransmitters and synapse function, inflammation and immune responses in the CNS, as well as epigenetic alterations and gene expression dysregulation. These mechanisms contribute synergistically to the detrimental effects of cadmium on neuronal structure and function. Understanding these underlying mechanisms is crucial for developing targeted therapeutic interventions and preventive strategies against cadmium-induced neurotoxicity.

4 Cadmium and Neurodegenerative Diseases: A Complex Connection Worth Exploring

Neurodegenerative diseases are debilitating conditions characterized by the progressive death of neurons in various regions of the nervous system. Recent research suggests that environmental factors, including heavy metal exposure, may contribute to their development and progression. Among these metals, cadmium has gained significant attention due to its widespread presence in the environment and its potential neurotoxic effects. This review aims to explore the connection between cadmium exposure and neurodegenerative disorders, specifically focusing on Alzheimer's disease, Parkinson's disease, and other related conditions.

(a) **Connection between cadmium exposure and neurodegenerative disorders**:

Several studies have investigated the association between cadmium exposure and the risk of developing neurodegenerative diseases. Research by Takeda et al. provided evidence suggesting that chronic exposure to cadmium may lead to the development of cognitive impairments and behavioral abnormalities, indicating a potential link to neurodegeneration. Moreover, studies conducted on animal models exposed to cadmium demonstrated an increase in oxidative stress markers and neuronal cell death.

(b) **Alzheimer's disease and cadmium involvement**:

Alzheimer's disease (AD) is the most common neurodegenerative disorder, characterized by the accumulation of amyloid-beta plaques and neurofibrillary tangles in the brain. Accumulating evidence suggests that cadmium exposure may contribute to the development and progression of AD. A study by Huang et al. revealed a positive association between blood cadmium levels and cognitive decline in AD patients. Additionally, research on cell cultures and animal models has shown that cadmium exposure can enhance amyloid-beta aggregation and promote neuroinflammation, key features of AD pathology.

(c) **Parkinson's disease and cadmium-induced neurotoxicity**:

Parkinson's disease (PD) is another prevalent neurodegenerative disorder, primarily characterized by the degeneration of dopaminergic neurons in the substantia nigra. Several studies have investigated the potential role of cadmium-induced neurotoxicity in the development of PD. Wu et al. found that chronic cadmium exposure could induce mitochondrial dysfunction and oxidative stress in experimental PD models, leading to the degeneration of dopaminergic neurons. Moreover, epidemiological studies have reported a positive association between occupational cadmium exposure and an increased risk of developing PD.

(d) **Other neurodegenerative diseases related to cadmium exposure**:

Besides AD and PD, cadmium exposure has been linked to other neurodegenerative diseases. For instance, amyotrophic lateral sclerosis (ALS), a progressive motor neuron disorder, has been associated with exposure to environmental toxins, including cadmium. Kim et al. demonstrated that cadmium exposure could induce neuronal cell death and motor function deficits in ALS animal models, suggesting a potential role of cadmium in ALS pathogenesis. Furthermore, cadmium exposure has also been implicated in the development of multiple sclerosis (MS), a chronic autoimmune demyelinating disorder of the central nervous system [5].

The evidence reviewed in this paper highlights the potential link between cadmium exposure and the development of neurodegenerative disorders, including Alzheimer's disease, Parkinson's disease, amyotrophic lateral sclerosis, and multiple sclerosis. The mechanisms by which cadmium exerts its neurotoxic effects are multifaceted, involving oxidative stress, mitochondrial dysfunction, neuroinflammation, and protein aggregation. Further research is essential to elucidate the exact pathways through which cadmium contributes to neurodegeneration and to develop preventive strategies for minimizing exposure to this hazardous heavy metal.

5 Detection and Assessment of Cadmium Neurotoxicity

Cadmium (Cd) is a heavy metal pollutant widely encountered in various industrial processes, causing significant concern due to its neurotoxic effects. Overexposure to cadmium has been linked to detrimental effects on the central nervous system (CNS), resulting in neurological impairments. Accurate detection and assessment of cadmium neurotoxicity are crucial for identifying at-risk populations and implementing effective preventive measures. This review aims to explore the biomarkers for cadmium exposure and neurotoxicity, techniques for measuring cadmium concentration in biological samples, and neurological assessment methods for cadmium toxicity.

(a) **Biomarkers for cadmium exposure and neurotoxicity**:

Biomarkers play a pivotal role in quantifying cadmium exposure and its neurotoxic effects. Several candidate biomarkers have been extensively studied, including metallothionein (MT), urinary *N*-acetyl-β-D-glucosaminidase (NAG), and urinary β2-microglobulin (β2-MG). Studies have shown that increased urinary concentrations of NAG and β2-MG significantly correlate with cadmium exposure and serve as sensitive biomarkers of renal toxicity. Moreover, MT, a small cysteine-rich protein, has been established as a reliable biomarker for cadmium-induced oxidative stress and neurotoxicity.

(b) **Techniques for measuring cadmium concentration in biological samples**:

Accurate quantification of cadmium levels in biological samples is essential for monitoring exposure and assessing toxicity. Various techniques have been developed for cadmium analysis, including atomic absorption spectrophotometry (AAS), inductively coupled plasma mass spectrometry (ICP-MS), and voltammetry. AAS is widely utilized due to its simplicity and sensitivity in quantifying cadmium, although interference from other metals can limit its accuracy. ICP-MS offers higher sensitivity and selectivity and allows simultaneous analysis of multiple heavy metals, including cadmium. Additionally, voltammetry techniques, such as differential pulse anodic stripping voltammetry (DPASV), have shown promising results for rapid and reliable cadmium detection in various biological matrices.

(c) **Neurological assessment methods for cadmium toxicity**:

Comprehensive neurological assessments are vital for evaluating the impact of cadmium exposure on the CNS. Neurobehavioral tests, including motor function assessments, cognitive tests, and sensory function evaluations, provide valuable insights into cadmium-induced neurotoxicity. The rotarod test, Morris water maze, and novel object recognition tests are commonly employed to assess motor coordination, spatial memory, and recognition memory, respectively. Electrophysiological techniques, such as electroencephalography (EEG), have also proved useful in detecting alterations in brain function caused by cadmium exposure. Furthermore, neuroimaging techniques, including magnetic resonance imaging (MRI) and positron emission tomography (PET), offer non-invasive means of visualizing structural and functional brain changes associated with cadmium neurotoxicity [5].

In the field of cadmium neurotoxicity detection and assessment, various biomarkers, techniques for measuring cadmium concentration, and neurobehavioral assessment methods have proven useful. The integration of these approaches facilitates reliable quantification of cadmium exposure and its associated neurotoxic effects, aiding in the identification of at-risk populations and the implementation of appropriate preventive measures. Further research and refinement of these methods are essential for advancing our understanding of cadmium neurotoxicity and ultimately safeguarding public health.

6 Conclusions and Future Perspectives

(a) Recap of Cadmium's Neurotoxic Effects

In conclusion, numerous studies have convincingly demonstrated the neurotoxic effects of cadmium exposure. Cadmium, a heavy metal present in various industrial and environmental sources, has been shown to cause significant damage to the central nervous system (CNS). It has been associated with impairments in cognitive function, memory, learning abilities, and motor coordination [10, 13]. Furthermore, cadmium exposure has been linked to an increased risk of neurodevelopmental disorders in children, such as attention deficit hyperactivity disorder (ADHD) and autism spectrum disorder (ASD) [35]. These neurological effects highlight the importance of understanding and addressing the issue of cadmium exposure.

(b) Importance of Further Research and Studies

Further research and studies on cadmium are essential for several reasons. Firstly, there is a need to unravel the underlying molecular mechanisms by which cadmium exerts its neurotoxic effects. In-depth knowledge about these mechanisms will aid in developing targeted therapeutic strategies for mitigating the damage caused by cadmium exposure. Secondly, continued research is crucial to identify vulnerable populations, I such as pregnant women and children, who may be more susceptible to cadmium-induced neurotoxicity [2]. This knowledge will enable the implementation of effective preventive measures and the development of tailored interventions to protect these vulnerable groups. Finally, ongoing research will contribute to the identification of biomarkers for cadmium exposure and neurotoxicity, allowing for early detection and timely intervention [1, 22]. Therefore, further research efforts are warranted to address the gaps in our understanding of cadmium's neurotoxic effects.

(c) Strategies for Reducing Cadmium Exposure and Protecting Neurohealth

Considering the detrimental impact of cadmium on neurohealth, it is crucial to implement strategies aimed at reducing cadmium exposure and protecting individuals from its neurotoxic effects. Firstly, stringent regulations and policies need to be enforced to limit cadmium emissions from industrial sources, such as power plants and manufacturing plants [27, 42]. This includes implementing and enforcing emission control measures and promoting the use of alternative, less toxic materials. Secondly, promoting sustainable agricultural practices can help reduce the accumulation of cadmium in food crops. This can be achieved through the use of alternatives to cadmium-containing fertilizers, soil remediation techniques, and agricultural practices that promote the uptake of essential nutrients while limiting the absorption of cadmium by plants [30]. Additionally, public awareness campaigns and education programs should be employed to inform individuals about potential sources of cadmium exposure, such as contaminated water and certain food products [9, 14]. This can empower individuals to take proactive measures to minimize their exposure.

Lastly, research and development should focus on the creation of effective chelating agents that can facilitate the removal of cadmium from the body, thereby reducing its toxic effects [21].

In summary, the neurotoxic effects of cadmium demand further research, especially in understanding the molecular mechanisms, identifying vulnerable populations, and developing biomarkers for early detection. Strategies for reducing cadmium exposure and protecting neurohealth include implementing regulations to limit industrial emissions, promoting sustainable agricultural practices, raising public awareness, and developing effective chelating agents. These efforts are crucial for mitigating the adverse effects of cadmium and safeguarding the neurological well-being of individuals.

References

1. Akaras, N., Ileriturk, M., Gur, C., Kucukler, S., Oz, M., & Kandemir, F. M. (2023). The protective effects of chrysin on cadmium-induced pulmonary toxicity; a multi-biomarker approach. *Environmental Science and Pollution Research*, 1–16.
2. Ali, N., et al. (2018). Environmental exposure to lead and manganese and neurodevelopmental outcomes in children. *Annals of Global Health, 84*(3), 386–394.
3. Arora, M., Reichenbach, A., & van de Peppel, J. (2017). Epigenetic states transregulated by developmental and environmental cues: Exploring models and datasets. *BioEssays, 39*(3), 1700047.
4. Bernard, A. (2008). Cadmium & its adverse effects on human health. *Indian Journal of Medical Research, 128*(4), 557–564.
5. Bodur, S., Erarpat, S., Kayın, İ., & Bakırdere, S. (2023). Cadmium determination at trace levels in lake water samples by cold vapor generation-atomic absorption spectrometry after magnetic dispersive solid phase extraction. *Chemical Papers*, 1–8.
6. Bonomi, W., Giordano, G., Tognolina, M., et al. (2018). Altered miRNA expression levels in neurotoxicity induced by acute cadmium in rats. *Toxicology Research, 4*(5), 1246–1257.
7. Casas, A. I., Dao, V. T. V., Daiber, A., Maghzal, G. J., Di Lisa, F., O'Donnell, V. B., et al. (2017). Reactive oxygen-related diseases: Therapeutic targets and emerging clinical indications. *Antioxidants & Redox Signaling, 29*(11), 1172–1185.
8. Collaborative on Health and the Environment. (2013). Neurodevelopmental impacts from low-level exposure to lead, mercury, and other environmental toxins: Project TENDR consensus statement. *Reviews on Environmental Health, 28*(3), 185–192.
9. Dobersztyn, C., et al. (2018). Cadmium-related neurotoxicity ingesting food: A public health problem. *Journal of Food Protection, 81*(2), 222–236.
10. Elemile, O. O., Gana, A. J., Ejigboye, P. O., Ibitogbe, E. M., Olajide, O. S., & Ibitoye, O. O. (2023). Analysis of potentially toxic elements from selected mechanical workshops using the geo-accumulation index and principal component analysis in Omu-Aran community, Nigeria. *Environmental Monitoring and Assessment, 195*(2), 276.
11. Farghaly, M. S., Arafa, M. H., Badawy, M. G., et al. (2020). Interactions between prenatal cadmium exposure and socioeconomic factors and their effects on autism spectrum disorder. *International Journal of Environmental Research and Public Health, 17*(12), 4295.
12. Flora, S. J. S. (2009). Structural, chemical and biological aspects of antioxidants for strategies against metal and metalloid exposure. *Oxidative Medicine and Cellular Longevity, 2*(4), 191–206.
13. Fowler, B., et al. (2009). Neurotoxicity of cadmium and lead. *Journal of Toxicology and Environmental Health, Part B, 12*(6), 587–614.

14. Gnatyshyna, L., Khoma, V., Martinyuk, V., Matskiv, T., Pedrini-Martha, V., Niederwanger, M., Stoliar, O., & Dallinger, R. (2023). Sublethal cadmium exposure in the freshwater snail *Lymnaea stagnalis* meets a deficient, poorly responsive metallothionein system while evoking oxidative and cellular stress. *Comparative Biochemistry and Physiology Part C: Toxicology & Pharmacology, 263*, 109490.
15. Hirsch, E. C., & Hunot, S. (2009). Neuroinflammation in Parkinson's disease: A target for neuroprotection? *The Lancet Neurology, 8*(4), 382–397.
16. Jung, H. A., & Jung, Y. J. (2015). Mitochondria as targets in cadmium-induced apoptosis in neuronal cells. *Biomedical Research International, 2015*, 545294.
17. Lafuente, A., Gonzalez-Carracedo, A., Romero, A., et al. (2016). Effects of chronic cadmium exposure on the neurobehavioral development of rats in a low level exposure model. *Neurotoxicology and Teratology, 54*, 66–73.
18. Miguel-Queralt, S., Espejo-Porras, F., & Garcia-Dominguez, I. (2018). Role of astrocytes and microglia in the regulation of synaptic plasticity: Implications for neurodevelopmental and neurodegenerative disorders. *Frontiers in Cellular Neuroscience, 12*, 222.
19. Nemmouche, S., Tritschler, L., Gabriel, C., et al. (2015). Cadmium induces apoptosis in a neuroblastoma cell line by caspase-dependent and p53-independent mechanisms. *Journal of Environmental Science and Health, Part A: Toxic/Hazardous Substances and Environmental Engineering, 50*(6), 646–654.
20. Niehoff, M. L., & Adkins, D. L. (2014). Age-dependent effects of chronic cadmium exposure on cognitive behavior. *Behavioural Brain Research, 263*, 171–179.
21. Prozialeck, W. C., et al. (2014). Cadmium toxicokinetics: The relationship between tissue accumulation, and biological effects and health risks of cadmium in humans. *Accounts of Chemical Research, 46*(3), 543–554.
22. Ratner, V., et al. (2018). The role of neurotoxicants and neurotoxicity mechanisms in neurodevelopmental disorders. *International Journal of Molecular Sciences, 19*(12), 3829.
23. Rocha, E. S., Ribeiro, A. P., Cerqueira, M. M. O. P., & Vieira, L. F. C. (2019). Cadmium-induced neurotoxicity: Targets and mechanisms of neurobehavioral disorders. *Environmental Toxicology and Pharmacology, 68*, 101–114.
24. Satarug, S., Garrett, S. H., Sens, M. A., & Sens, D. A. (2010). Cadmium, environmental exposure, and health outcomes. *Environmental Health Perspectives, 118*(2), 182–190.
25. Scandalios, J. G. (2011). Oxidative stress response genes and their regulation. *The Arabidopsis Book, 9*, e0153.
26. Shayan, M., Mehri, S., Razavi, B. M., & Hosseinzadeh, H. (2023). Minocycline protects PC12 cells against cadmium-induced neurotoxicity by modulating apoptosis. *Biological Trace Element Research, 201*(4), 1946–1954.
27. da Silva, A. O. F., Bezerra, V., Meletti, P. C., Simonato, J. D., & dos Reis Martinez, C. B. (2023). Cadmium effects on the freshwater teleost *Prochilodus lineatus*: Accumulation and biochemical, genotoxic, and behavioural biomarkers. *Environmental Toxicology and Pharmacology, 99*, 104121.
28. Soetan, K. O., Olaiya, C. O., & Oyewole, O. E. (2018). The importance of mineral elements for humans, domestic animals, and plants: A review. *African Journal of Food Science, 12*(3), 54–73.
29. Srivastava, A., Kumari, A., Jagdale, P., Ayanur, A., Pant, A. B., & Khanna, V. K. (2023). Potential of quercetin to protect cadmium induced cognitive deficits in rats by modulating NMDA-R mediated downstream signaling and PI3K/AKT—Nrf2/ARE signaling pathways in hippocampus. *NeuroMolecular Medicine*, 1–15.
30. Sun, H., et al. (2020). Emerging strategies for controlling cadmium uptake in rice. *Journal of Hazardous Materials, 393*, 122412.
31. Tang, Y., Xie, H., Chen, J., et al. (2017). Low-level prenatal exposure to cadmium and premature delivery: A birth cohort study in Wuhan, China. *PLoS ONE, 12*(3), e0173094.
32. Tinkov, A. A., Skalny, A. A., Cherkasov, S. V., et al. (2018). Impact of chronic exposure to cadmium on the blood-brain barrier integrity and dopamine metabolism in rats. *Food and Chemical Toxicology, 115*, 431–437.

33. Tinkov, A. A., Filippini, T., Ajsuvakova, O. P., Aaseth, J., Gluhcheva, Y. G., Ivanova, J. M., Bjørklund, G., Skalnaya, M. G., Gatiatulina, E. R., Popova, E. V., & Nemereshina, O. N., (2017). The role of cadmium in obesity and diabetes. *Science of the Total Environment*, 601–602, 741–755.
34. Vallejo, R., Toledo, R., Amma, L. L., et al. (2015). Ultrastructural changes and apoptosis in dentate gyrus of rat hippocampus induced by acute methylmercury intoxication. *Toxicology and Industrial Health, 31*(6), 616–629.
35. Wang, H., et al. (2012). Maternal cadmium exposure and neurodevelopmental outcomes in children aged 2–5 years. *Journal of Environmental Science and Health, Part A, 47*(7), 1032–1042.
36. Wang, M., Liu, W., Zhang, H., et al. (2019). Cadmium-induced neurotoxicity: Mechanisms and interpretations. *International Journal of Molecular Sciences, 20*(3), 569.
37. Wang, Y. X., Mao, L. J., Meng, Q. F., et al. (2020). Evaluation of cadmium, chromium, nickel, copper, and zinc concentrations in biological reference materials by inductively coupled plasma mass spectrometry. *BMC Chem., 14*(1), 9.
38. Wang, J., Pang, B., et al. (2021). Anxiety-like behavior induced by subchronic exposure to cadmium in mice: Role of oxidative stress, neuroinflammation, and neurotrophin factor impairment. *Chemosphere, 266*, 129206.
39. Xu, B., Huang, R., Huang, J., et al. (2016). Cadmium-induced neuron apoptosis and brain-borne encephalopathy. *Journal of Hygiene Research, 45*(4), 559–565.
40. Xu, Y., Zhang, H., Zhang, M., et al. (2019). Behavioral effects and mechanisms of neurodevelopmental cadmium exposure in rats. *Environmental Pollution, 248*, 928–935.
41. Yang, Y. J., Lee, E. Y., Choi, S. Y., et al. (2018). Cadmium-induced oxidative stress and apoptotic changes in the cerebellum of mice. *Journal of Veterinary Science, 19*(2), 246–252.
42. Zhai, Y., et al. (2019). Cadmium emissions from industrial sources and its control technologies. *Journal of Hazardous Materials, 365*, 416–425.
43. Zhang, J., Li, C., Yu, H., et al. (2017). Cadmium impairs cognitive function in rat frontal cortex and hippocampus. *Neurotoxicology, 62*, 1–9.

Sustainable Mitigation Strategies and Solutions

Phytoremediation of Cadmium: A Review

Iftikhar Ahmad, Hussein Alserae, Bo Zhu, Atif Zahoor, Zia Ur Rahman Farooqi, Adil Mihoub, Qurat Ul Ain, and Emanuele Radicetti

Abstract Cadmium (Cd) is a highly poisonous heavy metal that presents a considerable hazard to the environment and human health. It can move from water to soil, then food crops and accumulate in their edible parts. When it becomes part of the food chain, it causes toxic effects to human life via kidney damage to even cancers. Numerous methods have shown varied efficacies for its remediation from contaminated environments. One method that has shown promising results with sustainable Cd removal is phytoremediation, which involves phytoextraction, rhizofiltration, phytostabilization, and phytostimulation through diverse plant species. Based on Cd uptake in aerial part of plant three types of plants are existed and exploited

I. Ahmad (✉)
Department of Environmental Sciences, COMSATS University Islamabad, Vehari Campus, Vehari 61100, Pakistan
e-mail: iftikharahmad@cuivehari.edu.pk

H. Alserae
College of Agricultural Engineering Science, Baghdad University, Baghdad, Iraq

B. Zhu
Shanghai Yangtze River Delta Eco-Environmental Change and Management Observation and Research Station, Shanghai Cooperative Innovation Center for Modern Seed Industry, School of Agriculture and Biology, Shanghai Jiao Tong University, Shanghai, China

A. Zahoor
Department of Structures and Environmental Engineering, University of Agriculture, Faisalabad 38040, Pakistan

Z. U. R. Farooqi
Institute of Soil and Environmental Sciences, University of Agriculture, Faisalabad, Pakistan

A. Mihoub
Center for Scientific and Technical Research on Arid Regions, Biophysical Environment Station, Touggourt, Algeria

Q. U. Ain
Department of Environmental Science, COMSATS University Islamabad, Vehari Campus, Vehari 61100, Pakistan

E. Radicetti
Department of Chemical, Pharmaceutical and Agricultural Sciences, University of Ferrara, Via Luigi Borsari N. 46, 44121 Ferrara, Italy

A. K. Jha and N. Kumar (eds.), *Cadmium Toxicity in Water*, Springer Water,
https://doi.org/10.1007/978-3-031-54005-9_5

for phytoremdiation such as hyperaccumulator (concentrate substantial amount of Cd in shoot with BCF > 1), excluder (concentrate tiny amount of Cd in shoot with BCF < 1), and indicator plants (concentrate reasonable amount of Cd in shoot with BCF = 1). Numerous plant species have been exploited for remediation of Cd in water (*Eichhornia crassipes, Pistia stratiotes, Lemna minor*), whereas in soil (*Brassica* spp., *Sedum alfredii, Populus alba, Alyssum murale, Helianthus annuus, Thlaspi caerulescens, Zea mays*). This chapter critically reviewed the sources, and toxicity of Cd. Finally, selection criteria of plant, merits and demerits and types of phytoremediation elaborated in detail for remediation of Cd contaminated water and soil resources for achieving sustainable development.

Keywords Soil pollution · Bioaccumulation · Food chain · Heavy metals · Human health

1 Introduction

Cadmium (Cd) is a highly toxic heavy metal that show considerable dangers to human and ecosystem's health because of its extensive industrial usage and environmental persistence [1]. It can enter the environment through natural and anthropogenic sources. Anthropogenic sources of Cd in the environment include the extraction and processing of zinc (Zn), lead (Pb), and copper (Cu) ores, as well as the manufacturing and eventual disposal of Cd-containing products [2]. Natural sources of Cd include release of Cd from rocks through weathering process, forest fires, and volcanoes. The long-term environmental consequences of Cd include air, water, and soil pollution. Soil Cd accumulates and persists over time, posing a hazard to soil fertility and microbial activity. Ingestion of Cd-contaminated food and drinks, inhalation of polluted air, and skin contact with Cd-contaminated soil are all potential routes of exposure to Cd [3]. It is toxic because it interrupts cellular processes and causes organ damage i.e., kidneys, once it is absorbed. Renal failure, lung damage, heart disease, and cancer are only some of the health problems linked to long-term Cd exposure. Children and pregnant women are more vulnerable to the harmful effects of Cd exposure [4].

Plant life also faces similar consequences from small scale effects to plant death via disruption in photosynthesis, food absorption, and enzyme function. As a result, plant development slow down, develops chlorosis, and yields are diminished [5]. Because Cd may enter the food chain via contaminated crops, it is quite worrying when it accumulates in edible plants. Soil quality and fertility are negatively impacted by Cd pollution [6]. It lowers the soil pH, interferes with nitrogen (N) cycling, and inhibits the soil biodiversity and its functioning. The long-term impacts of bioaccumulation and biomagnification in the environment posed by Cd persistence in soil further amplifies its harmful effects [7, 8].

Due to the harmful consequences, and Cd poisoning on soil, plant and human health, its remediation should be done with adoption of efficient remediation strategies [9]. Commonly employed methods for Cd remediation from water, soil, and plants are soil excavation and its disposal which entails the removal of Cd-contaminated soil or sediment and the subsequent disposal of the material in a specially designed containment facility [10]. It is an efficient strategy, but it requires soil as disposal site and may cause Cd-contamination via its movement to other/ nearby regions. Physical or chemical washing agents or solvents are used in soil washing to remove Cd from contaminated soil [11]. The extracted Cd may be processed independently or disposed-off away. Soil washing is an effective remediation method for limited pollution, and it may be used in tandem with other methods to boost their effectiveness [11]. By including amendments that immobilize Cd, stabilization/solidification strives to decrease its mobility and bioavailability in soil. Lime, cement, and organic materials are often used as stabilizing agents because they react with Cd to generate its less-toxic species or stimulate chemical processes that diminish Cd toxicity [11]. Electrokinetic application is another method for Cd remediation. By applying an electric field over Cd-contaminated soil, electrokinetic remediation drives Cd ions to collect at targeted electrodes. This technique is useful for removing or immobilizing Cd from clayey soils only as they have more charges on their particles [12].

Using sorbents or resins designed to selectively bind Cd ions in solution is another strategy for Cd remediation [13], but use of plants for sustainable remediation is preferred. Phytoremediation uses plants and their associated microbes to remove, stabilize, or detoxify Cd from the environment, while ion exchange activities substitute Cd with less hazardous ions, effectively lowering Cd content in water or soil [8, 14]. Phytoremediation uses different techniques i.e., phytoextraction, rhizofiltration, and phytostabilization are all part of this eco-friendly method [7, 15, 16] (Rasool et al. 2023). Some plant species, called hyperaccumulators, are ideal for phytoextraction because of their propensity to store large amounts of Cd in their tissues [6]. Microorganisms are used in bioremediation to break down or convert Cd into less harmful forms. Microbial volatilization, biosorption, and enzymatic degradation are all mechanisms through which microbes might help lessen its hazardous effects [17].

Due to the environmental consequences, arousal of secondary problems and requirements of investment and labor, techniques other than phytoremediation are less preferred. Species of plants known as hyperaccumulators have been promising for phytoextraction due to their exceptional capacity to collect large quantities of Cd. The removal of Cd from water using rhizofiltration makes use of plants with deep root systems [7]. The goal of phytostabilization is to employ plants that immobilize Cd in soil to limit the Cd bioavailability and mobility in the soil [12, 15, 16]. Phytoremediation, the use of plants and the microbes that live with them, is a green and cheap way to clean up Cd [8, 18]. Phytoremediation has been proposed as a sustainable and effective method for Cd remediation [8]. This chapter briefly highlights the importance of phytoremediation in environmental cleanup from Cd from soil and water resources, while discussing their merits and demerits. In conclusion, the authors have highlighted the potential of using phytoremediation as a green method to reduce Cd contamination and revive polluted areas.

2 Cadmium in Soil–Water-Plant Systems: Sources and Effects

Both natural and anthropogenic sources exist for Cd entry into the water-soil–plant system. The main anthropogenic sources include industrial discharges, agricultural runoff and activities, atmospheric deposition, waste disposal, mining, and ore processing, while natural sources are weathering, volcanoes, and runoff/erosion [6, 12]. Mining and manufacturing, two major industrial sectors, both directly and indirectly emit Cd into water bodies and subsequently pollute the soil via deposition [5]. Cadmium may be introduced into the water supplies using Cd-containing agricultural inputs such as fertilizers, herbicides, and sewage sludge. Industrial and municipal waste incineration, as well as natural sources, contribute to Cd emissions into the atmosphere, where it remains until washed away by rainwater [19]. Contamination of soil and water may also result from improper disposal of waste items containing Cd. The mining and processing of Zn, Pb, and Cu ores results in the release of Cd into the environment [2]. The Cd levels in soil and water are impacted by the natural weathering of Cd-containing rocks and minerals where it moves from terrestrial surfaces to aquatic bodies through runoff and erosion [20, 21].

When Cd gets into the water bodies, it can stay there for a long time, endangering ecosystems, and people. Minimizing Cd poisoning and protecting soil and water resources requires efficient management tactics and remedial procedures [22]. To establish targeted remediation techniques and adopt preventative measures to decrease Cd entry into soil and water, knowing the sources and routes of Cd contamination is essential. Protecting ecosystems and human populations from negative effects of Cd on the environment requires addressing its origins and potential entrance points [23].

Cadmium contamination has significant impacts on soil and plant health. It can lead to soil degradation, reducing fertility and disrupting nutrient cycling. In plants, Cd toxicity inhibits growth and development, impairs nutrient uptake, induces oxidative stress, and alters plant physiology [24]. In plants, Cd disrupts plant growth via reducing photosynthesis, nutrient absorption, and generation of reactive oxygen species (ROS), leading to stunted plant growth, reduced biomass, and decreased crop yields [25]. Moreover, Cd accumulation in plant tissues poses a risk to human health through bioaccumulation into the food chain. Overall, Cd contamination has detrimental effects on soil fertility, plant growth, and plant health, emphasizing the need for effective remediation strategies and soil management practices to mitigate Cd contamination and ensure the production of safe and healthy food crops [26].

Moreover, Cd has devastating consequences on human health if consumed in contaminated food. Its build-up in a wide range of crops and is then ingested by humans cause its accumulation in kidneys and is known to severely impair renal functions. Chronic kidney disease, tubular dysfunction, and renal tubular necrosis are all possible outcomes of long-term Cd exposure via the diet [27]. The kidney's capacity to remove waste materials from the blood is impaired by exposure to Cd. Over time, Cd may also build up in the skeleton. Consumption of Cd-tainted food over a long period of time has been linked to low bone mineral density, an increased risk of osteoporosis, and skeletal deformities [28]. Bones are weakened by Cd because it competes with calcium during bone production, throwing off the delicate chemistry of bone remodeling. An increased risk of cardiovascular illnesses has been linked to cd contamination in food [29]. Hypertension, endothelial dysfunction, and an upped risk of cardiovascular disease may all result from exposure to dangerous levels of Cd. Damage to the cardiovascular system is accelerated by induction of Cd-produced oxidative stress, inflammation, and vascular damage [30]. Consumption of contaminated food might have an adverse influence on fertility. Reduced fertility, hormonal abnormalities, and poor reproductive function have all been associated to Cd exposure in both sexes [31]. Furthermore, prenatal exposure to Cd may cause growth retardation, cognitive impairments, and behavioral abnormalities in babies and children. Long-term use of Cd-contaminated food has been linked to an increased risk of developing many types of cancer as Cd is a tumor promoter; it speeds up the development and spread of cancer cells [32].

3 Cadmium Uptake in Plants: Plant Uptake Pathways, Bioaccumulation, and Hazards to Ecosystems

The Cd is entered in to the food production system via influencing plant health and toxicity to human health via the eating of Cd-contaminated foods is a serious issue [33]. In order to effectively develop ways to limit Cd accumulation and eliminate related health hazards, it is essential to understand the processes and variables controlling Cd absorption in plants [34]. The Cd is taken up by plant roots from soil that contains the metal. Many variables influence how readily available Cd is for absorption, including the content of Cd in the soil, its chemical speciation, the soil pH, and the surrounding environment [35]. Soluble ions, adsorbed to the soil particles, and organically bound Cd are all possible forms of Cd. Plants may easily absorb Cd ions in their soluble form, while the bioavailability of Cd in its adsorbed or complexed form may be lower [36]. It is normally bound to soil particles, rendering it inaccessible to plants. Its ions migrate from higher concentrations in the soil solution to lower concentrations in the root via a process called passive diffusion [33]. The Cd uptake also involves active uptake systems like ion channels and transporters. Because of its comparable chemical characteristics to important minerals like calcium, iron, and Zn, these transport mechanisms also govern the intake of Cd [35].

The xylem carries water and nutrients up the plant from the root, and Cd may go up it with them. Leaves, stems, and seeds are only some of the plant components that may accumulate Cd [37]. Some plants, called hyperaccumulators, are recognized for their remarkable capacity to store large amounts of Cd in their tissues without showing obvious signs of toxicity [38–40]. Plants' absorption of Cd from water works on the same principles as plant uptake from soil. Cd is taken up by the roots of aquatic plants and its availability is affected by factors such as the water chemistry, pH, and dissolved organic matter concentration [41]. Then Cd is transferred and cycled throughout aquatic environments, in part due to its buildup in aquatic plants. Plant health and production may suffer as a result of Cd absorption and buildup [42]. In plants, Cd disrupts processes including nutrition intake, enzyme function, photosynthesis, and resistance to oxidative stress [31]. Ingesting crops that have accumulated Cd in their edible components might be harmful to people. Several methods exist for reducing plant Cd absorption [43]. Changing the soil pH or adding amendments that generate less soluble Cd compounds are two ways to decrease Cd bioavailability in soil. Cd-tolerant or hyperaccumulator plant species may be used in phytoremediation processes to aid in the removal of Cd from polluted soil [44]. Cd transfer to crops may also be reduced by the use of water management procedures, such as lowering Cd concentrations in irrigation water or using water treatment methods [45].

Elevated levels of cadmium in ecosystems and the consequent transmission of this metal via the food chain are a major danger to ecosystem health and human wellbeing. Cd is a heavy metal that is not necessary to life but may have harmful consequences if it is released into the environment or if it accumulates in living entities [46]. In terrestrial ecosystems, plants are the primary sink for Cd buildup. Roots of plants may take up Cd from polluted soil. Plant tissues such as leaves, stems, and seeds may collect cadmium [42]. Hyperaccumulators are plant species that can store large amounts of Cd in their tissues without demonstrating obvious harm. It is transported to higher trophic levels in the food chain when animals ingest Cd-contaminated plants or plant-eating critters [47]. Larger trophic level species tend to have larger Cd contents due to biomagnification or bioaccumulation. Cd accumulates because it is not readily broken down or eliminated by the body. Higher trophic level organisms, such as predators, may collect more Cd than lower trophic level organisms [43].

The buildup of Cd in biological systems is harmful to a wide range of species. Reduced development, decreased photosynthesis, and nutritional imbalances are some of the negative effects that excessive Cd levels may have on plants [48]. Animals with high Cd levels may have impaired reproduction, weakened immune systems, organ damage, and even death [49]. The Cd toxicity is especially worrisome in humans since it is associated with kidney damage, skeletal abnormalities, and an increased risk of some malignancies when Cd-contaminated food is consumed [50]. In addition to the obvious health risks, Cd buildup also endangers the efficiency of ecosystems. Reduced plant production due to high Cd levels has consequences for the ecosystem's primary producers. Herbivores and higher trophic levels that depend on plants for food may be negatively affected by this disturbance [42]. The ecological balance and the provision of vital ecosystem services are vulnerable to changes in species abundance and distribution within the ecosystem. Soil health and microbial populations may potentially be adversely affected by Cd buildup [44]. Soil fertility and ecosystem function may be negatively impacted by Cd toxicity due to its ability to inhibit microbial activity and alter nutrient cycle mechanisms. Its pollution may also have negative effects on soil organisms like earthworms, which are essential to soil structure and nutrient cycling [51].

The risks associated with Cd buildup may be reduced by using remediation procedures and taking steps to decrease Cd discharge into the environment. Cleaner and sustainable industrial practices, better waste management, soil remediation methods, and crop monitoring for Cd levels are all part of the solution [52]. Effective management requires knowledge of the causes and effects of Cd buildup. These measures are used to lessen Cd contamination, shield ecosystems, and guarantee the security of the food supply. Taking action to reduce Cd buildup will reduce the hazards to ecosystem function, biodiversity, and human health [42].

4 Conventional Approaches for Cadmium Remediation

Common and conventional techniques for the Cd removal from contaminated sites (Fig. 1), include changing the soil pH, adding chemicals, soil washing, electrokinetics, phytoremediation, chemical stabilization, sedimentation, filtering, and chemical precipitation [53]. These procedures are designed to eliminate or significantly reduce the Cd pollution in both soil and water. The mobility and bioavailability of Cd may be decreased by adjusting the soil pH, making chemical amendments, or using chemical stabilization procedures [45].

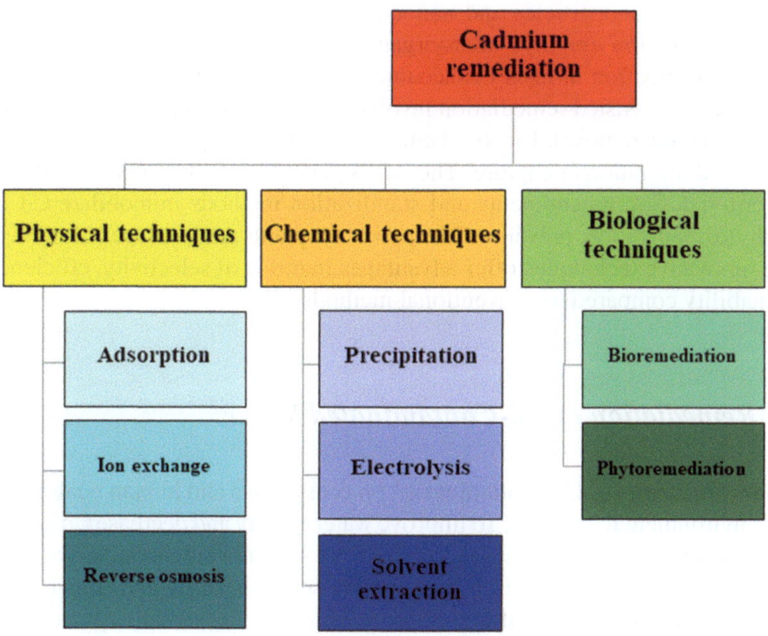

Fig. 1 Conventional and modern approached for the remediation of Cd-contaminated environs

In soil, Cd contamination may be remedied in two ways: either by cleaning the soil with a chemical agent or by applying an electric field. Plants are used in phytoremediation to take up and store Cd for later removal. Sedimentation and filtering methods are used to remove Cd from water by trapping or separating the Cd, while Cd removal from water include chemical precipitation, which yields in insoluble Cd molecules [52]. Considerations such as pollution levels, soil or water properties, and site-specific variables help determine which cleanup approach is best. To successfully remove Cd while limiting damage to the environment, it is essential to take into account the efficacy, cost, environmental implications, and long-term sustainability of the selected strategy [51].

5 Modern/Innovative Technique for Cadmium Remediation

Modern and innovative techniques for Cd remediation (Fig. 1), include nanotechnology, bioremediation, phytoextraction, chelation-assisted remediation, electrochemical remediation, advanced oxidation processes (AOPs), soil amendments and stabilization, and phytostabilization [24, 51]. These techniques offer effective approaches to enhance the efficiency and sustainability of Cd removal. Nanotechnology utilizes nanoparticles and nanocomposites to selectively bind and remove Cd. Bioremediation employs microorganisms to bioaccumulate or transform Cd, while phytoextraction utilizes Cd-accumulating plants to remove Cd from the soil [54]. Chelation-assisted remediation involves the use of chelating agents to increase Cd solubility for removal. Electrochemical remediation applies electric currents to facilitate Cd migration or capture. The AOPs generate reactive species to degrade or transform Cd. Soil amendments and stabilization methods immobilize Cd in soil, and phytostabilization prevents Cd uptake by plants and reduces leaching [45]. These innovative techniques offer advantages in terms of selectivity, efficiency, and sustainability compared to conventional methods [45].

5.1 Remediation of Cd-Contaminated Water

The harmful effects of Cd contamination on ecosystems and human health make it a major environmental problem. To improve water quality and decrease Cd contamination, corrective actions must be taken. There are several viable options for removing Cd from water bodies, and each has its own set of benefits and drawbacks [55]. Chemicals like aluminum sulfate or ferric chloride are added to water to create insoluble Cd precipitates via the coagulation and flocculation processes. The flocculation that occurs thereafter forms aggregates of the precipitates, making their subsequent removal by sedimentation or filtering more simpler [56]. These techniques are often used because they successfully lower Cd levels in water. Cd ions are bound preferentially by special resins or membranes in ion exchange procedures, and then exchanged for safer ions. As a result of its efficiency, this technique is often used in water purification systems to lower Cd levels [57]. Activated carbon, zeolites, and modified clays are only a few examples of sorbent materials with a high affinity for Cd that may be used in adsorption processes. These materials are very efficient in removing Cd from water, making them a viable cleanup choice [58]. To remove Cd ions from water, semi-permeable membranes are used in membrane filtration processes including reverse osmosis, nanofiltration, and ultrafiltration. However, these procedures may have substantial energy requirements and ongoing expenditures to eliminate Cd contamination [59, 60].

When plants can be safely introduced to Cd-contaminated water, a biological method called phytoremediation may be used. Water hyacinth and duckweed are two examples of aquatic plants that may absorb Cd from the water and significantly lower the concentration [61]. This approach is a greener and more long-lasting alternative to conventional cleanup methods. In order to create insoluble precipitates from Cd ions in water, chemical precipitation techniques include the addition of chemicals that react with the ions [56]. Among the most prevalent types of precipitating agents are lime, ferric chloride, and aluminum sulfate. Once the precipitates have formed, they may be filtered out or sedimented away [62]. These methods for removing Cd from water are simple and economical. Electrocoagulation and electrooxidation are two electrochemical techniques that use electric current or potential to remove Cd. These processes aid in the removal of Cd from the aqueous matrix by coagulation, oxidation, or electro-precipitation [63]. Potential benefits in terms of efficiency and controllability may be offered. Using microorganisms or biofilms, biological applications may clean up Cd-tainted water. Bioaccumulation and biotransformation by certain bacteria and fungi allow for significant concentrations of Cd to be reduced. However, there may be prerequisites and extended treatment times for biological therapies [64].

The level of Cd pollution, water quality criteria, financial concerns, and site-specific variables all play a role in deciding which remedial technique is best. Removing Cd from water supplies may need multi-pronged strategies that work together. In addition, these remedial solutions can only be kept in working order with consistent monitoring and continuing maintenance [56, 65].

5.2 Remediation of Cd-Contaminated Soil

Remediation of Cd pollution in soils is essential to mitigate the environmental and health risks associated with Cd contamination. Various effective strategies can be employed to reduce or remove Cd from contaminated soils, depending on the extent of contamination and site-specific conditions [66]. One approach is to adjust the soil pH through the application of lime or alkaline amendments. Increasing soil pH induces Cd precipitation or immobilization, reducing its availability for plant uptake and leaching into groundwater. Chemical amendments, such as phosphate-based compounds, can form less soluble Cd-phosphate complexes, reducing Cd mobility and bioavailability in the soil [66]. Soil washing is another remedial measure that involves the use of chemical agents or surfactants to extract Cd from contaminated soil. This method is particularly useful for localized contamination and allows for the separate treatment or disposal of the extracted Cd [67]. Electrokinetic remediation utilizes an electric field to induce Cd migration in the soil, the Cd ions move towards specific electrodes where they can be recovered. This technique is effective for reducing Cd concentrations in fine-grained soils [67]. Phytoremediation is a sustainable and environmentally friendly approach that utilizes plants to uptake, accumulate, and detoxify Cd from contaminated soil. Certain plant species, known as hyperaccumulators, can tolerate and accumulate Cd in their tissues. These plants can be harvested and removed, effectively reducing Cd levels in the soil [33].

Chemical stabilization involves the addition of stabilizing agents, such as lime, bentonite, or activated carbon, to immobilize Cd in the soil. These agents bind with Cd, reducing its mobility and bioavailability. Organic matter amendments, such as compost or manure, can enhance soil fertility and microbial activity, contributing to Cd immobilization [68]. Soil covering with impermeable barriers, such as plastic sheets or geomembranes, prevents the migration of Cd to surrounding areas, minimizing potential exposure, and spreading. In severe cases, excavation and removal of contaminated soil may be necessary, with replacement or off-site treatment [54, 67]. The selection of remedial measures depends on factors such as the extent of contamination, soil characteristics, and economic feasibility. In many cases, a combination of techniques may be required for optimal Cd reduction. Regular monitoring is crucial to assess the effectiveness of the chosen measures and ensure long-term soil quality improvement [68]. Implementing these remedial measures for Cd pollution in soils can contribute to the restoration of soil health, protect ecosystems, and minimize potential human health risks associated with Cd contamination [54]. It is important to consider site-specific factors and evaluate the feasibility and effectiveness of each approach to determine the most suitable remediation strategy (Fig. 1).

6 Phytoremediation of Cadmium Contaminated Environments: Techniques and Advantages

To clean up Cd polluted areas, phytoremediation is a method that is both safe for the environment and economical. Cadmium is removed, stabilized, or degraded with the help of plants and the microbes that live on and around them. Phytoextraction, rhizofiltration, phytostabilization, and phytovolatilization are only a few of the processes that may be used for phytoremediation. Phytoremediation typically employs three different kinds of plants: Cd hyperaccumulators (BCF $>$ 1), Cd excluders (BCF $<$ 1), and Cd indicators (BCF $=$ 1) (Fig. 2). Some plants such as *Eichhornia crassipes, Sedum alfredii, Populus alba, Alyssum murale, Helianthus annuus, Pteris vittate, Thlaspi caerulescens* are known as hyperaccumulators, may store higher Cd levels in their shoots compared to surrounding soil without displaying any ill effects and BCF should be more than 1 [69, 70]. There is no standard level of Cd accumulation in plants to qualify them for hyperaccumulator (100–500 mg/kg dry biomass), however, *T. caerulescens* shoots can accumulate > 2000 mg Cd per kg soil and tolerate Cd up to 30 μM in solution [69]. Cadmium excluders (*Glycine max, Triticum aestivum, Oryza sativa, Hordeum vulgare*) may store low levels of Cd in their tissues compared to surrounding soil and BCF should be less than 1. The process of exclusion involves binding of Cd on plant cell wall, lignification of cell wall to limit Cd entrance into plant, and efflux of Cd through ATP-dependent efflux pumps [69]. Cadmium indicator plants (*Zea mays*) may store equal level of Cd in their tissues compared to surrounding soil and BCF should be equal to 1. Phytoextraction is another technique of phytoremediation for Cd, in which plants can uptake Cd from contaminated soil and store it in their biomass. Rhizofiltration is a method for cleaning water systems of Cd by using plant roots as filters. Cadmium is less bioavailable and cannot leached into groundwater because it has been immobilized in the soil by phytostabilization [33], while the process by which Cd is converted into volatile molecules and then expelled into the atmosphere is called phytovolatilization [71].

6.1 Phytoextraction

Toxic heavy metals like Cd may be extracted from polluted soil with the help of plants via a phytoremediation process called phytoextraction. It is taken up by the plant roots and then transported to shoots and other aerial portions. This process requires root cell membranes to include specialized transporters that allow for the selective absorption of cadmium ions from the soil solution [72]. Cadmium ions, once within the plant, go to the leaves through the xylem vessels. Ion transporters and chelating agents help Cd travel across the plant. The sequestration of Cd ions inside plant tissues is an essential part of the phytoextraction process. Phytochelatins, tiny peptides generated by plants in response to metal stress, are metal-binding ligands that may bind Cd. Cadmium toxicity is mitigated, and it is stored in vacuoles and other cellular compartments due to the production of Cd-phytochelatin complexes. When the plants have reached a specific stage of development and Cd accumulation, the above-ground biomass may be collected. This biomass stores the accumulated Cd and may be recycled or disposed of to get rid of it [73]. *Sesbania sesban* can efficiently grow and accumulate 29% Cd more than control plants in an in vitro experiment at seedling stage [74], however, at this stage we should not recommend this plant for phytoextraction of Cd. In another study, [75] found greater accumulation of Cd in root and shoot of *S. sesban* because of biochemical homeostasis of plants under Cd stress. *E. crassipes* is a natural macrophyte that shows great accumulation of Cd in

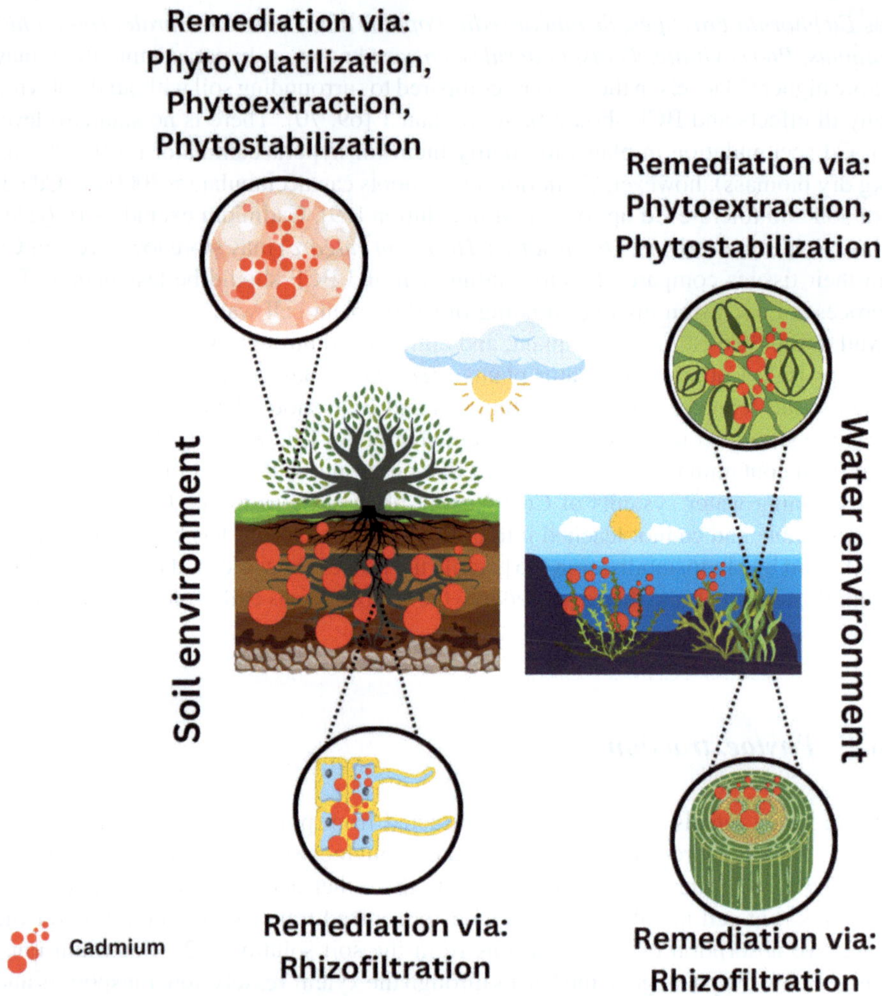

Fig. 2 Mechanisms of Cd-phytoremediation from contaminated soil and aquatic bodies

shoot, translocation factor for Cd was ≤ 2.34 while bioconcentration factors were 61.68 in roots and 21.94 in shoots (Rasool et al. 2023).

6.2 *Rhizofiltration*

Rhizofiltration is another method in phytoremediation for cleaning Cd-contaminated water by the action of plant roots. Cadmium ions are taken up by plant roots, transported via xylem vessels to above-ground plant parts, accumulated in plant tissues, filtered from water as it travels through the roots, and then collected when the plants are harvested. Hyperaccumulators are a special kind of plant that can store large amounts of Cd. Rhizofiltration, which makes use of plants inherent capacity to accumulate Cd, is a sustainable and inexpensive approach of cleaning up Cd-contaminated water [18]. *E. crassipes* concentrated most of Cd in roots from fresh, industrial, and urban wastewater, with respective bioconcentration factors of 19.41, 61.68 and 8.99 in roots (Rasool et al. 2023). Here, it is pertinent to mention that *E. crassipes* plant showed excellent rhizofiltration ability of Cd from industrial wastewater.

6.3 Phytostabilization

The phytostabilization reduces the mobility and bioavailability of Cd in the soil by immobilizing it. Cadmium ions are taken up by plant roots, and then the root exudates change the rhizosphere. These secretions may combine with Cd to produce complexes that limit its movement and make it less accessible to plants and soil organisms [76]. Adsorption of Cd onto soil particles has also been reported. Some plant species can prevent Cd from moving from their roots to their shoots and leaves. High-biomass plant growth is typically used in conjunction with phytostabilization to further enhance soil stability. Long-term immobilization of Cd in the soil is the result of root exudation, complexation, adsorption, and restricted translocation [77]. The danger of Cd contamination may be reduced in a sustainable way by phytostabilization, reducing the negative effects on ecosystems and human health [78]. Cadmium can be efficiently stabilized in soil–plant system with the application of different amendments, for instance addition of silicon in soil not only improves wheat growth also caused immobilization of Cd within soil-root system [38]. In another study, 2% zeolite reduced DTPA extractable Cd up to 87% in soil and decreased its availability to brinjal plant [39]. Organic manures application in soil induced organic acid production from maize roots, which can make stable complexes with Cd and decreased its availability to maize plant [76]. They further reported that positive significant correlation existed between Cd retention in maize roots and organic acid production in the rhizosphere. Application of rock phosphate, compost, and press mud synergistically decreased Cd uptake while increased growth and yield of maize [79], similarly, [80] found that application anaerobically composted municipal solid waste compost in sandy loam soil decreased Cd uptake in spinach and carrot, consequently these vegetables were found safe for human consumption without showing any risks to human health. Application of compost and biogas residues in soil also stabilized Cd in soil and thereby its uptake restricted to wheat and maize plants grown in Cd contaminated soil [15, 16, 81].

6.4 Phytostimulation

Phytostimulation is an innovative technique in which some plants or bacterial inoculations may promote growth of other plants or in themselves through diverse processes i.e., exopolysaccharide production [82]. Legumes, such as clover and beans, are examples of plants that are often employed in phytostimulation because they contain nitrogen-fixing bacteria in their root nodules. These bacteria transform air nitrogen into plant available nitrogen, enriching the soil and benefiting nearby plants [83]. Allelopathic plants, such as sunflowers and marigolds, produce chemical substances that encourage the development of surrounding plants [84]. Understanding and using these relationships may result in novel agricultural techniques that promote sustainable farming and environmental balance. Plant growth and resistance are aided by phytostimulants such as seaweed extracts, mycorrhizal fungi, and amino acids [85]. They boost nutrient intake, root growth, and pest and disease resistance. These natural remedies provide long-term alternatives to synthetic agricultural chemicals, promoting healthier crops and ecosystems [86].

6.4.1 Bacterial Assisted

Bacterial-assisted phytoremediation is a novel strategy to environmental cleaning in which specialized bacteria work with plants to remove, degrade, or immobilize Cd from soil, water, or air [87]. This method combines the natural ability of some plant species to remove toxins from the environment with the metabolic capabilities of bacteria that can detoxify or change to less toxic forms of pollutants. Some bacteria, such as *Pseudomonas* and *Bacillus* species, may convert Cd into less toxic forms [88]. These bacteria are administered to the polluted site or injected into the rhizosphere of phytoremediation-capable plants. The plant root system offers a home for the bacteria, which in turn improves the plant capacity to absorb and breakdown contaminants [87]. This mutually beneficial connection speeds up the remediation process, making it more effective and efficient than either phytoremediation or bacterial remediation alone. Furthermore, owing to their broad root systems and capacity to resist a wide range of environmental conditions, some plant species, such as willows and poplars, are especially well-suited for bacterial-assisted phytoremediation. These plants enhance the cleanup process by providing a conducive habitat for bacterial growth and activity [89]. The inoculation of *Bacillus* sp. and *Klebsiella* sp. stabilized Cd in soil and thereby its uptake lowered in wheat and maize plants grown in Cd contaminated soil [15, 16], however, maize plants was found more sensitive to Cd stress compared to wheat plants. In another study, root exudation from maize roots inoculated with *Pantoea* sp. decreased Cd uptake [76], this finding is further strengthening with strong positive correlation between organic acid (acetic and citric acids) and Cd retention in maize roots. Similarly, metal tolerant *Bacillus anthracis* inoculation to *S. sesban* increased plant growth and concentrate > 29% Cd in plants at seedling stage [74], whereas same bacteria induced metal stress tolerance in *S. sesban* through biochemical homeostasis under Cd stress and caused increase accumulation of Cd in plant [75].

6.4.2 Chemical Assisted Phytoremediation

Chemical-assisted phytoremediation is type of phytoremediation technology that combines the use of certain chemicals with plant-based cleaning procedures to successfully remove Cd from polluted settings [90]. Chelating agents, surfactants, fertilizers, pH-adjusting agents, microbial inoculants, and oxidizing agents are among the compounds used to aid in the absorption, destruction, or immobilization of Cd by plants. Chelators make Cd more available for plant absorption, surfactants make organic pollutants more soluble, and fertilizers improve plant growth and pollutant uptake capability. Plant–microbe interactions are improved by microbial inoculants, while pollutant degradation is aided by oxidizing agents [91]. When employing these compounds, proper management and awareness of environmental implications are critical. Chemical-assisted phytoremediation is a promising and ecologically benign way to cleaning up polluted places, combining natural plant powers with targeted chemical treatments to speed up the remediation process and restore ecological balance [87]. Soil application of silicon and zeolite improved wheat and brinjal growth due to Cd stabilization in soil [38, 39]. Although phytoremediation is a viable method for depleting soil and water Cd, it is important to keep an eye on the environmental and health effects of the released Cd compounds (Table 1).

7 Plants Selection for Phytoremediation

The selection of appropriate plant species is a critical aspect of phytoremediation for Cd removal from soil and water. The choice of plant species depends on their ability to tolerate Cd, accumulate it in their tissues, and adapt to specific environmental conditions [72].

Here are some key considerations for selecting plant species for Cd phytoremediation:

- Plants that can survive in or develop a tolerance for Cd-contaminated soil are favored. The harmful effects of Cd on these plants are negligible at the levels seen in nature. Metal
- Sequestration in vacuoles, improved antioxidant defenses, and effective root-to-shoot translocation are all possibilities among the potential mechanisms of tolerance. Plant species with natural Cd tolerance may be found via screening experiments.
- Many phytoremediation projects concentrate on hyperaccumulator plants because of their ability to store large amounts of Cd in their tissues. These plants are very

Table 1 Application and mechanisms of phytoremediation for management of cadmium contaminated soil and water

Mechanisms	Principles	Application	Species used	References
Phytoextraction	Extraction of contaminants in shoots using plants	Soil and water	*Brassica juncea, Solanum nigrum, Helianthus annuus*	[92]
Phytostabilization	Use of plants to immobilize or reduce the mobility of contaminants within soil-root system	Soil and water	*Lolium perenne, Festuca arundinacea, Zea mays, Triticum aestivum*	[15, 16, 93]
Phytostimulation	Use of stimulants i.e., bacteria and nutrients to enhance phytoremediation process	Soil and water	*Zea mays, Salix* sp. with *Pseudomonas* and *Bacillus* sp.	[94]
Rhizofiltration	Use of plant roots to remove pollutants from aqueous solution	Water	*Brassica juncea, Eichhornia crassipes*	[95], Rasool et al. (2023)

effective in absorbing Cd from their surrounding environment and storing it in their shoots and leaves.

- Slow-growing hyperaccumulators may not be suited for large-scale phytoremediation projects, thus it is vital to think about things like biomass production and growth rates.
- In terms of plant development, Cd phytoremediation often favors rapidly expanding plants with deep root systems. Strong root systems not only facilitate Cd absorption, but also facilitate the search for contaminated soil hotspots. Plants with extensive root systems may reach Cd buried below the surface. Plants with thick canopies also offer shade, which helps keep soil wet by lowering the rate at which water evaporates.
- Plant species that are native to or well-adapted to the environmental conditions of the polluted site have the upper hand. They have a greater potential for establishment and growth, increasing the likelihood of effective phytoremediation. Site-specific considerations to think about include soil pH, moisture, temperature, and more. To aid with ecological restoration and conservation efforts, it is best to choose native plant species that are already present in the area.
- Plant biomass is typically collected and handled for phytoremediation operations that attempt to eliminate Cd from the site. Plants that produce a lot of biomasses, can be harvested with little effort, and can be mined for metals efficiently are therefore preferable. Depending on legislation in the area and the ultimate purpose

of remediation, the collected biomass may be treated to either recover Cd or dispose of it.

- The ecological interactions between the chosen plant species and their associated microbes are an important factor to consider during phytoremediation. In certain cases, interactions between plants and microbes may improve Cd absorption, translocation, and tolerance in plants. Plant development and Cd accumulation may be aided by mutually beneficial symbiotic connections, such as mycorrhizal associations.
- For phytoremediation operations to be successful in removing Cd, monitoring and management are necessary. Therefore, it is preferable to focus on plant species that can be easily monitored and display obvious symptoms of Cd stress or buildup. Interactions with native plant and animal species, as well as other possible implications of phytoremediation on local ecosystems, should be carefully considered.

Based on the criteria, the most used plant species for phytoremediation of Cd were listed (Table 2), there species were either reported as hyperaccumulator, indicator or excluder of Cd from soil and plant resources and very helpful to detoxify the Cd from polluted environments.

Table 2 Types of plants based on Cd uptake and major plant species used for phytoremediation of cadmium contaminated soil and water

Plant type	Concept	Plant species	References
Hyperaccumulator	Accumulated high levels of Cd in the aerial part of plant compared to surrounding soil or water, BCF > 1	*Alyssum murale, Arabidopsis helleri, Eichhornia crassipes, Gnaphalium affine, Helianthus annuus, Malva rotundifolia, Pteris vittata, Sesbania sesban, Solanum nigrum, Salix viminalis, Setaria italica, Youngia erythrocarpa*	[47, 51, 92, 95, 97, 98]
Indicator	Accumulated levels of Cd in the aerial part of plant that reflect level of Cd in the surrounding soil or water, BCF = 1	*Cannabis sativa, Celosia argentea, Medicago sativa, Zea mays*	[15, 47, 95]
Excluder	Accumulated lower levels of Cd in the aerial part of plant compared to surrounding soil or water, BCF < 1	*Althaea rosea, Amaranthus spinosus, Avena sativa, Agrostis stolonifera, Agrostis capillaris, Arabidopsis thaliana, Pinus radiata, Silene maritime, Triticum aestivum,*	[16, 47, 93, 98, 99]

8 Merits and Demerits of Phytoremediation

8.1 *Merits*

Phytoremediation offers several advantages as a remediation approach for contaminated environments [41, 96]:

- It is a green remediation technique since it uses plants to clean up hazardous substances. Instead of using invasive and disruptive cleanup methods, it makes use of plants and related microbes to remove, decompose, or stabilize pollutants.
- It is often less expensive than conventional remediation strategies like soil replenishment or excavation and disposal. Large-scale cleanup initiatives may be undertaken because of the minimal capital investment and operating expenses necessitated by their use.
- Heavy metals, organic pollutants, and even radioactive elements are all amenable to phytoremediation. It is a flexible method for treating a broad variety of contaminated locations since different plant species and accompanying microbial communities may be chosen or created to target toxins.
- It provides a permanent and environmentally friendly way to clean up polluted areas. Once established, plants may keep growing and filtering out pollutants for a long time. Because of this, environmental cleaning activities can be sustained without continual upkeep and repair.
- It not only facilitates the removal of toxins, but it also encourages the restoration of habitat and the provision of ecosystem services. Plant growth has several benefits, including soil stabilization, erosion prevention, increased biodiversity, and new homes for animals. The aesthetic and practical value of a polluted site may be enhanced by phytoremediation, which contributes to the site's total ecological restoration.

It may be used in a wide range of contexts, from industrial sites and agricultural areas to brownfields and wetlands. Its adaptability to different sites and different degrees of pollution makes it a useful tool. Depending on the situation and desired outcome, several phytoremediation methods including phytoextraction, rhizofiltration, and phytostabilization may be used.

- Because of its perceived use of natural processes to clean up polluted environments, it has widespread popular support. When compared to conventional remediation techniques, such as the excavation, transportation, and disposal of polluted materials, this method is more acceptable to the public.
- Some phytoremediation plant species may collect very high pollutant concentrations in their biomass, which presents an opportunity for both resource consumption and recovery. Using methods like phytomining and biomass combustion, this biomass may be gathered and processed to extract important metals or turned into biofuels. Therefore, phytoremediation provides hope for resource reuse and recycling.

8.2 Demerits

While phytoremediation offers numerous advantages, it is important to consider its potential disadvantages and limitations [72, 96]:

- It is often a slow process that requires significant time to achieve significant reductions in contaminant levels. It may take several years or even decades to fully remediate a contaminated site. This prolonged timeline may not be suitable for sites requiring immediate remediation or those with time-sensitive constraints.
- While It is effective for many contaminants, it may have limitations for certain types of pollutants. For example, highly mobile contaminants or those with complex chemical structures may be challenging for plants to uptake or degrade. Additionally, it may not be as effective for contaminants that are strongly bound to soil particles or present in high concentrations.
- The selection of appropriate plant species is crucial for the success of phytoremediation. However, the availability and suitability of suitable plant species may be limited, especially for specific contaminants or challenging site conditions. Some contaminants may not have well-adapted hyperaccumulator plant species, making phytoremediation less feasible.
- In some cases, the harvested biomass from phytoremediation may contain high concentrations of contaminants. Proper management and disposal of this biomass are necessary to prevent secondary contamination or improper handling. The costs and logistics associated with biomass management can be a significant challenge.
- Phytoremediation success is influenced by various site-specific factors such as soil type, pH, moisture content, and climate conditions. Certain sites may not be suitable for phytoremediation due to extreme conditions or the presence of multiple contaminants that hinder plant growth or limit contaminant uptake.
- Phytoremediation is generally more suitable for smaller-scale or localized contamination areas. Implementing phytoremediation on a large scale may require extensive land availability, significant resources, and long-term monitoring and maintenance. It may not be feasible or cost-effective for larger contaminated sites.
- While phytoremediation aims to restore contaminated environments, there is a potential risk of unintended ecological effects. Certain hyperaccumulator plant species, for instance, may have allelopathic effects, inhibiting the growth of neighboring plants. The introduction of non-native plant species for remediation purposes can also pose risks to native biodiversity if not properly managed.
- Although phytoremediation is generally well-received by the public, there may still be skepticism or concerns regarding its long-term effectiveness and potential health risks associated with consuming plants grown in contaminated environments. Public perception and acceptance of phytoremediation projects can vary depending on cultural, social, and educational factors.

9 Conclusions

In summary, conventional methods of cadmium removal have limitations and drawbacks, including high costs, environmental disruption, and the generation of secondary waste. Phytoremediation, on the other hand, offers several advantages as an alternative approach. It is environmentally friendly, cost-effective, and provides a long-term solution for cadmium removal. Phytoremediation promotes ecological restoration, versatility, and public acceptance. However, it is important to consider site-specific factors and potential limitations, such as plant species selection and ongoing monitoring requirements. By harnessing the benefits of phytoremediation, we can effectively address cadmium contamination while promoting sustainability and ecosystem health.

References

1. Naik, M. R., Akash, Kumar, D., Sinam, V., & Siddique, A. (2022). Cadmium toxicity in plants: remedial strategy through seed priming. *Pollution Research.*
2. Sahoo, R., & Sahoo, S. (2014). Cadmium induced oxidative stress in plants and endogenous ameliorative mechanism—A review. *Trends in Biosciences, 7*, 1843–1852.
3. Peana, M., Pelucelli, A., Chasapis, C. T., Perlepes, S. P., Bekiari, V., Medici, S., & Zoroddu, M. A. (2023). Biological effects of human exposure to environmental cadmium. *Biomolecules, 13*, 36.
4. Hossain, S., Latifa, G. A., & Al Nayeem, A. (2019). Review of cadmium pollution in Bangladesh. *Journal of Health and Pollution, 9*, 190913.
5. Hocaoğlu-Özyiğit, A., & Genç, B. (2020). Cadmium in plants, humans and the environment.
6. Cui, X., Mao, P., Sun, S., Huang, R., Fan, Y., Li, Y., Li, Y., Zhuang, P., & Li, Z. (2021). Phytoremediation of cadmium contaminated soils by *Amaranthus hypochondriacus* L.: The effects of soil properties highlighting cation exchange capacity. *Chemosphere, 283*, 131067.
7. Raza, A., Habib, M., Kakavand, S. N., Zahid, Z., Zahra, N., Sharif, R., & Hasanuzzaman, M. (2020). Phytoremediation of cadmium: Physiological, biochemical, and molecular mechanisms. *Biology, 9*, 177.
8. Subašić, M., Šamec, D., Selović, A., & Karalija, E. (2022). Phytoremediation of cadmium polluted soils: current status and approaches for enhancing. *Soil Systems.*
9. Požgajová, M., Navrátilová, A., & Kovár, M. (2022). Curative potential of substances with bioactive properties to alleviate Cd toxicity: A review. *International Journal of Environmental Research and Public Health, 19*, 12380.
10. Huang, X., Zhao, H., Zhang, G., Li, J., Yang, Y., & Ji, P. (2020). Potential of removing Cd(II) and Pb(II) from contaminated water using a newly modified fly ash. *Chemosphere, 242*, 125148.
11. Wang, J. (2021). Remediation technology and effect evaluation of Cd pollution in farmland soil.
12. Zheng, Y., Xiao, C., & Chi, R. (2021). Remediation of soil cadmium pollution by biomineralization using microbial-induced precipitation: A review. *World Journal of Microbiology and Biotechnology, 37*, 208.
13. Bożęcka, A., Orlof-Naturalna, M., & Sanak-Rydlewska, S. (2016). Removal of lead, cadmium and copper ions from aqueous solutions by using ion exchange resin C 160. *Gospodarka Surowcami Mineralnymi-mineral Resources Management, 32*, 129–140.
14. Asnaoui, H., Khalis, M., Mlouk, L. A., & Ait Hou, M. (2018). Removal of heavy metal cadmium Cd (II) by adsorption onto biomass; adsorption models.

15. Ahmad, I., Akhtar, M. J., Asghar, H. N., Ghafoor, U., & Shahid, M. (2016). Differential effects of plant growth promoting rhizobacteria on maize growth and cadmium uptake. *Journal of Plant Growth Regulation, 35*, 303–315.
16. Ahmad, I., Akhtar, M. J., Zahir, Z. A., Naveed, M., Mitter, B., & Sessitsch, A. (2014). Cadmium-tolerant bacteria induce metal stress tolerance in cereals. *Environmental Science and Pollution Research, 21*, 11054–11065.
17. Huang, G., Zhang, Y., Tang, J. J., & Du, Y. (2020). Remediation of Cd contaminated soil in microbial fuel cells: Effects of Cd concentration and electrode spacing. *Journal of Environmental Engineering, 146*, 04020050.
18. Li, C., Yang, G., Liu, Z., & Cai, J. (2022). Overview of phytoremediation technology for heavy metal contaminated soil. In *E3S Web of Conferences.*
19. Cui, J., Wang, W., Peng, Y., Zhou, F., He, D., Wang, J., Chang, Y., Yang, J., Zhou, J., Wang, W., Yao, D., Du, F., Liu, X., & Zhao, H. (2019). Effects of simulated Cd deposition on soil Cd availability, microbial response, and crop Cd uptake in the passivation-remediation process of Cd-contaminated purple soil. *Science of The Total Environment, 683*, 782–792.
20. Iqbal, M. N., & Ashraf, A. (2018). Environmental pollution: Heavy metals removal from water sources.
21. Yu, J., Yu, H., Fang, H., Lei, M., Li, S., & Chi, J. (2014). Pollution characteristics of lead, zinc, arsenic, and cadmium in short-term storm water roof runoff in a suburban area. *Toxicological & Environmental Chemistry, 96*, 1034–1046.
22. Idrees, N., Tabassum, B., Abd Allah, E. F., Hashem, A., Sarah, R., & Hashim, M. (2018). Groundwater contamination with cadmium concentrations in some West U.P. Regions, India. *Saudi Journal of Biological Sciences, 25*, 1365–1368.
23. Zhang, C., Zang, X., Dai, Z., Zhang, X., & Ma, Z. (2021). Remediation techniques for cadmium-contaminated dredged river sediments after land disposal. *Sustainability, 13*, 6093.
24. Shabayev, V. P., Bocharnikova, E., & Ostroumov, V. (2020). Remediation of cadmium-polluted soil using plant growth-promoting Rhizobacteria and natural zeolite. *Eurasian Soil Science, 53*, 809–819.
25. Sabella, E., Aprile, A., Tenuzzo, B. A., Carata, E., Panzarini, E., Luvisi, A., De Bellis, L., Vergine, M. (2022): Effects of cadmium on root morpho-physiology of durum wheat. *Frontiers in Plant Science, 13.*
26. Haider, F. U., Liqun, C., Coulter, J. A., Cheema, S. A., Wu, J., Zhang, R., Wenjun, M., & Farooq, M. (2021). Cadmium toxicity in plants: Impacts and remediation strategies. *Ecotoxicology and Environmental Safety, 211*, 111887.
27. Bhattacharyya, K., Sen, D., Laskar, P., Saha, T., Kundu, G., Chaudhuri, A. G., & Ganguly, S. (2023). Pathophysiological effects of cadmium(II) on human health—A critical review. *Journal of Basic and Clinical Physiology and Pharmacology, 34*, 249–261.
28. Nordberg, G. F., Bernard, A., Diamond, G. L., Duffus, J. H., Illing, P., Nordberg, M., Bergdahl, I. A., Jin, T., & Skerfving, S. (2018). Risk assessment of effects of cadmium on human health (IUPAC Technical Report). *Pure and Applied Chemistry, 90*, 755–808.
29. Gunnar, N., Alfred, B., Gary, D., John, D., Paul, I., Monica, N., Ingvar, B., Taiyi, J., & Staffan, S. (2017). Risk assessment of effects of cadmium on human health. *Pure and Applied Chemistry.*
30. Kumar, S., & Sharma, A. (2019). Cadmium toxicity: Effects on human reproduction and fertility. *Reviews on Environmental Health, 34*, 327–338.
31. Suhani, I., Sahab, S., Srivastava, V., & Singh, R. P. (2021). Impact of cadmium pollution on food safety and human health. *Current Opinion in Toxicology, 27*, 1–7.
32. Zwolak, I. (2020). The role of selenium in arsenic and cadmium toxicity: An updated review of scientific literature. *Biological Trace Element Research, 193*, 44–63.
33. Yousefi, Z., Babanejad, E., Mohammadpour, R. A., Noori Esbokolaee, H. (2023). Evaluation of Cd phytoremediation by *Portulaca oleracea* irrigated by contaminated water. *Environmental Health Engineering and Management.*
34. Rezaei Kahkha, M. R., Salarifar, A.,& Rezaei Kahkha, B. (2022). Measurement of heavy metals in soil, plants and water samples based on MWCNTs modified with Bis(triethoxysilylpropyl)tetrasulfide by flame atomic absorption spectrophotometry. *Analytical Methods in Environmental Chemistry Journal.*

35. Yargholi, B., Azimi, A. A., Baghvand, A., Fariborz, A., Lyaghat, A., & Fardi, JA. (2010). Investigation of Cd adsorption and accumulation from contaminated soil in different parts of root crops.
36. Wu, Q., Wang, D., Zhang, J., Chen, C., Ge, H., Xu, H., Cai, D., & Wu, Z. (2022). Synthesis of iron-based carbon microspheres with tobacco waste liquid and waste iron residue for Cd(II) removal from water and soil. *Langmuir, 38*, 5557–5567.
37. Hu, Y., Zhou, X., Shi, A., Yu, Y., Rensing, C., Zhang, T., Xing, S., & Yang, W. (2023). Exogenous silicon promotes cadmium (Cd) accumulation in Sedum alfredii Hance by enhancing Cd uptake and alleviating Cd toxicity. *Frontiers in Plant Science, 14.*
38. Farooqi, Z. U. R., Murtaza, G., Bibi, S., Sabir, M., Owens, G., Saifullah, A. I., & Zeeshan, N. (2022). Immobilization of cadmium in soil-plant system through soil and foliar applied silicon. *International Journal of Phytoremediation, 24*, 1193–1204. https://doi.org/10.1080/15226514.2021.2024133
39. Farooqi, Z. U. R., Ahmad, I., Abdul Qadir, A., Murtaza, G., Rafiq, S., Jamal, A., Zeeshan, N., Murtaza, B., Javed, W., Radicetti, E., et al. (2022). Zeolite-assisted immobilization and health risks of potentially toxic elements in wastewater-irrigated soil under brinjal (*Solanum melongena*) Cultivation. *Agronomy, 2022*(12), 2433. https://doi.org/10.3390/agronomy12102433
40. Hussain, B., Li, J., Ma, Y., Tahir, N., & Ullah, A. (2020). Effects of Fe and Mn cations on Cd uptake by rice plant in hydroponic culture experiment. *PLoS ONE, 15*, e0243174.
41. Hussain, M. M., Farooqi, Z. U. R., Latif, J., Mubarak, M. U., & Younas, F. (2021). Bioremediation: a green solution to avoid pollution of the environment. *Soil Bioremediation: an Approach Towards Sustainable Technology*, 15–40.
42. Wu, F., Yang, W., Zhang, J., & Zhou, L.-Q. (2011). Growth responses and metal accumulation in an ornamental plant (*Osmanthus fragrans* var. *thunbergii*) submitted to different Cd Levels. *International Scholarly Research Notices, 2011*, 1–7.
43. Yang, J., Sun, H., Qin, J., Wang, X., & Chen, W. (2021). Impacts of Cd on temporal dynamics of nutrient distribution pattern of Bletilla Striata, a traditional Chinese medicine plant. Agriculture.
44. Huang, Q., Wan, Y., Luo, Z., Qiao, Y., Su, D.-c, & Li, H.-f. (2020). The effects of chicken manure on the immobilization and bioavailability of cadmium in the soil-rice system. *Archives of Agronomy and Soil Science, 66*, 1753–1764.
45. Guang-hu, X. (2015). Research advances in harmful effects of soil Cd pollution on crops. *Hunan Agricultural Sciences.*
46. Govind, P., & Madhuri, S. (2014). Heavy metals causing toxicity in animals and fishes.
47. Wu, M., Luo, Q., Liu, S., Zhao, Y., Long, Y., & Pan, Y. (2018). Screening ornamental plants to identify potential Cd hyperaccumulators for bioremediation. *Ecotoxicology and Environmental Safety, 162*, 35–41.
48. Tan, S., Huang, H., Zhu, M., Zhang, K., Xu, H., Wang, Z., Wu, X., & Zhang, Q. (2013). The negative effects of cadmium on Bermuda grass growth might be offset by submergence. *Environmental Science and Pollution Research, 20*, 7433–7440.
49. Chen, Z., Zhu, X., Lv, X., Huang, Y., Qian, W., Wang, P., Li, B., Wang, Z., & Cai, Z. (2019). Alleviative effects of C60 on the trophic transfer of cadmium along the food chain in aquatic environment. *Environmental Science & Technology, 53*, 8381–8388.
50. Khafaga, A. F., Abd El-Hack, M. E., Taha, A. E., Elnesr, S. S., & Alagawany, M. (2019). The potential modulatory role of herbal additives against Cd toxicity in human, animal, and poultry: A review. *Environmental Science and Pollution Research, 26*, 4588–4604.
51. Soubasakou, G., Cavoura, O., & Damikouka, I. (2022). Phytoremediation of cadmium-contaminated soils: A review of new cadmium hyperaccumulators and factors affecting their efficiency. *Bulletin of Environmental Contamination and Toxicology, 109*, 783–787.
52. Xu, Y., Yan, X., Fan, L., & Fang, Z.-q. (2016). Remediation of Cd(II)-contaminated soil by three kinds of ferrous phosphate nanoparticles. *RSC Advances, 6*, 17390–17395.
53. Singh, S., & Jawaid, D. S. M. A. (2015). Heavy metal removal from contaminated soil by soil washing—A review.

54. Changjia, J., Shuang, C., Qi, F., & Xiuli, S. (2021). Advances in Cd remediation techniques for contaminated soils. In *IOP Conference Series: Earth and Environmental Science 621*.
55. Li, Sp., Li, Jt., Kuang, J., Duan, H. N., Zeng, Y., & Shu, W. (2012). Effects of species richness on cadmium removal efficiencies of algal microcosms. *Journal of Applied Ecology, 49,* 261–267.
56. Padmaja, M., Bhavani, R., & Pamila, R. (2018). Adsorption of cadmium from aqueous solutions using low cost materials—A review. *International Journal of Engineering & Technology.*
57. Derbe, T. (2015). Cactus potential in heavy metal (Pb and Cd) Removal in water sample collected from rural area around Adigrat Town. *Chemistry and Materials Research, 7*, 84–92.
58. Visa, A., Maranescu, B., Lupa, L., Crisan, L., & Borota, A. (2020). New efficient adsorbent materials for the removal of Cd(II) from aqueous solutions. *Nanomaterials, 10*, 899.
59. Ansari, T. M., Shaheen, S., Manzoor, S., Naz, S., & Hanif, M. A. (2020). Litchi chinensis peel biomass as green adsorbent for cadmium (Cd) ions removal from aqueous solutions. *Desalination and Water Treatment, 173*, 343–350.
60. Lupa, L., Cocheci, L., Dobos, A. M., Onofrei, M. D., Negrea, P., & Filimon, A. (2022). Metal ions removal from contaminated water using membranes functionalized with ionic liquids. Water.
61. Lyu, R., Gu, B., Zhang, T., & Yang, Z. (2021). Simultaneous removal of Cd (II), Ni (II), and Pb (II) from water by a submerged macrophyte pondweed (*Potamogeton malaianus*). *Water Environment Research, 93*, 2637–2647.
62. Cheraghi, M., Sobhanardakani, S., Lorestani, B., & Zandipak, R. (2016). Tea wastes efficiency on removal of Cd(II) from aqueous solutions. *Archives of Hygiene Sciences, 5*, 184–191.
63. Fu, B., & Xie, F. (2020). Facile in situ synthesis of cellulose microcrystalline-manganese dioxide nanocomposite for effective removal of Pb(II) and Cd(II) from water. *Environmental Science and Pollution Research, 27*, 5108–5121.
64. Maleki, A., Mahvi, A. H., Zazouli, M. A., Izanloo, H., & Barati, A. H. (2011). Aqueous Cadmium removal by adsorption on Barley Hull and Barley Hull Ash.
65. Zaman, H. G., Baloo, L., Pendyala, R., & Kutty, S. R.M. (2020). Response surface methodology application in the optimization of Cd(II) removal from produced water by chitosan. In *2020 Second International Sustainability and Resilience Conference: Technology and Innovation in Building Designs* (51154) (pp. 1–4).
66. Shi, Y., Huang, Z., Liu, X., Imran, S., Peng, L., Dai, R., & Deng, Y. (2016). Environmental materials for remediation of soils contaminated with lead and cadmium using maize (*Zea mays* L.) growth as a bioindicator. *Environmental Science and Pollution Research, 23*, 6168–6178.
67. Adhikari, T., & Singh, M. V. (2008). Remediation of cadmium pollution in soils by different amendments: A column study. *Communications in Soil Science and Plant Analysis, 39*, 386–396.
68. Xian, L. (2015). Effects of chemical stabilization on soil Cd fractions and water spinach Cd contents after amendment with dredged river sediments. *Journal of Agro-Environment Science.*
69. Küpper, H., & Leitenmaier, B. (2013). Cadmium-accumulating plants. *Metal Ions in Life Sciences, 11*, 373–393. https://doi.org/10.1007/978-94-007-5179-8_12
70. Shu, R., & De, S. (2006). Characteristics of Cd pollution in soil and the mechanisms of phytoremediation for soil contamination. *Chinese Journal of Eco-agriculture.*
71. Mojiri, A. (2011). The potential of corn (*Zea mays*) for phytoremediation of soil contaminated with cadmium and lead. *Journal of Biological and Environmental Sciences*, 5.
72. Shakoor, A., Abdullah, M., Sarfraz, R., Altaf, M. A., & Batool, S. (2017). A comprehensive review on phytoremediation of cadmium (Cd) by mustard (*Brassica juncea* L.) and sunflower (*Helianthus annuus* L.).
73. Yu, G., Liu, J., Long, Y., Chen, Z., Sunahara, G. I., Jiang, P., You, S., Lin, H., & Xiao, H. (2020). Phytoextraction of cadmium-contaminated soils: Comparison of plant species and low molecular weight organic acids. *International Journal of Phytoremediation, 22*, 383–391.
74. Ali, J., Ali, F., Ahmad, I., Rafique, M., Munis, M. F. H., Hassan, S. W., Sultan, T., Iftikhar, M., & Chaudhary, H. J. (2021). Mechanistic elucidation of germination potential and growth of *Sesbania sesban* seedlings with *Bacillus anthracis* PM21 under heavy metals stress: An in vitro study. *Ecotoxicology and Environmental Safety, 208*, 111769. https://doi.org/10.1016/j.ecoenv.2020.111769

75. Ali, J., Wang, X., Rafique, M., Ahmad, I., Fiaz, S., Munis, M. F. H., & Chaudhary, H. J. (2021b). Phytoremediation of cadmium contaminated soil using *Sesbania sesban* L. in association with *Bacillus anthracis* PM21: A biochemical analysis. *Sustainability, 13*, 13529. https://doi.org/10.3390/su132413529
76. Tahir, M., Khan, M. B., Shahid, M., Ahmad, I., Khalid, U., Akram, M., Dawood, A., & Kamran, M. (2022). Metal-tolerant *Pantoea* sp. WP-5 and organic manures enhanced root exudation and phytostabilization of cadmium in the rhizosphere of maize. *Environmental Science and Pollution Research, 29*, 6026–6039.
77. Islam, M. S., Song, Z., Gao, R., Fu, Q., & Hu, H. (2022). Cadmium, lead, and zinc immobilization in soil by rice husk biochar in the presence of low molecular weight organic acids. *Environmental Technology, 43*, 2516–2529.
78. Gul, I., Manzoor, M., Hashim, N., Shah, G. M., Waani, S. P. T., Shahid, M., Antoniadis, V., Rinklebe, J., & Arshad, M. (2021). Challenges in microbially and chelate-assisted phytoextraction of cadmium and lead—A review. *Environmental Pollution, 287*, 117667.
79. Akhtar, M. J., Ali, Q., Javid, R., Asghar, H. N., Ahmad, I., Iqbal, M. Z., & Khaliq, A. (2019). Organic and inorganic amendments immobilized cadmium and improved maize growth and yield in Cd-contaminated soil. *International Journal of Agriculture and Biology, 22*, 1497–1506.
80. Shah, G. M., Tufail, N., Bakhat, H. F., Ahmad, I., Shahid, M., Hammad, H. M., Nasim, W., Waqar, A., Rizwan, M., & Dong, R. (2019). Composting of municipal solid waste by different methods improved the growth of vegetables and reduced the health risks of cadmium and lead. *Environmental Science and Pollution Research, 26*, 5463–5474. https://doi.org/10.1007/s11356-018-04068-z
81. Ahmad, I., Akhtar, M. J., Zahir, Z. A., & Mitter, B. (2015). Organic amendments: Effects on cereals growth and cadmium remediation. *International Journal of Environmental Science and Technology, 12*, 2919–2928.
82. Soni, R., & Keharia, H. (2021). Phytostimulation and biocontrol potential of Gram-positive endospore-forming Bacilli. *Planta, 254*, 49.
83. Fareleira, P., e Castro, I.V., Soares, R., Matos, S., Almeida, L., Barradas, A., & Nunes, A. P. (2022). Cover crops for the improvement of soil microbiological properties in intensive horticultural production systems. *Revista de Ciências Agrárias, 45*, 482–486.
84. Perisoara, A., Marinas, I. C., Geana, E. I., Constantin, M., Angheloiu, M., Pirvu, L., & Cristea, S. (2022). Phytostimulation and synergistic antipathogenic effect of *Tagetes erecta* extract in presence of Rhizobacteria. *Horticulturae, 8*, 779.
85. Geetha, N., Sunilkumar, C. R., Bhavya, G., Nandini, B., Abhijith, P., Satapute, P., Shetty, H. S., Govarthanan, M., & Jogaiah, S. (2022). Warhorses in soil bioremediation: Seed biopriming with PGPF secretome to phytostimulate crop health under heavy metal stress. *Environmental Research*, 114498.
86. Zhang, H., Yuan, X., Xiong, T., Wang, H., & Jiang, L. (2020). Bioremediation of co-contaminated soil with heavy metals and pesticides: Influence factors, mechanisms and evaluation methods. *Chemical Engineering Journal, 398*, 125657.
87. Sharma, P. (2021). Efficiency of bacteria and bacterial assisted phytoremediation of heavy metals: An update. *Bioresource Technology, 328*, 124835.
88. Choden, D., Pokethitiyook, P., Poolpak, T., & Kruatrachue, M. (2021). Phytoremediation of soil co-contaminated with zinc and crude oil using *Ocimum gratissimum* (L.) in association with *Pseudomonas putida* MU02. *International Journal of Phytoremediation, 23*, 181–189.
89. Danyal, Y., Mahmood, K., Ullah, S., Rahim, A., Raheem, G., Khan, A. H., & Ullah, A. (2023). Phytoremediation of industrial effluents assisted by plant growth promoting bacteria. *Environmental Science and Pollution Research, 30*, 5296–5311.
90. Hasan, M. M., Uddin, M. N., Ara-Sharmeen, I. F., Alharby, H., Alzahrani, Y., Hakeem, K. R., & Zhang, L. (2019). Assisting phytoremediation of heavy metals using chemical amendments. *Plants, 8*, 295.
91. Keshavarz, S., Ghasemi-Fasaei, R., Ronaghi, A., & Mousavi, A. A. (2021). Innovative assisted phytoremediation of multi-elements contaminated soil by ryegrass: An electro-bio-chemical approach. *Journal of soils and sediments, 21*, 2604–2618.

92. Asgari Lajayer, B., Khadem Moghadam, N., Maghsoodi, M. R., Ghorbanpour, M., & Kariman, K. (2019). Phytoextraction of heavy metals from contaminated soil, water and atmosphere using ornamental plants: Mechanisms and efficiency improvement strategies. *Environmental Science and Pollution Research, 26*, 8468–8484.
93. Shackira, A.,& Puthur, J. T. (2019). Phytostabilization of heavy metals: Understanding of principles and practices. *Plant-Metal Interactions*, 263–282.
94. Badawy, I. H., Hmed, A. A., Sofy, M. R., & Al-Mokadem, A. Z. (2022). Alleviation of cadmium and nickel toxicity and phyto-stimulation of tomato plant l. by endophytic micrococcus luteus and enterobacter cloacae. *Plants, 11*, 2018.
95. Bakshe, P., & Jugade, R. (2023). Phytostabilization and rhizofiltration of toxic heavy metals by heavy metal accumulator plants for sustainable management of contaminated industrial sites: A comprehensive review. *Journal of Hazardous Materials Advances*, 100293.
96. Farraji, H., Zaman, N. Q., Tajuddin, R. M., & Faraji, H. (2016). Advantages and disadvantages of phytoremediation : A concise review.
97. Janssen, J., Weyens, N., Croes, S., Beckers, B., Meiresonne, L., Van Peteghem, P., Carleer, R., & Vangronsveld, J. (2015). Phytoremediation of metal contaminated soil using willow: Exploiting plant-associated bacteria to improve biomass production and metal uptake. *International Journal of Phytoremediation, 17*, 1123–1136.
98. Mehes-Smith, M., Nkongolo, K., & Cholew, E. (2013). Coping mechanisms of plants to metal contaminated soil. *INTECH*. https://doi.org/10.5772/55124
99. Chinmayee, M. D., Mahesh, B., & Pradesh, S. (2012). The assessment of phytoremediation potential of invasive weed *Amaranthus spinosus* L. *Applied Biochemistry and Biotechnology, 167*, 1550–1559.

Microbial Strategies for Potential Cadmium Bioremediation: A Review

Iftikhar Ahmad, Qurat Ul Ain, Amna, Shehzad Mehmood, Muhammad Younis, Umar Daraz, and Bo Zhu

Abstract Cadmium (Cd) is a highly toxic heavy metal that poses serious environmental risks due to its persistence and bioaccumulation in various ecosystems. Traditional methods of Cd remediation, such as chemical precipitation and adsorption are not environment-friendly and expensive. In recent years, bioremediation has emerged as a promising and sustainable alternative for the removal of Cd from contaminated environments. This review explores the various microbial strategies (bioaugmentation, mycoremediation, phycoremediation, phytoremediation, nano bioremediation, microbial fuel cells, microbial consortia, and biofilm applications) employed for potential removal of Cd from soil and water resources, emphasizing their efficiency, applicability, and prospects. We also elaborated different mechanisms (such as biosorption, bioprecipitation, bioleaching, bioaccumulation, biotransformation, sequestration, phytoextraction, phytostabilization, rhizofiltration, and molecular mechanisms) responsible for Cd removal from wastewater. This review

I. Ahmad (✉) · M. Younis
Department of Environmental Sciences, COMSATS University Islamabad, Vehari Campus, Vehari 61100, Pakistan
e-mail: iftikharahmad@cuivehari.edu.pk

Q. U. Ain
Department of Environmental Science, COMSATS University Islamabad, Vehari Campus, Vehari 61100, Pakistan

Amna
Department of Botany, Rawalpindi Women University, 6th road Satellite Town, Rawalpindi 46300, Pakistan

U. Daraz
State Key Laboratory of Grassland Agro-Ecosystems, Center for Grassland Microbiome, College of Pastoral Agriculture Science and Technology, Lanzhou University, Lanzhou 730000, China

B. Zhu
Shanghai Yangtze River Delta Eco-Environmental Change and Management Observation and Research Station, Shanghai Cooperative Innovation Center for Modern Seed Industry, School of Agriculture and Biology, Shanghai Jiao Tong University, Shanghai, China

S. Mehmood
Department of Biotechnology, COMSATS University Islamabad Vehari Campus, Vehari 61100, Pakistan

A. K. Jha and N. Kumar (eds.), *Cadmium Toxicity in Water*, Springer Water,
https://doi.org/10.1007/978-3-031-54005-9_6

provides valuable insights into harnessing the power of microorganisms to combat Cd pollution effectively.

Keywords Soil pollution · Bioaccumulation · Food chain · Heavy metals · Human health

1 Introduction

Cadmium (Cd) contamination has become a major environmental concern due to its toxic nature and widespread occurrence in various ecosystems [9]. Cadmium pollution stems from industrial activities, mining operations, agricultural practices, and improper disposal of electronic waste, among other sources [5]. Its persistent presence in soil, water, and air poses significant risks to human health and ecosystem stability. Traditional remediation methods for Cd-contaminated sites, such as chemical precipitation and physical adsorption, have shown limited effectiveness and often require significant financial investments [9]. Moreover, these approaches may generate secondary pollutants, further complicating the cleanup process. In recent years, bioremediation has emerged as an eco-friendly and cost-effective alternative for removing Cd from polluted environments.

Bioremediation exploits the inherent capabilities of microorganisms, including bacteria, fungi, and algae, to sequester, transform, or remove contaminants from the environment. Microbes possess diverse strategies to mitigate Cd toxicity, such as binding the metal ions to their cell surfaces, sequestering them within intracellular compartments, and producing metal-chelating agents to neutralize its toxic effects. There are many other microbes that have been shown to be effective for bioremediation against Cd toxicity [9, 11]. Some of these microbes include *Bacillus* spp. [5, 9], *Enterobacter* spp. [3], *Pseudomonas aeruginosa* [25], *Enterobacter aerogenes* [78], *Rhodococcus erythropolis* [93], *Micrococcus luteus* and *Enterobacter cloacae* [14], *Trichoderma viride* [87], *Saccharomyces cerevisiae* [74], and *Trichoderma harzianum* [35] have been reported repeatedly for microbial assisted phytoremediation.

The choice of microbe will depend on the specific conditions of the contaminated site, such as the concentration of Cd^{+2} ions, the pH of the soil, and the presence of other pollutants [55]. Bioremediation is a promising approach for the removal of Cd toxicity from the environment. It is a relatively low-cost and environment-friendly way to treat contaminated sites. In a study *Baciullus anthracis* PM21 endured Cd stress through homeostasis of antioxidant activities, and positively impacted plant growth and biomass [9]. However, basic data on microbial species (bacteria, fungi, algae) is needed to optimize the bioremediation process and to develop new microbes that can be used to remove Cd from the environment. This comparative review aims to explore and analyze various microbial strategies and underlying mechanisms employed for potential Cd bioremediation. By assessing their efficacy, applicability, and limitations, this review will provide valuable insights into harnessing the power of microorganisms to combat Cd pollution effectively.

2 Sources and Environmental Impacts of Cd Pollution

Cadmium pollution is a significant environmental issue that arises from various human activities and natural processes. This toxic heavy metal can have detrimental effects on ecosystems, wildlife, and human health. Understanding the sources and environmental impact of Cd pollution is essential for developing effective mitigation strategies and protecting the environment. Industrial processes such as mining, smelting, battery manufacturing, and electroplating are major contributors to Cd pollution. These activities release Cd-containing dust, fumes, and wastewater into the environment [9, 88]. Normally Cd enters the soil with phosphate-based fertilizers and sewage sludge applied to farmlands [21, 49]. It can accumulate in crops and enter the food chain, posing a risk to human health through the consumption of contaminated food [85]. Improper disposal of municipal waste, electronic waste, batteries, and other consumer products containing Cd can lead to its leaching into soil and water bodies [29]. Cd is released into the environment during combustion of fossil fuels such as coal and oil and can eventually be deposited onto land and water surfaces through air pollution. Weathering of rocks and soil erosion can also release Cd into the environment [108]. These sources built up Cd in the environment and have following environmental impacts.

- Soil contamination: Cd is highly persistent in soil, where it can accumulate over time due to its low mobility [37]. Contaminated soil negatively impacts plant growth and can lead to the bioaccumulation of Cd in crops, potentially reaching humans through the food chain [104].
- Water contamination: Cd can leach into groundwater and surface water from contaminated soils, industrial discharges, and waste disposal sites. High Cd levels in water bodies can harm aquatic life and disrupt aquatic ecosystems [29].
- Bioaccumulation: Cd has a strong affinity for biological tissues and can bioaccumulate in organisms [7]. It can be biomagnified in the food chain, with predatory species accumulating higher levels of Cd from their prey [100].
- Toxicity to wildlife: Cd is highly toxic to a wide range of organisms, including fish, birds, and mammals. Chronic exposure to even low concentrations of Cd can impair reproduction, growth, and overall health [79].
- Human health concerns: Humans can be exposed to Cd through contaminated food, drinking water, and air. Chronic exposure to Cd can lead to kidney damage, respiratory issues, skeletal disorders, and an increased risk of certain cancers [34].
- Environmental persistence: Cd is not biodegradable and remains in the environment for long periods, contributing to its persistence and potential for long-term impacts [77]. The Cd pollution stems from anthropogenic activities, such as industrial processes and improper waste management, as well as natural processes. The environmental impact of Cd pollution includes soil and water contamination, bioaccumulation in the food chain, toxicity to wildlife, and potential health risks to humans. Implementing measures to control Cd pollution and adopting sustainable practices are crucial to safeguarding the environment and human health from its harmful effects.

3 Importance of Bioremediation in Cd Cleanup

Bioremediation is a process that uses living organisms, such as bacteria, fungi, plants, and microorganisms, to clean up and remove pollutants and contaminants from soil, water, air, and other environmental media. It is an eco-friendly and cost-effective approach to remediate contaminated sites and mitigate environmental damage [4, 9, 11]. The basic principle of bioremediation involves harnessing the natural abilities of these organisms to break down, transform, or remove harmful substances, converting them into less toxic or non-toxic forms [52]. The contaminants can include organic compounds like petroleum hydrocarbons, heavy metals, pesticides, solvents, and other toxic chemicals. Bioremediation plays a crucial role in the cleanup of Cd-contaminated environments due to its effectiveness, environmental sustainability, and potential for cost-efficiency. The importance of bioremediation in Cd cleanup can be understood through the following key points:

- Environmental-friendly: Unlike traditional methods that involve harsh chemicals or physical processes, bioremediation relies on the natural abilities of microorganisms, such as bacteria, fungi, and algae, to interact with and sequester Cd. This approach minimizes the generation of secondary pollutants and reduces the overall ecological impact during the cleanup process [8].
- Specificity and selectivity: Microorganisms can exhibit a high degree of specificity and selectivity towards Cd, targeting the metal contaminant while sparing essential elements. This ensures a more precise removal of Cd without causing unnecessary damage to the surrounding ecosystem [7, 41].
- Reduced cost and energy consumption: Bioremediation is often more cost-effective compared to traditional cleanup methods. It requires less energy and fewer resources since microorganisms can operate under ambient conditions, reducing the overall financial burden of Cd cleanup projects [4, 9, 94].
- Long-term effectiveness: Bioremediation offers long-term effectiveness by addressing the root cause of contamination. Microorganisms can transform Cd into less toxic forms, immobilize it in the soil matrix, or accumulate it within their cells, preventing its release back into the environment [112].
- Versatility and applicability: Bioremediation can be adapted to various contaminated environments, including soils, sediments, water bodies, and industrial effluents. Additionally, the approach can be employed in situ (on-site) or ex situ (off-site), providing flexibility in choosing the most appropriate remediation strategy for specific locations [27].
- Synergistic interactions: Microbial consortia and plant–microbe interactions can further enhance the efficiency of Cd cleanup. Co-culturing different microbial species can lead to cooperative mechanisms that result in improved Cd uptake and removal. Similarly, certain plants can act as hyperaccumulators, aiding in the phytoextraction of Cd with the help of associated microorganisms [59].

- Minimal disruption to surrounding ecosystems: Bioremediation minimizes the disturbance caused to the natural environment during the cleanup process. It allows for a gradual and controlled removal of Cd, which can be particularly beneficial in sensitive ecosystems where disruption could cause more harm than good [65].
- Potential for large-scale implementation: Bioremediation has the potential for large-scale implementation, making it applicable to widespread and severe Cd-contaminated sites. As technology advances and our understanding of microorganism-metal interactions improves, bioremediation strategies can be optimized for even greater efficiency [54]. Keeping in view of the above discussion, bioremediation offers a sustainable, effective, and environmentally friendly approach to Cd cleanup. Its specificity, long-term effectiveness, and potential for cost-efficiency make it a promising solution for addressing Cd contamination in various environmental settings. As we strive to protect ecosystems and human health from the harmful effects of heavy metal pollution, bioremediation stands out as a valuable tool in the remediation toolkit.

4 Microbial Strategies for Bioremediation of Cd

Microbial bioremediation strategies offer several advantages, including eco-friendliness, cost-effectiveness, and specificity in targeting Cd contaminants. As our understanding of bacterial interactions with Cd continues to grow, these strategies hold great promise for the remediation of Cd-contaminated environments, contributing to a cleaner and healthier ecosystem. Bacterial bioremediation strategies are effective and environmentally friendly approaches used to remove Cd contamination from various environments. The microbes have been shown to be effective in removing Cd from a variety of environments, including soil, water, and industrial wastewater. They can do this through a variety of mechanisms, including biosorption, bioaccumulation, and biotransformation. The use of microbes for Cd bioremediation is a promising technology with the potential to remove Cd from the environment in a safe and efficient manner. However, more research is needed to optimize the process and to develop cost-effective methods for large-scale application. Microorganisms such as bacteria, fungi and algae possess unique capabilities to interact with and sequester Cd through different mechanisms. Here are some key microbial bioremediation strategies for Cd cleanup in the environment.

4.1 Bioaugmentation

Cadmium bioremediation through bioaugmentation is a process that involves the introduction of specific microorganisms or microbial consortia to contaminated environments (Table 1) with the aim of enhancing the natural remediation of Cd or other heavy metals (Ali et al. 2023). Bioaugmentation is a type of bioremediation strategy that focuses on adding microorganisms with the ability to degrade, immobilize, or transform the target pollutants, such as Cd, into less toxic forms or sequester them, thereby reducing their environmental impact. In bioaugmentation, researchers isolate or engineer microorganisms that have the capacity to tolerate and interact with Cd [5, 3, 9, 11]. These microorganisms could include certain bacteria, fungi, or algae that possess metal-binding capabilities or enzymatic pathways to transform Cd into less toxic forms or precipitate it as insoluble compounds (Table 1). Once identified or engineered, these beneficial microorganisms are introduced into the contaminated site to establish or enhance their population and function. Key steps in the bioaugmentation process for Cd bioremediation may include:

1. Site Assessment: Conducting a thorough analysis of the contaminated site to determine the extent and distribution of Cd pollution and the site's environmental conditions. This step helps in designing an effective bioaugmentation strategy.
2. Microorganism Selection: Identifying or isolating microorganisms that are naturally present in the environment and capable of metabolizing or interacting with Cd. Alternatively, microbial strains may be genetically engineered to enhance their Cd bioremediation capabilities.
3. Microbial Cultivation: Growing and multiplying the selected or engineered microorganisms in the laboratory under controlled conditions to obtain enough for application.
4. Application: Introducing the cultured microorganisms into the contaminated site. This can be done through various methods, such as soil or water amendment, injection, or spraying.
5. Monitoring: Regularly monitoring the site to assess the effectiveness of bioaugmentation and the progress of Cd bioremediation. This step involves measuring Cd concentrations, microbial populations, and other relevant environmental parameters.
6. Maintenance: Ensuring appropriate environmental conditions, such as pH, temperature, and nutrient availability, to support the growth and activity of the introduced microorganisms.

Bioaugmentation is just one of several bioremediation approaches used to address Cd contamination. Other strategies include biostimulation, which involves adding nutrients to stimulate the activity of indigenous microorganisms, and phytoremediation, which utilizes plants to extract, stabilize, or immobilize Cd from the soil. It is important to note that the success of bioaugmentation largely depends on the specific characteristics of the contaminated site, the selected microorganisms, and the overall environmental conditions. As such, site-specific assessments and careful planning are essential for the effective application of Cd bioremediation through bioaugmentation.

Table 1 Microbial species (bacteria, fungi, algae, plants) and tolerance mechanisms involved in Cd remediation from soil, and water resources

Microorganism	Tolerance Mechanism	Niche	Reference
Bioaugmentation (Bacteria and Archaea)			
Natronolimnobius innermongolicus	Biosorption	Wastewater	[40]
Bacillus subtilis *Bacillus* spp.	Metal efflux, intracellular precipitation, complexation	Soil, Water, plants	[12, 19, 50]
Pseudomonas aeruginosa	Metal efflux, intracellular precipitation, complexation	Soil, water, plants	[48, 91]
Cupriavidus metallidurans	Metal efflux, intracellular precipitation, complexation	Soil, water	[67]
Shewanella oneidensis	Metal efflux, intracellular precipitation, complexation	Marine environments	[43]
Rhizobium leguminosarum	Metal efflux, intracellular precipitation, and complexation	Soil, plants	[92]
Enterobacter aerogenes	Intracellular Cd accumulation, Metal resistance	Water	[78]
Bacillus cereus	Biosorption, functional groups	Water	[106]
Mycoremediation (Fungi)			
Saccharomyces cerevisiae	Metal efflux, intracellular precipitation, complexation	Soil, plants	[74]
Pichia pastoris	Metal efflux, intracellular precipitation, complexation	Soil, plants	[38]
Trichoderma asperellum	Immobilization	Soil	[17]
Aspergillus niger	Bioaccumulation, Metal tolerance	Soil, plant	[19, 51]
Trichoderma viride	Bioaccumulation, Metal tolerance	Water	[87]

(continued)

Table 1 (continued)

Microorganism	Tolerance Mechanism	Niche	Reference
Amanita muscaria, Hypholoma fasciculare, Russula foetens, Boletus pinophilus	Uptake, Biomass, BAF	Soil	[10]
Paxillus involutus, Xerocomus badius, Amanita muscaria	Immobilization, Metal tolerance	Soil	[18]
Phycoremediation (Algae)			
Chlorella vulgaris	Metal chelation, Intracellular precipitation	Water	[47]
Scenedesmus quadricauda	Metal chelation, Intracellular precipitation	Water	[47]
Scenedesmus sp.	Bioaccumulation, Metal tolerance, Lipid rich biomass	Water	[100]
Cladophora sp.	Bioaccumulation, Detoxification	Water	[1]
Spirulina platensis	Metal chelation, Intracellular precipitation	Water	[23]
Bacterial assisted phytoremediation			
Bacillus anthracis associated with *Sesbania sesban*	Phytostabilization, Antioxidant activities, Biomass production	Soil, plant	[9]
Micrococcus luteus and Enterobacter cloacae associated with *Solanum lycopersicum*	Antioxidant activities, Osmoregulation, BAF, TF	Soil, plant	[14]
Pseudomonas fluorescens associated with *Sedum alfredii*	Phytoextraction, Indole acetic acid production, Lateral root formation	Soil, plant	[107]
Enterobacter aerogenes associated with *Oryza sativa*	Antioxidant activities, Metal resistance, Cd uptake	Soil, plant	[78]
Klebsiella sp. associated with *Triticum aestivum and Zea mays*	Phytostabilization, Biomass production, Cd tolerance	Soil, plant	[5]
Enterobacter sp. associated with *Zea mays*	Phytoextraction, Biomass production, Root colonization	Soil, plant	[3]

(continued)

Table 1 (continued)

Microorganism	Tolerance Mechanism	Niche	Reference
Rhodococcus erythropolis associated with *Mesembryanthemum crystallinum*	Metal resistance, Phytoremediation	Soil, plant	[93]
Bacillus subtilis associated with Lolium perenne	Cd immobilization, Biofilm sequestration, Intracellular accumulation	Soil, plant	[58, 56]
Glomus etunicatum and *Glomus mosseae associated with Lolium perenne*	AMF activates Cd transport channels, Cd influx, root uptake	Soil, plant	[39]

4.2 Mycoremediation

Mycoremediation is a specific type of bioremediation that utilizes fungi to clean up contaminated environments, including those contaminated with Cd and other heavy metals [17, 51]. Fungi possess unique capabilities to absorb and accumulate heavy metals from their surroundings, and some species are especially efficient at remediation processes (Table 1). Mycoremediation has gained attention as a sustainable and environmentally friendly approach to tackle Cd pollution and other toxic metal contamination [10]. Here's how mycoremediation works for Cd bioremediation:

1. Fungal Selection: Researchers identify and select fungal species that have a natural affinity for accumulating heavy metals, including Cd. Some fungi have specialized metal-binding proteins on their cell surfaces that facilitate the absorption of heavy metals.
2. Site Assessment: A thorough assessment of the contaminated site is conducted to determine the extent and type of Cd contamination, as well as the environmental conditions that could impact fungal growth and activity.
3. Fungal Inoculation: The selected fungi are introduced to the contaminated site. This can be done by applying fungal spores or mycelium directly to the soil or water, depending on the site conditions.
4. Metal Accumulation: Fungi grow and spread their mycelium, which is the thread-like vegetative part of the fungus. The mycelium acts as a network that extends into the contaminated soil or water, absorbing Cd, and other heavy metals along the way.
5. Metal Sequestration: Once absorbed, the Cd is often sequestered inside the fungal cells or bound to the fungal biomass. Fungi can immobilize heavy metals, reducing their bioavailability and potential for toxicity to other organisms.
6. Biomass Removal: After a certain period, the contaminated biomass can be harvested and removed from the site. The harvested fungal biomass, which now contains accumulated Cd, may require appropriate disposal to prevent recontamination.

7. Monitoring: Regular monitoring of the site is essential to assess the progress of mycoremediation. This includes measuring Cd concentrations, fungal population growth, and other relevant environmental parameters.

Mycoremediation offers several advantages as a bioremediation strategy for Cd and other heavy metals: (i) fungi are naturally occurring and have a widespread distribution in the environment, making them easily accessible for application, (ii) some fungal species can tolerate harsh environmental conditions and survive in polluted sites, (iii) mycoremediation is cost-effective and relatively easy to implement compared to traditional cleanup methods, and (iv) fungi can form symbiotic relationships with plants (mycorrhizal associations), enhancing plant growth and aiding in phytoremediation efforts. However, mycoremediation also has its limitations, such as potentially slower cleanup rates compared to chemical treatments, the need for site-specific fungal selection, and the risk of unintentional spreading of invasive fungal species. Overall, mycoremediation shows great promise as an eco-friendly and sustainable approach for Cd bioremediation, especially when combined with other remediation techniques to maximize its efficiency.

4.3 Phycoremediation

Phycoremediation is a bioremediation approach that uses photosynthetic microorganisms, such as algae, to clean up contaminated environments (Table 1), including those contaminated with Cd and other heavy metals. Algae can accumulate and transform heavy metals, making them potentially valuable candidates for remediation efforts [55]. Here's how phycoremediation works for Cd bioremediation:

1. Algae Selection: Researchers identify and select specific algae species or strains that have a high affinity for Cd uptake and accumulation. Some algae have metal-binding proteins or ligands on their cell surfaces, which facilitate the absorption of heavy metals [1].
2. Site Assessment: A thorough assessment of the contaminated site is conducted to determine the extent and type of Cd contamination, as well as the environmental conditions that could impact algal growth and metal uptake.
3. Algal Cultivation: The selected algae are cultured in the laboratory under controlled conditions to obtain enough biomass for application.
4. Algal Inoculation: The cultivated algae are introduced to the contaminated site, either by directly adding the algal biomass to the water or soil or by encouraging natural algal growth if suitable native species are already present.
5. Metal Uptake: Algae grow through photosynthesis, using sunlight as their energy source. During this process, they absorb water, and the dissolved nutrients present in it, including Cd. The accumulated Cd is stored within the algal cells.
6. Metal Sequestration: Once inside the algae, Cd is sequestered, reducing its availability in the environment, and minimizing its potential toxicity to other organisms.

7. Biomass Harvesting: After a certain period, the algal biomass can be harvested from the site. This harvested biomass contains accumulated Cd and may be further processed for metal recovery or proper disposal.
8. Monitoring: Regular monitoring of the site is essential to assess the progress of phycoremediation. This includes measuring Cd concentrations, algal population growth, and other relevant environmental parameters.

Phycoremediation offers several advantages as a bioremediation strategy for Cd and other heavy metals: (i) algae are efficient at removing heavy metals, including Cd, from the water and soil, (ii) algae can grow rapidly, providing a relatively fast remediation process compared to some other bioremediation methods, (iii) phycoremediation can be applied in various environments, including aquatic systems like ponds, lakes, and wastewater treatment facilities, and (iv) algae can be used in combination with other remediation methods for enhanced cleanup efficiency. However, phycoremediation also has some limitations, such as the potential to release accumulated metals back into the environment if not properly managed and the need for specific algae species adapted to the contaminated site conditions. Overall, phycoremediation shows promise as an eco-friendly and sustainable approach for Cd bioremediation, and ongoing research is continuously improving its effectiveness and practicality in real-world applications.

4.4 Phytoremediation

Phytoremediation is a bioremediation technique that uses plants to clean up contaminated environments or some time microorganisms assisted in phytoremediation of Cd polluted soil or water (Fig. 2, [9, 11]). This approach harnesses the natural abilities of certain plant species to absorb, accumulate, and detoxify pollutants from the soil, water, or air (Table 1). Phytoremediation is considered an eco-friendly and cost-effective alternative to traditional remediation methods [57]. Here's how phytoremediation works for Cd bioremediation:

1. Plant Selection: Researchers identify and select plant species that have a high affinity for Cd uptake and accumulation. Some plants, known as hyperaccumulators, have evolved to tolerate and thrive in metal-contaminated soils, and they can accumulate large concentrations of Cd and other heavy metals in their tissues.
2. Site Assessment: A thorough assessment of the contaminated site is conducted to determine the extent and type of Cd contamination, as well as the environmental conditions that could impact plant growth and metal uptake.
3. Planting: The selected plant species are cultivated and then planted in the contaminated area. This can be done using seeds, seedlings, or cuttings, depending on the plant species and site conditions.
4. Cd Uptake: As the plants grow, they absorb water and nutrients from the soil, including Cd. The Cd is taken up through the roots and transported to various parts of the plant, such as the leaves and stems.

5. Metal Sequestration: Once inside the plant, Cd is sequestered and stored in the plant's tissues. This process reduces the bioavailability of Cd in the environment and mitigates its potential toxicity to other organisms.
6. Biomass Harvesting: After a certain period, the plants are harvested from the site. This harvested plant material contains accumulated Cd and may need to be properly managed, either by disposing of it safely or by recycling the biomass through other processes to recover the metal.
7. Monitoring: Regular monitoring of the site is essential to assess the progress of phytoremediation. This includes measuring Cd concentrations in the soil and plant tissues, as well as monitoring plant health and growth.

Phytoremediation offers several advantages as a bioremediation strategy for Cd and other heavy metals: (i) phytoremediation is a sustainable and environmentally friendly approach that utilizes natural processes and does not require extensive site disruption, (ii) it can be applied in various environments, including both terrestrial and aquatic systems, (iii) phytoremediation can be cost-effective, especially in large-scale remediation projects, and (iv) it can be combined with other remediation methods for improved efficiency. However, phytoremediation also has some limitations, such as potentially slow cleanup rates, depending on factors such as plant growth rates, metal concentrations, and site conditions. Additionally, the success of phytoremediation relies on selecting appropriate plant species for the specific contaminants and site characteristics. Overall, phytoremediation shows promise as a viable and sustainable approach for Cd bioremediation, and ongoing research continues to refine and optimize this technique for various environmental cleanup scenarios.

4.5 Nano-bioremediation

Nano-bioremediation is an innovative approach that combines the principles of nanotechnology and bioremediation to address environmental contamination, including the cleanup of Cd and other heavy metals. This technique involves the use of nanoparticles or nanomaterials in conjunction with microorganisms (such as bacteria or fungi) to enhance the bioremediation process and improve the efficiency of removing pollutants from the environment [20]. Here's how nano-bioremediation works for Cd bioremediation:

1. Nanomaterial Synthesis: Nanoparticles or nanomaterials with specific Cd binding and sequestration properties are synthesized in the laboratory. These nanomaterials can be engineered to have a high surface area, functional groups that can bind with heavy metals like Cd, and other properties that enhance their interaction with contaminants.
2. Microorganism Selection: Suitable microorganisms, such as metal-resistant bacteria or metal-accumulating fungi, are selected based on their ability to interact with Cd and their compatibility with the chosen nanomaterials.

3. Nanoparticle-Cell Interaction: The synthesized nanomaterials are introduced to the selected microorganisms, creating a hybrid nanomaterial-microorganism system. The nanoparticles act as carriers or agents that facilitate the uptake and transport of Cd by the microbes [44].
4. Application: The hybrid nanomaterial-microorganism system can be applied to the contaminated site. This can be done by the amendment of soil or water with the engineered nanomaterials and microorganisms.
5. Metal Uptake and Sequestration: The microorganisms interact with Cd in the environment and absorb the metal into their cells or extracellular matrix. The nanoparticles aid to improve the efficiency of Cd uptake and its sequestration by increasing the surface area available for interaction.
6. Biomass Harvesting (Optional): Depending on the specific nano-bioremediation approach used, the metal-accumulating microorganisms and nanoparticles can be harvested after the remediation process for further processing or disposal.
7. Monitoring: Regular monitoring of the site is essential to assess the progress of nano-bioremediation, including measuring Cd concentrations, microbial population growth, and other relevant environmental parameters.

Nano-bioremediation offers several potential advantages for Cd and heavy metal bioremediation [98]: (i) nanoparticles can enhance the uptake and sequestration of Cd by microorganisms, leading to faster and more effective remediation, (ii) nanomaterials can be engineered with specific properties to target pollutants, increasing the selectivity of the remediation process, (iii) nano-bioremediation can be applied in various environments, including soils, sediments, and aqueous systems, and (iv) use of microorganisms and biodegradable nanomaterials may reduce long-term environmental impacts compared to traditional chemical treatments. However, it is essential to consider potential risks associated with the use of nanoparticles, such as unintended environmental effects and nanoparticle toxicity. As with any emerging technology, the safety and regulatory aspects of nano-bioremediation require careful consideration and further research. Overall, nano-bioremediation holds promise as a cutting-edge and potentially effective approach for Cd bioremediation, but further research and field trials are needed to fully assess its practicality and long-term impacts in real-world scenarios [70].

4.6 Microbial Fuel Cells

Microbial fuel cells (MFCs) are devices that utilize the metabolic activities of microorganisms to generate electrical energy from organic matter, such as wastewater, organic waste, or other carbon-rich substances. While MFCs are primarily designed for electricity generation, they can also play a role in Cd bioremediation, especially in environments where both electricity production and metal remediation are desirable outcomes [28]. Here's how microbial fuel cells can be integrated into Cd bioremediation:

1. Setup: A typical microbial fuel cell consists of an anode and a cathode separated by a proton exchange membrane. The anode is the chamber where microorganisms, typically bacteria or archaea, break down organic matter. During this process, electrons are released as byproducts.
2. Electron Transfer: The released electrons from the microbial metabolism travel through an external circuit from the anode to the cathode, generating electricity.
3. Cd Remediation: In the presence of Cd-contaminated water or soil, certain microbial species can interact with Cd. Some microorganisms have the ability to reduce Cd to less toxic forms or bind it to their cell surfaces.
4. Enhanced Bioremediation: The electrical current generated in the microbial fuel cell can create conditions that favor the growth and activity of specific metal-resistant microorganisms. This can enhance the bioremediation process, as the microbial community adapts to the presence of Cd and potentially increases its Cd-removal capabilities [61].
5. Electrochemical Processes: In some cases, the electrical current generated in MFCs can also induce electrochemical reactions that facilitate the removal of Cd through precipitation or electrodeposition.

Microbial fuel cells offer several advantages for Cd bioremediation [101]: (i) MFCs provide a dual benefit of electricity generation while also contributing to bioremediation efforts, making them a sustainable and environmentally friendly technology, (ii) MFCs can operate continuously as long as a constant supply of organic matter is provided, making them suitable for long-term bioremediation projects, and (iii) MFCs are relatively low-maintenance compared to other bioremediation methods, as the microorganisms can self-sustain and reproduce. However, some challenges and limitations of using microbial fuel cells for Cd bioremediation include (i) the bioremediation process in MFCs can be relatively slow compared to chemical treatments or other ex situ bioremediation techniques, (ii) MFCs require specific environmental conditions and optimal microbial consortia to effectively bioremediate Cd, and (iii) the electricity generated by MFCs may not be sufficient for large-scale remediation applications, especially in heavily contaminated sites. The microbial fuel cells show promise as a sustainable and energy-efficient approach for Cd bioremediation, but their practical application may require further research and optimization to address the specific challenges associated with heavy metal removal [33].

4.7 *Microbial Consortia*

Bacterial bioremediation strategies can be enhanced by using microbial consortia, which involve the co-culturing of different bacterial species. Each bacterial species within the consortium may possess distinct Cd uptake mechanisms, allowing for complementary interactions. Microbial consortia can facilitate the simultaneous uptake, sequestration, and precipitation of Cd, resulting in improved efficiency and effectiveness of the bioremediation process [16]. Microbial consortia in the context of Cd bioremediation refer to a group of different microorganisms, such as bacteria, fungi, or archaea, working together synergistically to efficiently degrade, immobilize, or transform Cd contaminants in the environment. These microbial communities often consist of species with complementary metabolic abilities that enhance the overall bioremediation process and improve the remediation efficiency of Cd and other pollutants [24]. Here's how microbial consortia work in Cd bioremediation:

1. Diversity and Specialization: Microbial consortia typically comprise a diverse array of microorganisms, each with unique capabilities. Some microorganisms in the consortium might possess specific Cd-binding proteins, while others can enzymatically transform Cd into less toxic forms [16].
2. Synergy: When these different microorganisms are working together in a consortium, they can create a synergistic effect. For example, one species might produce metabolites or substances that serve as growth factors for another species, leads to the improvement of overall performance.
3. Redundancy: Microbial consortia often exhibit redundancy, meaning that if one species is inhibited or becomes less active due to changing environmental conditions, another species with similar capabilities can take over the function, maintaining the stability of the remediation process [60].
4. Enhanced Adaptation: The presence of a diverse microbial community provides increased adaptability to changing environmental conditions. If the Cd concentration or other factors change, certain microorganisms within the consortium might become more dominant and better suited to address the altered conditions.
5. Improved Efficiency: Microbial consortia can be more efficient than individual microbial strains in tackling Cd contamination. By working together, they can handle a wider range of pollutants and ensure a more comprehensive degradation or immobilization of Cd.
6. Environmental Conditions: The success of microbial consortia in Cd bioremediation depends on various environmental factors such as pH, temperature, nutrient availability, and oxygen levels. Providing suitable conditions for the growth and activity of the microbial community is crucial for effective bioremediation. Microbial consortia can be developed naturally in contaminated environments through the selection and adaptation of microorganisms to Cd-rich conditions. Alternatively, researchers can engineer synthetic consortia by carefully selecting and combining specific microbial strains with desired bioremediation capabilities [73].

The use of microbial consortia for Cd bioremediation offers several advantages: (i) increased remediation efficiency: The combined abilities of different microorganisms lead to a more comprehensive and efficient remediation process, (ii) versatility: microbial consortia can adapt to various environmental conditions and different types of Cd-contaminated sites, and iii) sustainable and environmentally friendly: bioremediation using microbial consortia is a green and eco-friendly approach, producing fewer harmful byproducts compared to some traditional remediation methods. However, challenges in managing microbial consortia and controlling their behavior exist. Maintaining a balanced and stable consortium is essential for sustained bioremediation success, as the dominance of a single species might lead to inefficiencies or failure of the remediation process. Overall, microbial consortia present a promising and cutting-edge approach for Cd bioremediation, and ongoing research is continuously improving our understanding of these complex microbial communities and their applications in environmental cleanup [110].

4.8 Biofilm Applications

Bacterial biofilms are structured communities of microorganisms adhering to surfaces. Biofilms are known to enhance the resistance of bacteria to environmental stresses, including Cd exposure. Applying bacterial biofilms in Cd bioremediation can enhance the retention of bacteria at the contaminated site, extend the longevity of the bioremediation process, and increase the overall effectiveness of Cd removal. Biofilms have various applications in Cd bioremediation due to their unique properties and capabilities. Biofilms are complex communities of microorganisms that attach to surfaces and form a protective matrix of extracellular polymeric substances (EPS). They play a crucial role in bioremediation processes as they can facilitate the removal and immobilization of Cd and other pollutants [89]. Here are some biofilm applications in Cd bioremediation:

1. Fixed-Bed Reactors: Biofilms can be grown on solid supports in fixed-bed reactors. These reactors allow the development of a stable and dense microbial community, which can efficiently degrade or sequester Cd from contaminated water or wastewater.
2. Biofiltration: In biofiltration systems, biofilms develop on porous media, such as sand or gravel. As contaminated water passes through the biofilm-coated media, the microorganisms in the biofilm can remove Cd and other pollutants through sorption, precipitation, or enzymatic processes [68].
3. Bioaugmentation: Biofilms can be engineered or enriched with specific Cd-resistant or metal-accumulating microorganisms and then introduced into contaminated sites through bioaugmentation. These augmented biofilms can enhance the bioremediation capacity of indigenous microbial communities [69].
4. Biobarriers: Biofilms can be employed to create biobarriers, which act as physical barriers to prevent the spread of Cd and other contaminants in groundwater or soil. The biofilm matrix can also adsorb and immobilize Cd, reducing its mobility [86].
5. Microbial Fuel Cells (MFCs): In microbial fuel cells, biofilms form on the anode, where they oxidize organic matter to generate electricity. As the biofilm grows, it can sequester Cd, leading to both electricity generation and metal bioremediation [63].
6. Rhizoremediation: In rhizoremediation, biofilms develop around plant roots. The combination of plant roots and the associated biofilms can enhance Cd uptake and accumulation, increasing the efficiency of phytoremediation.

7. Biofilm-Enhanced Sediment Remediation: In aquatic environments, biofilms can form on sediment surfaces. These biofilms can trap and transform Cd, improving sediment bioremediation.

Biofilms offer several advantages for Cd bioremediation [68]: (i) biofilms provide a stable environment for microorganisms, ensuring prolonged and sustained remediation activity, (ii) biofilms can enhance the bioavailability of Cd by secreting chelating agents or enzymes that facilitate metal solubilization, (iii) EPS matrix in biofilms can shield microorganisms from adverse environmental conditions, enabling them to function effectively in challenging environments, and (iv) biofilms can adapt to changing environmental conditions, making them suitable for various Cd-contaminated sites. However, managing and controlling biofilms is essential, as uncontrolled growth may lead to biofouling and decreased bioremediation efficiency. Additionally, the specific microbial community within the biofilm can influence the bioremediation process, necessitating careful selection and management of the biofilm composition. The biofilm applications in Cd bioremediation offer promising and versatile solutions that can be tailored to specific contaminated environments, making them valuable tools in the field of environmental cleanup.

5 Microbial Mechanisms for Bioremediation of Cd

Microorganisms employ various mechanisms to interact with and uptake Cd from their surrounding environment. Understanding these microbial Cd uptake mechanisms is crucial for developing effective bioremediation strategies. Here are the key mechanisms through which microorganisms can uptake Cd.

5.1 Biosorption

Biosorption is a variant of sorption techniques. It's a passive process in which microbial cell surfaces, including bacterial cell walls and fungal hyphae, can possess functional groups such as carboxyl, amino, and sulfhydryl, that can bind Cd ions through electrostatic interactions, ion exchange, or coordination (Fig. 1). This surface binding, known as biosorption, allows microorganisms to remove Cd from the aqueous phase and accumulate it on their cell surfaces [99]. The effectiveness of biosorption depends on the surface properties of microbial cells, the availability and concentration of functional groups, and the physicochemical properties of Cd ions in the surrounding environment. Biosorption of Cd by microbes is generally mediated through various mechanisms like (i) exchange and adsorption through cell surfaces that includes the exchange of Cd ions (Cd_2+) by original metal ions (Mg_2+, Ca_2+) and/or by the interaction of chemical groups on the cell walls of microbes, (ii) extracellular enrichment and precipitation strategy by which they are able to form the complex and

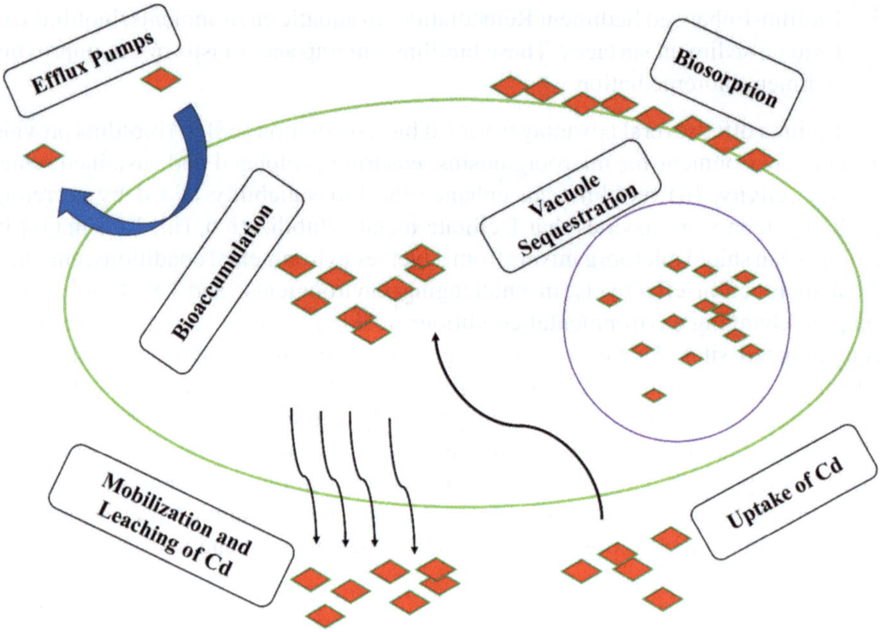

Fig. 1 Various mechanisms employed by microorganisms (bacteria, fungi, algae, biomass) responsible for Cd remediation in soil and water resources

precipitates the metals, (iii) intracellular transformation of metals with the help of internal auto-metabolic reactions of microbes like transportation and detoxification of metals. Beside these mechanisms, physical biosorption pathways are considered as dominant and effective one [12].

Following are the advantages of biosorption [40]: (i) biosorption can lead to a quick reduction of Cd concentration in the surrounding medium due to the high adsorption capacity of bacterial cell surfaces, (ii) the process does not require energy consumption by bacteria since it occurs passively through the affinity of cell surface functional groups for Cd ions, and (iii) simple and cost-effective: Biosorption is relatively simple to implement and does not require complex equipment, making it cost-effective for Cd bioremediation. However, biosorption has limitations, such as limited capacity and saturation points of functional groups on microbial cell surfaces. This may restrict the amount of Cd that can be removed through this process [99].

5.2 Bioaccumulation

The strategy behind the Cd bioremediation mainly involves extracellular biosorption and intracellular bioaccumulation of toxic ions (Fig. 1). These two strategies pose

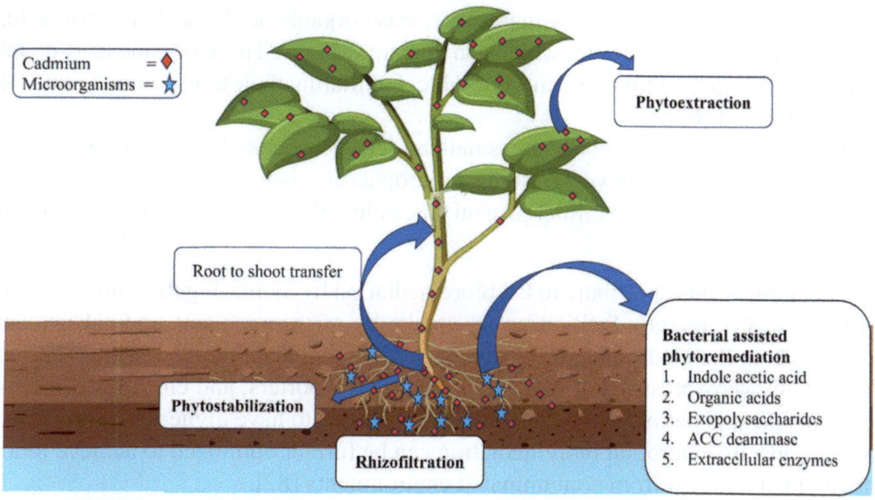

Fig. 2 Bacterial assisted phytoremediation of Cd and main mechanism employed by plants to extract, stabilize, and filter Cd from soil and water resources

great benefits over conventional physico-chemical remediation techniques. Bioaccumulation is an active process in which bacteria actively take up Cd ions from the surrounding environment into their intracellular compartments. This process relies on specific transport systems or channels on the bacterial cell membrane that facilitate the uptake of Cd against concentration gradients. Once inside the bacterial cell, Cd ions can be sequestered within various intracellular compartments, such as the cytoplasm, vacuoles, or specialized storage granules [83]. To counteract Cd toxicity, microbes evolved various resistance mechanisms like transformation, precipitation, efflux transport, and intracellular sequestration that is accompanied by the active production of thiol containing compounds and glutathione metallothionein, to combat the negative effects of heavy metal ions and efficiently proceed the intracellular accumulation [80]. Following are the advantages of bioaccumulation: (i) bioaccumulation allows microbes to concentrate higher levels of Cd within their cells compared to the external environment [13], (ii) bioaccumulation removes Cd from the surrounding medium, reducing its bioavailability and potential toxicity to the ecosystem [72], and (iii) bacteria can sequester and store Cd within their cells, preventing its release back into the environment over time [72]. However, bioaccumulation requires energy expenditure by bacteria to transport Cd ions into their cells, which may limit the efficiency of this process under certain environmental conditions.

Chelating agents are organic compounds that possess multiple coordination sites, enabling them to form stable complexes with metal ions. Microorganisms can produce chelating agents as a response to Cd exposure to enhance the removal and detoxification of the metal. Common chelating agents involved in Cd bioremediation include:

- Organic Acids: Some microorganisms release organic acids, such as citric acid, malic acid, and oxalic acid, which can chelate Cd ions. These organic acid-metal complexes are less toxic and more soluble, facilitating their transport and removal from the cell [62].
- Siderophores: Siderophores are small molecules secreted by microorganisms to scavenge iron. However, some siderophores also can bind Cd ions. By producing siderophores, microorganisms indirectly enhance Cd removal from the environment [6]

Chelating agents contribute to Cd bioremediation by increasing the solubility and mobility of Cd, thereby facilitating its uptake by microorganisms and subsequent removal from the environment. Researchers can manipulate the genetic makeup of bacteria to overexpress metal-binding proteins, transporters, and enzymes involved in Cd uptake and detoxification. Engineering bacteria to have higher Cd affinity and uptake capacity, microbial bioremediation can be further optimized to achieve more efficient Cd removal from contaminated environments [82].

5.3 *Biotransformation*

Biotransformation is a specific bioremediation process that involves the microbial-mediated conversion of Cd and other pollutants into less toxic or more easily removable forms. In the context of Cd bioremediation, biotransformation refers to the enzymatic or metabolic processes carried out by microorganisms to alter the chemical structure of Cd, reducing its toxicity and environmental impact [42]. There are several ways in which biotransformation can occur during Cd bioremediation:

1. Reduction: Certain microorganisms can reduce Cd ions (Cd2+) to Cd sulfide (CdS) or elemental Cd (Cd^0). These transformed forms of Cd are often less soluble and less bioavailable, reducing their potential harm to the environment [106].
2. Complexation: Microorganisms can produce extracellular compounds, such as organic acids and siderophores, that complex with Cd ions. These complexes reduce the free Cd concentration in the environment and limit its toxic effects (Yan et al. 2020).
3. Precipitation: Microorganisms can induce the precipitation of Cd in the form of insoluble compounds, such as Cd phosphate or Cd carbonate. Precipitated Cd becomes immobile and less likely to migrate through the environment [75].
4. Sequestration: Some microorganisms can sequester Cd within their cells or extracellular matrix, effectively removing it from the surrounding environment and reducing its bioavailability [50].
5. Methylation: In rare cases, certain anaerobic microorganisms can methylate Cd, transforming it into methyl-Cd. However, this process is of concern as methyl-Cd is highly toxic and poses greater risks to living organisms.

The success of biotransformation in Cd bioremediation depends on the specific microbial species present at the contaminated site and the environmental conditions. Factors such as temperature, pH, nutrient availability, and oxygen levels can influence the activity and efficiency of the microorganisms involved in the biotransformation process. Biotransformation is often one part of a comprehensive bioremediation strategy that may include other processes like biosorption, bioaccumulation, or biomineralization. The combination of these processes can maximize the overall effectiveness of Cd bioremediation. One of the advantages of biotransformation in Cd bioremediation is that it offers a natural and sustainable approach to reducing Cd toxicity without producing harmful byproducts. However, the use of biotransformation in practical bioremediation applications requires careful consideration of the specific microbial communities and conditions at the contaminated site to ensure successful and safe remediation outcomes.

5.4 *Bioleaching*

Bioleaching is a bioremediation technique that uses microorganisms to extract or leach metals, including Cd, from solid materials, such as contaminated soils or industrial waste, by employing their metabolic activities (Fig. 1). While bioleaching is commonly known for its application in the mining industry to recover valuable metals from ores, it can also be used in environmental remediation to reduce the levels of harmful metals like Cd in contaminated sites [71]. Here's how bioleaching can be applied to Cd bioremediation:

1. Microorganism Selection: Specific microorganisms, often acidophilic or acid-tolerant bacteria, are selected based on their ability to tolerate and thrive in acidic environments and their capacity to solubilize and extract Cd from solid matrices.
2. Environmental Conditions: The bioleaching process requires controlling the environmental conditions to create an acidic environment. This can be achieved through the addition of suitable acids or acidic waste materials to lower the pH of the medium.
3. Solid-Phase Contact: The contaminated solid material, such as soil or waste, is brought into contact with the selected microorganisms and the acidic medium. The microorganisms colonize the surface of the solid matrix and initiate the leaching process.
4. Metal Solubilization: The microorganisms produce organic acids and other metabolites that solubilize Cd and other metals from the solid phase, converting them into soluble metal ions.
5. Leachate Formation: As the metals are solubilized, leachate is formed, containing dissolved Cd and other metal ions.
6. Metal Recovery: The leachate is collected and processed to recover the dissolved Cd. This can involve various methods, such as precipitation, ion exchange,

or electrochemical processes, to isolate and concentrate the metals for further treatment or safe disposal.

Bioleaching offers several advantages for Cd bioremediation [30]: (i) bioleaching can target specific metals, allowing for the selective removal of Cd without disrupting other elements in the contaminated material, (ii) bioleaching is often less energy-intensive compared to traditional chemical extraction methods, (iii) bioleaching is a greener alternative to harsh chemical treatments, producing fewer toxic byproducts, and (iv) bioleaching can be applied in situ, reducing the need for excavation and transport of contaminated materials. However, bioleaching also has some limitations and challenges such as process optimization, long duration, and limited applicability. Overall, bioleaching shows potential as an innovative and eco-friendly bioremediation technique for Cd-contaminated environments, and ongoing research aims to improve its efficiency and applicability in real-world scenarios.

5.5 Bioprecipitation

Extracellular binding and precipitation are important mechanisms used by microorganisms to interact with and remove Cd from their surrounding environment during bioremediation. These processes involve the adsorption of Cd ions onto the outer surface of microbial cells and subsequent precipitation as insoluble compounds. Microorganisms, particularly bacteria, possess cell surfaces that contain functional groups, such as carboxyl (COOH), amino (NH_2), and sulfhydryl (SH) groups [52]. These groups can act as binding sites for metal ions like Cd. When Cd is present in the surrounding environment, it interacts with these functional groups on the microbial cell surfaces through electrostatic interactions or coordination bonds. Bacteria possess the capability to induce the precipitation of Cd as insoluble compounds. These bacteria can secrete extracellular substances, such as exopolysaccharides or proteins, which serve as nucleation sites for the formation of metal precipitates [66]. The extracellular binding of Cd to the microbial cell surface has several advantages [22]: (i) by binding Cd on their surfaces, microorganisms can prevent its release back into the environment, reducing the risk of Cd recontamination, (ii) extracellular binding allows microorganisms to concentrate Cd ions on their cell surfaces, increasing the efficiency of removal from the surrounding medium, (iii) extracellular binding is often the initial step before intracellular uptake and sequestration of Cd by microorganisms.

Once Cd ions are bound to the microbial cell surfaces, they can undergo a precipitation process. This involves the transformation of soluble Cd ions into insoluble compounds or complexes. Various factors influence the precipitation process, including pH, temperature, and the presence of other ions or ligands [102]. Microbial induced precipitation not only removes Cd from the environment but also immobilizes the metal in a less bioavailable form. The precipitation process can lead to the formation of metal sulfides, carbonates, phosphates, or metal-protein complexes, depending on the bacterial species and environmental conditions. Examples of Cd precipitation mechanisms include:

- Formation of metal sulfides: In environments with high concentrations of sulfide ions (S2-), Cd can precipitate as insoluble metal sulfides, such as Cd sulfide (CdS) or Cd metallothionein-sulfide complexes [58, 56].
- Carbonate precipitation: In alkaline conditions, Cd can form insoluble metal carbonate complexes, such as Cd carbonate ($CdCO_3$).
- Phosphate precipitation: In phosphate-rich environments, Cd can precipitate as less soluble metal phosphates, like Cd phosphate ($Cd_3(PO_4)_2$).

Precipitation is a key mechanism in permanently removing Cd from the environment, as the formed insoluble compounds are less likely to be released back into the ecosystem. Extracellular binding and precipitation are important components of microbial Cd bioremediation strategies. By leveraging the natural capabilities of microorganisms to bind and precipitate Cd, bioremediation offers an eco-friendly and effective approach for removing this toxic heavy metal from contaminated environments. Researchers continue to explore ways to optimize and harness these mechanisms for efficient and sustainable Cd cleanup [53].

5.6 Sequestration

Intracellular accumulation and sequestration are critical mechanisms used by microorganisms to uptake and store Cd within their cells during bioremediation processes (Fig. 1). These mechanisms help reduce the concentration of Cd in the surrounding environment and prevent its toxic effects. Here's a closer look at intracellular accumulation and sequestration in Cd bioremediation. Microorganisms possess specialized transport proteins or channels in their cell membranes that allow them to actively transport Cd ions from the surrounding environment into their cells [80]. These transporters can be selective for Cd or may have broader metal ion specificity. The active uptake of Cd is an energy-dependent process and often occurs against concentration gradients. Once inside the cell, Cd ions accumulate in various cellular compartments, including the cytoplasm, vacuoles, and vesicles. The accumulation of Cd within the cells helps reduce its concentration in the surrounding medium and enables microorganisms to tolerate higher levels of Cd than would be possible if the metal were present solely in the extracellular environment [42]. Intracellular sequestration refers to the process by which microorganisms store or sequester Cd within specific intracellular structures or compartments. The sequestration of Cd is a defense mechanism employed by microorganisms to reduce its toxic effects and protect cellular structures. Microorganisms can actively accumulate Cd within their cells to reduce its concentration in the surrounding environment. They possess transport proteins or channels that facilitate the uptake of Cd ions across their cell membranes. Once inside the cell, Cd can be sequestered in various intracellular compartments, such as vacuoles, vesicles, or specific storage granules [76]. Sequestration is an essential mechanism for microorganisms to cope with high Cd exposure and prevent the metal from interfering with vital cellular functions. Intracellular accumulation and sequestration of Cd by microorganisms contribute significantly to the success of bioremediation processes. By actively taking up Cd and storing it within their cells, microorganisms can reduce the bioavailability of Cd in the surrounding environment, preventing its accumulation in plants, animals, and humans. Moreover, these mechanisms enhance the overall effectiveness and sustainability of Cd cleanup efforts, making bioremediation a valuable and eco-friendly approach for addressing Cd contamination in various ecosystems [2]. Common intracellular compartments involved in Cd sequestration include:

- Vacuoles: Some microorganisms, especially yeast and plant cells, store Cd in vacuoles. Vacuoles are membrane-bound organelles that serve as storage sites for various substances, including Cd ions [95].
- Vesicles and inclusions: Microorganisms can form vesicles or inclusions that contain accumulated Cd. These compartments help sequester Cd away from critical cellular processes and minimize its harmful effects.

Metallothioneins (MTs) are small, cysteine-rich proteins present in various organisms, including bacteria, fungi, plants, and animals. They have a high affinity for binding metal ions, including Cd, copper, zinc, and mercury. MTs are inducible proteins, meaning their synthesis is upregulated in response to heavy metal exposure [97]. In the context of Cd bioremediation, MTs play the following roles:

- Cd binding: MTs can bind Cd ions through coordination with cysteine thiolate groups present in their amino acid sequence. This binding allows MTs to sequester Cd and reduce its bioavailability within the cell.
- Cd detoxification: By binding Cd, MTs serve as important detoxification agents. They prevent the accumulation of toxic Cd concentrations within the cytoplasm, protecting cellular structures and functions from Cd-induced damage.
- Metal ion homeostasis: MTs also play a role in maintaining metal ion homeostasis within the cell. They help regulate the concentrations of essential metal ions, such as zinc and copper, by binding excess ions and releasing them as needed.

5.7 Phytoextraction

Phytoextraction is a specialized phytoremediation technique used to remove Cd and other heavy metals from contaminated soils or water using plants (Fig. 2). In phytoextraction, certain plant species, known as hyperaccumulators, have the unique ability to absorb and accumulate high concentrations of Cd from the soil into their above-ground biomass. Here's how phytoextraction works for Cd bioremediation:

1. Hyperaccumulator Plant Selection: The first step is to identify and select plant species that are capable of hyperaccumulating Cd. These plants have developed

specific mechanisms to tolerate and accumulate high levels of heavy metals without being significantly harmed by their presence [107].

2. Planting and Growth: Hyperaccumulator plants are cultivated and planted in contaminated areas. As they grow, their roots take up water and nutrients, including Cd, from the soil.
3. Cd Uptake: The roots of hyperaccumulator plants absorb Cd from the soil and transport it to the above-ground parts of the plant, such as the leaves and stems.
4. Cd Accumulation: Once absorbed, the Cd is sequestered and stored within the plant's tissues. Hyperaccumulators can accumulate significantly higher concentrations of Cd than non-hyperaccumulator plant species.
5. Biomass Harvesting: After a certain period, the hyperaccumulator plants are harvested, and the above-ground biomass containing accumulated Cd is removed from the site.
6. Disposal or Metal Recovery: Depending on the goal of the remediation project, the harvested biomass can be properly disposed of in a controlled manner to prevent recontamination, or it can be processed to recover the Cd for recycling or other applications.
7. Repeating the Process: The process can be repeated over multiple growing seasons to continuously reduce Cd levels in the soil until the desired cleanup objectives are achieved. Phytoextraction offers several advantages for Cd bioremediation such as (i) phytoextraction is a natural and environmentally friendly remediation method that does not require the use of harsh chemicals or energy-intensive processes [32], (ii) once established, hyperaccumulator plants can grow and accumulate Cd with minimal additional inputs, making it a cost-effective remediation option, (iii) phytoextraction can be applied in situ, reducing the need for excavation and transport of contaminated soils [109], and (iv) phytoextraction not only removes Cd from the environment but also offers the potential for metal recovery, allowing for the recycling or reuse of the extracted Cd. However, phytoextraction also has some limitations such as slow process, site, and plant specific [96]. Overall, phytoextraction is a promising and sustainable approach for Cd bioremediation, especially in combination with other remediation methods to maximize its efficiency and effectiveness.

5.8 *Phytostabilization*

Phytostabilization is a bioremediation technique used to manage and stabilize Cd and other heavy metal-contaminated soils [58, 56]. Unlike phytoextraction, where plants actively take up and accumulate metals, phytostabilization focuses on reducing the mobility and bioavailability of Cd in the soil, thereby preventing its migration and uptake by plants, and minimizing its potential for causing harm to the environment (Fig. 2, [111]. Here's how phytostabilization works for Cd bioremediation:

1. Plant Selection: Specific plant species are selected for their ability to establish strong root systems and thrive in contaminated soil. These plants may not be hyperaccumulators but are effective in stabilizing the Cd in the soil matrix.
2. Root Development: The selected plants are cultivated and allowed to grow in the contaminated soil. As they establish root systems, the roots penetrate the soil and help bind the Cd and other metals.
3. Rhizosphere Effects: The rhizosphere, the zone of soil surrounding the plant roots, undergoes changes due to the presence of plant exudates and root activities. These changes can influence the soil properties and chemical reactions that reduce Cd mobility.
4. Metal-Binding Compounds: Some plants secrete organic compounds or release substances from their roots that have metal-binding capabilities. These compounds can bind with Cd, reducing its solubility and mobility in the soil.
5. Soil Amendments: In some cases, soil amendments, such as lime or phosphate-based materials, may be applied to further enhance the immobilization of Cd in the soil. These amendments can modify the soil pH or form stable metal precipitates, reducing Cd bioavailability.
6. Establishment and Maintenance: Successful phytostabilization requires proper plant establishment and regular maintenance to ensure the continued stability of the soil and the plant's health.

Phytostabilization offers several advantages for Cd bioremediation: (i) phytostabilization provides a long-term solution for managing Cd-contaminated soils, as it reduces the potential for Cd migration and exposure over time, (ii) once established, phytostabilization requires minimal maintenance and ongoing inputs, making it a cost-effective remediation option [26], and (iii) phytostabilization is a sustainable and environmentally friendly approach that avoids the use of chemicals and minimizes disruption to the site. However, phytostabilization also has some limitations such as limited cleanup, the success of phytostabilization depends on site-specific factors, such as soil properties, climate, and the selected plant species, and it may require an extended period to achieve full stabilization of the contaminated soil. Overall, phytostabilization is a valuable and sustainable approach for managing Cd-contaminated soils, particularly in scenarios where complete removal of Cd is not feasible or necessary. It can be integrated with other bioremediation techniques for a comprehensive remediation strategy based on the specific conditions of the contaminated site.

5.9 Rhizofiltration

Rhizofiltration is a specialized form of phytoremediation that focuses on using plant roots to remove contaminants, including Cd, from water sources [15, 81]. In rhizofiltration, certain plant species are chosen for their ability to take up and accumulate pollutants from water through their root systems (Fig. 2, [90]). This technique is particularly useful for remediating Cd-contaminated water and wastewater. Here's how rhizofiltration works for Cd bioremediation:

1. Plant Selection: Plant species with high metal uptake capabilities, particularly those that can efficiently take up Cd from water, are selected for rhizofiltration. These plants are often aquatic or semi-aquatic in nature [15].
2. Hydroponic Growth: The selected plants are grown hydroponically in a container or a constructed wetland system, with their roots submerged in the contaminated water [105].
3. Cd Uptake: As water flows through the hydroponic system, the plant roots take up Cd and other contaminants present in the water. Cd ions are absorbed into the root cells and transported to the above-ground parts of the plant.
4. Accumulation and Harvesting: Over time, the Cd accumulates in the plant biomass, especially in the roots. Periodically, the plants are harvested, and the Cd-rich biomass is removed from the system.
5. Repeated Application: Rhizofiltration can be repeated with new batches of plants to continue the process of Cd removal from the water source.

Rhizofiltration offers several advantages for Cd bioremediation in water sources: (i) rhizofiltration can efficiently remove Cd and other contaminants from water, as the plant roots have a high affinity for metal uptake, (ii) it is a sustainable and eco-friendly remediation method, as it relies on natural processes and plant-based systems, (iii) it can be applied in both natural and engineered systems, such as constructed wetlands, making it suitable for various water treatment scenarios, (iv) it is often a cost-effective method for Cd removal, especially when compared to traditional water treatment technologies, and (v) it can be combined with other water treatment methods to enhance overall remediation efficiency. However, rhizofiltration also has some limitations such as the success of rhizofiltration depends on selecting appropriate plant species with good metal-uptake capabilities, which may vary depending on the water quality and Cd concentration, it may take some time to achieve significant Cd removal from the water, especially in highly contaminated sources, and proper management and disposal of the Cd-rich plant biomass are necessary to prevent recontamination. Overall, rhizofiltration is a promising and sustainable technique for removing Cd from water sources, and ongoing research continues to optimize this method and explore its applications in various environmental cleanup scenarios.

5.10 Molecular Mechanisms

Some microorganisms possess active transport systems, such as ATP-dependent efflux pumps, which enable them to actively transport Cd ions out of their cells. These transporters play a role in Cd detoxification and maintaining cellular homeostasis [46, 64]. Microorganisms can regulate the expression of genes involved in Cd uptake and detoxification in response to environmental Cd levels. The activation of specific genes and regulatory pathways allows microorganisms to adapt and enhance their Cd uptake capabilities in contaminated environments [31]. It is important to note that the mechanisms of Cd uptake can vary among different microbial species and depend on their specific adaptations and environmental conditions. Moreover, microbial communities can exhibit synergistic effects, where different microorganisms work together to enhance Cd uptake and remediation efficiency. By harnessing the unique Cd uptake mechanisms of microorganisms, researchers can develop strategies to optimize bioremediation approaches, such as selecting suitable microbial strains, engineering genetic traits, and manipulating environmental conditions, to achieve effective and sustainable Cd cleanup in contaminated environments [36]. These microorganisms are all capable of surviving and growing in the presence of high concentrations of Cd. They do this by using a variety of mechanisms to tolerate the metal, including:

- Metal efflux: This is the process of pumping Cd out of the cell against its concentration gradient [45].
- Intracellular precipitation: This is the process of forming insoluble complexes of Cd with other molecules inside the cell [113].
- Complexation: This is the process of binding Cd to other molecules, such as chelating agents, which makes it less toxic to the cell.

The ability of these microorganisms to tolerate Cd makes them potential candidates for use in bioremediation. Bioremediation is the use of living organisms to remove or reduce pollutants from the environment. Cd-tolerant microorganisms could be used to clean up contaminated soil, water, or plants. It is important to note that the effectiveness of bioremediation using Cd-tolerant microorganisms will depend on several factors, including the type of pollutant, the concentration of the pollutant, and the environmental conditions [9, 11]. However, these microorganisms offer a promising new approach to the remediation of Cd-contaminated environments.

6 Economics and Constraints for Adoption the Microbial Strategies

The adoption of microbial strategies for Cd bioremediation comes with both economic opportunities and certain constraints. Understanding the economics and limitations associated with these approaches is crucial for their successful implementation and widespread adoption. Here are some key factors to consider:

6.1 Economics

1. Cost-Effectiveness: Microbial strategies can be cost-effective compared to traditional remediation methods that involve excavation, transport, and disposal of

contaminated materials. Bioremediation techniques, such as phytoremediation and bioleaching, may require lower operational and maintenance costs once established [4, 9, 11].
2. Sustainable and Green: Microbial strategies are often viewed as more sustainable and environmentally friendly compared to chemical treatments. As a result, they may align better with the growing demand for eco-friendly remediation solutions.
3. Potential for Resource Recovery: Some microbial strategies, such as phytoextraction and bioleaching, offer the potential for metal recovery, providing an economic incentive as extracted metals could be recycled or sold.
4. Long-Term Benefits: Microbial strategies can lead to long-term benefits, as the natural processes can continue to remediate the environment even after initial setup, reducing the need for ongoing interventions.

6.2 Constraints

1. Site-Specific Factors: The success of microbial strategies is influenced by site-specific factors such as soil conditions, climate, and the presence of suitable microorganisms. Some strategies might not be viable or efficient in all environments.
2. Slow Cleanup Rates: Microbial strategies, particularly plant-based ones like phytoremediation, can be slow compared to chemical methods. They may require more extended periods to achieve significant reductions in Cd levels.
3. Lack of Suitable Microbes: Some environments may lack the necessary metal-resistant or metal-accumulating microorganisms needed for effective bioremediation. In such cases, bioaugmentation might be necessary, which can add complexity and costs.
4. Uncertainty and Risk: Microbial strategies can be less predictable and may have variable outcomes based on the complexity of the microbial community, environmental conditions, and other factors. There may be uncertainties regarding their performance and success.
5. Land and Space Requirements: Some microbial strategies, such as phytoremediation, may require substantial land or space for implementation, which can limit their use in densely populated or urban areas.
6. Regulatory Approvals: The adoption of microbial strategies for bioremediation might require regulatory approvals and permits, which can add to the time and costs of implementing these technologies.
7. Public Perception: The acceptance and adoption of microbial strategies may face resistance or skepticism from the public or stakeholders who are unfamiliar with these innovative approaches.

To overcome these constraints and encourage the widespread adoption of microbial strategies for Cd bioremediation, ongoing research, development, and case studies demonstrating their effectiveness and economic viability are essential. Collaboration between researchers, regulators, industries, and communities can foster the integration of microbial strategies into sustainable and successful environmental cleanup practices.

7 Conclusion and Prospects

Bioremediation offers promising and sustainable solutions for mitigating the environmental impacts of Cd contamination. Various microbial strategies, including phytoremediation, bioleaching, rhizofiltration, and bioaugmentation have demonstrated their efficacy in removing, immobilizing, or transforming Cd in contaminated soil, water, and sediments. These bioremediation approaches leverage the metabolic capabilities of microorganisms and plants to achieve efficient and eco-friendly cleanup. Phytoremediation shows potential for long-term Cd remediation in soil, with hyperaccumulator plants offering the ability to accumulate high concentrations of Cd. Bioleaching and rhizofiltration have proven effective for Cd removal from water sources, providing economical and sustainable alternatives to traditional water treatment methods. Bioaugmentation and the use of microbial consortia enhance the bioremediation potential by introducing or enhancing the activities of specific metal-resistant or metal-accumulating microorganisms. Nano-bioremediation also presents a cutting-edge approach by combining nanotechnology with bioremediation to improve Cd uptake and sequestration. The prospects for Cd bioremediation are promising, with ongoing research and innovation continuously improving the effectiveness and practicality of these microbial strategies. Advancements in biotechnology, genomics, and molecular techniques allow for the development of genetically modified plants and microorganisms with enhanced metal uptake capabilities, further expanding the range of suitable species for bioremediation. The integration of different microbial strategies and their combination with other remediation techniques, such as physical and chemical methods, offer comprehensive and synergistic approaches to tackle Cd contamination in diverse environmental settings. The development of cost-effective and efficient processes, along with the optimization of microbial consortia and biofilm applications, can lead to more successful and scalable bioremediation practices.

References

1. Abioye, O. P., Ezugwu, B. U., Aransiola, S. A., & Ojeba, M. I. (2020). Phycoremediation of water contaminated with arsenic (As), cadmium (Cd) and lead (Pb) from a mining site in Minna. *Nigeria. European Journal of Biological Research, 10*(1), 35–44.
2. Abo-Alkasem, M. I., Hassan, N. H., & Abo Elsoud, M. M. (2023). Microbial bioremediation as a tool for the removal of heavy metals. *Bulletin of the National Research Centre, 47*, 31 (2023). https://doi.org/10.1186/s42269-023-01006-z

3. Ahmad, I., Akhtar, M. J., Asghar, H. N., Ghafoor, U., & Shahid, M. (2016). Differential effects of plant growth promoting rhizobacteria on maize growth and cadmium uptake. *Journal of Plant Growth Regulation, 35*, 303–315.
4. Ahmad, I., Akhtar, M. J., Jadoon, I. B. K., Imran, M., Imran, M., & Ali, S. (2017). Equilibrium modeling of cadmium biosorption from aqueous solution by compost. *Environmental Science and Pollution Research, 24*, 5277–5284.
5. Ahmad, I., Akhtar, M. J., Zahir, Z. A., Naveed, M., Mitter, B., & Sessitsch, A. (2014). Cadmium-tolerant bacteria induce metal stress tolerance in cereals. *Environmental Science and Pollution Research, 21*, 11054–11065.
6. Ahmed, E., & Holmström, S. J. (2014). Siderophores in environmental research: Roles and applications. *Microbial Biotechnology, 7*(3), 196–208. https://doi.org/10.1111/1751-7915.12117
7. Akhtar, M. J., Ali, Q., Javid, R., Asghar, H. N., Ahmad, I., Iqbal, M. Z., & Khaliq, A. (2019). Organic and inorganic amendments immobilized cadmium and improved maize growth and yield in Cd-contaminated soil. *International Journal of Agriculture and Biology, 22*(6), 1497–1506.
8. Al-Dhabi, N. A., Esmail, G. A., Mohammed Ghilan, A. K., & Valan Arasu, M. (2019). Optimizing the management of cadmium bioremediation capacity of metal-resistant *Pseudomonas* sp. strain Al-Dhabi-126 isolated from the industrial city of Saudi Arabian environment. *International Journal of Environmental Research and Public Health, 16*(23), 4788.
9. Ali, J., Ali, F., Ahmad, I., Rafique, M., Munis, M. F. H., Hassan, S. W., et al. (2021). Mechanistic elucidation of germination potential and growth of *Sesbania sesban* seedlings with *Bacillus anthracis* PM21 under heavy metals stress: An in vitro study. *Ecotoxicology and Environmental Safety, 208*, 111769.
10. Ali, A., Guo, D., Mahar, A., Wang, P., Shen, F., Li, R., & Zhang, Z. (2017). Mycoremediation of potentially toxic trace elements—A biological tool for soil cleanup: A review. *Pedosphere, 27*(2), 205–222.
11. Ali, J., Wang, X., Rafique, M., Ahmad, I., Fiaz, S., Munis, M.F.H., & Chaudhary, H. J. (2021b). Phytoremediation of cadmium contaminated soil using *Sesbania sesban* L. in association with bacillus anthracis PM21: A biochemical analysis. *Sustainability, 13*(24), 13529.
12. Alotaibi, B. S., Khan, M., & Shamim, S. (2021). Unraveling the underlying heavy metal detoxification mechanisms of Bacillus species. *Microorganisms., 9*(8), 1628. https://doi.org/10.3390/microorganisms9081628
13. Amna, Ali N., Masood S., Mukhtar T., Kamran M. A., Rafique M., et al. (2015). Differential effects of cadmium and chromium on growth, photosynthetic activity, and metal uptake of Linum usitatissimum in association with Glomus intraradices. Environ. Monit. Assess. https://doi.org/187:31110.1007/s10661-015-4557-8
14. Badawy, I. H., Hmed, A. A., Sofy, M. R., & Al-Mokadem, A. Z. (2022). Alleviation of cadmium and nickel toxicity and phyto-stimulation of tomato plant by endophytic *Micrococcus luteus* and *Enterobacter cloacae*. *Plants, 11*(15), 2018.
15. Bakshe, P., & Jugade, R. (2023). Phytostabilization and rhizofiltration of toxic heavy metals by heavy metal accumulator plants for sustainable management of contaminated industrial sites: A comprehensive review. *Journal of Hazardous Materials Advances*, 100293.
16. Bala, S., Garg, D., Thirumalesh, B. V., Sharma, M., Sridhar, K., Inbaraj, B. S., & Tripathi, M. (2022). Recent strategies for bioremediation of emerging pollutants: a review for a green and sustainable environment. *Toxics, 10*(8), 484. https://doi.org/10.3390/toxics10080484
17. Bandurska, K., Krupa, P., Berdowska, A., Jatulewicz, I., & Zawierucha, I. (2021). Mycoremediation of soil contaminated with cadmium and lead by *Trichoderma* sp. *Ecological Chemistry and Engineering S, 28*(2), 277–286.
18. Bandurska, K., Krupa, P., Berdowska, A., & Marczak, M. (2016). Adaptation of selected ectomycorrhizal fungi to increased concentration of cadmium and lead. *Ecological Chemistry and Engineering S., 23*(3), 483–491.

19. Bashir, S., Javed, S., Al-Anazi, K. M., Farah, M. A., & Ali, S. (2022). Bioremediation of cadmium toxicity in wheat (*Triticum aestivum* L.) plants primed with L-Proline, *Bacillus subtilis* and *Aspergillus niger*. *International Journal of Environmental Research and Public Health, 19*(19), 12683.
20. Bhatt, P., Pandey, S. C., Joshi, S., Chaudhary, P., Pathak, V. M., Huang, Y., Wu, X., Zhou, Z., & Chen, S. (2022). Nano-bioremediation: A sustainable approach for the removal of toxic pollutants from the environment. *Journal of Hazardous Materials, 427*, 128033.
21. Bogusz, A., & Oleszczuk, P. (2020). Effect of biochar addition to sewage sludge on cadmium, copper and lead speciation in sewage sludge-amended soil. *Chemosphere, 239*, 124719.
22. Bravo, D., & Braissant, O. (2022). Cadmium-tolerant bacteria: Current trends and applications in agriculture. *Letters in Applied Microbiology, 74*(3), 311–333.
23. Cepoi, L., & Zinicovscaia, I. (2020). Spirulina platensis as a model object for the environment bioremediation studies. In *Handbook of Algal science, technology and medicine* (pp. 629–640). Academic.
24. Chaudhary, P., Xu, M., Ahamad, L., Chaudhary, A., Kumar, G., Adeleke, B. S., Verma, K. K., Hu, D.-M., Širić, I., Kumar, P., et al. (2023). Application of synthetic consortia for improvement of soil fertility, pollution remediation, and agricultural productivity: A review. *Agronomy, 13*(3), 643. https://doi.org/10.3390/agronomy13030643
25. Chellaiah, E. R. (2018). Cadmium (heavy metals) bioremediation by *Pseudomonas aeruginosa*: A minireview. *Applied Water Science, 8*(6), 154.
26. Chen, Z. J., Tian, W., Li, Y. J., Sun, L. N., Chen, Y., Zhang, H., Li, Y.-Y., & Han, H. (2021). Responses of rhizosphere bacterial communities, their functions and their network interactions to Cd stress under phytostabilization by *Miscanthus* spp. *Environmental Pollution, 287*, 117663.
27. da Costa, F. P., Cipolatti, E. P., Furigo Junior, A., & Oliveira Henriques, R. (2022). Nanoflowers: A new approach of enzyme immobilization. *The Chemical Record, 22*(4), e202100293.
28. Cui, Y., Lai, B., & Tang, X. (2019). Microbial fuel cell-based biosensors. *Biosensors (Basel), 9*(3), 92. https://doi.org/10.3390/bios9030092
29. Deng, J., Fu, D., Hu, W., Lu, X., Wu, Y., & Bryan, H. (2020). Physiological responses and accumulation ability of *Microcystis aeruginosa* to zinc and cadmium: Implications for bioremediation of heavy metal pollution. *Bioresource Technology, 303*, 122963.
30. Dong, Y. et al. (2023) Bioleaching of heavy metals from metal tailings utilizing bacteria and fungi: mechanisms, strengthen measures, and development prospect. J. Environ. Manag.
31. Dong, X. B., Huang, W., Bian, Y. B., Feng, X., Ibrahim, S. A., Shi, D. F., Qiao, X., & Liu, Y. (2019). Remediation and mechanisms of cadmium biosorption by a cadmium-binding protein from *Lentinula edodes*. *Journal of Agricultural and Food Chemistry, 67*(41), 11373–11379.
32. Dou, X., Dai, H., Skuza, L., & Wei, S. (2022). Cadmium removal potential of hyperaccumulator *Solanum nigrum* L. under two planting modes in three years continuous phytoremediation. *Environmental Pollution, 307*, 119493.
33. Fang C, Achal V. (2019). The Potential of Microbial Fuel Cells for Remediation of Heavy Metals from Soil and Water—Review of Application. *Microorganisms, 7*(12):697. https://doi.org/10.3390/microorganisms7120697
34. Fatima, G., Raza, A. M., Hadi, N., Nigam, N., & Mahdi, A. A. (2019). Cadmium in human diseases: It's more than just a mere metal. *Indian Journal of Clinical Biochemistry, 34*(4), 371–378. https://doi.org/10.1007/s12291-019-00839-8
35. De Freitas Lima, A., Ferreira de Moura, G., Barbosa de Lima, M. A., Mendes de Souza, P., Alves da Silva, C. A., de Campos Takaki, G. M., & Do Nascimento, A. E. (2011). Role of the morphology and polyphosphate in *Trichoderma harzianum* related to cadmium removal. *Molecules, 16*(3), 2486–2500.
36. Gonzalez Henao, S., & Ghneim-Herrera, T. (2021). Heavy metals in soils and the remediation potential of bacteria associated with the plant microbiome. *Frontiers in Environmental Science, 15.*

37. Gul, I., Manzoor, M., Hashim, N., Shah, G. M., Waani, S. P. T., Shahid, M., Antoniadis, V., & Arshad, M. (2021). Challenges in microbially and chelate-assisted phytoextraction of cadmium and lead—A review. *Environmental Pollution, 287*, 117667.
38. Halecký, M., & Kozliak, E. (2020). Modern bioremediation approaches: Use of biosurfactants, emulsifiers, enzymes, biopesticides, GMOs. *Advanced Nano-bio Technologies for Water and Soil Treatment*, 495–526.
39. Han, Y., Zveushe, O. K., Dong, F., Ling, Q., Chen, Y., Sajid, S., Zhou, L., & de Dios, V. R. (2021). Unraveling the effects of arbuscular mycorrhizal fungi on cadmium uptake and detoxification mechanisms in perennial ryegrass (*Lolium perenne*). *Science of The Total Environment, 798*, 149222.
40. Hegazy, G. E., Soliman, N. A., Ossman, M. E., Abdel-Fattah, Y. R., & Moawad, M. N. (2023). Isotherm and kinetic studies of cadmium biosorption and its adsorption behaviour in multi-metals solution using dead and immobilized archaeal cells. *Science Report, 13*(1), 2550. https://doi.org/10.1038/s41598-023-29456-5
41. Hui, C. Y., Guo, Y., Liu, L., & Yi, J. (2022). Recent advances in bacterial biosensing and bioremediation of cadmium pollution: A mini-review. *World Journal of Microbiology and Biotechnology, 38*(1), 9.
42. Igiri, B. E., Okoduwa, S. I. R., Idoko, G. O., Akabuogu, E. P., Adeyi, A. O., & Ejiogu, I. K. (2018). Toxicity and bioremediation of heavy metals contaminated ecosystem from tannery wastewater: A review. *J Toxicol.* https://doi.org/10.1155/2018/2568038
43. Jaafar, R., & Al-Taee, A. (2020). Biosorption of lead and cadmium and the potential role of *Shewanella oneidensis*. In *IOP Conference Series: Materials Science and Engineering* (Vol. 928, No. 6, p. 062026). IOP Publishing.
44. Jadoun, S., Chauhan, N. P. S., Zarrintaj, P., et al. (2022). Synthesis of nanoparticles using microorganisms and their applications: A review. *Environmental Chemistry Letters, 20*, 3153–3197. https://doi.org/10.1007/s10311-022-01444-7
45. Janicka, M., Reda, M., Napieraj, N., Michalak, A., Jakubowska, D., & Kabała, K. (2022). Involvement of diamine oxidase in modification of plasma membrane proton pump activity in *Cucumis sativus* L. seedlings under cadmium stress. *International Journal of Molecular Sciences, 24*(1), 262.
46. Jebril, N., Boden, R., & Braungardt, C. (2022). Cadmium resistant bacteria mediated cadmium removal: a systematic review on resistance, mechanism and bioremediation approaches. In: *IOP Conference Series: Earth and Environmental Science* (Vol. 1002, No. 1, p. 012006). IOP Publishing.
47. Jose, S., Delilah, D., Mathew, N. M., Joseph, A., & Kumar, S. (2022). Bioremediation of cadmium, chromium and copper using green microalgae. *Research Journal of Agricultural Science, 2022*(13), 275–279.
48. Kaur, R., Mavi, G. K., & Raghav, S. (2019). Bioremediation of sludge using *Pseudomonas aeruginosa. International Journal of Current Microbiology and Applied Science, 8*, 69–79.
49. Khadim, M. U., Murtaza, G., Farooqi, Z. U. R., Hussain, T., Mahmood, N., & Hussain, S. (2023). An application of rock phosphate increased soil cadmium contamination and hampered the morphophysiological growth of *Brassica campestris* L. *Journal of Soil Science and Plant Nutrition*, 1–13.
50. Khan, A., Gupta, A., Singh, P., Mishra, A. K., Ranjan, R. K., & Srivastava, A. (2020). Siderophore-assisted cadmium hyperaccumulation in *Bacillus subtilis. International Microbiology, 23*, 277–286.
51. Khan, I., Aftab, M., Shakir, S., Ali, M., Qayyum, S., Rehman, M. U., Haleem, K. S., & Touseef, I. (2019). Mycoremediation of heavy metal (Cd and Cr)–polluted soil through indigenous metallotolerant fungal isolates. *Environmental Monitoring and Assessment, 191*, 1–11.
52. Khan, Z., Elahi, A., Bukhari, D. A., & Rehman, A. (2022). Cadmium sources, toxicity, resistance and removal by microorganisms—A potential strategy for cadmium eradication. *Journal of Saudi Chemical Society*, 101569.
53. Kubier, A., Wilkin, R. T., & Pichler, T. (2019). Cadmium in soils and groundwater: A review. *Applied Geochemistry, 108*, 1–16. https://doi.org/10.1016/j.apgeochem.2019.104388

54. Kumar, V., Dwivedi, S. K., & Oh, S. (2022). A review on microbial-integrated techniques as promising cleaner option for removal of chromium, cadmium and lead from industrial wastewater. *Journal of Water Process Engineering, 47*, 102727.
55. Kumar, M., Kushwaha, A., Goswami, L., Singh, A. K., & Sikandar, M. (2021). A review on advances and mechanism for the phycoremediation of cadmium contaminated wastewater. *Cleaner Engineering and Technology, 5*, 100288.
56. Li, W., Zhang, Y., & Achal, V. (2022). Mechanisms of cadmium retention on enzyme-induced carbonate precipitation (EICP) of Ca/Mg: Nucleation, chemisorption, and co-precipitation. *Journal of Environmental Chemical Engineering, 10*(3), 107507.
57. Li, G., Yan, L., Chen, X., Lam, S. S., Rinklebe, J., Yu, Q., Yang, Y., Peng, W., & Sonne, C. (2023). Phytoremediation of cadmium from soil, air and water. Chemosphere, 138058.
58. Li, Q., Xing, Y., Huang, B., Chen, X., Ji, L., Fu, X., Li, T., Wang, J., Chen, G., & Zhang, Q. (2022). Rhizospheric mechanisms of *Bacillus subtilis* bioaugmentation-assisted phytostabilization of cadmium-contaminated soil. *Science of the Total Environment, 825*, 154136.
59. Li, X., Lan, X., Feng, X., Luan, X., Cao, X., & Cui, Z. (2021). Biosorption capacity of Mucor circinelloides bioaugmented with *Solanum nigrum* L. for the cleanup of lead, cadmium and arsenic. *Ecotoxicology and Environmental Safety, 212*, 112014.
60. Liang, Y., Ma, A., & Zhuang, G. (2022). Construction of environmental synthetic microbial consortia: Based on engineering and ecological principles. *Frontiers in Microbiology, 13*, 829717. https://doi.org/10.3389/fmicb.2022.829717
61. Logan, B. (2009). Exoelectrogenic bacteria that power microbial fuel cells. *Nature Reviews Microbiology, 7*, 375–381. https://doi.org/10.1038/nrmicro2113
62. Lu, H., Qiao, D., Han, Y., Zhao, Y., Bai, F., Wang, Y. (2021). Low Molecular Weight Organic Acids Increase Cd Accumulation in Sunflowers through Increasing Cd Bioavailability and Reducing Cd Toxicity to Plants. *Minerals, 11*(3):243. https://doi.org/10.3390/min11030243
63. Malik, S., Kishore, S., Dhasmana, A., Kumari, P., Mitra, T., Chaudhary, V., Kumari, R., Bora, J., Ranjan, A., Minkina, T., et al. (2023). A perspective review on microbial fuel cells in treatment and product recovery from wastewater. *Water, 15*(2), 316. https://doi.org/10.3390/w15020316
64. Mandragutti, T., Dokka, M. K., Panchagnula, B., & Godi, S. (2021). Molecular characterization of marine bacterial isolates of Visakhapatnam coast—efficacy in dye decolorization and bioremediation of cadmium. *Journal of Genetic Engineering and Biotechnology, 19*(1), 1–11.
65. Mani, D., & Kumar, C. (2014). Biotechnological advances in bioremediation of heavy metals contaminated ecosystems: An overview with special reference to phytoremediation. *International Journal of Environmental Science and Technology, 11*, 843–872.
66. Manikandan, S. K., Pallavi, P., Shetty, K., Bhattacharjee, D., Giannakoudakis, D. A., Katsoyiannis, I. A., & Nair, V. (2023). Effective usage of biochar and microorganisms for the removal of heavy metal ions and pesticides. *Molecules, 28*(2), 719. https://doi.org/10.3390/molecules28020719
67. Minari, G. D., Saran, L. M., Constancio, M. T. L., da Silva, R. C., Rosalen, D. L., de Melo, W. J., & Alves, L. M. C. (2020). Bioremediation potential of new cadmium, chromium, and nickel-resistant bacteria isolated from tropical agricultural soil. *Ecotoxicology and Environmental Safety, 204*, 111038.
68. Mishra, S., Huang, Y., Li, J., Wu, X., Zhou, Z., Lei, Q., Bhatt, P., & Chen, S. (2022). Biofilm-mediated bioremediation is a powerful tool for the removal of environmental pollutants. *Chemosphere, 294*, 133609.
69. Muter O. (2023). Current Trends in Bioaugmentation Tools for Bioremediation: A Critical Review of Advances and Knowledge Gaps. *Microorganisms. 11*(3):710. https://doi.org/10.3390/microorganisms11030710
70. Najahi-Missaoui, W., Arnold, R. D., & Cummings, B. S. (2020). Safe nanoparticles: Are we there yet? *International Journal of Molecular Sciences, 22*(1), 385. https://doi.org/10.3390/ijms22010385

71. Nguyen, T. H., Won, S., Ha, M. G., Nguyen, D. D., & Kang, H. Y. (2021). Bioleaching for environmental remediation of toxic metals and metalloids: A review on soils, sediments, and mine tailings. *Chemosphere, 282*, 131108.
72. Nnaji, N. D., Onyeaka, H., Miri, T., et al. (2023). Bioaccumulation for heavy metal removal: A review. *SN Applied Science, 5*, 125. https://doi.org/10.1007/s42452-023-05351-6
73. Ojuederie, O. B., & Babalola, O. O. (2017). Microbial and plant-assisted bioremediation of heavy metal polluted environments: A review. *International Journal of Environmental Research and Public Health, 14*(12), 1504. https://doi.org/10.3390/ijerph14121504
74. Ozturk, M., Metin, M., Altay, V., De Filippis, L., Ünal, B. T., Khursheed, A., Gul, A., Hasanuzzaman, M., Nahar, K., Kawano, T., & Caparrós, P. G. (2021). Molecular biology of cadmium toxicity in *Saccharomyces cerevisiae. Biological Trace Element Research*, 1–15.
75. Palansooriya, K. N., Shaheen, S. M., Chen, S. S., Tsang, D. C., Hashimoto, Y., Hou, D., Bolan, N. S., & Ok, Y. S. (2020). Soil amendments for immobilization of potentially toxic elements in contaminated soils: A critical review. *Environment International, 134*, 105046.
76. Pande, V., Pandey, S. C., Sati, D., Bhatt, P., and Samant, M. (2022). Microbial interventions in bioremediation of heavy metal contaminants in agroecosystem. Front. Microbiol. 13:824084. https://doi.org/10.3389/fmicb.2022.824084
77. Peana, M., Pelucelli, A., Chasapis, C. T., Perlepes, S. P., Bekiari, V., Medici, S., & Zoroddu, M. A. (2022). Biological effects of human exposure to environmental cadmium. *Biomolecules, 13*(1), 36. https://doi.org/10.3390/biom13010036
78. Pramanik, K., Mitra, S., Sarkar, A., & Maiti, T. K. (2018). Alleviation of phytotoxic effects of cadmium on rice seedlings by cadmium resistant PGPR strain *Enterobacter aerogenes* MCC 3092. *Journal of Hazardous Materials, 351*, 317–329.
79. Qu, J., Wang, Q., Sun, X., & Li, Y. (2022). The environment and female reproduction: Potential mechanism of cadmium poisoning to the growth and development of ovarian follicle. *Ecotoxicology and Environmental Safety, 244*, 114029.
80. Raklami, A., Meddich, A., Oufdou, K., & Baslam, M. (2022). Plants-microorganisms-based bioremediation for heavy metal cleanup: Recent developments, phytoremediation techniques, regulation mechanisms, and molecular responses. *International Journal of Molecular Sciences, 23*(9), 5031. https://doi.org/10.3390/ijms23095031
81. Rasool, S., Ahmad, I., Jamal, A., Saeed, M. F., Zakir, A., Abbas, G., Seleiman, M. F., Caballero-Calvo, A. (2023). Evaluation of Phytoremediation Potential of an Aquatic Macrophyte (Eichhornia crassipes) in Wastewater Treatment. Sustainability , 15, 11533. https://doi.org/10.3390/su151511533
82. Raza, A., Habib, M., Kakavand, S. N., Zahid, Z., Zahra, N., Sharif, R., & Hasanuzzaman, M. (2020). Phytoremediation of cadmium: Physiological, biochemical, and molecular mechanisms. *Biology, 9*(7), 177.
83. Razzak, S. A., Faruque, M. O., Alsheikh, Z., Alsheikhmohamad, L., Alkuroud, D., Alfayez, A., Zakir Hossain, S. M., & Hossain, M. M. (2022). A comprehensive review on conventional and biological-driven heavy metals removal from industrial wastewater. *Environmental Advances, 7*, 100168.
84. Rehman, M. Z., Batool, Z., Ayub, M. A., Hussaini, K. M., Murtaza, G., Usman, M., Naeem, A., Khalid, H., Rizwan, M., & Ali, S. (2020). Effect of acidified biochar on bioaccumulation of cadmium (Cd) and rice growth in contaminated soil. *Environmental Technology & Innovation, 19*, 101015.
85. Rigby, H., & Smith, S. R. (2020). The significance of cadmium entering the human food chain via livestock ingestion from the agricultural use of biosolids, with special reference to the UK. *Environment International, 143*, 105844.
86. Roy, R., Tiwari, M., Donelli, G., & Tiwari, V. (2018). Strategies for combating bacterial biofilms: A focus on anti-biofilm agents and their mechanisms of action. *Virulence, 9*(1), 522–554. https://doi.org/10.1080/21505594.2017.1313372.PMID:28362216;PMCID: PMC5955472
87. Sahu, A., Mandal, A., Thakur, J., Manna, M. C., & Rao, A. S. (2012). Exploring bioaccumulation efficacy of *Trichoderma viride*: An alternative bioremediation of cadmium and lead. *National Academy Science Letters, 35*, 299–302.

88. Shah, G. M., Tufail, N., Bakhat, H. F., Ahmad, I., Shahid, M., Hammad, H. M., Nasim, W., Waqar, A., Rizwan, M., & Dong, R. (2019). Composting of municipal solid waste by different methods improved the growth of vegetables and reduced the health risks of cadmium and lead. *Environmental Science and Pollution Research, 26*(6), 5463–5474.
89. Sharma, P. (2022). Role and significance of biofilm-forming microbes in phytoremediation-a review. *Environmental Technology & Innovation, 25*, 102182.
90. Sharma, J. K., Kumar, N., Singh, N. P., & Santal, A. R. (2023). Phytoremediation technologies and their mechanism for removal of heavy metal from contaminated soil: An approach for a sustainable environment. *Frontiers in Plant Science, 14*, 1076876.
91. Shi, G. Y., Yan, Y. J., Yu, Z. Q., Zhang, L., Cheng, Y. Y., & Shi, W. L. (2020). Modification-bioremediation of copper, lead, and cadmium-contaminated soil by combined ryegrass (*Lolium multiflorum* Lam.) and *Pseudomonas aeruginosa* treatment. *Environmental Science and Pollution Research, 27*, 37668–37676.
92. Stedel, C., Efrose, R. C., & Rosu, C. M. (2019). Textile dye bioremediation potential of some rhizobial strains and their heavy-metal and high salinity tolerance. *Journal of Experimental and Molecular Biology, 20*(1–2), 15–26.
93. Supel, P., Śliwa-Cebula, M., Miszalski, Z., & Kaszycki, P. (2022). Cadmium-tolerant rhizospheric bacteria of the C3/CAM intermediate semi-halophytic common ice plant (*Mesembryanthemum crystallinum* L.) grown in contaminated soils. *Frontiers in Plant Science, 13*, 820097.
94. Tahir, M., Khan, M. B., Shahid, M., Ahmad, I., Khalid, U., Akram, M., Dawood, A., & Kamran, M. (2021). Metal-tolerant *Pantoea* sp. WP-5 and organic manures enhanced root exudation and phytostabilization of cadmium in the rhizosphere of maize. *Environmental Science and Pollution Research*, 1–14.
95. Tan, X., Li, K., Wang, Z., Zhu, K., Tan, X., Cao, J. (2019). A Review of Plant Vacuoles: Formation, Located Proteins, and Functions. Plants, 8(9):327. https://doi.org/10.3390/plants8090327
96. Teng, Y., Yu, A., Tang, Y. M., Jiang, Z. Y., Guan, W. J., Li, Z. S., Yu, H. Y., & Zou, L. Y. (2021). Visualization and quantification of cadmium accumulation, chelation and antioxidation during the process of vacuolar compartmentalization in the hyperaccumulator plant *Solanum nigrum* L. *Plant Science, 310*, 110961.
97. Thirumoorthy, N., Manisenthil Kumar, K. T., Shyam Sundar, A., Panayappan, L., & Chatterjee, M. (2007). Metallothionein: An overview. *World Journal of Gastroenterology, 13*(7), 993–996. https://doi.org/10.3748/wjg.v13.i7.993.PMID:17373731;PMCID:PMC4146885
98. Tian, S., Gong, Y., Ji, H., Duan, J., & Zhao, D. (2020). Efficient removal and long-term sequestration of cadmium from aqueous solution using ferrous sulfide nanoparticles: Performance, mechanisms, and long-term stability. *Science of the Total Environment, 704*, 135402.
99. Torres, E. (2020). Biosorption: A review of the latest advances. *Processes., 8*(12), 1584. https://doi.org/10.3390/pr8121584
100. Tripathi, S. et al. (2021) Elucidating the bioremediation mechanism of Scenedesmus sp. IITRIND2 under cadmium stress. Chemosphere. https://doi.org/10.1016/j.chemosphere.2021.131196
101. Verma, J., Kumar, D., Singh, N., et al. (2021). Electricigens and microbial fuel cells for bioremediation and bioenergy production: A review. *Environmental Chemistry Letters, 19*, 2091–2126.
102. Vidu, R., Matei, E., Predescu, A. M., Alhalaili, B., Pantilimon, C., Tarcea, C., & Predescu, C. (2020). Removal of heavy metals from wastewaters: A challenge from current treatment methods to nanotechnology applications. *Toxics, 8*(4), 101. https://doi.org/10.3390/toxics8040101
103. Wang, X., Li, D., Gao, P., Gu, W., He, X., Yang, W., & Tang, W. (2020). Analysis of biosorption and biotransformation mechanism of *Pseudomonas chengduensis* strain MBR under Cd (II) stress from genomic perspective. *Ecotoxicology and Environmental Safety, 198*, 110655.
104. Wang, J. F., Li, W. L., Ahmad, I., He, B. Y., Wang, L. L., He, T., Wang, F. P., Xu, Z.-M., & Li, Q. S. (2021). Biomineralization of Cd2+ and inhibition on rhizobacterial Cd mobilization function by *Bacillus cereus* to improve safety of maize grains. *Chemosphere, 283*, 131095.

105. Woraharn, S., Meeinkuirt, W., Phusantisampan, T., & Chayapan, P. (2021). Rhizofiltration of cadmium and zinc in hydroponic systems. *Water, Air, & Soil Pollution, 232*(5), 204.
106. Wu, H., Wu, Q., Wu, G., Gu, Q., & Wei, L. (2016). Cd-resistant strains of B. cereus S5 with endurance capacity and their capacities for cadmium removal from cadmium-polluted water. *PLoS One, 14;11*(4), e0151479. https://doi.org/10.1371/journal.pone.0151479
107. Wu, Y., Ma, L., Liu, Q., Vestergård, M., Topalovic, O., Wang, Q., Zhou, Q., Huang, L., Yang, Z., & Feng, Y. (2020). The plant-growth promoting bacteria promote cadmium uptake by inducing a hormonal crosstalk and lateral root formation in a hyperaccumulator plant *Sedum alfredii*. *Journal of Hazardous Materials, 395*, 122661.
108. Xia, X., Ji, J., Yang, Z., Han, H., Huang, C., Li, Y., & Zhang, W. (2020). Cadmium risk in the soil-plant system caused by weathering of carbonate bedrock. *Chemosphere, 254*, 126799.
109. Xu, L., Xing, X., Liang, J., Peng, J., & Zhou, J. (2019). In situ phytoremediation of copper and cadmium in a co-contaminated soil and its biological and physical effects. *RSC Advances, 9*(2), 993–1003.
110. Zhang, T, & Zhang, H. (2022). Microbial consortia are needed to degrade soil pollutants. *Microorganisms, 10*(2), 261. https://doi.org/10.3390/microorganisms10020261
111. Zhang, X., Yu, J., Huang, Z., Li, H., Liu, X., Huang, J., Zhuo, R.,Wu, Z., Qin, X., Gao, Y., Wang, M., & Zhu, Y. (2021). Enhanced Cd phytostabilization and rhizosphere bacterial diversity of *Robinia pseudoacacia* L. by endophyte *Enterobacter* sp. YG-14 combined with sludge biochar. *Science of the Total Environment, 787*, 147660.
112. Zhao, Q., Li, X., Wang, Y., Cheng, Y., & Fan, W. (2023). Long-term bioremediation of cadmium contaminated sediment using sulfate reducing bacteria: Perspective on different depths of the sediment profile. *Chemical Engineering Journal, 451*, 138697.
113. Zhao, F. J., & Wang, P. (2020). Arsenic and cadmium accumulation in rice and mitigation strategies. *Plant and Soil, 446*, 1–21.

Treatment Methods for Cadmium Removal from Wastewater

Iftikhar Ahmad, Rana Umair Asad, Laraib Maryam, and Marriam Masood

Abstract Wastewater produced from industrial and municipal processes has been reported for heavy metal contamination. Amongst them, cadmium (Cd) concentration has been reported (up to 0.02 mg/L) in Pakistan, (up to 0.03 mg/L) in China and (up to 0.07 mg/L) in India, which is quite higher than the standard limits of WHO (0.003 mg/L). It has serious implications for human health, getting into the tissues via direct ingestion, dermal contact, inhalation, and adsorption. Several methods have been explored for the extermination of Cd from the environment. Conventional methods (such as adsorption, chemical precipitation, membrane filtration, electrodialysis, photocatalysis) of metal removal are constrained by the processing problems, expenses, and the generation of toxic sludge, therefore more research is now focused on the use of advanced treatment methods (such as fluidized bed reactor, advanced oxidation process, nanotechnology, hybrid system method, chelation, solvent extraction, foam fractionation, cryogens, graphene, carbon nanotubes, activated carbon, zeolites, microbial fuel cells, and phytoremediation) for the removal of Cd metal ions from the wastewater. These advanced processes may be cost-effective, viable, more efficient and less time taking than the existing techniques and have intense opportunity.

Keywords Advanced techniques · Conventional methods · Heavy metal removal · Industrial wastewater · Wastewater treatment

1 Introduction

Heavy metals in industrial effluent constitute a serious threat to the environment and public health worldwide. Cadmium (Cd) distinguishes out among these heavy metals due to its toxicity and tenacity in the environment. In industrial effluents from mining, metal plating, battery manufacture, and other chemical processes, Cd

I. Ahmad (✉) · R. U. Asad · L. Maryam · M. Masood
Department of Environmental Science, COMSATS University Islamabad, Vehari Campus, Vehari 61100, Pakistan
e-mail: iftikharahmad@cuivehari.edu.pk

A. K. Jha and N. Kumar (eds.), *Cadmium Toxicity in Water*, Springer Water,
https://doi.org/10.1007/978-3-031-54005-9_7

is frequently present. One of the most dangerous and movable substances in the environment is Cd (e.g., [8, 96, 97]). Cd enters the human diet primarily through terrestrial channels, such as through vegetables, in contrast to other harmful elements like mercury (Hg) and arsenic (As). Rice bioaccumulated Cd, causing an enhanced daily Cd absorption in China, Korea, and Jamaica in locations with elevated soil and groundwater Cd concentrations from both geogenic and anthropogenic sources [75, 108]. The bioavailability of Cd is complicated, as evidenced by the fact that rice from China's southern region has more Cd than rice from its northern region. Possible causes include the production of rice in southern China that has a higher affinity for Cd accumulation and more acidic soils, an excessive use of nitrogen fertilizers, irrigation, and atmospheric deposition pollution, and more [22, 136]. Itai-itai sickness, also known as chronic Cd poisoning, was initially identified in Japan at the beginning of the twentieth century and is known to cause renal tubular dysfunction, osteomalacia, and osteoporosis because of competition with calcium and other minerals [10, 11, 55]. According to Khan et al. [55], Cd exposure is linked to heart failure, cerebral infarction, breast and lung cancer, diabetes, and abnormalities in glucose metabolism. Cd-contaminated wastewater discharged into water bodies has the potential to harm aquatic ecosystems and possibly infiltrate the food chain, posing a risk to human health. Figure 1 shows the sources and major effects of Cd in the environment. Therefore, it is essential to properly treat wastewater that contains Cd to protect both the environment and human health.

Numerous treatment techniques have been developed over time to remove Cd from industrial effluent. Many techniques, including membrane separation, ion exchange, coagulation, softening, solvent extraction, and adsorption, have been used for that purpose [33]. Substantial progress has been made recently in the field of treating industrial effluent contaminated with Cd. A modified zeolite was employed to remove Cd from wastewater [88], results revealed that ethylenediamine modified zeolite successfully removed Cd from wastewater samples with an astounding removal efficiency of 98.5%. Electrodes with graphene modifications were utilized in an electrochemical system that removed 99.8% Cd ions from wastewater [119]. Using a unique hybrid adsorbent made of bentonite and iron oxide nanoparticles, Rahman et al. [102] reported 97.6% of Cd from wastewater using this hybrid adsorbents. Li et al. [72] used nanofiltration to treat wastewater containing Cd from a metal plating facility in their industrial case study and they found nanofiltration system effectively removed Cd up to 95–98%. Innovative methods that can achieve greater removal efficiency, shorten treatment times, and limit the production of secondary waste have been the focus of research. These developments have the potential to transform the way of wastewater treatment contaminated with Cd and solve problems caused by traditional approaches.

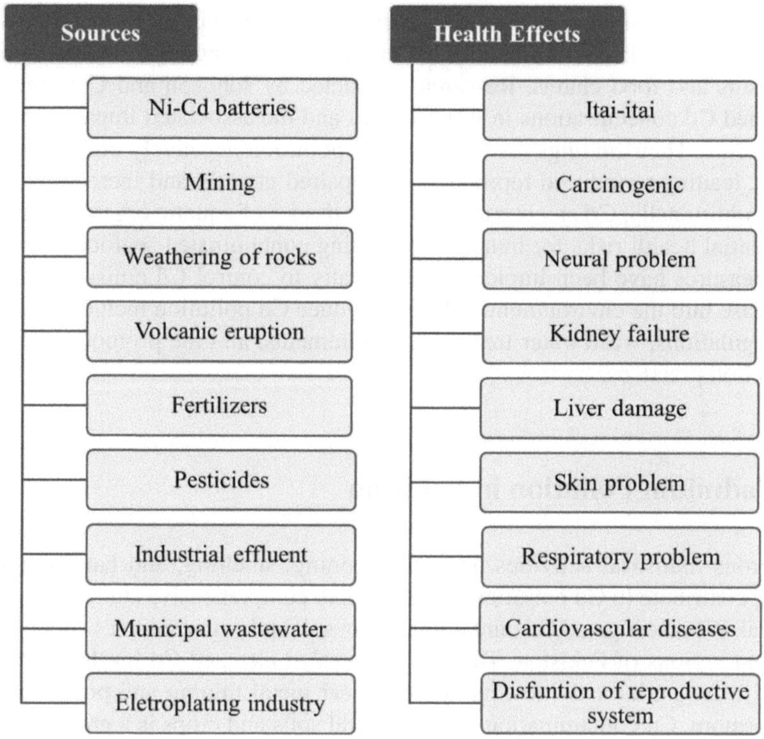

Fig. 1 Sources and health effects of Cd [2, 4, 37]

2 Status of Cd Pollution

Cd pollution is primarily attributed to human activities, particularly industrial processes such as mining, smelting, and manufacturing. These activities release Cd into the environment through emissions, wastewater discharges, and improper waste management practices. Additionally, Cd can also enter the soil and water using Cd-containing fertilizers, sewage sludge, and atmospheric deposition. Discharging sewage sludge into waterbodies would significantly increase organic load with a corresponding decrease in dissolved oxygen levels and nutrient enrichment since various pollutants in wastewater accumulate in sewage sludge (in the range of 80–90%) [91, 98]. Most of the time, methods used to treat wastewater do not ensure the quantitative removal of many contaminants, which means that after discharge, there may be additional environmental damage [15, 123]. The wastewater discharged may contain biological or chemical pollutants.

The widespread occurrence of Cd pollution has been documented in various regions around the world. A comprehensive study conducted by Smith et al. [115] assessed Cd contamination levels in soil, water, and food samples from multiple sites across different countries. The results indicated elevated Cd concentrations in areas with a history of industrial activities, particularly in proximity to metal mines and smelters. In aquatic environments, Cd pollution poses a significant threat to aquatic organisms and food chains. Research conducted by Johnson and Colleagues [52] examined Cd concentrations in water bodies and the associated impacts on aquatic ecosystems. Their findings revealed that Cd pollution negatively affects fish populations, leading to reduced reproduction, impaired growth, and increased mortality rates. Additionally, Cd can accumulate in the tissues of aquatic organisms, resulting in potential health risks for humans consuming contaminated seafood. The regulatory measures have been implemented globally to control Cd emissions and limit its release into the environment. Efforts to reduce Cd pollution include strict industrial regulations, wastewater treatment requirements, and the promotion of cleaner production practices.

3 Cadmium Pollution in Pakistan

Numerous industrial activities, including mining, smelting, and battery manufacturing, contribute to Cd pollution in Pakistan. A comprehensive study conducted by Ali et al. [7] investigated Cd contamination in soil and water samples collected from different regions of Pakistan. The results revealed elevated Cd levels in areas with intensive industrial activities, particularly near metal mining and processing sites. Furthermore, Cd contamination in agricultural soils and crops is a growing concern in Pakistan. Research conducted by Khan et al. [54] assessed Cd concentrations in soils and wheat crops from various agricultural regions. The study found significant Cd contamination in agricultural soils, mainly attributed to the application of Cd-containing fertilizers, irrigation with Cd-contaminated water, and atmospheric deposition. The Cd levels in wheat crops exceeded the permissible limits, raising concerns about potential health risks for consumers.

In addition to industrial and agricultural sources, Cd pollution in Pakistan is also associated with inadequate wastewater treatment practices. Untreated or poorly treated industrial and municipal wastewater containing Cd is discharged into water bodies, leading to contamination. A study by Ahmed et al. [6] investigated Cd levels in wastewater and surface water samples from different regions of Pakistan. The research revealed high Cd concentrations in the studied water bodies, indicating the presence of Cd pollution resulting from wastewater discharge. To mitigate Cd pollution in Pakistan, efforts have been made to strengthen environmental regulations, promote sustainable industrial practices, and improve wastewater treatment infrastructure. These initiatives aim to control Cd emissions, enhance pollution monitoring, and implement effective remediation strategies.

4 Treatment of Cd Contaminated Wastewater

The objectives of treating Cd from wastewater are primarily to reduce the concentration of Cd in the water to safe levels and prevent harm to human health and the environment. The objectives of implementing advanced Cd (Cd) treatment technologies in industrial wastewater are multifaceted, aiming to effectively remove Cd contaminants and ensure compliance with environmental regulations. Here are some objectives that can be achieved after the treatment of Cd contaminated wastewater.

Efficient Cd Removal: The primary objective is to achieve efficient removal of Cd from industrial or municipal wastewater, reducing its concentration to permissible levels. This involves employing advanced treatment technologies that can effectively capture and separate Cd ions from the water matrix.

Resource Recovery: Another objective is to recover and recycle Cd from the wastewater, where feasible. Advanced technologies, such as adsorption and precipitation methods, may offer opportunities for Cd recovery, allowing for its reuse or proper disposal, thus minimizing environmental impact.

Minimize Environmental Impact: The objective is to minimize the release of Cd into the environment, preventing its entry into natural water bodies and mitigating potential ecological risks. Advanced treatment technologies are designed to achieve high removal efficiencies, ensuring that treated wastewater meets the regulatory standards for Cd discharge.

Cost-Effectiveness: Implementing cost-effective Cd treatment technologies is an important objective for industries. Advanced techniques aim to optimize treatment processes, reducing operational costs, energy consumption, and chemical usage while maintaining high removal efficiencies. A study conducted by Li et al. [66, 69, 71] explored the objectives of Cd treatment in industrial wastewater using advanced technologies. The researchers highlighted the significance of efficient Cd removal, resource recovery, minimizing environmental impact, and cost-effectiveness as key objectives of implementing advanced treatment methods.

5 Conventional Techniques for Treatment of Cd Contaminated Wastewater

The various conventional techniques such as adsorption, precipitation, membrane filtration, electrodialysis, and photocatalysis are being used for Cd removal from wastewater (Table 1).

Table 1 Conventional techniques employed for treatment of Cd contaminated wastewater

Techniques	Mechanisms	Removal efficiency (%)	References
Adsorption	Physical adsorption	70–99	Cui and Yue [30], Liu et al. [83], Ma et al. [89]
Precipitation	Precipitation	≤ 90	Daraei et al. [31], Liu et al. [80], Sani et al. [107]
Membrane Filtration	Size exclusion	70–99	Kim and Park [56], Wang et al. [126]
Photocatalysis	Reactive radicals	70–99	Rajeshwar et al. [103], Sharma and Tripathi [111], Tian et al. [120]
Electrodialysis	Ion exchange	80–99	Avsec et al. [13], Pontié et al. [100], Ramos et al. [104], Zhu et al. [154]

5.1 Adsorption

Industrial wastewater contains the poisonous heavy metal Cd, which is dangerous for both the environment and human health. Adsorption has proven to be a successful way for removing Cd to solve this problem. It has been investigated how well different adsorbent substances can bind with Cd ions to remove them from wastewater. The interaction between Cd ions and the surface of the adsorbent material is the basis of the adsorption process used to remove Cd from wastewater. The Cd ions in the aqueous phase of the wastewater diffuse to the surface of the adsorbent when it meets the substance. The Cd ions are drawn to and bound by the active sites in the adsorbent material, such as functional groups or porous structures, through a variety of physicochemical factors, such as electrostatic interactions, ion exchange, and surface complexation. Due to the charge disparity between the Cd ions and the active sites of the adsorbent, electrostatic interactions take place. Cadmium ions are then adsorbed onto the adsorbent surface because of an attractive attraction created between species that have opposing charges. Based on variations in ion concentration and affinity, ion exchange entails replacing exchangeable ions on the adsorbent surface with Cd ions. When Cd ions and particular functional groups, such as hydroxyl (–OH) and carboxyl (–COOH) groups, are present on the surface of an adsorbent, the process is known as surface complexation.

Application of compost as adsorbent can remove 99% of Cd from spiked Cd-contaminated water at optimal pH, temperature, contact time and initial Cd concentration [3]. Due to its huge surface area and porosity, which offer plenty of binding sites for Cd ions, activated carbon is one of the most frequently studied adsorbents. For example, Liu et al. [84] showed that activated carbon made from agricultural waste had outstanding Cd adsorption efficiency, making it a promising choice for wastewater treatment. Other inexpensive adsorbents, besides activated carbon, have demonstrated the ability to remove Cd. As an illustration, a study by Sharma et al. [110]

investigated the usage of agricultural by-products, such as rice husk and sawdust, as efficient adsorbents for removing Cd from wastewater. Zhao et al. [147] investigated how metal–organic frameworks (MOFs) may be used as nanomaterials for effective Cd adsorption. Due to their high surface area and adjustable architecture, MOFs displayed excellent adsorption capability, potentially representing a technological advance in the removal of Cd. To increase the capacity of adsorbents to bind Cd, researchers have also investigated modifying them. According to a study by Chen et al. [20], immobilizing chelating agents like ethylenediaminetetraacetic acid (EDTA) onto activated carbon can greatly increase the effectiveness of Cd removal.

Chen et al. [19, 23–25] investigated using a modified adsorbent made from agricultural waste to remove Cd, which was successfully removed from wastewater by the adsorption technique, with removal rate of more than 90%. The modified adsorbent demonstrated a significant affinity for Cd, making it an advantageous option for real-world use in wastewater treatment. For instance, Zhang et al. [140, 141, 143, 144] investigated the removal of Cd using biochar made from agricultural waste and discovered that it was a viable adsorbent because of its high surface area and distinctive physicochemical features. In a different study, Gupta and Babu [40] investigated the potential of modified clay minerals for Cd adsorption and found that they were significantly more effective in absorbing the metal than raw clay minerals. Additionally, Li et al. [64, 68, 70, 73] examined the usage of zeolite-based materials for the removal of Cd and emphasized the need of optimizing operating conditions to increase the adsorption capacity. These investigations and results show that the adsorption technique, utilizing different adsorbent materials and modifications, provides a practical and effective means of removing Cd from wastewater, hence reducing the negative effects of this hazardous heavy metal on the environment and human health.

5.2 Chemical Precipitation

Chemical precipitation relies on the use of a precipitating agent to create insoluble precipitates that react with the Cd ions in the wastewater to create a solid phase that can be separated from the treated water. The following steps make up the mechanism.

The effluent is first treated by adding a suitable precipitating agent. Hydroxide-based substances like sodium hydroxide (NaOH) or lime (calcium hydroxide, $Ca(OH)_2$) are frequently employed precipitating agents for removing Cd. These substances cause the wastewater to have a higher pH, which causes hydrolysis to create hydroxide ions (OH^-).

Second, insoluble Cd hydroxide ($Cd(OH)_2$) precipitates are produced when the hydroxide ions combine with the dissolved Cd ions in the effluent. Maintaining the pH within an ideal range is essential for the efficient removal of Cd because this process is strongly pH dependent. When the pH is right, the precipitates of Cd hydroxide have limited solubility and a propensity to aggregate, making it easier to separate them from the water.

Thirdly, the precipitates of produced Cd hydroxide are removed from the cleaned wastewater. This can be done by using different techniques for separating solids from liquids, like centrifugation, filtration, or sedimentation.

$$Cd^{2+} + 2OH^{-} \rightarrow Cd(OH)_2$$

The use of lime (calcium hydroxide) as a precipitating agent was efficiently removed Cd from wastewater by bringing the pH into an ideal range, which led to the precipitation of insoluble Cd hydroxide [93]. The efficiency of calcium hydroxide-based chemical precipitation for the removal of Cd from industrial wastewater was examined in a study by Wang et al. [127]. The researchers found that the removal effectiveness was greatly impacted by the pH of the solution, with higher pH values resulting in faster precipitation rates. To achieve the best Cd removal efficiency, they also emphasized the need for pH optimization and precipitating agent dosing. In a different investigation, Daraei et al. [31] applied sodium hydroxide for the removal of Cd from synthetic wastewater. The researchers discovered that the chemical precipitation method was effective in bringing Cd concentrations down to below the permitted limits established by environmental standards. The addition of calcium hydroxide (lime), magnesium hydroxide and sodium sulfide improved the removal efficiency of Cd up to 95–98% from the wastewater by producing precipitates of either Cd hydroxide or Cd sulfide [9, 18, 21, 43, 122].

5.3 Membrane Filtration

Membrane filtration procedure makes use of semi-permeable membranes, which let water molecules pass while capturing pollutants like Cd ions. Frequently used membrane filtration techniques for Cd removal are reverse osmosis, nanofiltration, ultrafiltration, and microfiltration.

5.3.1 Reverse Osmosis

By applying pressure to the wastewater during the reverse osmosis process (RO), a thick membrane that discards Cd ions and other dissolved particles lets only clean water pass through. Zhang et al. [140, 141, 143, 144] investigated the use of RO to remove Cd from industrial effluent, they discovered that RO had a high removal effectiveness for Cd, effectively lowering its content to satisfy discharge regulations. According to the researchers, reverse osmosis had a phenomenal removal efficiency, eliminating over 99% of Cd. The small hole size of the membrane and pressure applied efficiently blocked the passage of Cd ions, producing filtered water with much lower Cd levels [16, 57].

5.3.2 Nanofiltration

Another membrane filtration technique is nanofiltration (NF), which offers the selective removal of divalent ions like Cd while permitting the passage of smaller ions and molecules. The removal of multivalent cations with this method is especially successful since it operates at a lower pressure than reverse osmosis. Nanofiltration works through several processes while removing Cd from wastewater. The Cd ions can interact with the charged surface of the nanofiltration membrane due to their charged nature. On the membrane surface, the negatively charged functional groups can attract the positively charged Cd ions (Cd^{2+}). Cd ions are more easily retained on the feed side of the membrane thanks to this electrostatic contact. Second, a big part is played by the size exclusion effect. Small, hydrated ions like Cd can pass through the tightly packed pores of the nanofiltration membrane, while larger particles and impurities are trapped. By separating the Cd ions from the wastewater stream using size-based selectivity, pure water can travel through the membrane while the Cd ions are captured. The elimination of Cd is also aided by the steric hindrance effect. The interactions of the Cd ions with the polymer matrix make it difficult for them to travel through the membrane's small pores, further promoting their retention.

Several studies showed that NF efficiently brought Cd concentrations down to below acceptable levels, demonstrating its promise as a successful Cd removal method from wastewater [61, 63, 148, 150, 151]. Another study, Huang et al. [45] found that NF had high exclusion rates up to 98% of Cd from wastewater. The combination of size exclusion and electrostatic repulsion between the negatively charged membrane surface and the Cd ions, which prevented their passage through the membrane, was credited with causing efficient removal. When compared to other traditional treatment methods, the membrane filtration method has various benefits, including the ability to function continuously, the production of high-quality treated water, and the need for less area. However, it is crucial to consider the potential for membrane fouling over time, since the buildup of pollutants on the membrane surface could decrease its effectiveness. Therefore, to guarantee the best performance, adequate pretreatment and frequent maintenance are required.

5.3.3 Ultrafiltration

Heavy metals like Cd and other pollutants can be removed from wastewater using the membrane-based separation technology known as ultrafiltration (UF). According to their molecular size and weight, pollutants are separated from water using this technique using a semi-permeable membrane with pores that range from 1 to 100 nm. While Cd ions and other bigger particles are retained and cleaned water is left behind, wastewater is pressurized and forced through the membrane during the ultrafiltration process. The hollow fiber UF membrane could remove Cd with exceptional removal efficiency that exceeded 95%. Additionally, they discovered that the UF method successfully eliminated additional suspended solids and colloidal particles, producing cleaner water appropriate for reusing or safe release [62]. Similarly, the researchers used a hollow fiber UF membrane with an appropriate molecular weight cutoff. They discovered that the UF approach removed 95% of Cd from industrial wastewater [149]. The ceramic UF membrane showed exceptional resistance to fouling, enabling stable and continuous Cd removal under various operating settings [64, 68, 70, 73]. The optimum feed flow rate and transmembrane pressure of ceramic UF membrane could remove up to 98% of Cd ions from wastewater [85]. The benefits of the UF technique include its effectiveness in removing Cd and other contaminants without the need of chemical additives, making it a greener process. In comparison to more sophisticated treatment methods, the procedure is also quite straightforward and uses little energy. To avoid fouling and guarantee consistent performance, it is crucial to monitor and maintain the membrane's integrity.

5.3.4 Microfiltration

Due to its effectiveness in effectively separating pollutants and suspended materials at the micro-scale level, microfiltration (MF) is a method for removing Cd from wastewater. This procedure involves passing the wastewater over a membrane with pores that typically have a size between 0.1 to 1.0 μm. These membranes serve as barriers that let water and smaller particles through while holding back larger particles, such as Cd ions. Wang et al. [126, 131, 132] studied the effectiveness of ceramic MF membranes in treating wastewater contaminated with heavy metals, the findings demonstrated that the MF method successfully reduced Cd concentration to levels that were safe for disposal or reusing. The process optimization of ceramic MF membrane such as pressure, flow rate, and membrane pore size enabled it to remove Cd up to 90% from industrial wastewater [113, 137, 139]. To increase the effectiveness of Cd removal, the researchers discovered that specific membrane types (such as ceramic MF and polymeric MF), elements (such as transmembrane pressure), pore size, and feed concentration had great impact on Cd removal from wastewater [48, 61, 63, 82]. The MF method is affordable and simple to operate, making it a feasible choice for extensive wastewater treatment applications [111]. Microfiltration is a workable and sustainable treatment for heavy metal-contaminated wastewater because it removes Cd ions efficiently and consistently by choosing the right membrane.

5.4 Electrodialysis

Electrodialysis is an efficient way to remove Cd from wastewater by using an ion-exchange membrane and an electric field to specifically transport Cd ions out of the contaminated water. The concentrate and the dilute compartments are created by applying a direct electric current across several ion-exchange membranes during the electrodialysis process. The ability to separate and transport ions according to their charge is made possible by the ion-exchange membranes' selective permeability to either cations or anions. Water is transported between cation-exchange membranes (CEM) and anion-exchange membranes (AEM), two different types of ion-exchange membranes, during the electrodialysis procedure. Positive ions, such as Cd ions (Cd^{2+}), flow through the cation-exchange membrane and towards the cathode when an electric field is produced, whereas negative ions move through the anion-exchange membrane and towards the anode. Fixed negatively charged functional groups in the CEM draw positively charged Cd ions while repelling other cations found in the effluent. Cadmium ions gather close to the cathode because of this selective transport, where they can be eliminated from the system. The overall charge balance in the system is maintained by the positively charged functional groups in the AEM, which draw in and eliminate the negative ions produced during the electrodialysis procedure. As the process goes on, the concentration of Cd ions close to the cathode rises, causing a progressive removal of the metal from the effluent. On the other side of the electrodialysis cell, a collection of the cleansed water with a reduced Cd concentration is made.

Electrodialysis has been studied for its potential to remove Cd. For instance, Liu et al. [76] evaluated the use of electrodialysis for the removal of Cd from simulated wastewater in their research. The results showed that electrodialysis successfully reduced Cd concentrations up to 99%. Yang et al. [138] used functionalized ion-exchange membranes to improve the efficiency of electrodialysis for the removal of Cd. To increase the adsorption capacity and selectivity for Cd ions, the researchers added certain functional groups to the membranes. In comparison to traditional ion-exchange membranes, the results showed higher removal efficiency. Many factors including feed flow rate, current density, and membrane materials, affect the effectiveness of electrodialysis procedure to remove Cd from wastewater [1, 58].

5.5 *Photocatalysis*

A promising technique for removing Cd and other heavy metals from wastewater using a sophisticated oxidation process is photocatalysis. Titanium dioxide (TiO_2), which is frequently used as a semiconductor photocatalyst, activates because of exposure to ultraviolet (UV) or visible light, and this is how photocatalysis works. On the photocatalyst's surface, this activation produces electron–hole pairs. Strong oxidation and reduction potentials are possessed by the electrons in the conduction band and by the holes in the valence band, respectively. When the photocatalyst is exposed to light to remove Cd, the excited electrons and holes engage in redox reactions with water and oxygen molecules in the wastewater. The holes create hydroxyl radicals (OH) when they interact with water molecules. These highly reactive radicals have the power to either precipitate Cd ions (Cd^{2+}) into solid metal oxides or oxidize them to less harmful forms. Conduction band electrons, meanwhile, take part in reduction processes with oxygen molecules to create superoxide radicals (O_2), which can also aid in the breakdown of organic pollutants or the conversion of Cd species into less dangerous forms. Cd and other pollutants are degraded and removed from the wastewater by the combined action of hydroxyl and superoxide radicals in the presence of the photocatalyst, which either causes their adsorption onto the photocatalyst's surface or causes them to change into less toxic compounds. This procedure successfully removes contaminants from wastewater and lowers the amount of dangerous Cd ions.

Shahryari et al. [109] investigated the effectiveness of TiO_2 photocatalysts in both suspended and immobilized forms. The paper highlights the photocatalytic mechanisms involved in Cd removal and offers insights into photocatalysis that might be improved as a viable technique for treating wastewater. Aslam et al. [12] investigated the effectiveness of TiO_2 nanoparticles in eliminating Cd from synthetic wastewater. The number of operating factors, including pH, light intensity, and catalyst dosage can influence Cd removal. The study also explored production of hydroxyl radicals during the removal of Cd and their function in oxidizing the pollutant. In another study, the outcomes showed that the photocatalyst based on TiO_2 could eliminate over 95% of the Cd ions in a reasonably short amount of time [134]. The combination of ultrafiltration and TiO_2 photocatalysis indicated considerable Cd removal with removal efficiencies reaching 90% [95].

6 Advanced Techniques for Treatment of Cd Contaminated Wastewater

The various conventional techniques such as fluidized bed reactors, advanced oxidation processes, nanotechnology, hybrid system method, chelation, solvent extraction, foam fractionation, graphene, and phytoremediation are being used for Cd removal from wastewater (Table 2).

6.1 Fluidized Bed Reactors

Cadmium removal from wastewater is accomplished using fluidized bed reactors (FBRs), a sort of advanced treatment method. A bed of solid particles, usually in the form of granules, is suspended and fluidized in FBRs by the upward flow of water or air. Due to highly effective contact that is made between the wastewater and the solid particles, Cd ions can be effectively adsorbate or precipitated. A cylindrical tank with a distribution system for injecting polluted wastewater and a gas or liquid flow system for fluidization constitutes a fluidized bed reactor. To withstand the demanding circumstances of wastewater treatment, the vessel is often composed of corrosion-resistant materials like stainless steel. The selection of medium is based on the Cd removal requirements. The high surface area and strong affinity for heavy metals like Cd of activated carbon, zeolites, or ion exchange resins make them popular

Table 2 Advanced techniques employed for treatment of Cd contaminated wastewater

Techniques	Mechanisms	Removal efficiency	References
Fluidized bed reactor	Adsorption and chemical reaction	Varies depending on reactor design and conditions	Miao et al. [92], Smith et al. [114], Zhang et al. [145]
Advanced oxidation process	Oxidation and degradation	Varies depending on the process and conditions	Chen et al. [19, 23–25], Wang et al. [130], Zhao et al. [148, 150, 151]
Nanotechnology	Nanoparticles-mediated removal	Varies depending on nanoparticle type and concentration	Wu et al. [133], Xu et al. [135], Zhang et al. [140, 141, 143, 144]
Hybrid system method	Combination of multiple techniques	Varies depending on the combination of techniques	Park et al. [99], Song et al. [117]
Chelation	Formation of soluble metal complexes	Varies depending on chelating agent and pH	Chen et al. [18, 21], Feng et al. [35]
Solvent extraction	Separation based on solubility differences	Varies depending on reactor design and conditions	Miao et al. [92], Zhang et al. [145]
Foam fractionation	Formation and removal of foam	Varies depending on the process and conditions	Wang et al. [130], Zhao et al. [148, 150, 151]
Cryogens	Precipitation and removal	Varies depending on nanoparticle type and concentration	Xu et al. [135], Zhang et al. [140, 141, 143, 144]
Graphene	Adsorption and electron transfer	Varies depending on graphene properties and dosage	Gao et al. [36], Liu et al. [77]
Carbon nanotubes	Adsorption and filtration	Varies depending on nanotube type and concentration	Jiang et al. [51], Lu et al. [87]
Activated carbon	Adsorption	Varies depending on carbon type and dosage	Shu et al. [112], Zhang et al. [142]
Zeolites	Ion exchange and adsorption	Varies depending on zeolite type and conditions	Hui et al. [46], Wang et al. [125, 129, 129]

(continued)

Table 2 (continued)

Techniques	Mechanisms	Removal efficiency	References
Microbial fuel cells	Microbial metabolism and electricity generation	Varies depending on microbial species and reactor design	Chen et al. [27], He et al. [42]
Phytoremediation	Uptake and accumulation by plants	Varies depending on plant species and conditions	Jiang and Xu [50], Li et al. [65, 72, 74], Rasool et al. [103]

choices for media. To create a bed, the chosen media is loaded into the reactor vessel. Typically, a porous plate or distributor at the bottom of the tank supports the bed and distributes the effluent flow uniformly. A liquid or gas, such as air or nitrogen, is injected from the reactor's bottom to fluidize the bed. To prevent entrainment or an excessive pressure decrease, the flow rate and velocity are carefully regulated to guarantee appropriate suspension of the solid particles. The distribution system introduces the contaminated wastewater into the fluidized bed reactor. Cd ions in the wastewater absorb onto the surface of the media particles as they flow through the bed by a variety of processes, such as ion exchange, chemical precipitation, or physical adsorption. A crucial parameter is how long the wastewater stays inside the reactor. It depends on elements like the fluidized bed's height and the wastewater's flow rate.

According to the findings, under ideal operating conditions, the FBR had a Cd removal effectiveness of above 90% [114]. The researchers believed that the increased interaction between the Cd ions and the sand particles in the fluidized bed was the cause of this high removal efficiency. Removal of Cd from wastewater using fluidized beds packed with microbeads offered a significant surface area for efficient Cd adsorption [64, 68, 70, 73]. The usage of modified zeolite particles in a FBR for the removal efficiently remove Cd from wastewater [140, 141, 143, 144]. High mass transfer rates, significant adsorption surface area, and the ability for medium regeneration are just a few benefits that fluidized bed reactors provide for the treatment of Cd contaminated wastewater. To guarantee efficient treatment, the reactor's design and operation should be adapted to the unique properties of the wastewater and the targeted removal efficiency.

6.2 Advanced Oxidation Processes

By producing extremely reactive hydroxyl radicals that oxidize and destroy the contaminant, Advanced Oxidation Processes (AOPs) are very successful at removing Cd from wastewater. The processes used by AOPs include photocatalysis, ozonation, electrochemical oxidation, and heterogeneous Fenton processes (Fig. 2).

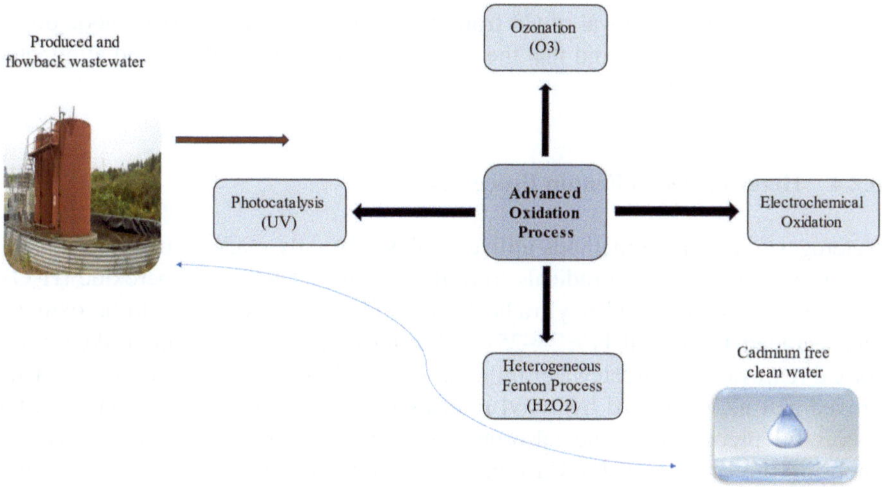

Fig. 2 Advanced oxidation processes used for the treatment of Cd contaminated wastewater

6.2.1 Photocatalysis

Photocatalytic procedures produce hydroxyl radicals from a catalyst and UV rays, which then react with the Cd ions in the effluent. Due to its powerful oxidation potential, titanium dioxide (TiO_2) is frequently used as a photocatalyst. TiO_2 nanoparticles were used to study the photocatalytic removal of Cd from wastewater by Kumar et al. [59]. They claimed that the production of Cd hydroxides and their subsequent photocatalytic breakdown resulted in a considerable reduction in Cd removal effectiveness.

6.2.2 Ozonation

Ozonation is the process of adding ozone (O_3) to wastewater, which causes water molecules to react to generate hydroxyl radicals. Cd ions are changed into less harmful forms or precipitates by the hydroxyl radicals. Manikandan et al. [90] investigated the use of ozonation to remove Cd from synthetic wastewater. Through oxidation and subsequent Cd hydroxide precipitation, they found effective Cd removal.

6.2.3 Electrochemical Oxidation

For removing Cd, electrochemical oxidation procedures use an electric current to produce hydroxyl radicals or other oxidizing species. This process can be used in electrochemical cells or with anodic oxidation. Mohanapriya et al. [94] investigated the electrochemical removal of Cd from wastewater using a diamond anode doped with boron. They discovered that the method successfully oxidized and eliminated Cd from the effluent.

6.2.4 Heterogeneous Fenton Processes

Heterogeneous Fenton methods utilize catalysts, usually made of iron-based materials to produce hydroxyl radicals from the reaction of hydrogen peroxide (H_2O_2) and the catalyst. The hydroxyl radicals subsequently cause the Cd to be oxidized and degraded. Chen et al. [19, 23–25] studied the heterogeneous Fenton-like process for Cd removal from wastewater utilizing nanoscale zero-valent iron (nZVI). They proved that nZVI created hydroxyl radicals in an efficient manner, leading to the successful precipitation and adsorption of Cd. It's important to remember that choosing the right AOP for Cd removal depends on various elements, including the Cd content, wastewater makeup, and process optimization. To identify the ideal operating conditions, pilot-scale investigations and optimization experiments are frequently carried out.

6.3 Nanotechnology

Due to the special characteristics and large surface area of nanoparticles, approaches based on nanotechnology have shown promise in the removal of Cd from wastewater.

6.3.1 Nanoscale Zero-Valent Iron (nZVI)

For the removal of Cd from wastewater, nanoscale zero-valent iron has received much research. Cd ions may effectively be reduced by the strong reactivity and substantial surface area of nZVI. Wu et al. [133] investigated the usage of nZVI for Cd removal from wastewater, they discovered that both adsorption and reduction processes were successful in lowering the concentration of Cd ions.

6.3.2 Metal Oxide Nanoparticles

Metal oxide nanoparticles, including iron oxide (Fe_3O_4), manganese oxide (MnO_2), and cerium oxide (CeO_2), have demonstrated remarkable adsorption properties for the elimination of Cd. According to Zhou et al. [152] Fe_3O_4 nanoparticles can remove Cd from wastewater, the interaction between the nanoparticles and Cd ions resulted in the high adsorption capacity.

6.3.3 Hybrid Nanocomposites

By mixing various nanomaterials, hybrid nanocomposites can increase the adsorption and removal efficiency of Cd from wastewater. Synergistic effects may be produced, for instance, when carbon-based nanomaterials are combined with metal oxide nanoparticles. MnO_2 nanoparticles were incorporated into multi-walled carbon nanotubes to create a hybrid nanocomposite for the elimination of Cd [79, 84]. Because both components worked together synergistically, they reported increased adsorption capacity compared to individual materials.

Effective Cd removal from wastewater is possible using nanotechnology-based techniques. To improve the synthesis of nanomaterials, comprehend their long-term stability, and evaluate any potential environmental effects of using nanomaterials in large-scale wastewater treatment applications, more study is nonetheless required.

6.4 Hybrid System Method

Different treatment techniques or materials are combined in hybrid system methods for the removal of Cd from wastewater to increase the total removal efficiency. To remove Cd from wastewater, these systems combine membrane filtration with adsorption techniques. Cd ions are attracted by adsorption to an adsorbent material, after which the wastewater is further purified by passing through a membrane. Utilizing a polymeric ultrafiltration membrane and activated carbon as an adsorbent, Giri et al. [39] created a hybrid system. The activated carbon sorbs the Cd ions, and the membrane provides additional separation. These systems utilize both membrane processes and catalytic oxidation or reduction to get rid of Cd from wastewater. Through a combination of manganese dioxide catalytic oxidation and adsorption on the ceramic membrane, Li et al. [67] successfully removed Cd ions from the solution. These systems combine membrane filtration and electrocoagulation, which is the process of creating coagulant species by applying an electric current. Cd ions are destabilized and aggregated by electrocoagulation before being eliminated by the membrane [65, 72, 74].

When compared to individual treatment processes, hybrid system approaches have better synergistic effects and elimination efficiency. Hybrid systems are chosen and optimized based on a variety of variables, including wastewater properties, the required level of removal efficiency and financial constraints.

6.5 Chelation

Chelation is a process in which a complex is created between a metal ion and a chelating agent, a molecule that can attach to the metal ion. Heavy metal intoxication, such as lead or mercury poisoning, is frequently treated in medicine with chelation therapy. However, its use in the removal of Cd from wastewater is restricted. The environment and human health may be negatively impacted by the poisonous heavy metal Cd. The removal of Cd ions from the water is the main objective of wastewater treatment to halt their release into water bodies or contamination of drinking water sources. Chelating substances like ethylenediaminetetraacetic acid (EDTA), diethylenetriaminepentaacetic acid (DTPA), and citric acid can create stable complexes with Cd ions, making it easier to remove them from water (Fig. 3). A chelating agent is typically added to the wastewater, given time to interact with the Cd ions, and then the complex that results is separated from the water. The use of EDTA chelating agent had significantly decreased Cd concentrations at operating condition of pH 6.0, chelating agent concentration of 5 mmol/L, and contact period of 60 min [22, 26, 106, 146]. In another study, citric acid, a typical chelating agent, had demonstrated good outcomes in the removal of Cd from wastewater [125, 128, 129].

Cd removal from wastewater has been successfully accomplished by chelation. It has many benefits, including being very efficient, economical, and able to handle enormous amounts of contaminated water. When using chelation to remove Cd, there are a few things to keep in mind, though. According to the unique properties of the wastewater, for instance, the selection of the chelating agent and its concentration should be carefully considered. The chelated Cd complexes must also be properly disposed of to prevent any negative effects on the environment.

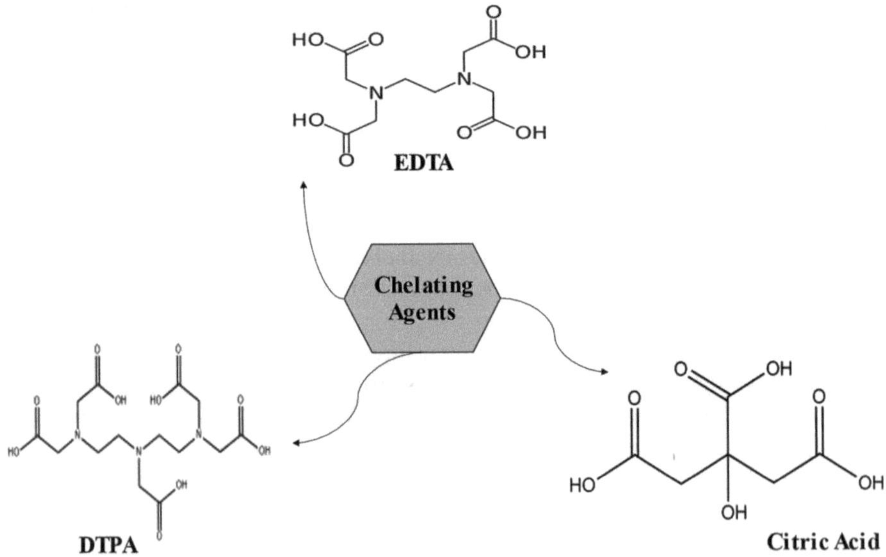

Fig. 3 Chelating agents (such as EDTA, DTPA and Citric acid) are being for treatment of Cd-contaminated wastewater

6.6 Solvent Extraction

The removal of heavy metals from wastewater, including Cd, is commonly accomplished using solvent extraction. Cadmium is a hazardous element that is frequently present in industrial wastewater. It must be removed or treated prior to disposed of, to prevent environmental contamination and safeguard public health. Effective Cd removal depends on the selection of the right extractant. Organic substances including di(2-ethylhexyl) phosphoric acid (D2EHPA), tributyl phosphate (TBP), and organophosphorus compounds are frequently utilized as Cd extractants. The organic phase is produced by dissolving the chosen extractant in an organic solvent. The wastewater compatibility, stability, and solubility of the solvent are all important considerations. To enable the transfer of Cd ions from the aqueous phase to the organic phase, the organic phase and the polluted wastewater are fully mixed. Usually, mechanical agitation or other mixing techniques are used to accomplish this. The mixture is allowed to settle after enough contact time, which makes it possible to separate the organic phase which contains the extracted Cd from the aqueous phase. Through settling, decantation, or centrifugation, phase separation can be accomplished. A stripping procedure is used to remove the Cd ions from the organic phase that is loaded with Cd and transfer them to a stripping solution, which affect the subsequent treatment procedures. The Cd-containing stripping solution is then properly processed to recover the metal for recycling or to ensure it is disposed of safely to prevent environmental contamination.

Organic solvent (D2EHPA) utilized to remove Cd from wastewater [153]. To obtain effective Cd extraction, the researchers adjusted several variables, including pH, extractant concentration, and contact time. They discovered that D2EHPA displayed high Cd ion extraction efficiency, with removal rates exceeding 95% under ideal circumstances. Liu et al. [78] examined the usage of Cyanex 272 and TBP (tri-n-butyl phosphate) as an extractant and efficiently remove Cd (up to 97%) from wastewater. The phenoxy-amino ligands 2-(((2-(diethylamino) ethyl) amino) methyl) phenol (L1) and 2-((2-mercaptoethyl) amino) methyl)-phenol (L2) were studied as model chelating agents for liquid–liquid extraction of Cd (II) cations from water using a dichloromethane-water biphasic system in a study conducted by Nyamato et al. [38]. For Cd (II), the binding effectiveness of the ligands L1 and L2 ranged from 76 to 92%.

Water contaminated with Cd can be treated effectively and economically by solvent extraction. It provides great selectivity, efficient Cd ion removal, and the capacity to produce low residual concentrations. To achieve efficient and economical operation, the process needs to have several parameters, including solvent concentration, pH, temperature, and contact time, carefully optimized. Additionally, the precise content and features of the wastewater play a role in the choice of an appropriate solvent and stripping agent.

6.7 Foam Fractionation

Cadmium removal from wastewater can be accomplished using the separation method known as foam fractionation. It is founded on the idea of selectively removing and concentrating a target substance—in this example, Cd ions from a liquid solution using a foam. The wastewater solution containing Cd ions is given a foam-generating substance. This ingredient may be a surfactant or foaming agent that aids in the production of stable foam. Through mechanical agitation or aeration, air or gas is often injected into the wastewater solution. On the liquid surface, a foam layer is produced by this process. Cd ions in wastewater solutions have a propensity to separate into liquid and foam phases. The Cd ions in the liquid are preferentially taken up by the foam, which functions as a selective adsorbent. The collected Cd ions-containing foam is separated from the liquid phase. This can be accomplished by employing strategies like overflow weirs, skimming, or vacuum-based systems. The concentrated Cd ions are subsequently released from the foam because of a mechanism that causes it to become unstable or collapse. A chemical agent that weakens the foam structure may be added, the pH may be changed, or the temperature may be raised.

Application of foam fractionation in industrial effluent significantly reduced the quantity of Cd in solution with the removal efficiency of above 90% under ideal circumstances [115], good results were also obtained of using this technique for the treatment of Cd-contaminated agricultural runoff [53], combination of foam fractionation in conjunction with nanofiltration effective for the removal of Cd from electroplating wastewater [19, 23–25]. Foam fractionation is a method that shows promise for treating wastewater that has been contaminated with Cd. It can be integrated into current treatment processes and delivers significant removal efficiency, particularly for diluted solutions. Numerous variables, such as pH, foam stability, foam formation method, surfactant choice, and process parameters, affect the efficiency of foam fractionation for Cd removal. To maximize the removal efficiency and reduce the loss of priceless metals throughout the process, it is essential to optimize these parameters. Furthermore, organic debris, oil, or surfactants in the wastewater might alter foam stability and lessen the process efficiency.

6.8 Cryogels

Cryogels are three-dimensional cross-linked structures made by reacting monomer or polymer solutions at low temperatures below the solvent's boiling point (Fig. 4). The gel system is considered a potential scaffold material for many applications because of the cryogel system's unique composition of highly porous properties, adequate osmotic stability, and mechanical strength. A procedure termed cryogelation is used to create cryogels, which can either be chemically or physically crosslinked. The procedure starts by creating a reaction mixture that includes monomers/precursors

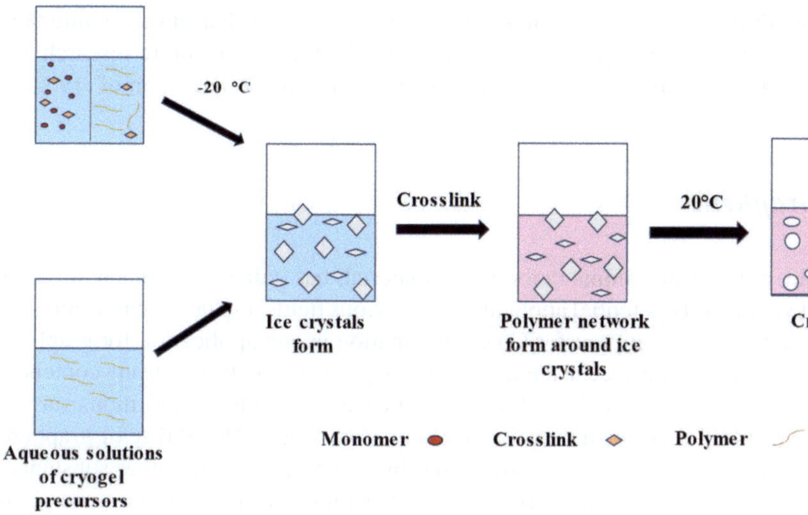

Fig. 4 Formation of cryogel for Cd treatment

for small molecules, precursors for polymers, or a combination of both, along with a solvent like water. The reagents typically make up only 5–20% of the mixture on average. An initiator may occasionally be introduced to kickstart the polymerization of the monomers or small molecule precursors. Thereafter, the solution is cooled below the freezing point of water, resulting in the formation of ice crystals. Within the structure, the water-based solvent hardens and forms solid solvent porogens. For the final cryogel's porous structure, these porogens serve as blueprints.

The removal of Cd ions was accomplished using cryogels and polymeric beads made of cysteine [47]. A functional monomer based on cysteine was used in two different molar ratios (1:1 and 2:1), with the greatest adsorption amounts of 76.35 and 98.8 mol/g, respectively. In addition, it was revealed that the cryogels had a 98.8% adsorption capability for Cd removal from synthetic wastewater. Cryogels made of ion-imprinted polymers are intriguing adsorbents in which *N*-methacryloyl-L-cysteine methyl ester (MAC) served as a functional monomer and 2-hydroxyethyl methacrylate (HEMA) served as a monomer. The cryogel network was then infused with Poly (HEMA-MAC) particles. According to their findings, these cryogels exhibited a high removal capacity and affinity for Cd ions (32.15 g/g of polymer) [118]. In another study, Baimenov et al. [14] used free-radical co-polymerization of acrylate-based precursors (methacrylic acid and 2-acrylamido-2-methyl-1-propansulfonic acid) with allylamine at various ratios to create amphoteric cryogels. Maximum sorption capacity was found to be between 132 and 249 mg/g according to equilibrium experiments. Cryogels have a lot of potential for removing Cd from wastewater. They offer a viable alternative to traditional treatment techniques because of their high adsorption capacity, selective adsorption capabilities, reusability, quick kinetics, and compatibility with a variety of wastewater sources. As well as investigating their scalability and long-term performance in practical applications, more research and development is required to improve the synthesis and functionalization of cryogels.

6.9 Graphene

Graphene, a two-dimensional carbon nanomaterial, has gained significant attention in recent years for its potential applications in various fields, including environmental remediation. Water treatment for Cd contamination is one application for graphene that holds promise. The special characteristics of graphene make it a strong contender for Cd elimination, such as lots of adsorption capacity due to its enormous surface area, makes it efficient for removal of Cd ions from water. The ability of graphene to absorb more substances can be increased by combining it with other substances to create composite structures. To optimize Cd removal and produce synergistic effects, these composites may also contain other nanomaterials, such as metal oxides, polymers, or carbon-based materials. It is available in various forms such as pristine graphene, graphene oxide and reduced graphene oxide.

According to a study by Mahmoudi et al. [32], graphene oxide has a very high capacity as an adsorbent because of its functional groups and interactions with other substances. However, by utilizing the synergistic effects of mixed-matrix composite, the adsorption capacity of graphene oxide can be further increased. The removal of Cd (II) from aqueous solutions in this study was accomplished using silica-decorated graphene oxide (SGO), a highly effective adsorbent. According to Freundlich isotherms, the greatest SGO adsorbent capacities for the efficient removal of Cd (II) were 43.45 mg/g. The amino-functionalized graphene oxide (GO-NH_2) was created by attaching (3-aminopropyl) triethoxysilane to the surface of graphene oxide (GO). The GO-NH_2, which has a large surface area and many active sites, is effective at absorbing Cd ions. The high Cd adsorption capability of GO-NH_2 (10.04 mg/g) suggests that this treatment method has a lot of potential for removing heavy metals from industrial wastewater [44]. In this regard, Ahmad et al. [5] reported high adsorption capacity (172–189 mg/g) of zeolitic imidazolate framework-L with graphene oxide for Cd removal from wastewater. The kinetic study revealed electrostatic attraction and chemisorption processes involved for Cd removal from wastewater, which conferred this adsorbent as good material to treat Cd-contaminated wastewater.

The treatment of wastewater contaminated with Cd has a lot of potential for graphene-based materials. Their excellent adsorption capacity, chemical stability, regenerability, selectivity, and scalability make them desirable candidates for efficient and long-lasting Cd removal. It is crucial to remember that additional study and development are still required to improve the synthesis processes, comprehend the mechanisms underlying Cd adsorption, and determine the long-term environmental effects of materials based on graphene. Overall, Cd pollution is a problem that can be solved and its negative impacts on the environment and human health can be mitigated using graphene-based materials.

6.10 Carbon Nanotubes

For the treatment of Cd-contaminated wastewater, carbon nanotubes (CNTs) have demonstrated remarkable promise. CNTs have special qualities that make them efficient at removing Cd ions from water, including their large surface area, strong mechanical strength, and excellent adsorption capacity. Due to these features, a significant amount of Cd ions can be absorbed, that eventually lowering the concentration of Cd ions in the effluent. The negatively charged CNTs and the positively charged Cd ions interact electrostatically, resulting in the adsorption process. The CNTs are regenerable and reusable after becoming saturated with Cd ions. The Cd ions can be removed from the CNTs using a variety of techniques, including altering the pH, temperature, or applying chemical agents. Cd ions that have been desorbed can be appropriately processed or disposed of. CNTs can be added to nanocomposites or functionalized with groups or coatings to increase their capacity for adsorption. For instance, adding amino, carboxyl, or thiol groups to CNTs can improve their affinity for Cd ions. The effectiveness and selectivity of Cd removal can be enhanced by these changes.

To remove Cd from industrial wastewater, Titus et al. [121] used nickel nanoparticles (NiNPs) supported on activated multi-walled carbon nanotubes (MWCNTs) as an adsorbent. The hydrophilic surface of MWCNTs-KOH was improved by the addition of NiNPs, leading to an increase in surface area, functional groups, and pore dispersion. The modified CNTs showed comparatively increased Cd adsorption capacities i.e. NiNPs (40.5%), MWCNTs-KOH (50.7%), and MWCNTs-KOH NiNPs (80.6%). The highest Cd (II) adsorption capacity attained using MWCNTs-KOH@NiNPs was 415.8 mg/g. In another study, Fard et al. [34] used ethylenediamine-surface-modified single-walled carbon nanotubes (EDA-SWCNTs) for the adsorption of Cd in wastewater. They discovered that at optimum Cd starting concentration, CNT dosage, duration, and pH the greatest adsorption capacity for Cd ions was 93.47%. The surface functionalization of CNTs with certain groups, such as carboxyl or amino, dramatically increased their ability to bind Cd [66, 69, 71]. The successful use of CNTs for Cd removal is dependent on several variables, including the pH of the wastewater, the length of the contact time, and the presence of other coexisting ions. Additionally, the possible toxicity of CNTs themselves as well as their effect on the environment must be considered. To ensure the safe and effective application of CNTs for Cd remediation in wastewater, more study and optimization are thus required.

6.11 Activated Carbon

Activated carbon is a common adsorbent for the treatment of heavy metals from wastewater, such as Cd because of its large surface area and high adsorption capacity. Selecting an appropriate raw material for the manufacturing of activated carbon is the first stage. Shells from coconuts, wood, peat, and coal are typical raw materials. The availability, pricing, and required qualities of the activated carbon influence the choice of raw material. To eliminate volatile chemicals and transform them into carbon, the raw material is heated in an oxygen-free environment. It is also known as pyrolysis or carbonization. Activated carbon's desired properties and the type of raw material determine the carbonization temperature and time. The carbonized material is then activated to produce a structure that is very porous. Activation is achieved by two methods which are chemical activation and physical activation. The carbonized material is impregnated with substances like potassium hydroxide (KOH) or phosphoric acid (H_3PO_4) during chemical activation. The impregnated material is subsequently heated in a controlled environment to a high temperature, usually between 800–900 °C. As the chemicals and carbon interact, pore structures are formed, and the surface area is increased. In the physical activation process, steam or carbon dioxide are present while the carbonized material is heated to a high temperature, usually exceeding 800 °C. Because of the high temperature and gas flow, the material expands and develops pores. The activated carbon is then rinsed with water to eliminate any impurities or leftover compounds after activation. After that, moisture is removed by drying. Activated carbon is typically milled or crushed to get the required particle size distribution. Different particle sizes could be needed for different purposes. Finally, the activated carbon is packed and made ready for usage. The activated carbon that has been created can be utilized in several ways to remove Cd from wastewater. It can be utilized in batches or continuous stirred-tank reactors as a powdered form, or as a granular material in packed-bed filters. When wastewater and activated carbon come into contact, the Cd ions in the water are adsorbed onto the surface of the carbon particles. Following that, the treated water and activated carbon are separated using filtration or another method.

Activated carbon with moso and ma bamboo efficiently removed 100% Cd with once activated maa bamboo and 91.7% with twice activated maa bamboo. However, the removal efficiency decreased for once activated moso bamboo and twice activated moso bamboo by 20–77% [86]. The regeneration of activated carbon that had been exposed to Cd was examined using several desorption methods [60]. They discovered that the recovery of Cd and the restoration of the adsorption capacity of activated carbon by thermal desorption and chemical regeneration processes made this material a viable and affordable option for the long-term treatment of wastewater. This study clarifies the viability of reusing activated carbon, which lowers overall operating costs and has a smaller negative impact on the environment. For the treatment of Cd-contaminated wastewater, activated carbon is a popular and extremely effective material. Its outstanding adsorption capabilities and renderability make it the best option for removing Cd ions from wastewater, improving water quality, and preserving the environment.

6.12 Zeolite

Minerals with a distinct crystalline structure, zeolites, can be created artificially or naturally. The ion-exchange and adsorption processes result in a high affinity for heavy metals like Cd. Clinoptilolite, chabazite, and mordenite are a few of the zeolites that are frequently employed to remove heavy metals because of strong affinity for Cd ions. Prior to use, the zeolite should be adequately prepared. The zeolite is typically crushed into small particles, and any pollutants or impurities are normally adsorbed on its surface. This can be accomplished using a variety of techniques, including adding zeolite directly to the wastewater or circulating the wastewater through a bed or column containing zeolite particles. The zeolite particle's surfaces pick up the Cd ions that are present in the wastewater. Cd can be absorbed at many different places thanks to the porous structure and high surface area of zeolites. Additionally, zeolites have the capacity for ion exchange, which enables them to swap Cd ions for other cations like sodium or calcium. This characteristic makes it possible to effectively remove Cd from wastewater. The Cd-loaded zeolite particles can be extracted from the cleaned wastewater following the adsorption procedure. Filtration or other separation methods can accomplish this. For numerous Cd removal cycles, the zeolite can be renewed and used. Using a suitable solution to desorb the Cd ions from the zeolite is a common step in regeneration techniques.

The potential of these composite nanofibers for Cd was then examined by Rad et al. [101] after synthesizing NaX nano zeolite using a microwave heating method and creating polyvinyl acetate polymer/NaX nanocomposite nanofibers by electrospinning. With 80% removal at an ideal pH of 5, the maximum reported adsorption capacity was 838.7 mg/g. To create a net negative charge, Choi et al. [29] modified zeolite by substituting the Si (IV) and Al (III) sites in the lattice with exchangeable cations like sodium, magnesium, potassium, or calcium. In comparison to unmodified adsorbents, zeolite modified with magnesium has some advantages, including non-toxicity, low cost, abundance (and thus availability), and large pore sizes of 40–50 nm. At an ideal pH of 7, this Mg-modified adsorbent can remove more than 98% of Cd. The transformed fly ash into an amorphous aluminosilicate zeolite adsorbent, reporting a maximum adsorption capacity of 26.246 mg/g with 84% Cd removal at an ideal pH of 5 [49], whereas Visa [123] transformed fly ash into zeolite through a hydrothermal procedure utilizing sodium hydroxide for the removal of Cd. This product exhibits more than 80% Cd elimination at an ideal pH of 7–8. It is crucial to keep in mind that the efficacy of zeolites for removing Cd may rely on elements including pH, contact time, starting Cd concentration, and the presence of other competing ions in the effluent. When assessing the viability of zeolite-based treatment systems, additional factors including the scope of application, cost, and particular needs of the wastewater treatment process should be considered.

6.13 Microbial Fuel Cells

Heavy metals like Cd and other pollutants can be treated in wastewater using microbial fuel cells (MFCs), a sustainable and effective technique. MFCs produce power and clean the environment at the same time by using the metabolic activity of microorganisms. MFCs are made up of an anode and a cathode compartment that are divided by a selectively permeable membrane (Fig. 5). They have the unique capacity of some microorganisms to oxidize organic materials. Microorganisms that can oxidize organic substances often inhabit the anode, whereas the cathode promotes reduction processes. It is necessary to develop a sufficient microbial population in the anode chamber to handle Cd. Bacteria that can oxidize organic materials and deliver electrons to the anode should be a part of this population. Electric currents can be produced by several bacteria, including *Geobacter sulfurreducens* and *Shewanella* species. Through several MFC methods, Cd can be removed from wastewater. Some bacteria can bind metals, allowing them to generate extracellular polymeric substances (EPS) or adsorb Cd onto their cell surfaces. By using enzymatic or chemical mechanisms, specific microbes can decrease heavy metals, including Cd. These bacteria can help to reduce Cd ions toxicity in solution if there is a enough electron donor present. Cd may precipitate as Cd hydroxide because of pH shifts and the generation of hydroxide ions brought on by electrochemical conditions inside the MFC. This aids in immobilizing the metal, thereby lowering its toxicity and mobility. To maximize MFC performance for Cd treatment, several aspects should be considered. These include changing the pH, temperature, level of organic matter, and electrode materials used during operation. Optimization ensures effective microbial activity and improves Cd removal effectiveness. For assessing the efficiency of the system, it is essential to regularly monitor MFC performance, including current production, Cd content, and microbial community dynamics. Long-term performance is supported by proper maintenance, which includes cleaning the electrodes and replenishing the organic substrates.

A dual-chamber MFC system, with one chamber acting as the anode and the other as the cathode, was created and used for Cd treatment. A mixed microbial culture that can oxidize organic material and reduce Cd ions was introduced into the anode chamber, which had removed up to 86% Cd after 80 days of incubation [64, 68, 70, 73]. In a subsequent work, Cheng et al. [28] investigated the contribution of several electrode materials to improving Cd removal in MFCs. The performance of carbon fabric, graphite, and activated carbon as anode materials was compared. They discovered that activated carbon had a clearance rate of more than 90% for Cd. Liu et al. [81] showed the importance of electrochemical interactions in the removal of Cd by identifying specific microbial species. They clarified that the microbial-driven mechanisms can play important role in the Cd treatment process utilizing MFCs. While MFCs present a potential strategy for Cd treatment, it's critical to remember that their efficacy is dependent on several variables, including the initial Cd concentration, the presence of other co-contaminants, and the makeup of the wastewater matrix. MFCs must be improved for large-scale applications and any potential constraints by doing additional research and development.

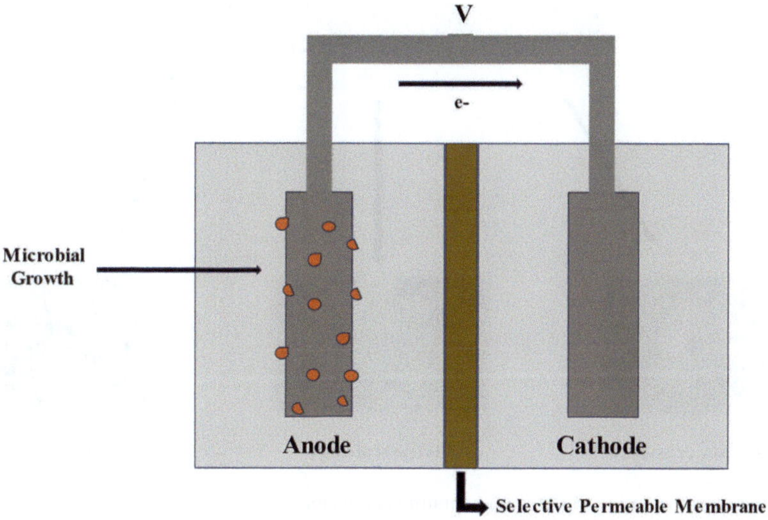

Fig. 5 Schematic diagram showing the working of MFCs for treatment of Cd-contaminated wastewater

6.14 Phytoremediation

Phytoremediation is a sustainable and environmentally friendly approach used for the treatment of Cd-contaminated wastewater. Cd is a toxic heavy metal that poses serious risks to human health and the environment, making its removal from wastewater essential. To lower the content of heavy metals in the water, it entails using plants to ingest, collect, and detoxify chemicals like Cd. Different plant species have been investigated for their potential to remove Cd, and their effectiveness depends on elements such metal uptake effectiveness, tolerance to high Cd levels, and biomass output. The ability of plant species to absorb and withstand high amounts of Cd is carefully considered when choosing plants for phytoremediation. These plants are grown in artificial wetlands, floating mats, or other aquatic systems where they are exposed to contaminated water. These plants absorb the Cd ions from the water as it passes through their root zone and transfers them to their shoots and leaves. When Cd builds up in the tissues of plants, it can be harvested and appropriately disposed of, eliminating the contaminant from wastewater. Some plant species, referred to as hyperaccumulators, have a remarkable capacity to concentrate heavy metals in their tissues, including Cd, making them particularly useful in phytoremediation efforts (Fig. 6).

Pistia stratiotes, *Lemna minor*, *Eichhornia crassipes,* and *Sedum alfredii* have been reported as potential candidates for phytoremediation of Cd-contaminated wastewater [17, 19, 22–26, 105]. According to these findings, these plant species accumulate considerable levels of Cd in their shoots and successfully lower the Cd content in wastewater by up to 80%. In a further study, Gupta et al. [41] reported

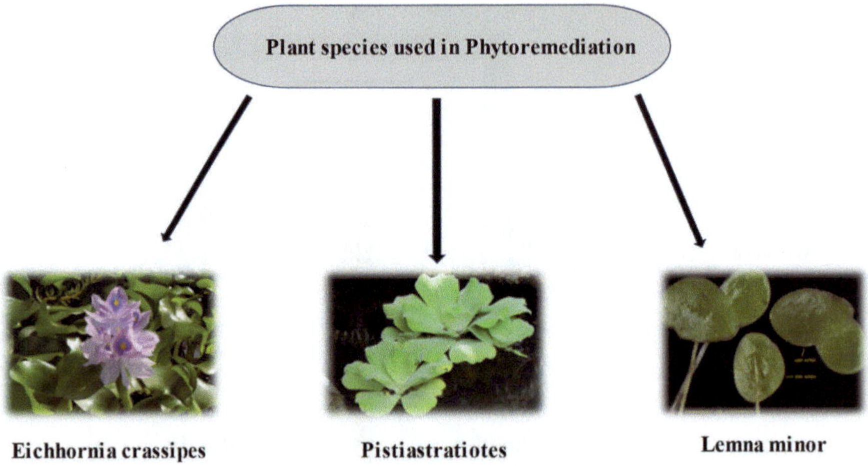

Fig. 6 Phytoremediation of Cd from contaminated water

that employing phytoremediation in conjunction with traditional treatment methods increased the efficacy of removing Cd from the environment compared to using either method by itself. The impact of microbial interactions in boosting phytoremediation effectiveness has also been investigated. Zhuang et al. [155] showed that the bacteria considerably increased the ability of vetiver to remove Cd, indicating the possibility of synergistic interactions between plants and microbes in phytoremediation processes. Water treatment for Cd can be done via phytoremediation, which is environmentally safe and sustainable. The effectiveness of this approach can be greatly increased by choosing the right plant species, taking advantage of microbial interactions, and using genetic engineering approaches. The successful treatment of wastewater contaminated with Cd may benefit from continued study and development in this area.

References

1. Adamczyk, Z., Robak, Z., Dąbrowska, A., & Ochowiak, M. (2019). Cd recovery from industrial waste using electrodialysis. *Materials, 12*(5), 815. https://doi.org/10.3390/ma1205 0815
2. Ahmad, I., Akhtar, M. J., Asghar, H. N., Ghafoor, U., & Shahid, M. (2016). Differential effects of plant growth promoting rhizobacteria on maize growth and cadmium uptake. *Journal of Plant Growth Regulation, 35*, 303–315.
3. Ahmad, I., Akhtar, M. J., Jadoon, I. B. K., Imran, M., & Ali, S. (2017). Equilibrium modeling of cadmium biosorption from aqueous solution by compost. *Environmental Science and Pollution Research, 24*, 5277–5284.
4. Ahmad, I., Akhtar, M. J., Zahir, Z. A., Naveed, M., Mitter, B., & Sessitsch, A. (2014). Cadmium-tolerant bacteria induce metal stress tolerance in cereals. *Environmental Science and Pollution Research, 21*, 11054–11065.

5. Ahmad, S., Salleh, W., Ismail, N., Razali, N., Hamdan, R., & Ismail, A. (2021). Effects of operating parameters on Cd removal for wastewater treatment using zeolitic imidazolate framework L/graphene oxide composite. *Journal of Environmental Chemical Engineering, 9*(5). https://doi.org/10.1016/j.jece.2021.106139
6. Ahmed, M., Bashir, S., Khan, M. N., & Akhtar, S. (2018). Cd accumulation in surface water and wastewater-irrigated vegetables and its potential human health risk. *Environmental Science and Pollution Research, 25*(20), 20068–20077. https://doi.org/10.1007/s11356-018-2016-3
7. Ali, A., Khan, S., & Malik, R. N. (2019). Spatial distribution and contamination assessment of heavy metals in soil and water of industrial city Sialkot, Pakistan. *Human and Ecological Risk Assessment: An International Journal, 25*(5–6), 1262–1284. https://doi.org/10.1080/10807039.2018.1566665
8. Alloway, B. J., & Jackson, A. P. (1991). The behavior of heavy metals in sewage sludgeamended soils. *Science of the Total Environment, 100*, 151–176.
9. Anjum, R., Khan, Z. M., Baig, U., & Siddique, M. (2018). Simultaneous removal of multiple heavy metals from wastewater by chemical precipitation method. *Journal of Environmental Chemical Engineering, 6*(1), 1220–1227.
10. Aoshima, K. (2016). Itai-itai disease: Renal tubular osteomalacia induced by environmental exposure to cadmium—historical review and perspectives. *Soil Science and Plant Nutrition, 62*, 319–326.
11. Arain, M. B., Kazi, T. G., Baig, J. A., Afridi, H. I., Sarajuddin, B. K. D., Panhwar, H., & Arain, S. S. (2015). Co-exposure of arsenic and cadmium through drinking water and tobacco smoking: Risk assessment on kidney dysfunction. *Environmental Science and Pollution Research, 22*, 350–357.
12. Aslam, M., Iqbal, M., Iftikhar, A., & Abbas, F. (2020). Photocatalytic remediation of cadmium-contaminated water using TiO_2 nanoparticles: Optimization, kinetics, and mechanistic study. *Catalysts, 10*(3), 335. https://doi.org/10.3390/catal10030335
13. Avsec, J., Bricelj, M., Kosjek, T., & Heath, E. (2017). Electrodialysis for removal of heavy metal ions and humic acid from different water matrices. *Journal of Membrane Science, 529*, 173–180.
14. Baimenov, A., Berillo, D. A., Poulopoulos, S. G., & Inglezakis, V. J. (2020). A review of cryogels synthesis, characterization, and applications on the removal of heavy metals from aqueous solutions. *Advances in Colloid and Interface Science, 276*, 102088. https://doi.org/10.1016/j.cis.2019.102088
15. Cantinho, P., Matos, M., Trancoso, M. A., & Correia dos Santos, M. M. (2016). Behaviour and fate of metals in urban wastewater treatment plants: A review. *International Journal of Environmental Science and Technology, 13*, 359–386.
16. Chelliapan, S., Wilby, T., & Sallis, P. (2017). Removal of cadmium from industrial wastewater using reverse osmosis. *Desalination, 406*, 1–7.
17. Chen, H. P., Tang, Z., Wang, P., & Zhao, F. J. (2018). Geographical variations of cadmium and arsenic concentrations and arsenic speciation in Chinese rice. *Environmental Pollution, 238*, 482–490.
18. Chen, C., Lu, X., Yang, J., Li, H., Wu, W., & Li, L. (2017). Precipitation of cadmium from wastewater by sulfide as affected by pH and sulfide concentration. *Journal of Chemistry.*
19. Chen, J., Chen, J., Zheng, Y., & Pan, X. (2019). Efficient removal of cadmium(II) from aqueous solutions using a novel modified adsorbent: Equilibrium, kinetics, and thermodynamics studies. *Environmental Science and Pollution Research, 26*(15), 14789–14800.
20. Chen, L., Zhao, R., Ren, X., Wang, X., Wang, Y., Zhang, L., et al. (2020). Immobilization of EDTA on activated carbon for efficient cadmium removal from wastewater. *Journal of Environmental Management, 260*, 110136.
21. Chen, W., Xiao, C., Shi, J., Zheng, X., & Gao, Y. (2017). Chelating agents for heavy metal removal from contaminated water: A review. *Materials, 10*(4), 424.
22. Chen, X., Ding, S., & Dong, S. (2018). Optimization of Cd removal from wastewater by EDTA chelation. *Water Science and Technology, 77*(9), 2306–2314.

23. Chen, X., et al. (2019). Heterogeneous Fenton-like process using nanoscale zero-valent iron (nZVI) for Cd removal from wastewater. *Journal of Environmental Sciences, 84*, 89–98.
24. Chen, X., et al. (2019). Integration of foam fractionation and nanofiltration for efficient removal of Cd from electroplating wastewater. *Separation and Purification Technology, 222*, 146–153.
25. Chen, Y., Han, Y., Cui, S., Sun, Y., Sun, G., & Du, L. (2019). Cd removal from wastewater by three aquatic plants: Pistia stratiotes, Lemna minor, and Eichhornia crassipes. *Water, 11*(6), 1123.
26. Chen, Y., Luo, X., Li, L., Cao, X., Liao, X., & Li, X. (2018). Phytoremediation of Cd-contaminated soil: A review. *Ecotoxicology and Environmental Safety, 144*, 430–440.
27. Chen, Z., Yang, J., & Zhang, L. (2021). Electricity generation and cadmium removal by Rhodoferax ferrireducens in microbial fuel cells: Mechanisms and applications. *Journal of Hazardous Materials, 411*, 125080.
28. Cheng, X., Liu, H., Li, J., Zhang, Z., Zhang, H., & Su, Z. (2021). High removal efficiency of Cd in microbial fuel cells with activated carbon as the anode material. *Environmental Science and Pollution Research, 28*(31), 42939–42949.
29. ChoiH, J., YuS, W., & KimK, H. (2016). Efficient use of Mg-modified zeolite in the treatment of aqueous solution contaminated with heavy metal toxic ions. *Journal of the Taiwan Institute of Chemical Engineers, 63*, 482–489.
30. Cui, Y., & Yue, Q. (2018). Removal of heavy metal ions from wastewater by adsorption technology. *The Open Biotechnology Journal, 12*, 94–106.
31. Daraei, H., Dalvand, A., Mahvi, A. H., & Dehghani, M. H. (2016). Cadmium removal from aqueous solutions by chemical precipitation onto sodium hydroxide. *Desalination and Water Treatment, 57*(25), 11665–11673.
32. Mahmoudi, E., Azizkhani, S., Wahab Mohammad, A., Ng, L. Y., Benamor, A., Ang, W. L., & Ba-Abbad, M. (2020). Simultaneous removal of Congo red and cadmium (II) from aqueous solutions using graphene oxide–silica composite as a multifunctional adsorbent. *Journal of Environmental Sciences, 98*, 151–160. ISSN 1001-0742. https://doi.org/10.1016/j.jes.2020.05.013
33. Al-Khaldi, F. A., Abu-Sharkh, B., Abulkibash, A. M., & Atieh, M. A. (2015). Cadmium removal by activated carbon, carbon nanotubes, carbon nanofibers, and carbon fly ash: A comparative study. *Desalination and Water Treatment, 53*(5), 1417–1429. https://doi.org/10.1080/19443994.2013.847805
34. Fard, E. M., Parvareh, A., & Moravaji, M. K. (2022). Optimization of removal of lead and cadmium from industrial wastewater by ethylenediamine-modified single-walled carbon nanotubes. *International Journal of Environmental Science and Technology, 19*, 2747–2760.
35. Feng, L., Wang, M., & Xu, S. (2019). A comprehensive review of removal of heavy metals by acid mine drainage treatment. *Journal of Cleaner Production, 215*, 386–403.
36. Gao, Z., Wang, X., & Yang, H. (2017). Removal of heavy metals from aqueous solution using graphene oxide nanosheets: A review. *Frontiers of Environmental Science & Engineering, 11*(4), 8.
37. Genchi, G., et al. (2020). Nickel: Human health and environmental toxicology.
38. Nyamato, G. S., Wambugu, K., Kiratu, J., & Ojwach, S. O. (2022). Liquid-liquid extraction of copper (II), zinc (II), cadmium (II), and lead (II) from aqueous solution and sewage effluent using phenoxy-amino ligands. *Water Science and Technology, 85*(10), 2993–3001. https://doi.org/10.2166/wst.2022.164
39. Giri, B., et al. (2021). Adsorption–membrane hybrid system for effective removal of Cd ions from wastewater. *Environmental Technology & Innovation, 22*, 101413.
40. Gupta, A., & Babu, B. V. (2018). A review on adsorption of heavy metals from aqueous solutions by modified clay minerals. *Environmental Science and Pollution Research, 25*(1), 1003–1024.
41. Gupta, S., Kaithwas, G., & Agrawal, S. (2020). Phytoremediation of heavy metals from industrial wastewater: A review on emerging technologies. *Journal of Environmental Management, 263*, 110396.

42. He, Y., Wang, Z., & Guo, L. (2018). Bioremediation of Cd(II)-polluted water by Microbial Fuel Cell with simultaneous recovery of bioelectricity. *Journal of Hazardous Materials, 343*, 159–167.
43. Huang, H., Liu, Y., Liu, J., Cai, J., & Wang, X. (2015). Removal of cadmium from industrial wastewater via a chemical precipitation method. *Desalination and Water Treatment, 56*(11), 3071–3077.
44. Huang, H., Wang, Y., Zhang, Y., Niu, Z., & Li, X. (2020). Amino-functionalized graphene oxide for Cr(VI), Cu(II), Pb(II) and Cd(II) removal from industrial wastewater. *Open Chemistry, 18*(1), 97–107. https://doi.org/10.1515/chem-2020-0009
45. Huang, K., Zeng, Q., Xu, J., Luo, J., Zhang, L., & Chen, H. (2019). Cadmium removal from wastewater using nanofiltration membranes: Mechanisms and influencing factors. *Journal of Membrane Science, 573*, 530–539.
46. Hui, K., Sun, S., & Hui, K. (2017). Recent progress in the use of zeolite-based materials as adsorbents for environmental protection. *Environmental Science and Pollution Research, 24*(3), 2222–2238.
47. Huseynli, S., Bakhshpour, M., Qureshi, T., Andac, M., & Denizli, A. (2020). Composite polymeric cryogel cartridges for selective removal of Cd ions from aqueous solutions. *Polymers (Basel), 12*, 1–15. https://doi.org/10.3390/polym12051149
48. Jafari, S., Kargari, A., & Rezakazemi, M. (2018). Application of microfiltration process for heavy metal removal from industrial wastewater. *Journal of Environmental Chemical Engineering, 6*(1), 1254–1262. https://doi.org/10.1016/j.jece.2018.01.029
49. Javadian, H., Ghorbani, F., Tayebi, H. A., & Asl, S. H. (2015). Study of the adsorption of Cd (II) from aqueous solution using zeolite-based geopolymer, synthesized from coal fly ash; kinetic, isotherm and thermodynamic studies. *Arabian Journal of Chemistry, 8*(6), 837–849.
50. Jiang, J., & Xu, R. (2019). Cadmium accumulation and growth of water lettuce under the conditions of various Cd concentrations and soil water contents. *Chemosphere, 224*, 625–631.
51. Jiang, L., Han, Y., & Zhao, Y. (2016). Carbon nanotubes in heavy metal removal from wastewater: A review. *Chemical Engineering Journal, 306*, 1116–1137.
52. Johnson, B., Thompson, C., Davis, M., & Brown, K. (2020a). Impacts of Cd pollution on aquatic ecosystems: A review. *Environmental Science and Pollution Research, 27*(18), 22165–22178. https://doi.org/10.1007/s11356-020-08876-5
53. Johnson, C. D., et al. (2020b). Pilot-scale evaluation of foam fractionation for the removal of Cd from agricultural runoff. Journal of Hazardous Materials, 396, 122693
54. Khan, S., Khan, M. A., Khan, A., Alam, M., & Khan, M. S. (2020). Cd contamination in soil and wheat crop in the vicinity of tanneries and its potential health risk. *Environmental Monitoring and Assessment, 192*(2), 88. https://doi.org/10.1007/s10661-020-8066-7
55. Khan, M. A., Khan, S., Khan, A., & Alam, M. (2017). Soil contamination with cadmium, consequences and remediation using organic amendments. *Science of the Total Environment, 601*, 1591–1605.
56. Kim, H., & Park, Y. (2017). Recent advances in membrane technologies for water treatment and reuse. *Environmental Engineering Research, 22*(2), 131–139.
57. Kim, J., Phuntsho, S., Kim, H., & Shon, H. K. (2017). Cadmium removal by reverse osmosis: Process behavior and membrane fouling. *Journal of Hazardous Materials, 332*, 125–132.
58. Kowalska, I., Konieczny, K., Koter, S., & Yel, E. (2017). Electrodialysis for the recovery of cadmium from industrial wastewater. *Separation Science and Technology, 52*(8), 1419–1425. https://doi.org/10.1080/01496395.2017.1282882
59. Kumar, A., et al. (2021). Photocatalytic removal of Cd from wastewater using TiO_2 nanoparticles. *Environmental Technology & Innovation, 21*, 101309.
60. Kumar, S., Deepika, K., Kothiyal, N. C., & Saravanan, P. (2020). Regeneration of cadmium-loaded activated carbon: A study on the recovery of adsorption capacity for sustainable wastewater treatment. *Journal of Environmental Management, 255*, 109886.
61. Lee, J., et al. (2019). Microfiltration for the removal of cadmium ions from synthetic wastewater using polymeric membranes. *Separation and Purification Technology, 215*, 186–194.

62. Lee, S., Choi, K., Choi, S., Kim, J., & Kim, J. (2017). Application of ultrafiltration for the treatment of cadmium-contained wastewater in the metal finishing industry. *Separation and Purification Technology, 172*, 205–212.
63. Lee, S., Kim, J., Kim, H., Hong, S., Lee, S., Hong, S., et al. (2019). Nanofiltration membranes with uniform and vertically oriented cylindrical pores for highly efficient water purification. *Nature Communications, 10*(1), 1–9.
64. Li, H., Wang, J., Chen, J., & Ma, Y. (2019). Zeolite-based materials for adsorption of heavy metals from wastewater: A review. *Frontiers of Environmental Science & Engineering, 13*(5), 1–18.
65. Li, H., Ye, Y., Li, Y., & Dong, B. (2016). Application of nanofiltration in treating wastewater containing cadmium from a metal plating facility. *Desalination and Water Treatment, 57*(28), 13090–13096.
66. Li, J., Wang, Y., Zhang, H., Zhang, Y., & Chen, L. (2020). Advanced technologies for Cd removal from industrial wastewater: Objectives, challenges, and perspectives. *Chemosphere, 244*, 125526. https://doi.org/10.1016/j.chemosphere.2019.125526
67. Li, N., et al. (2018). Ceramic membrane assisted by manganese dioxide for efficient Cd removal. *Separation and Purification Technology, 200*, 58–65.
68. Li, X., Zhao, Y., Wang, J., Zhao, L., & Ma, J. (2019). Ceramic ultrafiltration membrane for efficient cadmium removal from wastewater: Performance and mechanisms. *Chemosphere, 233*, 466–473.
69. Li, X., Zhao, Y., Yang, Y., & Gao, H. (2020). Effective adsorption of Cd (II) from wastewater by amino carboxylate modified multi-walled carbon nanotubes. *Chemical Engineering Journal, 382*, 122862. https://doi.org/10.1016/j.cej.2019.122862
70. Li, X., Zhu, N., Wei, L., & Sun, G. (2019). Simultaneous electricity generation and Cd removal using a dual-chamber microbial fuel cell. *Journal of Hazardous Materials, 367*, 318–325.
71. Li, Y., Feng, C., Wang, J., Guo, Y., Li, G., & Li, C. (2020). Highly efficient removal of Cd^{2+} from wastewater by new attapulgite/Fe_3O_4/graphene composite through foam fractionation. *Environmental Science and Pollution Research, 27*(25), 31420–31429.
72. Li, Y., Jin, Y., & Huang, X. (2016). Phytoextraction of cadmium by water hyacinth [*Eichhornia crassipes* (Mart.)]: A review. *International Biodeterioration & Biodegradation, 110*, 212–220.
73. Li, Y., Liu, X., Zhang, Z., & Zhu, H. (2019). Removal of Cd from wastewater using fluidized bed packed with microbeads. *Environmental Science and Pollution Research, 26*(6), 5901–5910.
74. Li, Z., et al. (2016). Electrocoagulation-membrane hybrid system for efficient removal of Cd from wastewater. *Journal of Environmental Chemical Engineering, 4*(4), 3862–3868.
75. Liu, Y. Z., Xiao, T. F., Perkins, R. B., Zhu, J. M., Zhu, Z. J., Xiong, Y., & Ning, Z. P. (2017). Geogenic cadmium pollution and potential health risks, with emphasis on black shale. *Journal of Geochemical Exploration, 176*, 42–49.
76. Liu, B., Zhang, Q., Li, Y., Lu, Y., Li, Y., Li, X., et al. (2019). Electrodialysis removal of cadmium ions from simulated wastewater: Effects of current density and initial cadmium concentration. *Journal of Water Process Engineering, 30*, 100608.
77. Liu, D., Wang, J., & Liu, C. (2020). Removal of cadmium from wastewater using graphene: A review. *Journal of Environmental Management, 258*, 110018.
78. Liu, H., Leng, F., Wang, Q., & Liu, Y. (2019). Extraction of Cd ions from wastewater using a mixture of Cyanex 272 and tri-n-butyl phosphate. *Journal of Chemical Engineering of Japan, 52*(6), 524–530.
79. Liu, X., et al. (2018). Removal of Cd from aqueous solution using MnO_2 nanoparticles loaded on multi-walled carbon nanotubes. *Journal of Environmental Management, 211*, 351–359.
80. Liu, X., Qu, F., Wang, Y., Wu, Z., Cai, H., & Zhou, Q. (2014). Heavy metal removal from wastewater using chemical precipitation and membrane filtration. *Journal of Hazardous Materials, 280*, 36–43.
81. Liu, X., Wang, Y., Li, Z., Li, H., Wang, H., & Huang, Y. (2022). Mechanisms of Cd removal from wastewater using microbial fuel cells. *Environmental Science and Pollution Research, 29*(5), 5432–5442.

82. Liu, X., Zhang, Y., Wang, Q., Liu, Y., & Chen, J. (2019). Microfiltration performance in heavy metal removal from wastewater: A review. *Chemosphere, 225*, 74–84. https://doi.org/10.1016/j.chemosphere.2019.03.189
83. Liu, Y., Lv, B., & Xu, J. (2019). Adsorption and recovery of cadmium from wastewater by hybrid adsorbents: A review. *Chemical Engineering Journal, 358*, 1283–1297.
84. Liu, Y., Zhang, Y., Wang, X., Wang, J., & Wang, S. (2018). Removal of cadmium ions from wastewater by activated carbon derived from agricultural waste. *Journal of Cleaner Production, 172*, 3651–3661.
85. Liu, Y., Zhao, L., Zhang, Y., Quan, X., & Chen, S. (2019). High-efficient cadmium removal from wastewater by ceramic ultrafiltration membrane. *Journal of Hazardous Materials, 367*, 447–455.
86. Lo, F., Wang, S. Y., Tsai, M. J., & Lin, L. D. (2012). Adsorption capacity and removal efficiency of heavy metal ions by Moso and Ma bamboo activated carbons. *Chemical Engineering Research and Design, 90*(9), 1397–1406.
87. Lu, J., Zhang, Y., & Li, Z. (2021). Removal of cadmium from wastewater using carbon nanotubes: A review. *Journal of Hazardous Materials, 402*, 123502.
88. Luo, L., Xu, Y., Xu, X., Chen, L., & Qiu, H. (2019). Removal of cadmium from wastewater using an ethylenediamine-functionalized zeolite. *Journal of Environmental Chemical Engineering, 7*(4), 103192.
89. Ma, F., Li, Y., & Shi, Y. (2018). Adsorption techniques for the removal of heavy metals from wastewater: A review. *Journal of Environmental Management, 217*, 56–70.
90. Manikandan, S., et al. (2020). Ozonation process for the removal of Cd from synthetic wastewater. *Journal of Environmental Chemical Engineering, 8*(5), 104320.
91. McCasland, M., Trautmann, N., Porter, K., & Wagenet, R. (2008). *Nitrate: Health effects in drinking water*. Available online: http://pmep.cee.comell.edu/facts.slides-self/facts/nit-heefgrw85.html. Accessed on July 20, 2020.
92. Miao, L., Zhang, X., Zhao, Y., & Yang, Q. (2018). Removal of cadmium(II) from wastewater using polymer adsorption enhanced advanced fluidized bed reactor. *Chemical Engineering Journal, 351*, 341–351.
93. Mohammadi, M., Mahdavian, M., Ghorbanpour, M., & Saeidi, M. (2021). Cadmium Removal from Wastewater by Precipitation Technique. *Water Environment Research, 93*(9), 1568–1575.
94. Mohanapriya, K., et al. (2018). Electrochemical removal of Cd from wastewater using boron-doped diamond anode. *Journal of Environmental Management, 210*, 357–365.
95. Muthukumaran, C., & Swaminathan, M. (2018). Hybrid photocatalytic membrane system for effective removal of cadmium ions from industrial wastewater. *Journal of Environmental Management, 210*, 353–360. https://doi.org/10.1016/j.jenvman.2017.12.036
96. Nies, D. H. (1999). Microbial heavy-metal resistance. *Applied Microbiology and Biotechnology, 51*, 730–750.
97. Nies, D. H. (2003). Efflux-mediated heavy metal resistance in prokaryotes. *FEMS Microbiology Reviews, 27*, 313–339.
98. Okoh, A. T., Odjadjare, E. E., Igbinosa, E. O., & Osode, A. N. (2007). Wastewater treatment plants as a source of microbial pathogens in receiving water sheds. *African Journal of Biotechnology, 6*, 2932–2944.
99. Park, S., Lee, S., Kang, H., Yang, J., & Lee, H. (2018). Hybrid treatment of metal-containing wastewater using chemical precipitation and microbial reduction. *Water Research, 138*, 232–240.
100. Pontié, M., Lazar, M. D., Mendret, J., & Brosillon, S. (2017). Electrodialysis for desalination of metal containing solutions and recovery of valuable metals and acid. *Water Research, 109*, 1–15.
101. Rad, L. R., Momeni, A., Ghazani, B. F., Irani, M., Mahmoudi, M., & Noghreh, B. (2014). Removal of Ni^{2+} and Cd^{2+} ions from aqueous solutions using electrospun PVA/zeolite nanofibrous adsorbent. *Chemical Engineering Journal, 256*, 119–127.

102. Rahman, M. S., Rahman, M. M., & Rahman, M. M. (2019). A novel hybrid adsorbent composed of bentonite and iron oxide nanoparticles for the removal of cadmium from wastewater. *Journal of Environmental Management, 240*, 302–310.
103. Rajeshwar, K., Osugi, M. E., Chanmanee, W., Chenthamarakshan, C. R., & Mohamed, H. N. (2016). Heterogeneous photocatalytic treatment of organic dyes in air and aqueous media. *Journal of Photochemistry and Photobiology C: Photochemistry Reviews, 31*, 1–35.
104. Ramos, A. M., & Borba, C. E. (2018). An overview on the applications of electrodialysis and electrodialysis reversal for industrial wastewater treatment. *Separation and Purification Technology, 194*, 145–161.
105. Rasool, S., Ahmad, I., Jamal, A., Saeed, M. F., Zakir, A., Abbas, G., Seleiman, M. F., & Caballero-Calvo, A. (2023). Evaluation of phytoremediation potential of an aquatic macrophyte (*Eichhornia crassipes*) in wastewater treatment. *Sustainability, 15*, 11533. https://doi.org/10.3390/su151511533
106. Razi, A. A., et al. (2018). Cd removal from synthetic wastewater using EDTA. *Iranian Journal of Health, Safety, and Environment, 5*(2), 1050–1055.
107. Sani, R. K., & Aminabhavi, T. M. (2018). Chemical precipitation for heavy metal ions removal from wastewater. *Journal of Environmental Chemical Engineering, 6*(1), 192–212.
108. Sebastian, A., & Prasad, M. N. V. (2014). Cadmium minimization in rice. A review. *Agronomy for Sustainable Development, 34*, 155–173.
109. Shahryari, A., Mousavi, M. F., & Karimi, S. (2020). Photocatalytic removal of cadmium ions from aqueous solutions using TiO_2 and immobilized TiO_2. *Environmental Management and Sustainable Development, 9*(3), 935–943. https://doi.org/10.5296/emsd.v9i3.17427
110. Sharma, R. K., Thakur, N., Kumari, A., & Kumar, V. (2019). Low-cost adsorbents for heavy metals uptake from contaminated wastewater: A review. *Journal of Environmental Management, 246*, 347–366.
111. Sharma, Y. C., & Tripathi, M. (2018). Photocatalytic treatment of organic dyes in aqueous media. *Journal of Environmental Chemical Engineering, 6*(2), 2799–2823.
112. Shu, X., Li, J., & Liu, S. (2015). Adsorption behavior of cadmium ions from aqueous solution by activated carbon microspheres. *Chemical Engineering Journal, 260*, 291–298.
113. Singh, A., & Srivastava, V. C. (2019). Treatment of industrial wastewater containing heavy metals using ceramic microfiltration membrane. *International Journal of Environmental Science and Technology, 16*(12), 7587–7596. https://doi.org/10.1007/s13762-019-02305-2
114. Smith, A., Johnson, B., & Williams, C. (2017). Treatment of Cd-contaminated wastewater using a fluidized bed reactor. *Environmental Engineering Research, 25*(3), 286–292.
115. Smith, A. B., et al. (2018). Foam fractionation for the removal of Cd from industrial wastewater. *Environmental Science and Technology, 52*(10), 5891–5898.
116. Smith, J., Johnson, R., Anderson, L., & Martinez, E. (2021). Global assessment of Cd pollution: Sources, impacts, and mitigation strategies. *Environmental Pollution, 250*, 123–135. https://doi.org/10.1016/j.envpol.2021.112345
117. Song, M., Jin, J., Guo, Z., Liu, Y., Cao, H., & Liu, X. (2016). Hybrid removal of heavy metals from wastewater by magnetic separation and electrocoagulation. *Chemical Engineering Journal, 289*, 465–473.
118. Tabakli, B., Topçu, A. A., Döker, S., & Uzun, L. (2015). Particle-assisted ionimprinted cryogels for selective Cd II ion removal. *Industrial and Engineering Chemistry Research, 54*, 1816–1823. https://doi.org/10.1021/ie504312e
119. Tang, W., Cui, Y., Wang, Y., Xing, L., Du, J., & Wang, H. (2018). Electrochemical removal of cadmium using a graphene-modified electrode. *Water Science and Technology, 77*(7–8), 1973–1981.
120. Tian, F., Zhang, J., Zheng, M., Yan, W., Qiao, L., Yu, H., & Chen, Z. (2016). Photocatalytic degradation of organic pollutants over TiO_2 based photocatalysts: A review. *Journal of Environmental Sciences, 51*, 90–109.
121. Titus, C. E., Michael, C. E., Jimoh, O. T., Saheed, M., Ambali, S. A., Abdulsalami, S. K., Vida, K., Andrei, V., Michal, W., & Grzegorz, L. (2022). Activated multi-walled carbon nanotubes decorated with zero valent nickel nanoparticles for arsenic, cadmium, and lead adsorption

from wastewater in a batch and continuous flow modes. *Journal of Hazardous Materials, 423*, Part B, 126993. ISSN 0304-3894.
122. Tran, H. N., You, S. J., & Chao, H. P. (2016). Precipitation of cadmium from wastewater: A review. *Journal of Environmental Management, 166*, 387–399.
123. Tytła, M. (2019). Assessment of heavy metal pollution and potential ecological risk in sewage sludge from municipal wastewater treatment plant located in the most industrialized region in Poland—Case Study. *International Journal of Environmental Research and Public Health, 16*, 2430.
124. Visa. (2016). Synthesis and characterization of new zeolite materials obtained from Fly ash for heavy metals removal in advanced wastewater treatment. *Powder Technology, 294*, 338–347.
125. Wang, D., Yin, S., Zhang, J., & Li, X. (2020). Extraction of Cd from wastewater using ionic liquids. *Journal of Environmental Chemical Engineering, 8*(2), 103728.
126. Wang, J., Smith, A. L., Johnson, L. A., & Chen, V. (2019). Removal of heavy metals from wastewater using ceramic microfiltration membranes. *Journal of Environmental Engineering, 145*(10), 04019048. https://doi.org/10.1061/(ASCE)EE.1943-7870.0001591
127. Wang, L., Wang, X., & Zhang, L. (2018). Precipitation of cadmium from industrial wastewater by a hydroxide precipitation method. *Separation Science and Technology, 53*(17), 2780–2790.
128. Wang, W., Zhang, L., & Chen, X. (2020). Removal of cadmium (II) ions from aqueous solutions using modified zeolites: A review. *Journal of Hazardous Materials, 386*, 121880.
129. Wang, X., et al. (2020). Chelation characteristics of citric acid for Cd removal from wastewater. *Environmental Science and Pollution Research, 27*(11), 11551–11560.
130. Wang, Y., Wang, H., Zhao, J., & Li, D. (2021). Comparative study on removal of cadmium from wastewater by three advanced oxidation processes: H_2O_2/UV, H_2O_2/Fe^{2+}/UV and H_2O_2/Fe^{2+}. *Journal of Hazardous Materials, 402*, 123798.
131. Wang, Y., Wu, X., Wu, Z., & Huang, Q. (2019). Membrane filtration for water and wastewater treatment: A comprehensive review. *Chemical Engineering Journal, 378*, 122222.
132. Wang, Y., Wu, X., Wu, Z., & Huang, Q. (2019). Microfiltration membranes for heavy metal removal from wastewater: A review. *Journal of Cleaner Production, 213*, 149–159.
133. Wu, K., et al. (2019). Adsorption and reduction of Cd ions in aqueous solution using nanoscale zero-valent iron. *Journal of Nanoscience and Nanotechnology, 19*(4), 2293–2299.
134. Xu, C., Chen, X., Hu, J., Chen, G., & Xing, W. (2017). Photocatalytic removal of cadmium from aqueous solution by titanium dioxide modified with sodium dodecyl sulfate. *Journal of Water Process Engineering, 18*, 46–55. https://doi.org/10.1016/j.jwpe.2017.02.010
135. Xu, J., Li, J., Wu, Y., Zhang, H., & Wang, J. (2019). Nanomaterials in wastewater treatment: An overview. *Chemical Engineering Journal, 359*, 898–906.
136. Yang, Y. J., Xiong, J., Chen, R. J., Fu, G. F., Chen, T. T., & Tao, L. X. (2016). Excessive nitrate enhances cadmium (Cd) uptake by up regulating the expression of OsIRT1 in rice (*Oryza sativa*). *Environmental and Experimental Botany, 122*, 141–149.
137. Yang, C., et al. (2018). Efficient removal of cadmium ions from wastewater by microfiltration using a ceramic membrane. *Journal of Membrane Science, 563*, 148–157.
138. Yang, X., Cao, Y., Liu, X., Xu, T., Jin, Z., & Jin, X. (2020). Efficient removal of cadmium from wastewater by electrodialysis using functionalized ion-exchange membranes. *Separation and Purification Technology, 235*, 116203.
139. Yang, Y., Tang, C., & Luo, Z. (2018). Current status and future perspective of solvent extraction for the purification of rare earth metals. *Hydrometallurgy, 178*, 147–160.
140. Zhang, C., Chen, X., Xu, W., Li, J., & Ma, J. (2017). Removal of heavy metals from industrial wastewater by reverse osmosis: Optimization using response surface methodology. *Desalination and Water Treatment, 74*, 366–374.
141. Zhang, C., Wang, X., & Li, Y. (2017). Cadmium removal from wastewater using biochar: A review. *Journal of Environmental Management, 186*, 685–694.
142. Zhang, D., & Huang, J. (2021). Efficient removal of cadmium and lead from wastewater using activated carbon derived from sewage sludge by K_2CO_3 activation. *Journal of Environmental Management, 292*, 112877.

143. Zhang, J., Zhu, W., & Ma, F. (2017). Applications of nanotechnology in wastewater treatment—A review. *Chinese Journal of Chemical Engineering, 25*(3), 421–430.
144. Zhang, Q., Zhang, Z., Hu, X., & Jiang, L. (2017). Removal of Cd from wastewater using a fluidized bed reactor with modified zeolite particles. *Journal of Environmental Sciences, 58*, 194–200.
145. Zhang, X., Yang, Q., Zhao, Y., Miao, L., & Wang, L. (2020). Simultaneous adsorption and transformation of Cd(II) from wastewater using ion-exchange resin enhanced advanced fluidized bed reactor. *Chemical Engineering Journal, 381*, 122668.
146. Zhang, Y., et al. (2019). Simultaneous removal of Cd, lead, and copper from industrial wastewater using modified natural adsorbents. *Chemosphere, 227*, 298–306.
147. Zhao, D., Su, J., & Ren, J. (2020). Metal-organic frameworks for highly efficient removal of cadmium ions from water. *Journal of Environmental Chemical Engineering, 8*(1), 103626.
148. Zhao, H., Li, W., & Gao, Y. (2019). High-efficient removal of cadmium from wastewater by advanced oxidation processes. *Journal of Environmental Sciences, 75*, 270–282.
149. Zhao, L., Liu, Y., Quan, X., Chen, S., & Zhang, Y. (2017). Performance and mechanism of a novel nanofiltration-ultrafiltration hybrid process for cadmium removal from industrial wastewater. *Water Research, 126*, 177–185.
150. Zhao, L., Su, Y., & Zheng, M. (2019). Efficient recovery of Cd and Zn from wastewater using cryogenic process coupled with precipitation. *Chemical Engineering Journal, 359*, 1041–1048.
151. Zhao, Y., Liu, X., Li, Y., Zhang, W., & Han, X. (2019). Cadmium removal from groundwater by nanofiltration. *Water Science and Technology: Water Supply, 19*(4), 1205–1213.
152. Zhou, J., et al. (2016). Adsorption of Cd from aqueous solution using Fe_3O_4 nanoparticles as adsorbents. *Water Science and Technology, 73*(3), 569–577.
153. Zhu, J., Zhu, L., Li, J., & Wu, L. (2018). Solvent extraction of Cd (II) from wastewater using di(2-ethylhexyl) phosphoric acid. *Separation Science and Technology, 53*(9), 1425–1435.
154. Zhu, L., Wang, J., Lv, Y., Wang, Y., Wang, S., Sun, Y., et al. (2018). Cadmium removal from wastewater by electrodialysis: Process optimization and mechanism analysis. *Journal of Membrane Science, 549*, 133–142.
155. Zhuang, P., Shu, W., Luo, Y., Cai, P., Li, Z., & Huang, Z. (2020). Combined effects of Cd-resistant bacteria and vetiver on phytoremediation of Cd-contaminated water. *Environmental Science and Pollution Research, 27*(32), 40246–40253.

Cadmium Removal from Aqueous Solutions Using Differents Biosorbents

Ismael Acosta, Adriana Rodríguez, Juan Fernando Cárdenas, Víctor Manuel Martínez, and María Eugenia Torre

Abstract Cadmium is a compound toxic, which can induce adverse effects on the different organs like heart, kidneys, liver, lung, and their exposition has been linked with cancer in humans, hence important remove it of sites contaminated with it. Some methods are used to eliminate this metal from differents contaminated places. The objective of the present work, was analyze the remediation of the heavy metal with nine natural obtained from fungal strains fungi from samples of soil contaminated of zone high-rish, and one of a place near the Universitary zone of San Luis Potosí, S.L.P., México, finding that the biomasses of *Purpureocillim lilacinium, Aspergillus flavus, Penicillium* sp-1, and *Aspergillus terreus,* had the best removal rates of the metal (with dithizone at 518 nm), with a removal between 76 and 99%, in the conditions analyzed. These biomasses are a great alternative for the elimination of this and other contaminants from the different high-risk sites, and in this study, the hability of eliminate this contaminant by environmental polluting fungus strain of *Purpureocillim lilacinium,* was analysed.

Keywords Contamination · Cadmium · Removal · Fungal biomasses

I. Acosta (✉) · M. E. Torre
Laboratorio de Micología Experimental, Facultad de Ciencias Químicas, Universidad Autónoma de San Luís Potosí, San Luís Potosí, México
e-mail: iacosta@uaslp.mx

A. Rodríguez · J. F. Cárdenas
Centro de Investigación y Extensión de La Zona Media El Balandran, Universidad Autónoma de San Luís Potosí, Río Verde, San Luís Potosí, San Luís Potosí, México

V. M. Martínez
Instituto de Ciencias Agropecuarias, Área Académica de Medicina Veterinaria y Zootecnia, Instituto de Ciencias Agrícolas, Universidad Autónoma del Estado de Hidalgo, Hidalgo, México

A. K. Jha and N. Kumar (eds.), *Cadmium Toxicity in Water*, Springer Water,
https://doi.org/10.1007/978-3-031-54005-9_8

1 Introduction

Between the main contaminants and more dangerous of aquifers due to its high toxicity, persistence, and mobility, are found the heavy metals. Not being biodegradable, they are highly polluting, can become toxic to vertebrates and invertebrates, and represent a great danger to the health of living beings in all over the planet. These risks are related with the physicochemical properties of these elements and/or compounds, mostly by its accumulation in plants and its transfer to humans, plants, and animals [36]. Their distribution in the different environments is highly complex and involves different factors, among which are: oxide reduction potential, pH, the presence of organic matter, and other properties, groundwater level and its fluctuations [33], and there are different investigations for carry out to determine the contamination of this chemical elements in the environment, like: lead, cobalt, chromium (VI), mercury, cadmium, and others, for example: Evaluation of contaminants in medlars (*Eriobotrya japonica*) a place from in the urban of Quito, Ecuador [58], the pollution by this type of contaminants in palm residues of the Peruvian Amazon [9], heavy metals bioaccumulation in aquatic animals from Mayabeque, Cuba) [43], in the marine coastal zone in Nuevitas maritime terminal (Cuba) [28], and their presence in Mexican Duck [20]. Too in México, there are some reports like: cadmium contamination in the micro-watershed of the Pixquiac river, Veracruz [42], heavy metals in caudal scales of *Crocodylus moreletii* in the Lacandon jungle (Chiapas) [8], their bioaccumulation in *Cucumber* crops [50], presence of lead, cadmium, and manganese in soils of Tulancingo, Hidalgo [53], the heavy metals pollutions in coastal ecosystems of Southeast [19], and to analyze the levels of some metallic ions in liver, kidney, brain, heart, spleen and muscle of the Golden Cownose Ray *Rhinoptera steindachneri* [12].

Some heavy metals, like Cadmium (Cd), are present in the atmosphere because of natural processes, such as volcanic emissions or erosion of rocks and minerals, as well as industrial applications in metallurgy, burning of fossil fuels, incineration of organic waste, stabilizer of plastics, pigment manufacturing, sewage sludge and for its agricultural use in phosphate-based fertilizers [33]. In water and air, the soluble forms of cadmium are the most mobile, while in soil it is much less mobile and can occur in soluble and insoluble forms, complexing with both organic and inorganic soil constituents. In this way, said metal can contaminate agricultural crops and can accumulate in terrestrial and marine animals [8]. Organisms such as crustaceans and fungi are natural accumulators of cadmium. Therefore, in the last link of the trophic chain, people are exposed to this metal when they eat vegetables and the meat of terrestrial and aquatic animals that contain cadmium [47]. Cadmium maintains properties similar toxicities among all its chemical species, but this does not make it less dangerous. The agency of Environmental Protection of the United States of America established that children should not use or drink water with more than 0.00004 g of cadmium per liter, since after 11 days of exposure they will present reactions adverse [59]. Its final deposition is usually the kidneys, where it accumulates for up to 30 years, which increases the risk of kidney damage [13]. It has been reported that even a small amount of this metal in the blood (0.41 μg/L) can cause renal failure in adults [17]. In the case of poisoning acute, cadmium is associated with liver damage, kidney stone formation, iron deficiency, water accumulation in the lungs (edema lung), impaired calcium absorption, bone weakness, DNA alteration and malformations in fetuses [35], breast cancer [41], and cancers of various organs in humans [38].

There are numerous remediation technologies form contaminated soils and can be grouped into 3 types: (a) biologica (bioremediation, biostimulation, phytoremediation, biotill, etc.), where the metabolic activities of certain organisms permits the transformation, degradation, removal of pollutants to harmless products; (b) physicochemical (electroremediation, washing, solidification/stabilization, etc.), here take advantage of the properties chemical and physical characters of pollutants to separate, destroy, or keep pollution; and (c) thermal (incineration, vitrification, thermal desorption, etc.), in which heat is used to promote volatilization, burn, decompose or immobilize the [36]. Additionally, the bioremediation process can use organisms that are native to the contaminated site or to other places (exogenous), and it can be done in situ or ex situ, aerobically (with oxygen present) or anaerobically (without oxygen). The bioremediation procedures have been employed successfully to clean soils, sludge, and sediments contaminated with a variety of compounds that jeopardise the health of living things, even if not all compounds are amenable to biodegradation [57], among the different natural biomasses, the fungi are a great tool for these technologies, for ezample: the bioremediation of cadmium (II) and lead (II) by fungal strains from a tailings pond [6, 7], the removal of this heavy metal by fungal strains in a river from Indian state [23], the biosorption of some heavy metals by contaminant ambiental fungi [63], the removal of the same metal for 15 different fungal biomasses [1], the elimination of the same contaminant the *Aspergillus versicolor* biomass [55], *Aspergillus terreus* and *Penicillium chrysogenum* tolerant to toxic metals, isolated from the contaminated zone from a community of an north state in México [22, 60], the isolation of 23 fungal strains tolerant to cadmium isolated from differents plants crops [34], some contaminants fungi, which removal different contaminants [26], the tolerance and removal capacity of *Cunninghamella bertholletiae* [52], two filamentous fungi with tolerance to cadmium [2], the isolate of different fungi in effluent industries in Zanjan province, Iran [25], and a yeast tolerant to the heavy metal from industrial samples [51]. Therefore, our objective was study the biosorption of cadmium (II) in aqueous solution, by some fungal strains isolated from contaminated samples from a place of high risk in San Luis Potosi, México, and we determined the removal capacity for a fungal strain of *Purpureocillum lilacinum.*

2 Material and Methods

We collected different samples (5 g), from the contaminated sites from a place with high traffic flow, near the University zone, San Luis Potosi, México. Subsequently, 1 g of sample was sowing in solid and modified Lee's minimal medium (LMM) [40], added with 500 mg/L of lead nitrate [$Pb(NO_3)_2$], pH of 5.3. The fungal strain of *P. lilacinum*, was isolated through open petri dishes with the same LMM [40] [contained 500 mg/L $K_2Cr_2O_7$, pH 7.0] [16]. All the simples were incubated at 28 °C for 7 days, and the fungal strain was purified by successive replanting in Malt Extract Agar, and was identified by their macro and microscopic properties [39].

2.1 Analysis of Resistant by Growth in Dry Wet

The test of tolerant analysis to Cadmium (II)-tolerant of the fungal strains (1×10^6 spores/mL), were carried out on 100 mL of liquid LMM, with diverse amountss of the heavy metal (like $CdCl_2$), incubated to 28 °C, seven days, and 100 rpm. Afterwards, the supernatant was removed by centrifugation in previously weighed tubes, and the precipitate was incubated at 60 °C for 72 h in a bacteriological oven, and the dry weight of the samples was determined by weight difference.

2.2 Obtaining the Fungal Biomasses

We inoculate 1×10^6 spores/600 mL in a sterile liquid medium, and it incubate 7 days at 30 °C, with constant stirringin. After, we separate the fungal biomass obtained by filtration in Whattman paper No. 1, and was washed with 500 mL of sterile trideionized water, dried for 72 h at 60 °C, ground and stored in an amber bottle until use.

2.3 Removal of Cadmium (II) in Solution by Fungal Biomass

Different solutions of 50 mg/L of cadmium ($CdCl_2$), solutions are prepared, adjusting to desired pH with different acids and/or salts, according to the requirements, and this solution was added to Erlenmeyer flask of 250 mL, which contain 1 g/100 mL of of the biomass to be analyzed, and at different times, 5 mL aliquots are taken, removing biomass by centrifuga (2000 rpm/10 min), and in the supernatant is determined the cadmium (II) concentration by the Dithizona colorimetric method, forming a color complex pink red, which can be extracted with chloroform, and reading the absorbance of the samples at a wavelength of 518 nm. This determination, can detect a concentration of cadmium up 0.5 μg/15 mL of dithizone solution [27].

2.4 Biorremediation of Contaminated Soils

In special containers, which containing 195 mL of wastewater [100 mg/L of cadmium (II) (adjusted)] of a wastewater lagoon (Tenorio Tank, ponient zone of this capital city, were incubated with 5 g of the fungal biomass to be analysed, in the optimal laboratory conditions previously, and every 24 h, the concentration of the heavy metal in the samples was analysed. All analysis was performed three times and in duplicate.

3 Results and Discussion

3.1 Study of Isolated Cadmium II Resistant Fungi

In this work, 10 different fungi were identified (Table 1), and, these strains isolated, presented different macro and microscopic characteristics, like: pigmentation, and the characteristics of the crops [39]. Too, the fungi have different tolerance concentrations of cadmium (II), which may be resistant to it, and the different cadmium resistance could be explained by genetic and physiological differences among the isolated fungi [33] (Table 1).

This results are like for some of the similar studies reported, evidencing the high degree of contamination of the sites analyzed, like: the isolaton of *Aspergillus niger* and *Rhizopus* sp., which growth at 3 mg/L of the heavy metal in study, from a tailing pond [6, 7], for 15 fungal isolates obtained from Bharalu River, Assam [23], 15 different fungal biomasses, with differents percentages of removal [1], *A. versicolor* tolerant to cadmium (II) isolated from from dumping area of plant [55], *A. terreus*

Table 1 Fungi identified in this study

Fungi	Cadmium (II) concentration (mg/L)	Growth (mg of dry weight)
Purpureocillium lilacinum	200	13
Aspergillus flavus	200	12
Penicillium sp.	200	8
Aspergillus terreus	175	8
Paecelomyces sp.	150	3
Mucor sp.	150	2.8
Alternaria sp.	150	2.4
Trichoderma sp.	100	2.0
Cladosporium	100	1.6
Scopulariopsis sp.	75	1.2

which growth at 250 mg/L of this metal [60], the cadmium tolerance (1 g/L) of *P. chrysogenum* FMS2 isolated from contaminated sites [22], different fungal strains isolated from soybean (*G. max*) and barley (*H. vulgare*) tolerant to 100 μg/mL of $Cd(NO_3)_2$ [34], *A. terreus* LCM8, which growth in presence 2 g/L of cadmium (II) [31], for *C. bertholletiae*, with resistance to the same heavy metal [52], with different fungi which tolerate 50 mg Cd/100 mL [2], for 16 fungal strains could tolerate 100 mg/L of the metal [25], and the yeast *C. tropicalis* isolated from wastewater samples, and growth in 2.8 g/L of the metalic ion [51].

3.2 Tolerance of Purpureocillium Lilacinum to Different Cadmium (II) Concentrations

Purpureolillium lilacinum growth up 350 mg/L of Cadmium (II), although the growth is higher than in the control without metal (Fig. 1).

These results are like previous reports, such as for 15 fungal isolates obtained from Bharalu River, Assam, which tolerance between 100 and 400 mg/L of the metal [23], the cadmium tolerance (1 g/L) for *P. chrysogenum* FMS2 [22], *A. terreus* LCM8, can tolerate up to 2 g/L of cadmium (II) [31], for 19 fungi, most of which tolerate up to 50 mg/L of the metal [2], for 16 fungal strains could tolerate between 1000 and 4000 mg/L of the metal [25], and for *C. tropicalis,* which tolerate 2800 mg/L of cadmium (II), 2300 mg/L of copper (II), 2000 mg/L of chromium (VI), and 2400 mg/L of mercury (II) [51]. But are better for *A. niger* and *Rhizopus* sp., which growth at 3 mg/L of cadmium (II) and lead (II) [6, 7], for some fungal strains, which have minimum inhibitory concentrations from 290 to 2400 μg/mL [34].

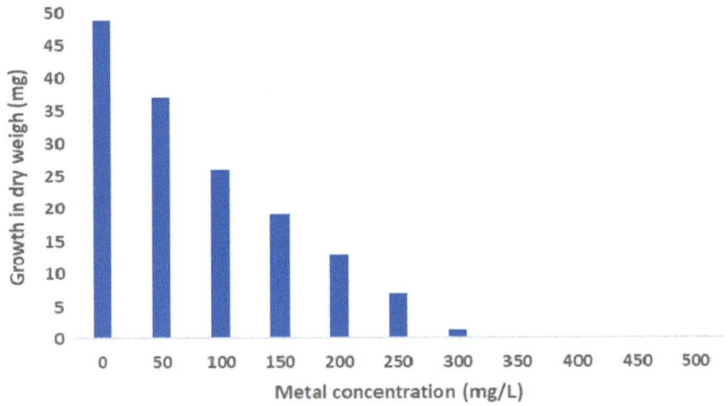

Fig. 1 Growth of *Purpureolillium lilacinum* in MLL. 1×10^6 Spores/mL, with different cadmium concentrations (mg/L). 28 °C. 1 week. Constant agitation. pH 5.6

3.3 *Biosorption of Lead (II) by the Biomass of Some Species of Penicillium*

3.3.1 Incubation Time

The biomasses removal 56.3% of the heavy metal, 32 h of incubation [50 mg/L cadmium (II)], at 28 °C, pH 6.0, 1.0 g of dry cells and 100 rpm (Fig. 2).

These results are like for *Rhizopus* sp., which eliminate 69.23% of cadmium (II) (100 mg/L) at 58 h, 28 °C, pH 5.0, and 90 rpm [6, 7], for fungal biomass of *Mucor rouxii* IM-80, with a removal of 63% of the metal, at 24 h of incubation, 1 mg/200 mL, 80 mg of dry cells, pH 5.0, 100 rpm, 28 °C [1].

But, but these results do not coincide with the dead biomass of *A. versicolor* tolerant to cadmium (II), in which the maximum biosorption was 27.56 mg/g, at 1.5 h, and pH 4.0 [55], for the 49% of removal for *P. chrysogenum*, isolated from soil within 15 days of incubation [22], for fungal strains of *B. bassiana* T7 (58.7% of removal of cadmium) in similar conditions [34], for the biosynthetic melanin from *A. terreus* LCM8, whit a removal of 40% of metalic ion [31], a time of 84 h for *C. bertholletiae*, with an elimination of 4 mg/L, pH 4.0, and 6 g of biomass [52], for *A. Fumigatus*, 98% of cadmium (II), removal (10 mg/L) [2], 10 days for the removal of cadmium (II) (100 mg/L) by some environmental polluting fungus [25], 144 h for the biosorption of 80% with *C. tropicalis*, in 5 mL of liquid culture, pH 7.2–75, 30 °C [51].

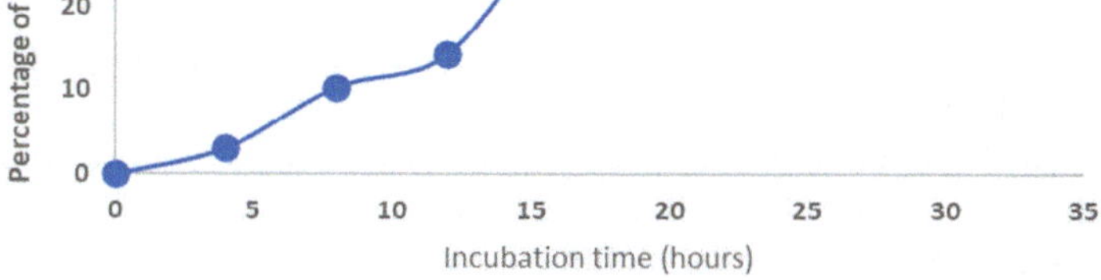

Fig. 2 Effect of the incubation time on the biosorption of cadmium (II) (50 mg/L) at 100 rpm. Fungal biomass (1 g). Incubation Temp: 28 °C and pH 6.0

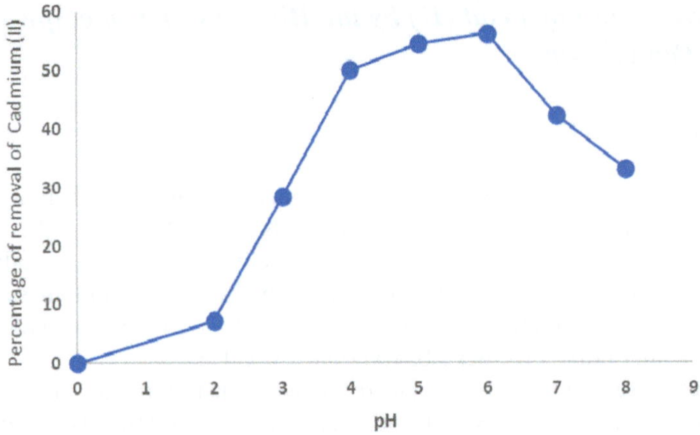

Fig. 3 Effect of pH on the biosorption of cadmium (II) (50 mg/L) at 100 rpm. Fungal biomass (1 g). Incubation temp: 28 °C and incubation duration: 32 h

3.3.2 Effect of the pH

The pH optimum was at 6.0 (Fig. 3).

These results are similar for *M. rouxii* IM-80, with a removal of 63% of the metal, same pH (6.0) [1], for the dried biomass of *A. versicolor* which reached at a pH of 6 [55], too pH 6.0, for the removal of cadmium (II) (100 mg/L) by different fungal strains isolated from contaminated sites [25], for a strain of *Penicillium amphipolaria*, XK11 [37], a pH between 6.0 and 7.0 for the removal of this metal by *Trichoderma atrobrunneum* (800 mg/L), in 7 days of incubation [46]. The results obtained in this work, do not match what was reported for *Rhizopus* sp., which eliminate 69.23% at pH 5.0 [6, 7], for dead biomass of *A. versicolor* tolerant to cadmium (II), in which the optimum pH maximum biosorption was 4.0 [55], a pH of 4.0 for the removal of 4 mg/L by *C. bertholletiae* [52], for *A. Fumigatus*, a pH of 5.0, with 98% of cadmium (II) removal (10 mg/L) after 90 min of contact [2], a pH between 7.2 and 7.5, with *C. tropicalis* [51], a pH of 5.0 with *Rhizopus* sp biomass, at a initial metal concentration of 50 ppm, with different percdentage of elimination [6, 7].

3.3.3 Effect of the Temperature

In this determination, 28 °C was the most effective (56.3% of removal) (Fig. 4).

This coincides with what was reported for *M. rouxii* IM-80, with a removal of 63% of the metal, at 28 °C, but if increase this parameter, decreases removal efficiency [1], 30 °C for the biosorption of 80% [100 mg/L], with *C. tropicalis*, [51], *Saccharomyces cerevisiae*, biosorbs more metals, including cadmium (II) at temperatures close to 25 °C [45], a temperature of 25 °C with the dry cells of *Rhizopus* sp., and initial metal concentration of 50 ppm [6, 7], with *T. atrobrunneum* (800 mg/L), if increased the

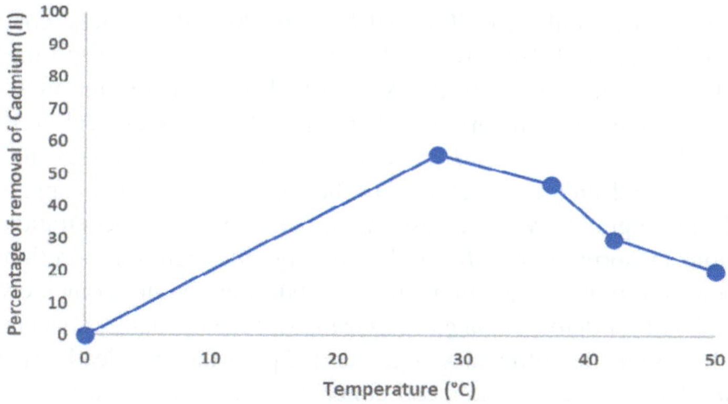

Fig. 4 Effect of the temperature of incubation on the removal of cadmium (II) (50 mg/L) at 100 rpm. Fungal biomass (1 g), incubation pH 6.0 and incubation duration 32 h

temperature, decrease the biosorption of the heavy metal [46], too with cadmium ions biosorption onto *Trichoderma* fungi, the highest absorption occurred at 28 °C [44], and the same temperature for the elimiknation of the same metal with *Eichhornia crassipes* biomass [30]. But are different for *C. bertholletiae,* whith 35 °C [52], and with *A. terreus* BCRC 32068, there is no obvious difference for metal uptake by the biomass among different temperatures [56].

3.3.4 Effect of Different Initial Concentration of Cadmium (II)

At a 50 mg/L of the metal, we obtain 56.3% of biosorption, but if we increase the heavy metal concentration (400 mg/L) the removal decreases to 2.2% (Fig. 5).

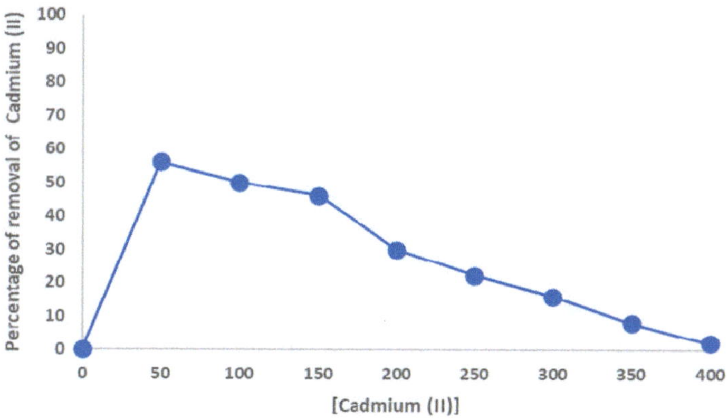

Fig. 5 Effect of the initial concentration of the metal on the elimination. 1 g of dry cells, pH 6.0, 32 h of incubation, 28 °C, 100 rpm

These results are similar for *M. rouxii* IM-80, which if the metal concentration is increased, the removal decrease of 63% (1 mg/200 mL) to 45% (2 mg/200 mL) [1], for differente fungi from riverbank, in which if increased the metal concentration decrease the removal capacity of the fungal biomasses [21], too using the *Rhizopus* sp. biomass [6, 7], for *Trichoderma simmonsii*, if increases the cadmium concentration, too decrease the removal of the [62], for *Penicillium* spp. XK10, the removal decrease gradually if increase the initial cadmium concentration [32], for cadmium ions biosorption onto *Trichoderma* fungi, the highest absorption occurred at lower concentrations of cadmium (II) [44], but, they do not coincide with what was reported with *Colletotrichum gloeosporioides*, because at high concentrations of the metal the elimination efficiency is greater [23], for the live, dead, dried biomass of *A. versicolor* in which in the metal concentration is increased too increase the removal efficiency [55], too for *C. bertholletiae*, similar results are obtained, in similar conditions [52], with the fungi *A. terreus* BCRC, 320, *Penicillium chrysogenum* and *Cephalotheca foveolate*, *Saccharomyces cerevisiae*, and *Piriformospora indica*, is better the elimination at higher metal concentrations [4, 45, 56, 62].

3.3.5 Effect of the Initial Concentration of Fungal Biomass

With 1 g of fungal biomass we obtain a biosorption of 56.3%, while with 5 g, the elimination is total in 24 h (Fig. 6).

These results are like for the 100 mg/L of *A. fumigatus* biomass, with 10 mg/100 mL of the metal [2], for different biosobents like: commercial chitosan biopolymer, *E. crassipes*, *A. niger*, where if increase the biomass concentration

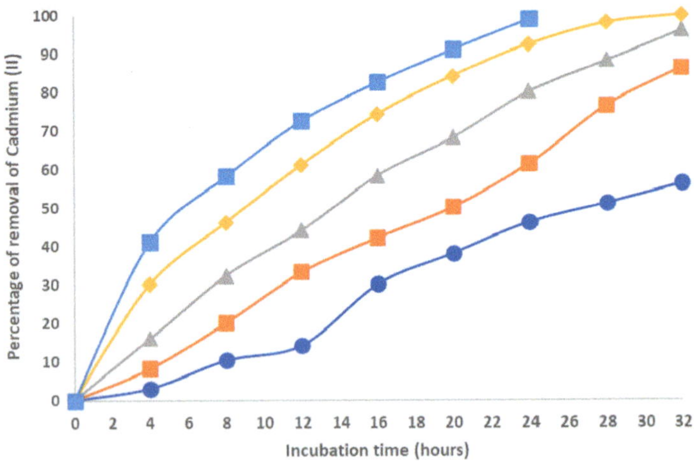

Fig. 6 Effect of the initial fungal biomass concentration on the elimination of 50 mg/L of cadmium (II). pH 6.0, 32 h of incubation, 28 °C, 100 rpm •---• 1g ■---■ 2g ▲----▲ 3g ♦---♦ 4g ■---■ 5g

too increase the removal of the metal [15, 29, 30]. Too, similar results have been reported for corn leaves [10] and *Leucaena leucocephala* [18], ash residues [49], for the biomass of *Lentinula edodes* modified, since it increases the elimination of cadmium (II) and copper (II) by increasing the biomass concentration [61], too for diferrent heavy metals with *Lactarius Scrobiculatus* [5]. But they are different for *Cunninghamella bertholletiae*, in which it has been reported if increase the biomass concentration, decrease the biosorption [52], and the 2 g/L of the biomass of the alga *Spirulina platensis* eliminate 60 mg/L of Cd (II) [3].

4 Removal of Cadmium (II) from a Wastewater Lagoon Contaminated

Industrial, municipal and manufacturing plants can contain toxic heavy metals in their wastewater, affecting downstream processes, water reuse, the environment and the well-being of the local community. These metals include cadmium, cobalt, copper, iron, mercury, nickel, silver, tin, and zinc, and with the aim of improve the elimination of heavy metals in order to reduce residual levels, something that is not always achieved with traditional methods, the elimination of 100 mg/L cadmium (II) of wastewater was analyzed, using 5 g of the different biomasses obtained, under the conditions described above, finding percentages of biosorpion of the metal of 60.1%, 52.3%, 48.1%, 46.2%, of the samples analysed, for *P. lilacinum*, *A. flavus*, *Penicillium* sp., and *A. terreus*, respectively (Fig. 7; Table 2).

These results are similar for the boremediation of lead (II) and cadmium (II) from a tailings pond of by fungal strains [6, 7], for the purification of the waters and soils of

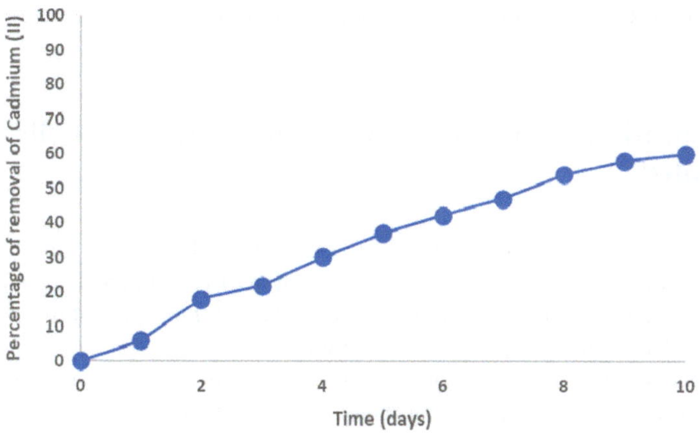

Fig. 7 Biosroption of Cadmium (II) (100 mg/L, adjusted), from samples contaminated of the Tenorio Tank, by *P. lilacinum*. pH 6.0 (adjusted), 5 g of biosorbent, 28 °C, 100 rpm

Table 2 Biosroption of Cadmium (II) from samples contaminated of the Tenorio Tank, by the different fungal biomasses*

Fungal strain	Percentage of removal (%)
Purpureocillium lilacinum	60.1
Aspergillus flavus	52.3
Penicillium sp	48.1
Aspergillus terreus	46.2
Paecelomyces sp	44.2
Mucor sp	43.1
Alternaria sp	40.4
Trichoderma sp	32.1
Cladosporium sp	26.3
Scopulariopsis sp	18.1

*100 mg/L of cadmium (II), pH 6.0 (adjusted), 5 g of biodsorbent, 10 days of incubation, 28 °C, 100 rpm

diferent heavy metals with an yeast and *Trichoderma* sp. [11, 45], the yeast *C. tropicalis* eliminate 56% and 73%, after 6 and 12 days of incubation, respectively, of the metal from contaminated industrial effluents [51], too, with the dry cells of *A. terreus* BCRC 32068 [56], 52% of efficiency by a loofa sponge of *P. chrysogenum* [4], the studies fin the RMoch6 point, Moche River, La Libertad, Perú, with a commercial chitosan biopolymer [29], the removal of chromium (VI) with some fungi isolated from industrial effluent contaminated area [48], with the macrofungus *Pleurotus* sp., in polluted river waters [54], the mycoremediation of soil contaminated with this metal by *Trichoderma* sp. [11], the bioremediation of mining waste with *Pleurotus ostreatus* [14], the bioremediation of soils contaminated with cadmium (II) by filamentous fungus *Penicillium chrysogenum* [22], and the elimination of cadmium (II) from Bharalu River, Assam, by autoctonous fungi [23].

5 Biosorption of Cadmium (II) by the Fungal Dry Cells Isolated

Finally, the cadmium (II) removal of 10 fungal dry cells isolated was studied, and these biosorbents eliminate the metal in solution in different percentages described, being the most efficient: *P. lilacinum*, *A. flavus*, *Penicillium* sp., and *A. terreus*, with 99%, 87%, 81%, and 76%, respectively (Table 3).

The results obtained in this work are similar to those reported for fungal biomass of A. *niger* and *Rhizopus* sp., which removal cadmium (II) and lead (II) in different percentages from a tailing pond [6, 7], for 15 fungal isolates obtained from Bharalu River, Assam, which biosorb different concentrations of the metal [23], the removal of different heavy metals for environmental contaminating fungi [63], for a *M. rouxii* IM-80 wild type, *M. rouxi* mutant, *A. flavus*-I, and *Helmintosporium* sp., isolated fron

Table 3 Biosorptio of 50 mg/L cadmium (II), with the biosorbent isolated*

Fungal biomass	Percentage of removal (%)
Purpureocillium lilacinum	99
Aspergillus flavus	87
Penicillium sp.	81
Aspergillus terreus	76
Paecelomyces sp.	61
Mucor sp.	54
Alternaria sp.	48
Trichoderma sp.	38
Cladosporium sp.	36
Scopulariopsis sp.	33

*28 °C, 100 rpm, pH 6.0, 5 g of biosorbent, 32 h of incubation

different contaminates sites [1], for the bioremediation of cadmium by *P. chrysogenum* and *Cunninghamella bertholletiae* [22, 52], the isolation of highly cadmium tolerant fungy [25], *A. terreus* tolerance to toxic metals [60], for the concentration of contaminantes in a zone from a north state, of our country [24], and the concentration of metalic ions in organs and tissues of *Rhinoptera steindachneri* [12].

6 Conclusion

1. We isolate 10 species of fungal strains, whit tolerance to cadmium (II).
2. *Purpureocillium lilacinum*, biosorb notably this metal, 99% in the conditions analysed.
3. This biosorbent biosorb between 33 and 99% of cadmium (II) of wastewater.
4. The bosorbents analysed show efficient elimination percentages, and *P. lilacinum, Aspergillus flavus, Penicillium* sp., and *A. terreus,* presented the best removal efficiencies.

References

1. Acosta, I., Moctezuma Zárate, M. G., Cárdenas, J. F., & Gutiérrez, C. (2007). Bioadsorción de Cadmio (II) en solución acuosa por biomasas fúngicas. *Revista Información Tecnológica de Chile, 18*(1), 9–14. https://doi.org/10.4067/S0718-07642007000100003
2. Al-Garni, S. M., Ghanem, K. M., & Bahobail, A. D. (2009). Biosorption characteristics of *Aspergillus fumigatus* in removal of cadmium from an aqueous solution. *African Journal of Biotechnology., 8*(17), 4163–4172. https://doi.org/10.5897/AJB2009.000-9401

3. Al-Homaidan, A. A., Alabdullatif, J. A., Al-Hazzani, S. A. A., Al-Ghanayem, A. A., & Alabbad, A. F. (2015). Adsorptive removal of cadmium ions by *Spirulina platensis* biomass. *Saudi Journal of Biological Sciences., 22*(6), 795–800. https://doi.org/10.1016/j.sjbs.2015.06.010
4. Ali, E. A. M., Sayed, M. A., Abdel-Rahman, T. M. A., & Hussein, R. (2021). Fungal remediation of Cd(II) from wastewater using immobilization techniques. *RSC Advances, 11*, 4853–4863. https://doi.org/10.1039/d0ra08578b
5. Anayurt, R. A., Sari, A., & Tuzen, M. (2009). Equilibrium, thermodynamic and kinetic studies on biosorption of Pb(II) and Cd(II) from aqueous solution by macrofungus (*Lactarius Scrobiculatus*) biomass. *Chemical Engineering Journal, 151*, 255–261. https://doi.org/10.1016/j.cej.2009.03.002
6. Apaza Aquino, H., & Valderrama Valencia, M.R. 2020. Aislamiento de cepas fúngicas de una poza de relaves para la biorremediación de plomo y cadmio. UVC Hacer. *Revista de Investigación y Cultura. Universidad Cesar Vallejo, 9*(4), 1–8. https://orcid.org/0000-0002-1515-3195
7. Apaza Aquino, H., Valderrama Valencia, M. R. (2020). Eficiencia del la biomasa de *Rhizopus* sp., en la remoción de cadmio (II). Universidad Nacional del Altiplano. *Revista de Investigaciones de la Escuela de Posgrado, 9*(4), 11805–1816. https://doi.org/10.26788/riepg.2020.4.192
8. Aranda-Coello, J. M., Mendoza-Velázquez, O. M., & Gutiérrez-Olvera, C. (2023). Metales pesados en escamas caudales de *Crocodylus moreletii* en la porción sur de la selva lacandona, Chiapas, México. *Revista Internacional de Contaminación Ambiental, 39*, 85–94. https://doi.org/10.20937/RICA.54454
9. Arbaiza-Peña, A. K., Grober Panduro-Isco, G., Díaz-Zúñiga, E. J., Guadalupe-Baylon, N. K., Angulo-García, N., & Iannacone, J. (2022). Composición elemental y de metales pesados en los residuos de palma en la Amazonia Peruana. *Tropical and Subtropical Agroecosystems, 25*(083), 1–11. https://doi.org/10.56369/tsaes.3967
10. Astudillo, S., Vera, L., Astudillo, J., & Castro, C. (2020). Evaluación del poder biosorbente de la hoja de maíz en la remoción de metales pesados. *AFINIDAD, LXXVII, 591*, 182–188.
11. Bandurska, K., Krupa, P., Berdowska, A., Jatulewicz, I., & Zawierucha, I. (2021). Mycoremediation of soil contaminated with cadmium and lead by *Trichoderma* sp. *Ecological Chemistry Engineering Sciendo, 28*(2), 277–286. https://doi.org/10.2478/eces-2021-0020
12. Benítez Colio, L. R. (2023). Concentración de As, Cd, Cu, Pb y Zn en órganos y tejidos de *Rhinoptera steindachneri* (Everman y Jenkins, 1981) capturado en la zona costera de marismas nacionales. México. Tesis que para obtener el grado de Maestria en Ciencias en Recursos Acuáticos en el Área de Manejo Sustentable de Recursos Costeros. Facultad de Ciencias del Mar. Maestría en Ciencias en Recursos Acuaticos. Mazatlán, Sinaloa, México. http://repositorio.uas.edu.mx/jspui/handle/DGB_UAS/191
13. Bustamante-Montes, L. P., Flores-Polanco, J. A., Isaac-Olivé, K., Hernández-Téllez, M., Campuzano-González, M. E., & Ramírez-Durán, N. (2018). Estudio exploratorio sobre la asociación de metales pesados y la nefropatía de etiología desconocida en el poniente del estado de México. *Revista Internacional de Contaminación Ambiental, 34*(4), 555–564. https://doi.org/10.20937/RICA.2018.34.04.01
14. Calderón García, L. R. (2022). Análisis de las aplicaciones de *Pleurotus ostreatus* como alternativa en la biorremediación de residuos mineros. Tesina. Para la obtención del Diploma de Especialidad en Gestión Ambiental de Residuos. Facultad de Ciencias Biológicas. Univdersidad Autónoma del Estado de Morelos. Cuernavaca, Morelos, México. http://riaa.uaem.mx/handle/20.500.12055/3367
15. Cárdenas González, J. F., Acosta Rodríguez, I., Rodríguez Pérez, A. S., Martínez Juárez, V. M., Moctezuma Zárate, M. G., Pacheco Castillo, N. C., & Muñoz Morales, A. (2019). Bioremoval of heavy metals by the native strain *Aspergillus niger*. *Modern Concept & Developments in Agronomy, 5*(2), 525–534. https://doi.org/10.31031/MCDA.2019.05.000610
16. Cárdenas González, J. F., Acosta Rodríguez, I., Terán Figueroa, Y., Lappe Oliveras, P., Martínez Flores, R., & Rodríguez Pérez, A. S. (2021). Biotransformation of Chromium (VI) via a reductant activity from the fungal strain *Purpureocillium lilacinum*. *Journal of Fungi, 7*(1022), 1–14. https://doi.org/10.3390/jof7121022. 29 November 2021.

17. Chávez-Gómez, N., Cabello-López, A., Gopar-Nieto, R., Aguilar-Madrid, G., Marín-López, K., Aceves-Valdez, M., Jiménez-Ramírez, C., Cruz-Angulo, M., & Juárez-Pérez, C. (2017). Enfermedad renal crónica en México y su relación con los metales pesados. *Revista Médica del Instituto Mexicano del Seguro Social, 55*(6), 725–734. Medigraphic. https://www.medigraphic.com › new › resumen
18. Cimá-Mukul, C. A., Abdellaoui, Y., Abatal, M., Vargas, J., Santiago, A. A., & Barrón-Zambrano, J. A. (2019). Eco-efficient biosorbent based on *Leucaena leucocephala* residues for the simultaneous removal of Pb(II) and Cd(II) ions from water system: Sorption and mechanism. *Bioinorganic Chemical and Applications, 2019*, 13. Article ID 2814047. https://doi.org/10.1155/2019/2814047
19. Cruz Casanova, F. E. (2014). Efecto de la contaminación por metales pesados en los ecosistemas costeros del sureste de México. Kuxulkab, 19(37), 65–68. https://doi.org/10.19136/kuxulkab.a19n37.351
20. Delgado Díaz, D. M., Zarza Villanueva, H., Flores Hernández, N., Cruz Monterrosa, R., Rayas Amor, A., Díaz Ramírez, M., Jiménez Guzmán, J., García Garibay, M., Miranda de la Lama, G., & Hernández Flores, L. I. (2019). Presencia de metales pesados en Pato Mexicano (*Anas diazi* Ridgway). *Agro Productividad, 12*(11), 89–92. https://doi.org/10.32854/agrop.vi0.1504
21. Dey, P., Gola, D., Mishra, A., Malika, A., Kumar, P., Kumar Singh, D., Patel, N., von Bergend, M., & Jehmlich, N. (2016). Comparative performance evaluation of multi-metal resistant fungal strains for simultaneous removal of multiple hazardous metals. *Journal of Hazardous Materials, 318*(2016), 679–685. https://doi.org/10.1016/j.jhazmat.2016.07.025
22. Din, G., Hassan, A., Dunlap, J., Ripp, S., & Shah, A. A. (2022). Cadmium tolerance and bioremediation potential of filamentous fungus *Penicillium chrysogenum* FMS2 isolated from soil. *International Journal of Environmental Science and Technology, 19*, 2761–2770. https://doi.org/10.1007/s13762-021-03211-7
23. Dipannita, D., Pinky, B., Devi, N. N., & Mayuri, Ch. (2021). Biosorption of cadmium by fungi isolated from Bharalu river, Assam. *Journal of Tropical Life Science, 11*(3), 279–284 https://doi.org/10.11594/jtls.11.03.04
24. Domínguez-Zúñiga, L. I., Puente-Valenzuela, C. T., Estrada-Arellano, J. R., Aguirre-Acosta, E., & Aguillón-Gutiérrez, D. R. (2022). Concentración de metales pesados en hongos de la zona metropolitana de la Comarca Lagunera, Mexico. *Scientia Fungorum, 22*(e1389), 1–10. https://doi.org/10.33885/sf.2021.52.1389
25. Fazli, M. M., Soleimani, N., Mehrasbi, M., Darabian, D. S., Mohammadi, J., & Ramazani, A. (2015). Highly cadmium tolerant fungi: Their tolerance and removal potential. *Journal of Environmental Health Science & Engineering, 13*(19), 1–9. https://doi.org/10.1186/s40201-015-0176-0
26. Ghosh, S., Rusyn, I., Dmytruk, O. V., Dmytruk, K. V., Onyeaka, H., Gryzenhout, M., & Gafforov, Y. (2023). Filamentous fungi for sustainable remediation of pharmaceutical compounds, heavy metal and oil hydrocarbons. *Frontiers in Bioengineering and Biotechnology, 11*(1106973), 1–20. https://doi.org/10.3389/fbioe.2023.1106973
27. Greenberg, A. E., Clesceri, L. S., & Eaton, A. D. (1998). Standard methods for the examination of water and wastewater. 18a. Ed. American Public Health Association. Washington DC
28. Guillama Barroso, G., Ramos Delgado, N. A., Sanjuan Galindo, R., Herrera Mendoza, R., Rivera Haro, J. A., & Quevedo Álvarez, O. (2022). Evaluación de la contaminación por As, Ni, Cu, Pb, Zn y Cr en sedimentos de la zona marino-costera asociada a la terminal marítima de Nuevitas, Cuba. *Revista Internacional de Contaminación Ambiental, 38*, 81–94. https://doi.org/10.20937/RICA.54080
29. Gutiérrez, L. (2022). Remoción de metales pesados del punto RMoch6 del río Moche mediante biopolímero quitosano commercial. *Revista Ciencia y Tecnología, 18*(12), 205–214. https://doi.org/10.17268/rev.cyt.2022.01.15
30. Gutiérrez, P., Aradillas, D., & Acosta, I. (2021). Aplicación de la biomasa de *Eichhornia crassipes* en la remoción de Cd^{+2} en aguas contaminadas por desechos industriales. *Avances en Ciencias e Ingeniería, 12*(3), 17–29. 20 de septiembre. http://www.executivebs.org/publishing.cl/category/revista-aci/

31. Hayat, R., Din, G., Farooqi, A., et al. (2022). Characterization of melanin pigment from *Aspergillus terreus* LCM8 and its role in cadmium remediation. *International Journal of Environmental Science and Technology, 20*, 3151–3160. https://doi.org/10.1007/s13762-022-04165-0
32. He, Y., Li, C., Sun, Z., Zhang, W., He, J., Zhao, Y., Xu, Z., & Zhao, W. (2023). *Penicillium* spp. XK10, fungi with potential to repair cadmium and antimony pollution. *Applied Sciences, 13*(1228), 1–13. https://doi.org/10.3390/app13031228
33. Hernández-Caricio, C., Ramírez, V., Martínez, J., Quintero-Hernández, V., Baez, A., Munive, J. A., & Rosas-Murrieta, N. (2022). Los metales pesados en la historia de la humanidad, los efectos de la contaminación por metales pesados y los procesos biotecnológicos para su eliminación: El caso de *Bacillus* como bioherramienta para la recuperación de suelos. *AyTBUAP, 7*(27), 1–68. https://doi.org/10.5281/zenodo.7091407
34. Ignatova, L., Kistaubayeva, A., Brazhnikova, Y., Omirbekova, A., Mukasheva, T., Savitskaya, I., Karpenyuk, T., Goncharova, A., Egamberdieva, D., & Sokolov, A. (2021). Characterization of cadmium-tolerant endophytic fungi isolated from soybean (*Glycine max*) and barley (*Hordeum vulgare*). *Heliyon, 7*(11), 1–9. https://doi.org/10.1016/j.jwliyon.2021.e08240
35. Jaishankar, M., Tseten, T., Anbalagan, N., Mathew, B. B., Krishnamurthy, N., & Beeregowda, N. (2014). Toxicity, mechanism and health effects of some heavy metals. *Interdisciplinary Toxicology, 7*(2), 60–72. https://doi.org/10.2478/intox-2014-0009
36. Jeyakumar, P., Debnath, Ch., Vijayaraghavan, R., & Muthuraj. (2023). Review trends in bioremediation of heavy metal contaminations. *Environmental Engineering Research, 28*(4), 1–19, 220631. https://doi.org/10.4491/eer.2021
37. Jingyi, D., Chaoyang, L., Yu, S., Yunlin, Z., Huimin, H., Yingzi, M., & Zhenggang, X. (2023). Adsorption capacity of *Penicillium amphipolaria* XK11 for cadmium and antimony. *Archives of Microbiology, 205*. Article number: 139. https://doi.org/10.1007/s00203-023-03484-1
38. Khan, Z., Elahi, A., Bukhari, D. A., & Rehman, A. (2022). Cadmium sources, toxicity, resistance and removal by microorganisms. A potential strategy for cadmium eradication. *Journal of Saudi Chemical Society, 26*(6), 1–18, 101569. https://doi.org/10.1016/j.jscs.2022.101569
39. Kirk, M. P., Cannon, F. P., David, C. J., & Stalpers, A. J. (2001). *Dictionary of the fungi*. CABI Publishing.
40. Lee, K., Buckley, L., & Campbell, C. C. (1975). An aminoacid liquid synthetic medium for the development of mycelial and yeast forms of *Candida albicans*. *Journal of Medical Veterinary Mycology, 13*(32), 148–153. https://doi.org/10.1080/00362177585190271
41. Mehari, T. F., Rorie, Ch. J., Godfrey, S. S., Minor, R. C., Sayo O. & Fakayode, S. O. (2017). Influence of Arsenic (III), Cadmium (II), Chromium (VI), Mercury (II), and Lead (II) ions on human triple negative breast cancer (HCC1806) cell cytotoxicity and cell viability. *Journal of Chemical Health Risks, 7*(1), 1–17. https://doi.org/10.22034/JCHR.2017.544160
42. Menchaca Dávila, S., & Ríos Fuentes, L. M. (2020). Análisis diacrónico de la contaminación por cadmio en la microcuenca del río Pixquiac, Veracruz. *UVserva, 9*, 8–19. https://doi.org/10.25009/uvs.v0i9.2644
43. Mesa Pérez, M. A., Díaz Rizo, O., García Acosta, H., Alarcón Santos, O. A., Tavella, M. J., Bagué, D., Sánchez Pérez, J. M., Guerrero Domínguez, L., Hernández Rodríguez, D. H., & Díaz Almeida, C. M. (2021). Heavy metals bioaccumulation and risk estimation in edible freshwater fish from Pedroso Reservoir (Matabeque, Cuba). *Revista Internacional de Contaminación Ambiental, 37*, 527–537. https://doi.org/10.20937/RICA.53850
44. Mohammadi, L., Sadeghi Malvajerdi, M., Rahdar, A., & Kyzas, G. Z. (2022). Optimization of cadmium ions biosorption onto *Trichoderma* fungi. *Biointerfase Research in Applied Chemistry, 12*(3), 3316–3331. https://doi.org/10.33263/BRIAC123.33163331
45. Moreno-Rivas, S. C., & Clamont-Montfort, G. R. (2018). Descontaminación de arsénico, cadmio y plomo en agua por biosorción con *Saccharomyces cerevisiae*. TIP. *Revista Especializada en Ciencias Químico-Biológicas, 2*(2), 51–69. https://doi.org/10.22201/fesz.23958723e.2018.0.155
46. Mushtaq, S., Bareen, F., & Tayyeb, A. (2023). Equilibrium kinetics and thermodynamic studies on biosorption of heavy metals by metal-resistant strains of *Trichoderma* isolated from tannery

solid waste. *Environmental Science and Pollution Research, 30*(4), 10925–10954. https://doi.org/10.1007/s11356-022-22860-w

47. Naranjo-Jiménez, C., & WingChing-Jones, R. (2023). Arsénico, cadmio, mercurio y plomo en alimentos importados para mascotas en Costa Rica. *Agronomía Mesoamericana, 34*(1), 1–13. Artículo 48399. https://doi.org/10.15517/am.v34i1.48399
48. Narolkar, S., Jain, N., & Mishra, A. (2022). Biosorption of chromium by fungal strains isolated from industrial effluent contaminated area. *Pollution, 8*(1), 159–168. https://doi.org/10.22059/POLL.2021.326818.1132
49. Olabemiwo, F. A., Tawabini, B. S., Patel, F., Oyehan, T. A., Khaled, M., & Laoui, T. (2017). Cadmium removal from contaminated water using polyelectrolyte-coated industrial waste fly ash. *Bioinorganic Chemical and Applications, 2017*, 13. Article ID 7298351. https://doi.org/10.1155/2017/7298351
50. Ortíz Olguín, A. M., Lander Valenzuela, N., Edith Hernández Nataren, E., Lara Viveros, F., Granados Echegoyen, C. A., & Andrade Hoyos, P. (2023). Bioaccumulation of heavy metals in *Cucumber* plants inoculated with *Trichoderma* strains (Hypocreaceae) in the Mezquital Valley, Hidalgo, México. *Revista Internacional de Contaminación Ambiental, 39*, 117–126. https://doi.org/10.20937/RICA.54525
51. Rehman, A., & Anjum, M. S. (2010). Cadmium uptake by yeast, *Candida tropicalis*, isolated from industrial effluents and its potential use in wastewater clean-up operations. *Water Air Soil Pollution, 205*, 149–159. https://doi.org/10.1007/s11270-009-0062-4
52. Ren, Y. Y., Zhu, Z. Y., Dong, F. Y., & Song, Q. Y. (2018). Screening, characteristics and mechanism of Cd-tolerance *Cunninghamella bertholletiae. Journal of Cleaner Production, 191*(2018), 480–489. https://doi.org/10.1016/j.jclepro.2018.04.225
53. Romero-Oliva, O. J., Acevedo-Sandoval, O. A., Prieto-García, F., & Prieto-Méndez, J. (2023). Riesgo toxicológico por plomo, cadmio y manganese en suelos DR028, Tulancingo, Hidalgo, México. *Ibn Sina–Revista electrónica semestral en Ciencias de la Salud, 14*(1), 1–12. http://revistas.uaz.edu.mx/index.php/ibnsina
54. Santos, M. P. O., Santos, M. V. N., Matos, R. S., Van Der Maas, A. S., Faria, M. C. S., Batista, B. L., Rodrigues, J. L., & Bomfeti, C. A. (2021). *Pleurotus* strains with remediation potential to remove toxic metals from Doce River contaminated by Samarco dam mine. *International Journal of Environmental Science and Technology, 19*, 6625–6638. https://doi.org/10.1007/s13762-021-03597-4
55. Soleimani, N., Mohammadian Fazli, M., Ramazani, A., & Mehrasbi, M. R. (2016). Application of live, dead, and dried biomasses of *Aspergillus Versicolor* for cadmium biotreatment. *Journal of Human, Environment Health Promotion, 1*(2), 87–98. https://doi.org/10.29252/jhehp.1.2.87
56. Sun, Y. M., Horng, C. Y., Chang, F. L., Cheng, L. C., & Tian, W. X. (2010). Biosorption of lead, mercury, and cadmium ions by *Aspergillus terreus* immobilized in a natural matrix. *Polish Journal of Microbiology, 59*(1), 37–44. PMID: 20568528.
57. Tarfeen, N., Nisa, K. I., Hamid, B., Bashir, Z., Yatoo, A. M., Dar, M. A., Mohiddin, F. A., Amin, Z., Ahmad, R. A., & Sayyed, R. Z. (2022). Microbial remediation: A promising tool for reclamation of contaminated sites with special emphasis on heavy metal and pesticide pollution: A review. *Processes, 10*(1358), 1–27. https://doi.org/10.3390/pr10071358
58. Tulcan Rivera, M. I., & Coyago Cruz, E. (2022). Evaluación de metales pesados en nísperos (*Eriobotrya japonica*) cultivados en la sección urbana del distrito metropolitano de Quito. Tesis Posgrado. Maestría en Recursos Naturales Renovables con Mención en Remediación y Restauración. Universidad Politécnica Saleciana. http://dspace.ups.edu.ec/handle/123456789/25016
59. U.S. Environmental Protection Agency. Deposition of air pollutants to the great waters. First Report to Congress. EPA-453/R-93-055. Office of Air Quality Planning and Standards, Research Triangle Park, NC, 1994. https://www.epa.gov ›default›files›documents.
60. Villalba-Villalba, A. G., & González-Méndez, B. (2021). Evaluating *Aspergillus terreus* tolerance to toxic metals. *Revista Chapingo Serie Ciencias Forestales y del Ambiente, 27*(3), 449–464. https://doi.org/10.5154/r.rchscfa.2020.07.047

61. Xie, H., Zhao, Q., Zhou, Z., Wu, Y., Wang, H., & Xu, H. (2015). Efficient removal of Cd(II) and Cu(II) from aqueous solution by magnesium chloride-modified *Lentinula edodes. Royal Society Chemistry Advances, 5*, 33478–33488. https://doi.org/10.1039/C4RA17272H
62. Yaghoubian, Y., Siadat, S. A., Moradi Telavat, M. R., Pirdashti, H., & Yaghoubian, I. (2019). Bio-removal of cadmium from aqueous solutions by filamentous fungi: *Trichoderma* spp. and *Piriformospora indica. Environmental Science and Pollution Research, 26*, 7863–7872. https://doi.org/10.1007/s11356-019-04255-6
63. Zaki, A., Hasanien, Y. A., & Abdel-Razek, A. S. (2022). Biosorption optimization of lead(II) and cadmium(II) ions by two novel nanosilica-immobilized fungal mutants. *Journal of Applied Microbiology, 33*(2), 987–1000. https://doi.org/10.1111/jam.15624

Microbial Native Soil Bacteria Against Cadmium Toxicity

Prasann Kumar and Debjani Choudhury

Abstract Cadmium (Cd) is a toxic heavy metal that poses a significant threat to the environment and human health due to its persistence and ability to accumulate in various ecosystems. In recent years, the use of microbial bioremediation strategies to mitigate the adverse effects of cadmium toxicity has gained considerable attention. Native soil bacteria have emerged as promising candidates for bioremediation due to their adaptability to diverse soil environments and ability to interact with heavy metals. This abstract highlights the potential of microbial native soil bacteria in combating cadmium toxicity. Firstly, it explores how these bacteria can alleviate cadmium toxicity, including metal sequestration, enzymatic detoxification, and bioaccumulation. Native soil bacteria possess various physiological and genetic adaptations that enable them to survive in cadmium-contaminated soils and tolerate high levels of cadmium exposure. The abstract discusses the interactions between native soil bacteria and plants in the context of cadmium remediation. Certain soil bacteria have been found to form symbiotic associations with plants, enhancing their cadmium tolerance through mechanisms such as phytoextraction, rhizodegradation, and rhizofiltration. These interactions hold great potential for developing efficient and sustainable strategies for cadmium bioremediation. The abstract also discusses the challenges of applying native soil bacteria for cadmium bioremediation. Factors such as microbial competition, nutrient availability, pH, and temperature can influence the effectiveness of microbial remediation approaches. Therefore, optimizing bacterial growth and activity conditions is crucial for maximizing their remediation potential. Utilizing microbial native soil bacteria represents a promising approach for mitigating cadmium toxicity. Their unique adaptations and interactions with plants provide potential solutions for remediating cadmium-contaminated soils and reducing the associated environmental and health risks. Further research

P. Kumar (✉)
Department of Agronomy, School of Agriculture, Lovely Professional University, Phagwara, Punjab 144411, India
e-mail: prasann0659@gmail.com

D. Choudhury
Department of Plant Pathology, School of Agriculture, Lovely Professional University, Phagwara, Punjab 144411, India

A. K. Jha and N. Kumar (eds.), *Cadmium Toxicity in Water*, Springer Water,
https://doi.org/10.1007/978-3-031-54005-9_9

and development in this field are necessary to optimize the efficacy of microbial bioremediation strategies and facilitate their practical implementation on a larger scale.

Keywords Crop · Cadmium · Toxicity · Microbes · Bioremediation · Bacteria

1 Introduction

Cadmium (Cd), a toxic heavy metal, is a ubiquitous environmental pollutant that significantly threatens ecosystems and human health [32, 43, 83, 95]. It is released into the environment through industrial activities, mining operations, and phosphate fertilizers, among other sources. Due to its long half-life and ability to accumulate in various environmental compartments, cadmium contamination persists and can harm soil quality, plant growth, and overall ecosystem health [6, 16, 32, 34, 43, 51, 54, 83, 95, 100].

Traditional approaches to cadmium remediation, such as physical and chemical methods, have shown limited effectiveness and often come with high costs and potential environmental risks. In recent years, microbial bioremediation strategies have emerged as a promising alternative for combating cadmium toxicity. Among the various microbial agents, native soil bacteria have garnered considerable attention due to their adaptability to soil environments and inherent capacity to interact with heavy metals [9, 65, 84, 98].

Microbial native soil bacteria have evolved in cadmium-contaminated soils and have developed unique mechanisms to tolerate and detoxify this toxic metal. These bacteria possess a range of physiological and genetic adaptations that allow them to survive and thrive under high cadmium concentrations. One such adaptation is the expression of metal-binding proteins or peptides, such as metallothioneins, which sequester cadmium ions and prevent their interaction with cellular components. Additionally, native soil bacteria can produce enzymes, such as metal-reducing enzymes, which facilitate the enzymatic detoxification of Cadmium [12, 67, 68, 78].

Furthermore, the interactions between microbial native soil bacteria and plants play a crucial role in cadmium remediation. Certain soil bacteria can form symbiotic associations with plants, providing mutual benefits. These plant–microbe interactions can enhance the tolerance of plants to cadmium through various mechanisms. For example, some bacteria promote phytoextraction, facilitating the uptake and accumulation of cadmium in plant tissues, thus reducing its concentration in the soil. Other bacteria engage in rhizodegradation, breaking down cadmium compounds in the rhizosphere by producing chelating agents or metal-reducing enzymes. Additionally, rhizofiltration, the process by which bacteria immobilize or degrade cadmium in the root zone, has shown the potential to reduce cadmium levels in contaminated groundwater [7, 17, 23, 36, 46, 59, 106, 109].

Despite the promising potential of microbial native soil bacteria in mitigating cadmium toxicity, several challenges must be addressed. Microbial competition, limited nutrient availability, suboptimal pH, and temperature fluctuations can significantly impact the effectiveness of bacterial remediation strategies. Therefore, understanding the ecological dynamics of native soil bacteria and optimizing the environmental conditions for their growth and activity is critical for maximizing their cadmium remediation potential [3, 81, 91].

In conclusion, microbial native soil bacteria represent a promising avenue for combating cadmium toxicity in contaminated soils. Their unique adaptations and interactions with plants offer potential solutions for the remediation of cadmium-contaminated environments. However, further research is needed to explore and optimize the mechanisms underlying bacterial cadmium tolerance and detoxification. Additionally, field-scale studies are necessary to evaluate the feasibility and long-term efficacy of microbial bioremediation strategies for cadmium-contaminated sites. Ultimately, harnessing the capabilities of native soil bacteria holds great promise for sustainable and eco-friendly cadmium remediation practices [4, 55, 69, 86, 102].

2 Research Gap and Possible Solution

While the potential of microbial native soil bacteria in mitigating cadmium toxicity is promising, several research gaps must be addressed. One significant research gap is the limited understanding of the diversity and functional capabilities of native soil bacteria in different cadmium-contaminated environments. Identifying and characterizing specific bacterial species or highly efficient strains in cadmium tolerance and detoxification would significantly enhance the development of targeted bioremediation strategies [33, 75, 103, 105, 113].

Furthermore, the mechanisms by which native soil bacteria interact with cadmium and facilitate its immobilization or degradation in the soil are still not fully understood. Elucidating the molecular and biochemical processes underlying these interactions would provide valuable insights into the potential of microbial bioremediation for cadmium-contaminated sites [25, 44, 74, 111].

Another research gap lies in optimizing environmental conditions for microbial growth and activity in cadmium-contaminated soils. Factors such as pH, temperature, nutrient availability, and co-contamination with other heavy metals can influence the effectiveness of microbial remediation strategies. Understanding the optimal ranges of these parameters and their impact on bacterial performance would aid in designing and implementing efficient and sustainable bioremediation approaches [21, 66, 76, 104, 110].

Possible Solution: To address the research gaps in microbial native soil bacteria against cadmium toxicity, several approaches can be undertaken. Firstly, comprehensive metagenomic and metatranscriptomic studies can be conducted to explore the diversity and functional potential of native soil bacterial communities in cadmium-contaminated environments. This would involve using high-throughput sequencing technologies to identify critical bacterial taxa and assess their genetic repertoire for cadmium tolerance and detoxification [8, 24, 72, 85, 87]. By correlating the presence and abundance of specific bacterial genes or pathways with cadmium concentrations, researchers can gain insights into the most effective microbial candidates for bioremediation. In parallel, experimental studies can be performed to elucidate the mechanisms by which native soil bacteria interact with cadmium and mediate its immobilization or degradation [8, 35, 60, 82, 92]. These studies may use advanced analytical techniques, such as proteomics and metabolomics, to identify the proteins and metabolites involved in cadmium binding, sequestration, and detoxification. Genetic and molecular tools can be employed to manipulate the expression of essential genes or pathways in native soil bacteria, further elucidating their role in cadmium remediation [20, 63, 77, 89, 108]. To optimize the environmental conditions for microbial bioremediation, integrated approaches can be employed. This may include the manipulation of pH and temperature regimes, supplementing essential nutrients, and using amendments or additives that enhance bacterial growth and activity in cadmium-contaminated soils. Field-scale trials should be conducted to assess these optimization strategies' long-term efficacy and practical feasibility, considering the complex and dynamic nature of real-world contaminated sites [28, 40, 79, 88, 97, 107]. Adressing the research gaps in microbial native soil bacteria against cadmium toxicity requires a multidisciplinary approach combining metagenomics, molecular biology, and environmental engineering. By investigating the diversity, functional capabilities, and mechanisms of native soil bacteria and optimizing the environmental conditions for their activity, researchers can develop targeted and efficient bioremediation strategies for cadmium-contaminated environments. Such advancements would contribute to conserving cadmium pollution and protect ecosystems and human health (Table 1).

3 Biochemical Mechanisms of Rhizospheric Phytoextraction Under Cadmium Stress

Phytoextraction, a phytoremediation technique, utilizes plants and associated microorganisms to extract and accumulate heavy metals from contaminated soils. Rhizospheric phytoextraction, specifically focusing on the interactions between plants and soil microorganisms in the root zone, has gained significant attention as an effective strategy for cadmium remediation [1, 15, 19, 30, 41, 47, 96]. Understanding the biochemical mechanisms underlying rhizospheric phytoextraction and the microbial diversity involved is crucial for optimizing this process. Additionally, metabolomics analyses offer valuable insights into the metabolic responses of plants and microorganisms to cadmium stress, providing a comprehensive understanding of the phytoextraction process. Biochemical mechanisms of rhizospheric phytoextraction: Rhizospheric phytoextraction involves the coordinated efforts of plants and rhizospheric microorganisms to facilitate cadmium uptake and accumulation. Several biochemical mechanisms contribute to this process. Firstly, plants release various root exudates, such as organic acids, phenolics, and enzymes, which

Table 1 Microbial species and it is targeted metals and possible mechanism

Microbial species	Metal(s) targeted	Bioremediation mechanisms
Pseudomonas putida	Cadmium, Copper, Chromium, Zinc	Metal sequestration, enzymatic detoxification, biosorption
Bacillus subtilis	Lead, Cadmium, Chromium, Mercury	Metal chelation, precipitation, biosorption, reduction/oxidation
Arthrobacter sp.	Uranium, Chromium, Copper	Metal reduction, chelation, biosorption
Cupriavidus metallidurans	Cadmium, Copper, Zinc, Silver	Metal resistance, metal accumulation, reduction/oxidation
Shewanella oneidensis	Uranium, Chromium, Technetium	Metal reduction, electron transfer, precipitation, biosorption
Desulfovibrio desulfuricans	Uranium, Chromium, Mercury	Metal reduction, sulfate reduction, methylation
Rhodococcus erythropolis	Polychlorinated biphenyls (PCBs)	PCB degradation, enzymatic detoxification, bioaccumulation
Trichoderma harzianum	Cadmium, Copper, Lead, Zinc	Biosorption, enzymatic detoxification, chelation
Saccharomyces cerevisiae	Lead, Cadmium, Copper, Uranium	Metal biosorption, intracellular accumulation, reduction/oxidation
Geobacter sulfurreducens	Uranium, Chromium, Technetium	Metal reduction, electron transfer, biomineralization
Candida sp.	Chromium, Lead, Cadmium, Mercury	Metal biosorption, biotransformation, reduction/oxidation
Rhizobium leguminosarum	Nickel, Zinc, Copper	Metal sequestration, plant–microbe symbiosis, chelation
Methylobacterium sp.	Copper, Chromium, Nickel	Metal volatilization, biosorption, enzymatic detoxification
Penicillium chrysogenum	Cadmium, Lead, Copper, Uranium	Metal biosorption, extracellular precipitation, enzymatic detoxification
Thiobacillus ferrooxidans	Copper, Uranium, Arsenic	Metal oxidation, acid mine drainage remediation, extracellular precipitation
Glomus intraradices	Cadmium, Zinc, Lead	Metal uptake, enhancement of plant metal tolerance, rhizosphere remediation
Rhizophagus irregularis	Cadmium, Copper, Zinc	Metal sequestration, mycorrhizal association with plants, enhancement of plant metal tolerance

(continued)

Table 1 (continued)

Microbial species	Metal(s) targeted	Bioremediation mechanisms
Acidithiobacillus ferrooxidans	Copper, Uranium, Arsenic	Metal oxidation, acid mine drainage remediation, extracellular precipitation, biosorption
Stenotrophomonas maltophilia	Cadmium, Chromium, Nickel	Metal sequestration, enzymatic detoxification, biosorption
Deinococcus radiodurans	Uranium, Chromium, Plutonium	Metal resistance, DNA repair mechanisms, biosorption
Burkholderia cepacia	Chromium, Cadmium, Nickel	Metal chelation, reduction/oxidation, biosorption
Exiguobacterium oxidotolerans	Uranium, Chromium, Plutonium	Metal reduction, biosorption, enzymatic detoxification
Chlorella vulgaris	Cadmium, Copper, Zinc	Phytoremediation, metal biosorption, intracellular accumulation
Magnetospirillum magneticum	Cadmium, Iron, Manganese	Magnetic separation, biomineralization, bioremediation

Source based on the review of literature

significantly mobilize and solubilize cadmium in the soil. These exudates enhance the availability of cadmium ions for uptake by plant roots [1, 19, 61, 64, 73, 96].

Microbial diversity in rhizospheric phytoextraction: The rhizosphere harbours a diverse microbial community that interacts closely with plants and influences their metal uptake and tolerance. Microorganisms in the rhizosphere can enhance phytoextraction by facilitating the soil's mobilization, solubilization, and transformation of cadmium. Various factors, including plant species, soil characteristics, and cadmium concentrations, influence the microbial diversity in the rhizosphere. Understanding the composition and dynamics of the rhizospheric microbial community is crucial for optimizing phytoextraction efficiency [26, 39, 57, 70, 71, 90, 112].

Metabolomics analyses in rhizospheric phytoextraction: Metabolomics, a powerful analytical approach, provides a comprehensive snapshot of the metabolites present in plant tissues and associated microorganisms. Metabolomic analyses enable identifying and quantifying various metabolites, including organic acids, sugars, amino acids, and secondary metabolites. In rhizospheric phytoextraction under cadmium stress, metabolomics analyses can elucidate the metabolic changes occurring in plants and microorganisms, highlighting the critical biochemical pathways involved in cadmium uptake, translocation, and detoxification [26, 29, 31, 38, 39, 57, 70, 90, 101, 112].

By comparing metabolite profiles between plants exposed to cadmium stress and control conditions, metabolomics can identify specific metabolites associated with cadmium tolerance and phytoextraction efficiency. Furthermore, metabolomics analyses of the rhizospheric microbial community can reveal the metabolic adaptations and interactions between microorganisms and plants in response to cadmium stress [26, 29, 31, 38, 39, 57, 70, 90, 101, 112]. These analyses provide valuable insights into the functional roles of specific microorganisms and their contribution to cadmium remedsiation.

The biochemical mechanisms underlying rhizospheric phytoextraction and the microbial diversity involved play crucial roles in successfully remedying cadmium-contaminated soils. Metabolomics analyses provide a powerful tool to decipher the metabolic responses of plants and microorganisms to cadmium stress, offering valuable insights into the critical biochemical pathways involved in phytoextraction. By integrating microbial diversity analyses and metabolomics, researchers can comprehensively understand the complex interactions between plants, microorganisms, and cadmium in the rhizosphere. This knowledge can be leveraged to optimize rhizospheric phytoextraction strategies, enhancing the efficiency and sustainability of cadmium remediation efforts (Table 2).

4 Cadmium Has a Hazardous Impact; Its Incoming Sources in the Environment

Cadmium (Cd) is a toxic heavy metal that poses significant risks to human, microbial, and soil health. High levels of Cadmium exposure can lead to severe health problems and environmental damage [13, 18, 29, 45, 58, 62, 94]. Here, we discuss the severe hazardous impacts of cadmium on these aspects:

1. Human Health: Cadmium toxicity is a significant concern for human health. It can enter the body through various routes, including inhalation, ingesting contaminated food and water, and dermal contact. Cadmium has a long half-life and accumulates in the body over time, mainly in the kidneys and liver. Some of the severe health effects associated with cadmium exposure include:

    - Kidney Damage: Cadmium is a nephrotoxic metal that can cause severe damage to the kidneys. Prolonged exposure to high levels of cadmium can lead to chronic kidney disease, impaired kidney function, and potentially kidney failure.
    - Bone Disorders: Cadmium interferes with calcium metabolism and can lead to the loss of bone mineral density. Chronic exposure to cadmium increases the risk of osteoporosis and bone fractures.
    - Respiratory and Cardiovascular Problems: Inhalation of cadmium-contaminated air can cause lung damage, leading to respiratory disorders such as chronic bronchitis and emphysema. Cadmium exposure is also associated with an increased risk of cardiovascular diseases, including hypertension and heart disease.

Table 2 Biochemical mechanism during the cadmium removal process

Mechanism	Description	Examples
Biosorption	Microbes or their components (e.g., cell wall, extracellular polymeric substances) bind heavy metals to their surfaces, removing them from the solution	*Bacillus subtilis*, *Pseudomonas putida*
Bioaccumulation	Microbes uptake and store heavy metals within their cells	*Cupriavidus metallidurans*, *Saccharomyces cerevisiae*
Extracellular precipitation	Microbes produce substances that induce the precipitation of heavy metals as insoluble compounds	*Shewanella oneidensis*, *Rhodococcus erythropolis*
Intracellular precipitation	Heavy metals are sequestered and precipitated within microbial cells	*Geobacter sulfurreducens*, *Desulfovibrio desulfuricans*
Enzymatic detoxification	Microbial enzymes degrade or transform heavy metals into less toxic or less mobile forms	*Arthrobacter* sp., *Rhizopus oryzae*
Reduction/Oxidation	Redox reactions involving microbes can convert heavy metals into less toxic or insoluble forms	*Shewanella putrefaciens*, *Acidithiobacillus ferrooxidans*
Chelation	Microbes produce chelating agents that bind heavy metals, forming stable complexes	*Pseudomonas fluorescens*, *Penicillium chrysogenum*
Methylation	Microbes methylate heavy metals, converting them into more volatile and potentially less toxic forms	*Methylobacterium* spp., *Desulfovibrio desulfuricans*
Volatilization	Microbes convert heavy metals into volatile compounds, releasing them into the atmosphere	*Thiobacillus ferrooxidans*, *Pseudomonas aeruginosa*
Ion exchange	Microbes exchange metal ions with other ions present in the environment	*Bacillus cereus*, *Bacillus megaterium*
Complexation	Microbes produce organic molecules that form stable complexes with heavy metals	*Arthrobacter* sp., *Pseudomonas putida*
Biomineralization	Microbes induce the formation of mineral compounds that incorporate heavy metals	*Arthrobacter oxydans*, *Bacillus megaterium*
Rhizodegradation	Root exudates and microbial activity in the rhizosphere facilitate the degradation of heavy metal compounds	*Rhizobium* sp., *Trichoderma harzianum*
Phytoextraction	Plants and associated microbes facilitate the uptake and accumulation of heavy metals in plant biomass	*Glomus intraradices*, *Populus* sp.
Rhizofiltration	Microbes associated with plant roots filter and remove heavy metals from contaminated water	*Azolla caroliniana*, *Salix viminalis*

(continued)

Table 2 (continued)

Mechanism	Description	Examples
Rhizostabilization	Microbes and plants work together to immobilize heavy metals in the rhizosphere, preventing their migration	*Pseudomonas putida, Festuca arundinacea*
Electrobioremediation	Microbes participate in electrochemical reactions to enhance the removal or recovery of heavy metals	*Geobacter sulfurreducens, Shewanella oneidensis*
Nanoparticle synthesis	Microbes are used to synthesize nanoparticles that can efficiently bind and remove heavy metals	*Shewanella* sp., *Pseudomonas aeruginosa*

- Reproductive and Developmental Effects: Cadmium can affect reproductive health in both males and females. It may cause infertility, hormonal imbalances, and adverse effects on fetal development, leading to congenital disabilities.

2. Microbial Health: Cadmium contamination harms microbial health and the overall microbial ecosystem [11, 14, 27, 42, 53, 99, 114]. Microbes play vital roles in nutrient cycling, soil fertility, and the degradation of organic matter. Cadmium disrupts these processes and can lead to the following impacts:

   - Microbial Diversity Loss: High levels of cadmium in soil can reduce microbial diversity, as certain microbial species are more sensitive to cadmium toxicity than others. This loss of microbial diversity can disrupt ecological balance and compromise soil functions.
   - Impaired Soil Nutrient Cycling: Cadmium toxicity inhibits the activities of microbial enzymes involved in nutrient cycling processes, such as nitrogen fixation, nitrification, and organic matter decomposition. This can lead to nutrient imbalances, reduced soil fertility, and decreased plant productivity.
   - Disrupted Microbial Interactions: Cadmium can alter microbial community interactions and disrupt symbiotic relationships, such as mycorrhizal associations. This disruption can impact plant–microbe interactions, nutrient uptake, and soil health.

3. Soil Health: Cadmium contamination profoundly impacts soil health, crucial for sustaining plant growth and ecosystem functioning. The presence of cadmium in the soil leads to the following adverse effects:

   - Soil Contamination: Cadmium accumulates in the soil, persisting for long periods due to its low mobility. The contaminated soil becomes unfit for cultivation, posing risks to food safety and human health.
   - Decreased Soil Fertility: Cadmium toxicity hampers the biological activity in the soil, leading to reduced microbial biomass and enzyme activities. This results in impaired nutrient cycling decreased organic matter decomposition, and diminished soil fertility.

- Ecological Imbalance: Cadmium contamination can disrupt the soil microbial community structure and impact the balance of soil ecosystems. This imbalance can lead to the dominance of stress-tolerant or cadmium-resistant microorganisms, affecting the overall ecological functions of the soil.

Addressing the hazardous severe impacts of cadmium on human, microbial, and soil health requires implementing effective pollution control measures, promoting sustainable agricultural practices, and adopting appropriate remediation techniques. Minimizing cadmium pollution, improving soil management practices, and raising awareness about the risks associated with cadmium exposure are crucial steps toward safeguarding human health and preserving the integrity of our environment [2, 5, 10, 37, 49, 56, 93].

4.1 Sources of Cadmium Contamination in Soil, Environment, and Ecosystem

Cadmium (Cd) contamination in soil and the wider environment is a significant concern due to its detrimental effects on human health, ecosystem health, and the overall balance of ecosystems. Cadmium contamination can arise from various sources, both natural and anthropogenic [16, 22, 48, 50, 52, 100]. Understanding these sources is crucial for effective prevention and remediation efforts. Here, we discuss the primary sources of cadmium contamination:

1. Natural Sources:

   - Geological Deposits: Cadmium is naturally present in the Earth's crust and can be released into the environment through weathering and erosion of cadmium-rich rocks and minerals. Geological processes can contribute to the background levels of cadmium in soils.
   - Volcanic Activity: Volcanic eruptions release gases and ash containing cadmium into the atmosphere. These emissions can deposit cadmium onto soils and other environmental compartments.
   - Weathering and Erosion: Natural weathering and erosion processes can release cadmium from rocks and minerals, gradually increasing its presence in soil.

2. Anthropogenic Sources:

   - Industrial Activities: The mining, smelting, and refining metals, particularly zinc, lead, and copper, can release cadmium into the environment as a byproduct. Industries that produce batteries, pigments, plastics, and fertilizers are notable sources of cadmium contamination.
   - Agricultural Practices: Cadmium-containing fertilizers, especially phosphate fertilizers derived from cadmium-rich phosphate rock, can introduce cadmium into agricultural soils. Prolonged use of such fertilizers can result in significant cadmium accumulation.

- Sewage Sludge and Wastewater: Disposal of sewage sludge, which may contain cadmium from industrial and domestic sources, on agricultural land can contribute to cadmium contamination. Similarly, untreated industrial wastewater discharged into water bodies can contaminate soils through irrigation and surface runoff.
- Atmospheric Deposition: Cadmium can be transported long distances in the atmosphere and deposited onto soils and water bodies through rainfall, air pollution, and atmospheric dust. This deposition can be derived from natural sources and anthropogenic activities, such as the combustion of fossil fuels and waste incineration.
- Contaminated Water: Industrial discharges, improper waste disposal, and accidental spills can contaminate water bodies with cadmium. Irrigation with cadmium-contaminated water can result in cadmium accumulation in agricultural soils.

3. Ecosystem Impacts:

- Plant Uptake: Plants can take up cadmium primarily through the roots and accumulate in various plant tissues. This can lead to the transfer of cadmium through the food chain, impacting animals and humans that consume cadmium-contaminated plants or plant-eating organisms.
- Soil Microbial Communities: Cadmium contamination can disrupt the microbial community structure and function in soil. Certain microorganisms are sensitive to cadmium toxicity, which can lead to biodiversity loss and impaired soil microbial processes vital for nutrient cycling and ecosystem functioning.
- Wildlife and Aquatic Organisms: Cadmium contamination in soil and water bodies can negatively impact wildlife and aquatic organisms. It can accumulate in the tissues of organisms, leading to toxic effects, reproductive issues, and overall ecological imbalance.

Addressing cadmium contamination requires implementing pollution control measures, sustainable agricultural practices, proper waste management, and remediation strategies such as soil and water treatment [6, 9, 16, 23, 34, 46, 51, 54, 78, 81, 91, 95, 98]. Additionally, raising awareness about the sources and risks of cadmium contamination is vital for mitigating its environmental and health impacts (Table 3).

5 Bioengineered Microbial Strains Versus Contamination

Bioengineered microbial strains have emerged as a promising approach for detoxifying toxic environmental pollutants. These genetically modified microorganisms are designed to possess enhanced abilities to degrade or transform harmful contaminants, contributing to environmental remediation efforts [14, 16, 22, 25, 30, 51, 63, 81, 83, 85, 95, 98, 101]. The field of bioengineered microbial strains encompasses various

Table 3 Sources of cadmium contamination and its description

Source of Cadmium contamination	Description
Industrial activities	– Mining, smelting, and refining of metals such as zinc, lead, and copper that release cadmium as a byproduct
	– Manufacturing processes involving cadmium, such as battery production, electroplating, and pigment manufacturing
Agricultural practices	– Use of cadmium-containing fertilizers, particularly those derived from phosphate rock
	– Application of sewage sludge or biosolids on agricultural land, which may contain cadmium from industrial and domestic sources
Atmospheric deposition	– Emission of cadmium-containing particulate matter from industrial processes, fossil fuel combustion, waste incineration, and volcanic eruptions
	– Long-range transport of cadmium in the atmosphere, leading to its deposition onto soils and water bodies through rainfall, dust, and air pollution
Contaminated water	– Industrial wastewater discharges that contain cadmium from various manufacturing processes
	– Improper disposal of industrial waste and effluents, leading to the contamination of water bodies and subsequent cadmium uptake by soil and organisms
Geological sources	– Weathering and erosion of cadmium-rich rocks and minerals, leading to the release of cadmium into the environment
	– Natural occurrences of cadmium in soil and water due to geological deposits
Urban and industrial runoff	– Stormwater runoff from urban areas and industrial sites, which can carry cadmium and other contaminants into water bodies and soil
Consumer products	– Improper disposal of cadmium-containing products such as batteries, electronics, and plastics, which can leach into the environment
Occupational exposure	– Workers in industries involved in cadmium-related processes, such as mining, smelting, battery production, and metal finishing
Historical contamination	– Legacy contamination from past industrial activities and improper waste disposal practices
	– Historical use of cadmium-based pesticides, insecticides, and fungicides

techniques, including genetic engineering, metabolic engineering, and synthetic biology, to develop microorganisms with specific properties and functionalities.

The use of bioengineered microbial strains offers several advantages in the detoxification of environmental pollutants:

1. Enhanced Biodegradation: Introducing specific genes or gene pathways into microbial strains can significantly enhance their capacity to degrade or transform toxic pollutants. These bioengineered strains can produce enzymes or metabolic intermediates that break down complex pollutants into less harmful forms, facilitating their removal from the environment.
2. Targeted Degradation: Bioengineered strains can target specific contaminants or pollutants, allowing for tailored and efficient detoxification approaches. By manipulating the genetic makeup of microorganisms, they can be equipped with enzymes or pathways that specifically target and degrade the pollutants of concern.
3. Expanded Substrate Range: Microorganisms can be engineered to expand their substrate range, enabling them to degrade a more comprehensive range of environmental pollutants. This flexibility allows for the remediation of complex contaminated sites with multiple pollutants.
4. Synergistic Interactions: Bioengineered microbial strains can be designed to work in synergy with other microbes or natural processes to enhance pollutant removal. For example, engineered microorganisms can produce enzymes that stimulate the growth of indigenous microorganisms or facilitate the breakdown of pollutants cooperatively.
5. Site-Specific Adaptation: Microbial strains can be engineered to thrive in specific environmental conditions, such as extreme temperatures, pH levels, or high pollutant concentrations. This adaptability increases their efficiency and survival in challenging environmental settings.

However, the use of bioengineered microbial strains for environmental detoxification also presents challenges and considerations:

1. Biosafety and Ecological Impact: Before the deployment of bioengineered strains, rigorous biosafety assessments are necessary to evaluate their potential ecological impact and ensure containment. These assessments help prevent unintended consequences, such as the spread of engineered genes to indigenous microbial populations or the creation of new ecological imbalances.
2. Long-Term Stability and Persistence: Bioengineered strains should be designed to maintain stability and functionality over extended periods. This ensures their continued effectiveness in detoxification processes without genetic modifications being lost or degraded.
3. Regulatory Considerations: Using bioengineered microbial strains for environmental remediation may be subject to regulatory frameworks. Compliance with applicable regulations and guidelines is crucial to ensure these technologies' safe and responsible use.

4. Public Perception and Acceptance: The public perception and acceptance of bioengineered microbial strains play a significant role in their implementation. Transparent communication, public engagement, and education efforts are essential to foster understanding and address concerns regarding their use.

Bioengineered microbial strains offer a promising avenue for detoxifying toxic environmental pollutants [6, 22, 29, 34, 48, 58, 72, 85, 87, 94, 100]. Continued research, development, and responsible implementation of these technologies can contribute to effective and sustainable environmental remediation efforts, helping to safeguard ecosystems, human health, and the well-being of future generations (Table 4).

Table 4 Bioengineered microbial strain

Bioengineered microbial strains	Target pollutants	Detoxification mechanisms	Applications and benefits
Pseudomonas putida	Aromatic compounds, hydrocarbons	Production of oxygenases and monooxygenases	Bioremediation of oil spills, contaminated soils, and groundwater
Escherichia coli	Heavy metals, organic pollutants	Metal-binding proteins, enzymatic degradation pathways	Removal of heavy metals from industrial wastewater and contaminated sites
Saccharomyces cerevisiae	Heavy metals, organic pollutants	Biosorption, intracellular accumulation	Remediation of metal-contaminated soils and industrial effluents
Rhodococcus erythropolis	Polycyclic aromatic hydrocarbons	Production of oxygenases and lignin-degrading enzymes	Biodegradation of persistent organic pollutants and hydrocarbon-contaminated sites
Alcaligenes eutrophus	Polyethylene terephthalate (PET)	Enzymatic degradation of PET	Recycling and biodegradation of PET plastics
Cupriavidus necator	Polyhydroxyalkanoates (PHA)	Production of PHA from organic waste	Bioplastic production and waste management
Deinococcus radiodurans	Radioactive contaminants	DNA repair mechanisms, metal resistance proteins	Remediation of radioactive waste and contaminated sites
Methylococcus capsulatus	Volatile organic compounds (VOCs)	Methane oxidation, enzymatic degradation	Biofiltration of VOCs in air and wastewater treatment
Trichoderma reesei	Cellulosic materials	Production of cellulases and hemicellulases	Bioconversion of agricultural waste and lignocellulosic biomass
Synechocystis sp.	Nitrogen and phosphorus compounds	Nutrient uptake and assimilation	Eutrophication control and wastewater treatment

6 Cadmium Absorption by the Root of Plants

Cadmium (Cd) absorption by plant roots is essential in understanding plants' uptake, transport, and accumulation of this toxic heavy metal. Cadmium absorption by plant roots involves several mechanisms and processes, which can vary depending on plant species, environmental conditions, and cadmium availability in the soil [9, 32, 43, 80, 95]. Here is a detailed explanation of cadmium absorption by plant roots:

1. Root Uptake Mechanisms: (a) Passive Uptake: Cadmium can passively diffuse through the plasma membrane of root cells due to its chemical similarity to calcium (Ca^{2+}). This passive uptake occurs when cadmium concentrations in the soil solution are high. (b) Active Uptake: Using specific transporters, plants can actively transport cadmium ions into root cells. These transporters, such as ZIP (ZRT/IRT-like Protein) and NRAMP (Natural Resistance-Associated Macrophage Protein), facilitate cadmium uptake from the soil into the root cells against the concentration gradient.
2. Factors Affecting Cadmium Absorption: (a) Cadmium Concentration: Higher concentrations of cadmium in the soil increase its availability for absorption by plant roots. However, excessive levels of cadmium can lead to toxicity in plants. (b) Soil pH: Cadmium uptake is influenced by soil pH. Acidic soils (low pH) tend to increase cadmium availability, facilitating its absorption by plant roots. (c) Soil Organic Matter: High levels of organic matter in the soil can reduce cadmium bioavailability by binding with cadmium ions, limiting their uptake by plant roots. (d) Nutrient Interactions: Interactions with other nutrients like iron (Fe), zinc (Zn), and calcium (Ca) can affect cadmium uptake. For instance, high levels of zinc or calcium can compete with cadmium for uptake, reducing its absorption by plant roots.
3. Root-to-Shoot Translocation: Once plant roots take up cadmium, they can be translocated to the aboveground plant parts through the xylem, primarily driven by the plant's transpiration stream. The cadmium translocation from roots to shoots occurs via the symplastic and apoplastic pathways.
4. Binding and Sequestration: Inside plant cells, cadmium ions can bind to various ligands, such as phytochelatins (PCs), glutathione (GSH), and metallothioneins (MTs). These ligands help detoxify cadmium by sequestering it in vacuoles or forming complexes that reduce its toxicity and prevent further movement within the plant.
5. Plant Responses and Tolerance: Plants have evolved various mechanisms to tolerate or cope with cadmium stress. This includes activating antioxidative systems, synthesis of chelating compounds (e.g., PCs), and altered gene expression related to cadmium detoxification. These responses help plants minimize cadmium-induced damage and limit its impact on growth and development.

It is important to note that excessive accumulation of cadmium in plants can lead to phytotoxicity and pose risks to human and animal health if consumed through the food chain. Therefore, efforts are made to develop plant-based strategies for phytoremediation or the use of plants to remove or mitigate cadmium contamination in soils.

7 Microbially and Chelate-Assisted Phytoextraction of Cadmium

Microbially and chelate-assisted phytoextraction of cadmium is a promising approach for the remediation of cadmium-contaminated soils. However, several challenges need to be addressed to optimize the efficiency and effectiveness of this remediation technique [9, 12, 43, 65, 68, 98]. Here are some key challenges:

1. Toxicity and Bioavailability: Cadmium is highly toxic to plants, microorganisms, and other organisms. Its high soil bioavailability can result in significant plant uptake, leading to potential phytotoxic effects. The challenge lies in selecting plant species that can tolerate high cadmium concentrations and effectively translocate and sequester cadmium without suffering from toxicity symptoms.
2. Microbial Activity and Community Dynamics: The success of microbially-assisted phytoextraction depends on the presence and activity of cadmium-resistant and plant growth-promoting microorganisms in the rhizosphere. However, soil conditions, such as pH, nutrient availability, and organic matter content, can impact microbial activity and diversity, influencing the efficacy of microbial-assisted remediation.
3. Chelator Selection and Application: Chelating agents enhance the phytoextraction of cadmium by increasing its solubility and availability in the soil. However, the choice of chelators is critical, as they should have a high affinity for cadmium and be environmentally friendly and biodegradable. Improper chelator application or excessive use can lead to cadmium leaching and contamination of groundwater or surface water bodies.
4. Mobility and Leaching: Cadmium is highly mobile in soils, particularly under acidic conditions. This mobility can result in cadmium leaching into deeper soil layers or contaminating groundwater, posing risks to ecosystems and human health. Proper soil and water management practices, including using barriers or liners, are essential to prevent cadmium leaching during phytoextraction.
5. Long-Term Effectiveness: Phytoextraction is a long-term remediation strategy that requires sustained plant growth and root development over an extended period. Maintaining plant health and vigour, especially in contaminated soils, can be challenging due to nutrient deficiencies, competition with other plants, and pest/disease pressure. Ensuring the long-term effectiveness of phytoextraction requires careful monitoring and management.
6. Economic Viability: Implementing microbially and chelate-assisted phytoextraction can involve significant costs associated with plant selection, microbial inoculation, chelator application, monitoring, and maintenance. Assessing the

economic feasibility of this technique and evaluating cost-effective strategies are crucial for its practical application and large-scale implementation.

Addressing these challenges requires interdisciplinary research, including plant biology, soil science, microbiology, and environmental engineering [4, 69, 86, 102]. It is essential to optimize plant–microbe interactions, develop efficient chelators, enhance microbial activity, and consider environmental, economic, and social factors to successfully implement microbially and chelate-assisted phytoextraction of cadmium in contaminated sites (Table 5).

Table 5 Challenges during the management of cadmium toxicity

Challenges	Description
Cadmium toxicity	Cadmium is highly toxic to plants, microorganisms, and other organisms. Its toxicity can affect plant health and limit the success of phytoextraction
Selection of Cadmium-tolerant plants	Identifying plant species that can tolerate high cadmium concentrations without suffering from phytotoxic effects is crucial for effective phytoextraction
Microbial activity and diversity	Soil conditions can impact microbial activity and diversity, which can affect the efficacy of microbially-assisted phytoextraction
Chelator selection and application	Choosing appropriate chelators with a high affinity for cadmium and environmentally friendly properties is essential for effective phytoextraction
Cadmium mobility and leaching	Cadmium's mobility in soils can result in leaching to deeper soil layers or contaminating groundwater, posing risks to ecosystems and human health
Long-term effectiveness	Maintaining plant health and root development over an extended period is essential for sustained phytoextraction and long-term effectiveness
Economic viability	The economic feasibility of microbially and chelate-assisted phytoextraction, considering costs and practical implementation, is a significant challenge
Environmental impact and regulations	Ensuring that phytoextraction techniques comply with environmental regulations and minimize potential negative impacts is essential for their acceptance

8 Effects of Cadmium on Antibiotic Resistance Genes in Rhizosphere Soil

Cadmium, a toxic heavy metal, has been found to affect antibiotic resistance genes (ARGs) in rhizosphere soil. The rhizosphere, the soil environment surrounding plant roots, is a hotspot for microbial activity and interactions [8, 24, 80, 85]. Here are some effects of cadmium on antibiotic-resistance genes in the rhizosphere soil:

1. Increase in ARG Abundance: Studies have shown that cadmium contamination in the rhizosphere soil can increase the abundance of antibiotic-resistance genes. Cadmium stress may induce a cellular stress response in bacteria, leading to the upregulation of ARGs as a defence mechanism.
2. Co-Selection of Resistance: Cadmium exposure can contribute to the co-selection of antibiotic resistance in the rhizosphere soil. This occurs when bacteria possessing resistance genes for both cadmium and antibiotics are selected, leading to the persistence and spread of antibiotic resistance in the microbial community.
3. Horizontal Gene Transfer (HGT): Cadmium can promote the horizontal transfer of antibiotic-resistance genes among bacteria in the rhizosphere soil. HGT is a process by which genetic material, including ARGs, is transferred between different bacterial species, potentially leading to the dissemination of antibiotic resistance.
4. Influence on Microbial Community Composition: Cadmium exposure can alter the microbial community structure in the rhizosphere soil. Certain bacteria, including those carrying ARGs, may become more dominant, while others may decline. This shift in microbial composition can impact the abundance and diversity of ARGs in the rhizosphere.
5. Co-Occurrence of Metal Resistance and Antibiotic Resistance: Cadmium and antibiotic resistance genes may co-occur in bacterial populations. The presence of cadmium resistance genes on mobile genetic elements can facilitate the co-transfer of metal and antibiotic resistance genes, exacerbating the problem of multidrug resistance.

It is important to note that the effects of cadmium on antibiotic-resistance genes in the rhizosphere soil can be complex and influenced by various factors, including the concentration and duration of cadmium exposure, microbial community dynamics, and other environmental stressors. Further research is needed to understand these interactions better and develop strategies to mitigate the spread of antibiotic resistance in the rhizosphere soil exposed to cadmium contamination (Table 6).

Table 6 Antibiotic resistance gene class with cadmium association

Antibiotic resistance gene class	Examples of specific genes	Cadmium association
β-lactamase	blaTEM, blaCTX-M, blaSHV, blaOXA	Potential co-occurrence
Tetracycline	tetA, tetB, tetM, tetW, tetX	Potential co-occurrence
Macrolide-Lincosamide-Streptogramin B	ermB, ermC, ermF, ermG	Limited evidence of association
Sulfonamide	sul1, sul2, sul3	Limited evidence of association
Quinolone	qnrA, qnrB, qnrS, qepA, oqxAB	Limited evidence of association
Aminoglycoside	aadA, aacC, aphA, strA, strB, rmtB	Limited evidence of association
Chloramphenicol	catI, catII, cmlA, floR	Limited evidence of association
Glycopeptide	vanA, vanB, vanC	Limited evidence of association
Others	Various ARGs belonging to different classes	Variable association depending on the gene

Source based on a review of literature

9 Sulphur Can Be Used for the Mitigation of Cadmium from the Soil

Sulfur plays a significant role in plants' uptake and translocation of cadmium. Adequate sulfur availability can influence the uptake and accumulation of cadmium by plants. Studies have shown that sulfur-deficient plants accumulate more cadmium levels than sulfur-sufficient plants [6, 24, 37, 49, 54, 56, 100]. This suggests that sulfur nutrition can modulate cadmium uptake and potentially reduce toxicity.

Sulfur and Metallothioneins: Metallothioneins (MTs) are a class of low-molecular-weight proteins with a high affinity for binding metals, including cadmium. Sulfur-containing amino acids, such as cysteine, are essential components of MTs. Adequate sulfur nutrition can enhance the synthesis of MTs, which in turn can sequester and detoxify cadmium ions. The binding of cadmium to MTs prevents its interaction with essential cellular components, reducing its toxic effects.

Sulfur and Glutathione: Glutathione (GSH) is a tripeptide crucial in cellular redox balance and detoxification. Sulfur is a critical component of GSH, and it is involved in synthesizing and regulating this antioxidant molecule. GSH participates in cadmium detoxification by binding to the metal ion and forming complexes that are transported and stored in vacuoles. Sulfur availability influences the levels of GSH and, consequently, the capacity of plants to combat cadmium-induced oxidative stress [48–50].

Sulfur and Phytochelatins: Phytochelatins (PCs) are small peptides of sulfur-rich amino acids, such as cysteine and glutamate. PCs have a high affinity for binding cadmium and other heavy metals, forming complexes that facilitate their sequestration and detoxification. Sulfur availability influences the synthesis of PCs, with sulfur-deficient plants often exhibiting lower levels of PCs and reduced ability to cope with cadmium toxicity.

Sulfur and Antioxidant Defense: Cadmium exposure induces oxidative stress in plants, generating reactive oxygen species (ROS). Sulfur nutrition influences the antioxidant defense system in plants, including the activity of enzymes such as superoxide dismutase, catalase, and peroxidase. Adequate sulfur availability can enhance the antioxidant capacity of plants, helping to mitigate the oxidative damage caused by cadmium toxicity.

Sulfur and Plant Growth: Cadmium toxicity negatively affects plant growth and development. Sulfur nutrition plays a vital role in supporting overall plant growth and can contribute to alleviating the inhibitory effects of cadmium on various physiological processes. Adequate sulfur availability promotes healthy root development, nutrient uptake, photosynthesis, and biomass accumulation, which can indirectly help plants tolerate and minimize the toxic effects of cadmium. It is important to note that the specific effects of sulfur on cadmium toxicity can vary depending on plant species, sulfur availability, cadmium concentrations, and other environmental factors. Further research is needed to gain a more comprehensive understanding of the underlying mechanisms and to develop strategies for optimizing sulfur nutrition to mitigate cadmium toxicity in different agricultural and environmental settings [39, 71, 73, 100].

10 Microbial Biofilm and Its Application for Soil Health

In recent years, there has been a growing recognition of the critical role that microbial biofilms play in shaping the health and productivity of plants and soil ecosystems. These complex communities of microorganisms, including bacteria, fungi, and archaea, adhere to surfaces and form biofilms that act as dynamic and interactive microhabitats. Harnessing the potential of microbial biofilms as climate-smart strategies for improving plant and soil health has emerged as a promising avenue for sustainable agriculture and ecosystem management [26, 38, 57, 70, 90, 112].

The conventional practices of chemical fertilizers, pesticides, and intensive tilling have long dominated agricultural systems, often leading to detrimental consequences such as soil degradation, nutrient loss, and increased greenhouse gas emissions. Recognizing the need for alternative approaches, scientists and researchers are now exploring the potential of microbial biofilms to promote climate-smart and sustainable solutions [11, 14, 27, 42, 99].

One of the critical advantages of prospecting microbial biofilms is their ability to enhance nutrient cycling and availability in soil ecosystems. Certain microbial species within biofilms can solubilize minerals, fix atmospheric nitrogen, and release plant growth-promoting substances. These biofilms can help plants access essential nutrients, improve their nutrient use efficiency, and ultimately enhance crop productivity while reducing the need for synthetic fertilizers [14, 27, 42, 45, 52, 53].

Furthermore, microbial biofilms exhibit remarkable resilience to environmental stressors, including drought, salinity, and disease. Colonizing plant roots can establish a symbiotic relationship, creating a protective barrier that enhances plant resilience to abiotic and biotic stress factors. This phenomenon, known as rhizosphere engineering, can reduce plants' vulnerability to climate change impacts and enhance their adaptability to changing environmental conditions [1, 6, 16, 34, 51, 54, 100].

In addition to their direct effects on plants, microbial biofilms play a crucial role in soil health. Biofilms improve soil structure and stability by promoting aggregation, thus preventing erosion and maintaining soil moisture. They also contribute to organic matter decomposition and the cycling of carbon, nitrogen, and other essential nutrients, thereby influencing soil fertility and long-term carbon sequestration.

Harnessing the potential of microbial biofilms as climate-smart strategies requires a multidisciplinary approach, integrating knowledge from microbiology, ecology, agronomy, and biotechnology. Advances in molecular techniques, such as metagenomics and metatranscriptomics, have revolutionized our ability to study and manipulate these complex microbial communities, paving the way for targeted interventions and biofilm-based solutions [10, 15, 26, 30, 41, 47, 99, 114].

In conclusion, prospecting microbial biofilms as climate-smart strategies for improving plant and soil health holds immense promise for sustainable agriculture and ecosystem management. By leveraging the inherent capabilities of biofilms to enhance nutrient availability, promote plant resilience, and improve soil quality, we can pave the way for a more resilient and productive agricultural system better equipped to mitigate and adapt to the challenges posed by climate change (Table 7).

11 Conclusion

In conclusion, the role of microbial native soil bacteria in mitigating cadmium toxicity is significant and offers promising avenues for environmental remediation. Cadmium, a toxic heavy metal, poses severe threats to ecosystems and human health due to its persistence and bioaccumulation potential. However, the presence of microbial native soil bacteria provides a natural defence mechanism against cadmium toxicity. Microbial native soil bacteria play a crucial role in cadmium detoxification through various mechanisms. These bacteria can accumulate and sequester cadmium ions within their cells, reducing the concentration of free cadmium in the soil. Additionally, certain bacteria can convert cadmium into less toxic forms or facilitate its immobilization, preventing its uptake by plants and other organisms. Moreover, microbial native soil bacteria exhibit a remarkable capacity for producing metal-binding proteins and enzymes, such as metallothioneins and metalloproteins, aiding cadmium detoxification. These biomolecules can effectively bind to cadmium ions, forming stable complexes and preventing their interaction with cellular components. Furthermore,

Table 7 Microbial biofilm and its use in agriculture for cadmium removal

Microbial biofilm	Target pest/disease	Mode of action	Benefits
Bacillus subtilis	Various plant pathogens	Production of antimicrobial compounds, competition for resources, induction of systemic resistance	Effective biocontrol agent against fungal, bacterial, and viral pathogens. It enhances plant growth and reduces disease incidence
Pseudomonas fluorescens	Soil-borne pathogens, such as Fusarium spp.	Production of antifungal metabolites, induction of systemic resistance	Suppresses root diseases caused by Fusarium and other soil-borne pathogens. Enhances plant growth and nutrient uptake
Trichoderma spp.	Various plant pathogens	Mycoparasitism, production of antifungal enzymes, competition for resources	Biocontrol agents against fungal pathogens, including Fusarium, Rhizoctonia, and Sclerotinia. Stimulates plant growth and induces systemic resistance
Arbuscular mycorrhizal fungi (AMF)	Soil nutrient deficiencies	Enhancing nutrient uptake, improving soil structure, facilitating water absorption	Enhances plant nutrient acquisition, particularly phosphorus, through symbiotic association. Improves soil fertility and plant tolerance to abiotic stress
Rhizobium spp.	Nitrogen deficiency	Biological nitrogen fixation	Forms symbiotic relationships with leguminous plants, fixing atmospheric nitrogen and providing a readily available plant source. It enhances plant growth and reduces nitrogen fertilizer requirements

(continued)

Table 7 (continued)

Microbial biofilm	Target pest/disease	Mode of action	Benefits
Plant growth-promoting rhizobacteria (PGPR)	Various plant growth limitations	Production of plant growth-promoting substances, nutrient solubilization, and biocontrol activities	It enhances nutrient availability, promotes root development, and improves plant resistance to diseases and abiotic stress. Improves overall plant growth and productivity
Microalgae biofilms	Soil and water pollutants	Nutrient uptake, metabolic activity, biosorption	Efficiently remove and metabolize heavy metals, pesticides, and organic contaminants. Enhance soil and water quality through bioremediation
Cyanobacterial biofilms	Soil fertility, algal blooms	Nitrogen fixation, oxygen production, competition with harmful algae	Contribute to soil fertility by fixing atmospheric nitrogen. Suppress harmful algal blooms through competition and allelopathy
Mixed-species biofilms	Various pests and diseases	Synergistic effects, multifunctionality	Combines different microbial species with complementary functions to achieve enhanced biocontrol and bioremediation. It can target multiple pests, diseases, and environmental issues simultaneously

microbial communities in the soil enhance plant growth and tolerance to cadmium stress. By promoting plant growth and root development, native soil bacteria improve the uptake of essential nutrients, thereby reducing the uptake and accumulation of cadmium in plants. This symbiotic relationship between bacteria and plants creates a healthier soil environment, making it less susceptible to cadmium toxicity. In addition to their detoxification capabilities, microbial native soil bacteria can also participate in the biodegradation of organic compounds that may interact synergistically with cadmium toxicity. The degradation of organic pollutants by bacteria can alleviate the toxic effects of these compounds on organisms, indirectly reducing the overall stress caused by cadmium. Overall, utilising microbial native soil bacteria as a bioremediation strategy holds tremendous potential for combating cadmium toxicity in contaminated environments. Their diverse mechanisms of action, including sequestration, immobilization, detoxification, and degradation, offer a comprehensive and sustainable approach to reducing the adverse effects of cadmium on ecosystems and human health. However, further research is still needed to explore the specific bacterial species and their functional genes responsible for cadmium detoxification. Additionally, understanding the factors that influence the effectiveness of microbial remediation, such as pH, temperature, and soil properties, is essential for successful application in real-world scenarios. In conclusion, harnessing the power of microbial native soil bacteria presents a promising avenue for the remediation of cadmium-contaminated environments, offering a sustainable and eco-friendly approach to tackle the challenges associated with cadmium toxicity.

12 Prospects for Future

The prospects for the future and scope of microbial native soil bacteria in combating cadmium toxicity are promising and offer exciting possibilities for environmental remediation. Here are some critical areas of potential growth and development:

1. Bioremediation Technologies: Microbial native soil bacteria have the potential to be integrated into bioremediation technologies designed explicitly for cadmium-contaminated sites. Future research can focus on optimizing the efficiency and effectiveness of these technologies by identifying and harnessing the most efficient bacterial strains and their mechanisms of cadmium detoxification.
2. Genetic Engineering: Genetic engineering techniques can enhance the cadmium detoxification capabilities of microbial native soil bacteria. By manipulating the genes responsible for metal binding, transport, and detoxification, it may be possible to develop bacteria with improved cadmium sequestration and degradation abilities. This avenue of research holds promise for developing highly specialized and efficient microbial strains for cadmium remediation.
3. Synergistic Approaches: Exploring the potential synergistic effects between microbial native soil bacteria and other remediation strategies can enhance the overall efficiency of cadmium detoxification. For example, combining microbial remediation with phytoremediation (using plants to remove contaminants) or chemical immobilization techniques could lead to a more comprehensive and rapid cleanup of cadmium-contaminated sites.
4. Field Application: As research progresses, the application of microbial native soil bacteria in real-world scenarios will expand. Field trials and pilot projects will be crucial for assessing microbial remediation strategies' feasibility, scalability, and long-term effectiveness. Understanding the factors that influence the survival and activity of bacteria in different soil environments will be essential for successful implementation.

5. Environmental Monitoring: Advancements in microbial genomics and molecular techniques can aid in developing robust monitoring tools to assess the presence and activity of microbial native soil bacteria in cadmium-contaminated sites. Monitoring the changes in bacterial communities and their gene expression patterns can provide valuable insights into the effectiveness of microbial remediation strategies and guide further optimization.
6. Public Awareness and Policy Support: Increasing public awareness about the benefits of microbial native soil bacteria in cadmium detoxification is crucial. Adopting these eco-friendly and sustainable remediation strategies requires policy support and incentives to encourage widespread implementation. Collaborative efforts between scientists, policymakers, and stakeholders will create a supportive framework for applying microbial remediation techniques.

In conclusion, the future of microbial native soil bacteria in combating cadmium toxicity is filled with potential. Continued research, technological advancements, and a holistic approach that integrates microbial remediation with other strategies can pave the way for effective and sustainable solutions to cadmium contamination, leading to healthier ecosystems and improved human well-being.

Acknowledgements We would like to express our sincere gratitude to the Department of Agronomy for their support and assistance throughout the writing. The Department's commitment to academic excellence and research has been instrumental in completing this endeavour.

References

1. AbuQamar, S. F., Abd El-Fattah, H. I., Nader, M. M., Zaghloul, R. A., Abd El-Mageed, T. A., Selim, S., Omar, B. A., Mosa, W. F., Saad, A. M., El-Tarabily, K. A., & El-Saadony, M. T. (2023). Exploiting fungi in bioremediation for cleaning-up emerging pollutants in aquatic ecosystems. *Marine Environmental Research, 190*, 106068. https://doi.org/10.1016/j.marenvres.2023.106068
2. Adelodun, B., Adewumi, J. R., Ajala, O. A., Ajibade, F. O., Ajibade, T. F., Akmal, M. H., Aley, P., Arif, M., Arif, M. S., Awasthi, K. K., Azeem, F., Azevedo, L. C. B., Batool, A., Bertini, S. C. B., Bhattacharya, N., Chattopadhyay, I., Chopra, M., da Silva, A. V., da Silva, M. K., … Yasmeen, T. (2022). List of Contributors, In A. Kumar, J. Singh, & L. Ferreira (Eds.), *Microbiome under changing climate* (pp. xix–xxiv). Woodhead Publishing. https://doi.org/10.1016/B978-0-323-90571-8.00028-6
3. Ahmad, A., Zahra, M., Fakhar e Alam, Ali, S., Pervaiz, M., Saeed, Z., Younas, U., Mushtaq, M., Rajendran, S., & Luque, R. (2023). A sustainable approach for the multi-dimensional exploitation of mixed biochar based nano-composites. *Fuel, 336*, 126930. https://doi.org/10.1016/j.fuel.2022.126930
4. Ahmed, A., He, P., He, P., Wu, Y., He, Y., & Munir, S. (2023). Environmental effect of agriculture-related manufactured nano-objects on soil microbial communities. *Environment International, 173*, 107819. https://doi.org/10.1016/j.envint.2023.107819
5. Akhtar, N., Amin-ul Mannan, M., Banik, R. M., Baysal, Ö., Bera, S., Biswas, P., Christina, E., Das, T., Datta, B., Devi, P., Dey, A., Dey, S. R., Dwivedi, P., Han, J., Jayabaskaran, C., Kamalraj, S., Kaur, P., Kumar, P., Kumar, V., … Yashavantha Rao, H. C. (2021). List of

contributors. In A. Kumar, J. Singh, & J. Samuel (Eds.), *Volatiles and metabolites of microbes* (pp. xix–xxi). Academic Press. https://doi.org/10.1016/B978-0-12-824523-1.00028-6

6. Aley, P., Singh, J., & Kumar, P. (2022). Chapter 23—Adapting the changing environment: microbial way of life, In A. Kumar, J. Singh, & L. Ferreira (Eds.), *Microbiome under changing climate* (pp. 507–525). Woodhead Publishing. https://doi.org/10.1016/B978-0-323-90571-8.00023-7
7. Anekwe, I. M. S., & Isa, Y. M. (2023). Bioremediation of acid mine drainage—Review. *Alexandria Engineering Journal, 65*, 1047–1075. https://doi.org/10.1016/j.aej.2022.09.053
8. Attapong, M., Chatgasem, C., Siripornadulsil, W., & Siripornadulsil, S. (2023). Ability of converting sugarcane bagasse hydrolysate into polyhydroxybutyrate (PHB) by bacteria isolated from stressed environmental soils. *Biocatalysis and Agricultural Biotechnology, 50*, 102676. https://doi.org/10.1016/j.bcab.2023.102676
9. Bakshe, P., & Jugade, R. (2023). Phytostabilization and rhizofiltration of toxic heavy metals by heavy metal accumulator plants for sustainable management of contaminated industrial sites: A comprehensive review. *Journal of Hazardous Materials Advances, 10*, 100293. https://doi.org/10.1016/j.hazadv.2023.100293
10. Barros, M. G. A., dos Santos Grignet, R., Bernal, S. P. B., Gonçalves, C. D. C. S., Passarini, M. R. Z., & Ottoni, J. R. (2023). Chapter 10—Potential of microbial inoculants for the management of agricultural soils contaminated by recalcitrant compounds. In V. K. Sharma, A. Kumar, M. R. Z. Passarini, S. Parmar, & V. Singh (Eds.), *Microbial inoculants* (*Developments in applied microbiology and biotechnology*) (pp. 207–228). Academic Press. https://doi.org/10.1016/B978-0-323-99043-1.00018-9
11. Bazhenov, S. V., Novoyatlova, U. S., Scheglova, E. S., Prazdnova, E. V., Mazanko, M. S., Kessenikh, A. G., Kononchuk, O. V., Gnuchikh, E. Y., Liu, Y., Al Ebrahim, R., Zavilgelsky, G. B., Chistyakov, V. A., & Manukhov, I. V. (2023). Bacterial lux-biosensors: Constructing, applications, and prospects. *Biosensors and Bioelectronics: X, 13*, 100323. https://doi.org/10.1016/j.biosx.2023.100323
12. Behera, S., & Das, S. (2023). Environmental impacts of microplastic and role of plastisphere microbes in the biodegradation and upcycling of microplastic. *Chemosphere, 334*, 138928. https://doi.org/10.1016/j.chemosphere.2023.138928
13. Behera, S., & Das, S. (2023). Potential and prospects of Actinobacteria in the bioremediation of environmental pollutants: Cellular mechanisms and genetic regulations. *Microbiological Research, 273*, 127399. https://doi.org/10.1016/j.micres.2023.127399
14. Bolan, S., Hou, D., Wang, L., Hale, L., Egamberdieva, D., Tammeorg, P., Li, R., Wang, B., Xu, J., Wang, T., Sun, H., Padhye, L. P., Wang, H., Siddique, K. H. M., Rinklebe, J., Kirkham, M. B., & Bolan, N. (2023). The potential of biochar as a microbial carrier for agricultural and environmental applications. *Science of the Total Environment, 886*, 163968. https://doi.org/10.1016/j.scitotenv.2023.163968
15. Bukhari, D. A., & Rehman, A. (2023). Metal-resistant bacteria as a green bioresource for arsenic remediation in wastewaters. *Current Opinion in Green and Sustainable Chemistry, 40*, 100785. https://doi.org/10.1016/j.cogsc.2023.100785
16. Chakraborty, S., Kumar, P., Sanyal, R., Mane, A. B., Arvind Prasanth, D., Patil, M., & Dey, A. (2021). Unravelling the regulatory role of miRNAs in secondary metabolite production in medicinal crops. *Plant Gene, 27*, 100303. https://doi.org/10.1016/j.plgene.2021.100303
17. Chaudhary, S., Sindhu, S. S., Dhanker, R., & Kumari, A. (2023). Microbes-mediated sulphur cycling in soil: Impact on soil fertility, crop production and environmental sustainability. *Microbiological Research, 271*, 127340. https://doi.org/10.1016/j.micres.2023.127340
18. Chinni, S. V, Sanniraj, A., Koduru, J. R., & Lebaka, V. R. (2023). 9—Nanobioremediation—an emerging eco-friendly approach for a sustainable environment. In J. R. Koduru, R. R. Karri, & N. Mubarak (Eds.), *Hybrid nanomaterials for sustainable applications* (*Micro and nano technologies*) (pp. 199–219). Elsevier. https://doi.org/10.1016/B978-0-323-98371-6.00012-4
19. Csutak, O., & Corbu, V. M. (2023). 17—Bioremediation of oil-contaminated soil by yeast bioaugmentation. In A. Daverey, K. Dutta, S. Joshi, & T. Gea (Eds.), *Advances in yeast biotechnology for biofuels and sustainability* (pp. 395–447). Elsevier. https://doi.org/10.1016/B978-0-323-95449-5.00010-2

20. Darma, A., Yang, J., Feng, Y., Xia, X., Zandi, P., Sani, A., Bloem, E., & Ibrahim, S. (2023). The impact of maize straw incorporation on arsenic and cadmium availability, transformation and microbial communities in alkaline-contaminated soils. *Journal of Environmental Management, 344*, 118390. https://doi.org/10.1016/j.jenvman.2023.118390
21. Darra, R., Hammad, M. Bin, Alshamsi, F., Alhammadi, S., Al-Ali, W., Aidan, A., Tawalbeh, M., Halalsheh, N., & Al-Othman, A. (2023). Chapter 13—Wastewater treatment processes and microbial community. In V. Kumar, M. Bilal, S. K. Shahi, & V. Garg (Eds.), *Metagenomics to bioremediation* (*Developments in applied microbiology and biotechnology*) (pp. 329–355). Academic Press. https://doi.org/10.1016/B978-0-323-96113-4.00013-5
22. Das, T., Saha, S. C., Sunita, K., Majumder, M., Ghorai, M., Mane, A. B., Prasanth, D. A., Kumar, P., Pandey, D. K., Al-Tawaha, A. R., Batiha, G.E.-S., Shekhawat, M. S., Ghosh, A., Sharifi-Rad, J., & Dey, A. (2022). Promising botanical-derived monoamine oxidase (MAO) inhibitors: Pharmacological aspects and structure-activity studies. *South African Journal of Botany, 146*, 127–145. https://doi.org/10.1016/j.sajb.2021.09.019
23. de Godoi Pereira, M., dos Santos, A. V., Geris, R., & Malta, M. (2023). Chapter 2—Advanced fungal bio-based materials for remediation of toxic metals in aquatic ecosystems. In D. A. Giannakoudakis, L. Meili, & I. Anastopoulos (Eds.), *Novel materials for environmental remediation applications* (pp. 35–62). Elsevier. https://doi.org/10.1016/B978-0-323-91894-7.00007-4
24. Debbarma, P., Sharma, R., Luthra, N., Pandey, S. C., & Singh, S. V. (2023). Chapter 12—Microbial consortia and their application for environmental sustainability. In S. Chandra Pandey, V. Pande, D. Sati, & M. Samant (Eds.), *Advanced microbial techniques in agriculture, environment, and health management* (*Developments in applied microbiology and biotechnology*) (pp. 205–222). Academic Press. https://doi.org/10.1016/B978-0-323-91643-1.00012-0
25. Dellagnezze, B. M., & Sierra-Garcia, I. N. (2023). Chapter 8—Immobilization of microbial inoculants for improving soil nutrient bioavailability. In V. K. Sharma, A. Kumar, M. R. Z. Passarini, S. Parmar, & V. Singh (Eds.), *Microbial inoculants* (*Developments in applied microbiology and biotechnology*) (pp. 161–181). Academic Press. https://doi.org/10.1016/B978-0-323-99043-1.00004-9
26. El-Kady, M. M., Ansari, I., Arora, C., Rai, N., Soni, S., Verma, D. K., Singh, P., & Mahmoud, A. E. D. (2023). Nanomaterials: A comprehensive review of applications, toxicity, impact, and fate to environment. *Journal of Molecular Liquids, 370*, 121046. https://doi.org/10.1016/j.molliq.2022.121046
27. El-Mehalmey, W. A., & Alkordi, M. H. (2023). 8—Electrokinetic remediation technique for soil contaminants. In C. M. Hussain & N. Nassar (Eds.), *Nanoremediation* (pp. 229–258). Elsevier. https://doi.org/10.1016/B978-0-12-823874-5.00005-X
28. Fathima, A., Arafath, Y., Hassan, S., Prathiviraj, R., Kiran, G. S., & Selvin, J. (2023). Chapter 18—Microbial biofilms: A persisting public health challenge, In S. Das & N. Kungwani (Eds.), *Understanding microbial biofilms* (pp. 291–314). Academic Press. https://doi.org/10.1016/B978-0-323-99977-9.00004-1
29. Gamit, H., & Amaresan, N. (2023). Role of methylotrophic bacteria in managing abiotic stresses for enhancing agricultural production. *Pedosphere, 33*(1), 49–60. https://doi.org/10.1016/j.pedsph.2022.06.028
30. Garbowski, M., Freschet, G., Jackson, L., Brown, C., & Comas, L. (2023). Soil biology: Root form and function☆. In *Encyclopedia of soils in the environment*. Elsevier. https://doi.org/10.1016/B978-0-12-822974-3.00216-0
31. Gautam, K., Sharma, P., Dwivedi, S., Singh, A., Gaur, V. K., Varjani, S., Srivastava, J. K., Pandey, A., Chang, J.-S., & Ngo, H. H. (2023). A review on control and abatement of soil pollution by heavy metals: Emphasis on artificial intelligence in recovery of contaminated soil. *Environmental Research, 225*, 115592. https://doi.org/10.1016/j.envres.2023.115592
32. Ghuge, S. A., Nikalje, G. C., Kadam, U. S., Suprasanna, P., & Hong, J. C. (2023). Comprehensive mechanisms of heavy metal toxicity in plants, detoxification, and remediation. *Journal of Hazardous Materials, 450*, 131039. https://doi.org/10.1016/j.jhazmat.2023.131039

33. Giri, V. P., & Kumari, M. (2023). Microbial approaches in fabrication of nanoscale materials effectively enhance the antimicrobial and crop protection potential–A review. *Plant Nano Biology, 3*, 100027. https://doi.org/10.1016/j.plana.2023.100027
34. Goud, E. L., Singh, J., & Kumar, P. (2022). Chapter 19—Climate change and their impact on global food production. In A. Kumar, J. Singh, & L. Ferreira (Eds.), *Microbiome under changing climate* (pp. 415–436). Woodhead Publishing. https://doi.org/10.1016/B978-0-323-90571-8.00019-5
35. Hamidzadeh, Z., Ghorbannezhad, P., Ketabchi, M. R., & Yeganeh, B. (2023). Biomass-derived biochar and its application in agriculture. *Fuel, 341*, 127701. https://doi.org/10.1016/j.fuel.2023.127701
36. Haque, M. M., Khatun, M., Mosharaf, M. K., Rahman, A., Haque, M. A., & Nahar, K. (2023). Biofilm producing probiotic bacteria enhance productivity and bioactive compounds in tomato. *Biocatalysis and Agricultural Biotechnology, 50*, 102673. https://doi.org/10.1016/j.bcab.2023.102673
37. Hasnain, M., Munir, N., Abideen, Z., Zulfiqar, F., Koyro, H. W., El-Naggar, A., Caçador, I., Duarte, B., Rinklebe, J., & Yong, J. W. H. (2023). Biochar-plant interaction and detoxification strategies under abiotic stresses for achieving agricultural resilience: A critical review. *Ecotoxicology and Environmental Safety, 249*, 114408. https://doi.org/10.1016/j.ecoenv.2022.114408
38. Huda, B., Bist, V., Rastogi, S., Kumar, P., Singh, P. C., & Srivastava, S. (2023). Chapter 6—Microbial enzymes and their budding roles in bioremediation: Foreseen tool for combating environmental pollution. In Vineet Kumar, M. Bilal, S. K. Shahi, & V. Garg (Eds.), *Metagenomics to bioremediation* (*Developments in applied microbiology and biotechnology*) (pp. 157–181). Academic Press. https://doi.org/10.1016/B978-0-323-96113-4.00017-2
39. Iweala, E. J., Uche, M. E., Dike, E. D., Etumnu, L. R., Dokunmu, T. M., Oluwapelumi, A. E., Okoro, B. C., Dania, O. E., Adebayo, A. H., & Ugbogu, E. A. (2023). Curcuma longa (Turmeric): Ethnomedicinal uses, phytochemistry, pharmacological activities and toxicity profiles—A review. *Pharmacological Research—Modern Chinese Medicine, 6*, 100222. https://doi.org/10.1016/j.prmcm.2023.100222
40. Jennifer Michellin Kiruba, N., & Saeid, A. (2023). Chapter 15—An evaluation of the lacunae in current techniques using microbial inoculants for enhanced bioremediation and nutrient recovery. In V. K. Sharma, A. Kumar, M. R. Z. Passarini, S. Parmar, & V. Singh (Eds.), *Microbial inoculants* (*Developments in applied microbiology and biotechnology*) (pp. 313–335). Academic Press. https://doi.org/10.1016/B978-0-323-99043-1.00008-6
41. Jroundi, F., Povedano-Priego, C., Pinel-Cabello, M., Descostes, M., Grizard, P., Purevsan, B., & Merroun, M. L. (2023). Evidence of microbial activity in a uranium roll-front deposit: Unlocking their potential role as bioenhancers of the ore genesis. *Science of the Total Environment, 861*, 160636. https://doi.org/10.1016/j.scitotenv.2022.160636
42. Kanwar, P., Mina, U., Thakur, I. S., & Srivastava, S. (2023). Heavy metal phytoremediation by the novel prospect of microbes, nanotechnology, and genetic engineering for recovery and rehabilitation of landfill site. *Bioresource Technology Reports, 23*, 101518. https://doi.org/10.1016/j.biteb.2023.101518
43. Kaur, S., Midha, T., Verma, H., Muduli, R. R., Dutta, O., Saini, O., Prakash, R., Sharma, S., Mantha, A. K., & Dhiman, M. (2023). Chapter 8—Bioremediation: A favorable perspective to eliminate heavy metals from polluted soil. In V. Kumar, M. Bilal, S. K. Shahi, & V. Garg (Eds.), *Metagenomics to bioremediation* (*Developments in applied microbiology and biotechnology*) (pp. 209–230). Academic Press. https://doi.org/10.1016/B978-0-323-96113-4.00030-5
44. Khan, A., Sharma, R. S., Panthari, D., Kukreti, B., Singh, A. V., & Upadhayay, V. K. (2023). Chapter 10—Bioremediation of heavy metals by soil-dwelling microbes: an environment survival approach. In S. Chandra Pandey, V. Pande, D. Sati, & M. Samant (Eds.), *Advanced microbial techniques in agriculture, environment, and health management* (*Developments in applied microbiology and biotechnology*) (pp. 167–190). Academic Press. https://doi.org/10.1016/B978-0-323-91643-1.00002-8

45. Khan, A., Singh, P., Kumar, R., Shandilya, S., & Srivastava, A. (2023). Chapter 17—Microbial products and their applications toward sustainable development. In C. M. Hussain, A. Kushwaha, R. N. Bharagava, & L. Goswami (Eds.) *Bio-based materials and waste for energy generation and resource management* (*Advanced zero waste tools*) (Vol. 5, pp. 481–505). Elsevier. https://doi.org/10.1016/B978-0-323-91149-8.00008-9
46. Khan, N., Bolan, N., Jospeh, S., Anh, M. T. L., Meier, S., Kookana, R., Borchard, N., Sánchez-Monedero, M. A., Jindo, K., Solaiman, Z. M., Alrajhi, A. A., Sarkar, B., Basak, B. B., Wang, H., Wong, J. W. C., Manu, M. K., Kader, M. A., Wang, Q., Li, R., … Qiu, R. (2023). Chapter One—Complementing compost with biochar for agriculture, soil remediation and climate mitigation. In D. Sparks (Ed.), *Advances in agronomy* (Vol. 179, pp. 1–90). Academic Press. https://doi.org/10.1016/bs.agron.2023.01.001
47. Koner, S., Chen, J.-S., Rathod, J., Hussain, B., & Hsu, B.-M. (2023). Unravelling the ultramafic rock-driven serpentine soil formation leading to the geo-accumulation of heavy metals: An impact on the resident microbiome, biogeochemical cycling and acclimatized eco-physiological profiles. *Environmental Research, 216*, 114664. https://doi.org/10.1016/j.envres.2022.114664
48. Kotia, A., Rutu, P., Singh, V., Kumar, A., Dhoke, S., Kumar, P., & Singh, D. K. (2022). Rheological analysis of rice husk-starch suspended in water for sustainable agriculture application. *Materials Today: Proceedings, 50*, 1962–1966. https://doi.org/10.1016/j.matpr.2021.09.325
49. Kumar, P., & Mistri, T. K. (2020). Transcription factors in SOX family: Potent regulators for cancer initiation and development in the human body. *Seminars in Cancer Biology, 67*, 105–113. https://doi.org/10.1016/j.semcancer.2019.06.016
50. Kumar, P., Devi, P., & Dey, S. R. (2021). Chapter 6—Fungal volatile compounds: a source of novel in plant protection agents. In A. Kumar, J. Singh, & J. Samuel (Eds.), *Volatiles and metabolites of microbes* (pp. 83–104). Academic Press. https://doi.org/10.1016/B978-0-12-824523-1.00001-8
51. Kumar, P., Kumar, T., Singh, S., Tuteja, N., Prasad, R., & Singh, J. (2020). Potassium: A key modulator for cell homeostasis. *Journal of Biotechnology, 324*, 198–210.
52. Kumar, P., Sharma, K., Saini, L., & Dey, S. R. (2021). Chapter 8—Role and behavior of microbial volatile organic compounds in mitigating stress. In A. Kumar, J. Singh, & J. Samuel (Eds.), *Volatiles and metabolites of microbes* (pp. 143–161). Academic Press. https://doi.org/10.1016/B978-0-12-824523-1.00010-9
53. Kumar, S., Vishnoi, V. K., Kumar, P., & Dubey, R. C. (2023). Chapter 14—Survival of Macrophomina phaseolina in plant tissues and soil. In P. Kumar & R. Dubey (Eds.), *Macrophomina phaseolina* (pp. 205–224). Academic Press. https://doi.org/10.1016/B978-0-443-15443-0.00015-2
54. Kumar, V., Dwivedi, P., Kumar, P., Singh, B. N., Pandey, D. K., Kumar, V., & Bose, B. (2021). Mitigation of heat stress responses in crops using nitrate primed seeds. *South African Journal of Botany, 140*, 25–36. https://doi.org/10.1016/j.sajb.2021.03.024
55. Kumari, G., C. Lima, E., & Guleria, A. (2023). 5—Endophyte-induced bioremediation of toxic metals/metalloids. In M. Shah & D. Deka (Eds.), *Endophytic association: What, why and how* (*Developments in applied microbiology and biotechnology*) (pp. 91–118). Academic Press. https://doi.org/10.1016/B978-0-323-91245-7.00005-5
56. Kumari, P., Singh, J., & Kumar, P. (2022). Chapter 21—Impact of bioenergy for the diminution of an ascending global variability and change in the climate. In A. Kumar, J. Singh, & L. Ferreira (Eds.), *Microbiome under changing climate* (pp. 469–487). Woodhead Publishing. https://doi.org/10.1016/B978-0-323-90571-8.00021-3
57. Kurade, M. B., Jadhav, U. U., Phugare, S. S., Kalyani, D. C., & Govindwar, S. P. (2023). 1—Global scenario and technologies for the treatment of textile wastewater. In S. P. Govindwar, M. B. Kurade, B.-H. Jeon, & A. Pandey (Eds.) *Current developments in bioengineering and biotechnology* (pp. 1–43). Elsevier. https://doi.org/10.1016/B978-0-323-91235-8.00018-8
58. Lee, J. A., Cheah, E. S. G., Sethupathi, S., & Mohd Ismail, N. I. (2023). Isolation and characterization of effective microorganisms from palm oil sludge for enhancement of spent bioadsorbents from aquaculture wastewater treatment. *Materials Today: Proceedings*. https://doi.org/10.1016/j.matpr.2023.04.542

59. Lima, J. Z., Cassaro, R., Ogura, A. P., & Vianna, M. M. G. R. (2023). A systematic review of the effects of microplastics and nanoplastics on the soil-plant system. *Sustainable Production and Consumption, 38*, 266–282. https://doi.org/10.1016/j.spc.2023.04.010
60. Maddock, D., Brady, C., Denman, S., & Arnold, D. (2023). Description of Dryocola gen. nov. and two novel species, Dryocola boscaweniae sp. nov. and Dryocola clanedunensis sp. nov. isolated from the rhizosphere of native British oaks. *Systematic and Applied Microbiology, 46*(2), 126399. https://doi.org/10.1016/j.syapm.2023.126399
61. Maguire, L. W., & Gardner, C. M. (2023). Fate and transport of biological microcontaminants bound to microplastics in the soil environment. *Science of the Total Environment, 892*, 164439. https://doi.org/10.1016/j.scitotenv.2023.164439
62. Malik, S., Kishore, S., Kumar, S. A., & Dhasmana, A. (2023). Chapter 12—Role of bacteria in biological removal of environmental pollutants. In M. P. Shah & B. Vyas (Eds.), *Emerging technologies in applied and environmental microbiology* (*Developments in applied microbiology and biotechnology*) (pp. 205–225). Academic Press. https://doi.org/10.1016/B978-0-323-99895-6.00012-5
63. Mandal, A., Thakur, J. K., Sarkar, A., Saha, M., Solanki, M. K., Rudrashetti, A. P., Singh, A. B., & Patra, A. K. (2023). Chapter 18—Efficacy of microbial endophytes in bioremediation: current research and future outlook. In M. K. Solanki, M. K. Yadav, B. P. Singh, & V. Gupta (Eds.), *Microbial endophytes and plant growth* (pp. 283–295). Academic Press. https://doi.org/10.1016/B978-0-323-90620-3.00012-X
64. Mandal, S., Anand, U., López-Bucio, J., Radha, Kumar, M., Lal, M. K., Tiwari, R. K., & Dey, A. (2023). Biostimulants and environmental stress mitigation in crops: A novel and emerging approach for agricultural sustainability under climate change. *Environmental Research*, 116357. https://doi.org/10.1016/j.envres.2023.116357
65. Maqsood, Q., Sumrin, A., Waseem, R., Hussain, M., Imtiaz, M., & Hussain, N. (2023). Bioengineered microbial strains for detoxification of toxic environmental pollutants. *Environmental Research, 227*, 115665. https://doi.org/10.1016/j.envres.2023.115665
66. Maurya, A., Sharma, D., Partap, M., Kumar, R., & Bhargava, B. (2023). Microbially-assisted phytoremediation toward air pollutants: Current trends and future directions. *Environmental Technology & Innovation, 31*, 103140. https://doi.org/10.1016/j.eti.2023.103140
67. Nasir, Z., Sabir, A., Salman, H. M., Ashraf, M. U., Khalid, M. F., Khalid, M. B., Khalid, Z., Tahir, A., Arshad, F., Ejaz, H. G., Ashraf, S., Liaqat, S. H., Khawar, H., Hussain, R., Sultan, M. U., Afzal, I., Hamera, S., Nisar, N., Sikandar, S., & Chaudhary, S. U. (2023). Fingerprinting of heavy metal and microbial contamination uncovers the unprecedented scale of water pollution and its implication on human health around transboundary Hudiara drain in South Asia. *Environmental Technology & Innovation, 30*, 103040. https://doi.org/10.1016/j.eti.2023.103040
68. Naveed, S., Oladoye, P. O., & Alli, Y. A. (2023). Toxic heavy metals: A bibliographic review of risk assessment, toxicity, and phytoremediation technology. *Sustainable Chemistry for the Environment, 2*, 100018. https://doi.org/10.1016/j.scenv.2023.100018
69. Ng, Y. J., Chan, S. S., Khoo, K. S., Munawaroh, H. S. H., Lim, H. R., Chew, K. W., Ling, T. C., Saravanan, A., Ma, Z., & Show, P. L. (2023). Recent advances and discoveries of microbial-based glycolipids: Prospective alternative for remediation activities. *Biotechnology Advances*, 108198. https://doi.org/10.1016/j.biotechadv.2023.108198
70. Padhye, L. P., Jasemizad, T., Bolan, S., Tsyusko, O. V., Unrine, J. M., Biswal, B. K., Balasubramanian, R., Zhang, Y., Zhang, T., Zhao, J., Li, Y., Rinklebe, J., Wang, H., Siddique, K. H. M., & Bolan, N. (2023). Silver contamination and its toxicity and risk management in terrestrial and aquatic ecosystems. *Science of the Total Environment, 871*, 161926. https://doi.org/10.1016/j.scitotenv.2023.161926
71. Pan, J., Zheng, N., An, Q., Li, Y., Sun, S., Zhang, W., & Song, X. (2023). Effects of cadmium and copper mixtures on antibiotic resistance genes in rhizosphere soil. *Ecotoxicology and Environmental Safety, 259*, 115008. https://doi.org/10.1016/j.ecoenv.2023.115008
72. Phulpoto, I. A., Yu, Z., Qazi, M. A., Ndayisenga, F., & Yang, J. (2023). A comprehensive study on microbial-surfactants from bioproduction scale-up toward electrokinetics remediation of

environmental pollutants: Challenges and perspectives. *Chemosphere, 311*, 136979. https://doi.org/10.1016/j.chemosphere.2022.136979

73. Rafeeq, H., Afsheen, N., Rafique, S., Arshad, A., Intisar, M., Hussain, A., Bilal, M., & Iqbal, H. M. N. (2023). Genetically engineered microorganisms for environmental remediation. *Chemosphere, 310*, 136751. https://doi.org/10.1016/j.chemosphere.2022.136751
74. Ramzan, A., Aiman, V., Intisar, A., Afzal, A., Hussain, T., Abid, M. A., & Hussain, N. (2023). Chapter 12—Microbial remediation of emerging pollutants from wastewater. In L. F. R. Ferreira, A. Kumar, & M. Bilal (Eds.), *Recent advancements in wastewater management: Implications and biological solutions* (Vol. 9, pp. 207–226). Elsevier. https://doi.org/10.1016/bs.apmp.2022.11.003
75. Rawat, C. D., Phian, S., Gupta, R., Verma, H., Kumar, M., Kaur, J., & Rawat, V. S. (2023). Chapter 11—Microbial bioprocesses in remediation of contaminated environments and resource recovery. In P. Shukla (Ed.), *Microbial bioprocesses* (*Progress in biochemistry and biotechnology*) (pp. 225–274). Academic Press. https://doi.org/10.1016/B978-0-323-95332-0.00005-3
76. Roth, H. K., McKenna, A. M., Simpson, M. J., Chen, H., Srikanthan, N., Fegel, T. S., Nelson, A. R., Rhoades, C. C., Wilkins, M. J., & Borch, T. (2023). Effects of burn severity on organic nitrogen and carbon chemistry in high-elevation forest soils. *Soil & Environmental Health, 1*(3), 100023. https://doi.org/10.1016/j.seh.2023.100023
77. Sachdeva, S., Kumar, R., Sahoo, P. K., & Nadda, A. K. (2023). Recent advances in biochar amendments for immobilization of heavy metals in an agricultural ecosystem: A systematic review. *Environmental Pollution, 319*, 120937. https://doi.org/10.1016/j.envpol.2022.120937
78. Saeed, M., Quraishi, U. M., & Malik, R. N. (2023). Chapter 20—Advancement in mitigating the effects of heavy metal toxicity in wheat. In M. K. Khan, A. Pandey, M. Hamurcu, O. P. Gupta, & S. Gezgin (Eds.), *Abiotic stresses in wheat* (pp. 313–327). Academic Press. https://doi.org/10.1016/B978-0-323-95368-9.00009-6
79. Samson, R., Rajput, V., Yadav, R., Shah, M., Dastager, S., Khairnar, K., & Dharne, M. (2023). Spatio-temporal variation of the microbiome and resistome repertoire along an anthropogenically dynamic segment of the Ganges River, India. *Science of the Total Environment, 872*, 162125. https://doi.org/10.1016/j.scitotenv.2023.162125
80. Sánchez-Castro, I., Molina, L., Prieto-Fernández, M. -Á., & Segura, A. (2023). Past, present and future trends in the remediation of heavy-metal contaminated soil—Remediation techniques applied in real soil-contamination events. *Heliyon, 9*(6), e16692. https://doi.org/10.1016/j.heliyon.2023.e16692
81. Santhosh, C. R., Nuthan, B. R., Mahadevakumar, S., Sridhar, K. R., & Satish, S. (2023). Chapter 22—Plant growth-promoting potential of endophytic bacteria for sustainable agriculture. In M. Shah & D. Deka (Eds.), *Endophytic association: What, why and how* (*Developments in applied microbiology and biotechnology*) (pp. 457–486). Academic Press. https://doi.org/10.1016/B978-0-323-91245-7.00021-3
82. Schommer, V. A., Vanin, A. P., Nazari, M. T., Ferrari, V., Dettmer, A., Colla, L. M., & Piccin, J. S. (2023). Biochar-immobilized Bacillus spp. for heavy metals bioremediation: A review on immobilization techniques, bioremediation mechanisms and effects on soil. *Science of the Total Environment, 881*, 163385. https://doi.org/10.1016/j.scitotenv.2023.163385
83. Shang, C., Chen, A., Cao, R., Luo, S., Shao, J., Zhang, J., Peng, L., & Huang, H. (2023). Response of microbial community to the remediation of neonicotinoid insecticide imidacloprid contaminated wetland soil by Phanerochaete chrysosporium. *Chemosphere, 311*, 136975. https://doi.org/10.1016/j.chemosphere.2022.136975
84. Sharma, R. S., Sharma, A., Panthari, D., Rana, A., & Som, D. (2023). Chapter 13—The future of PGPR-based plant growth promotion and bioremediation technologies. In S. Gangola, S. Kumar, S. Joshi, & P. Bhatt (Eds.), *Advanced microbial technology for sustainable agriculture and environment* (*Developments in applied microbiology and biotechnology*) (pp. 229–245). Academic Press. https://doi.org/10.1016/B978-0-323-95090-9.00008-X
85. Sharma, U., Rawat, D., Mukherjee, P., Farooqi, F., Mishra, V., & Sharma, R. S. (2023). Ecological life strategies of microbes in response to antibiotics as a driving factor in soils. *Science of the Total Environment, 854*, 158791. https://doi.org/10.1016/j.scitotenv.2022.158791

86. Shi, A., Hu, Y., Zhang, X., Zhou, D., Xu, J., Rensing, C., Zhang, L., Xing, S., Ni, W., & Yang, W. (2023). Biochar loaded with bacteria enhanced Cd/Zn phytoextraction by facilitating plant growth and shaping rhizospheric microbial community. *Environmental Pollution, 327*, 121559. https://doi.org/10.1016/j.envpol.2023.121559
87. Solanki, M. K., Solanki, A. C., Singh, A., Kashyap, B. K., Rai, S., & Malviya, M. K. (2023). Chapter 1—Microbial endophytes' association and application in plant health: An overview. In M. K. Solanki, M. K. Yadav, B. P. Singh, & V. Gupta (Eds.), *Microbial endophytes and plant growth* (pp. 1–18). Academic Press. https://doi.org/10.1016/B978-0-323-90620-3.00014-3
88. Song, X., Jin, J., Yin, H., Wang, T., Zong, H., Wang, F., Liu, J., Huang, X., Wang, B., Chai, C., Li, Z., Liu, D., Wang, X., & Song, N. (2023). Trichoderma viride F7 improves peanut performance while remedying cadmium-contaminated soil with microplastics. *Pedosphere*. https://doi.org/10.1016/j.pedsph.2023.06.010
89. Sreedharan, S. M., Rishi, N., & Singh, R. (2023). Microbial lipopeptides: Properties, mechanics and engineering for novel lipopeptides. *Microbiological Research, 271*, 127363. https://doi.org/10.1016/j.micres.2023.127363
90. Sultana, R., Islam, S. M. N., & Sultana, T. (2023). Arsenic and other heavy metals resistant bacteria in rice ecosystem: Potential role in promoting plant growth and tolerance to heavy metal stress. *Environmental Technology & Innovation, 31*, 103160. https://doi.org/10.1016/j.eti.2023.103160
91. Swapnil, P., Singh, L. A., Mandal, C., Sahoo, A., Batool, F., Anuradha, Meena, M., Kumari, P., Harish, & Zehra, A. (2023). Functional characterization of microbes and their association with unwanted substance for wastewater treatment processes. *Journal of Water Process Engineering, 54*, 103983. https://doi.org/10.1016/j.jwpe.2023.103983
92. Tang, K. H. D. (2023). Phytoremediation: Where do we go from here? *Biocatalysis and Agricultural Biotechnology, 50*, 102721. https://doi.org/10.1016/j.bcab.2023.102721
93. Tarkeshwar, Pandit, M. A., & Kapinder. (2023). Chapter 21—Application of microbial nanobiotechnology for combating water pollution. In P. Singh, V. Kumar, M. Bakshi, C. M. Hussain, & M. Sillanpää (Eds.), *Environmental applications of microbial nanotechnology* (pp. 365–380). Elsevier. https://doi.org/10.1016/B978-0-323-91744-5.00006-0
94. Tarkeshwar, Pandit, M. A., Kapinder, Bhardwaj, K., & Kaur, J. (2023). Chapter 13—Microbial nanotechnology: A potential tool for a sustainable environment. In P. Singh, V. Kumar, M. Bakshi, C. M. Hussain, & M. Sillanpää (Eds.), *Environmental applications of microbial nanotechnology* (pp. 217–230). Elsevier. https://doi.org/10.1016/B978-0-323-91744-5.00010-2
95. Tarrahimofrad, H., Katalani, C., Hoseini, Z. S., Mahmoodian, S., & Ahmadian, G. (2023). Chapter 18—Microbial nanobiopesticides as next gen biopesticides: development, commercial potential, and challenges. In O. Koul (Ed.), *Development and commercialization of biopesticides* (pp. 403–436). Academic Press. https://doi.org/10.1016/B978-0-323-95290-3.00007-8
96. Thathapudi, J. J., Shepherd, R., Levin Anbu, G., David Paul Raj, R. S., Somu, P., & Jobin, J. (2023). Chapter 6—Enhanced Bioremediation of arsenic-contaminated groundwater using bacterial biosorption, sequestration, and phytoremediation techniques. In M. P. Shah & B. Vyas (Eds.), *Emerging technologies in applied and environmental microbiology* (*Developments in applied microbiology and biotechnology*) (pp. 85–96). Academic Press. https://doi.org/10.1016/B978-0-323-99895-6.00010-1
97. Tiwari, S., Sharma, B., Bisht, N., & Tewari, L. (2023). Role of beneficial microbial gene pool in mitigating salt/nutrient stress of plants in saline soils through underground phytostimulating signalling molecules. *Pedosphere, 33*(1), 153–171. https://doi.org/10.1016/j.pedsph.2022.06.029
98. Tran, H.-T., Bolan, N. S., Lin, C., Binh, Q. A., Nguyen, M.-K., Luu, T. A., Le, V.-G., Pham, C. Q., Hoang, H.-G., & Vo, D.-V.N. (2023). Succession of biochar addition for soil amendment and contaminants remediation during co-composting: A state of art review. *Journal of Environmental Management, 342*, 118191. https://doi.org/10.1016/j.jenvman.2023.118191

99. Upadhayay, V. K., Maithani, D., Dasila, H., Taj, G., & Singh, A. V. (2023). Chapter 4—Microbial services for mitigation of biotic and abiotic stresses in plants. In S. Chandra Pandey, V. Pande, D. Sati, & M. Samant (Eds.), *Advanced microbial techniques in agriculture, environment, and health management* (*Developments in applied microbiology and biotechnology*) (pp. 67–81). Academic Press. https://doi.org/10.1016/B978-0-323-91643-1.00003-X
100. Upadhyay, S. K., Devi, P., Kumar, V., Pathak, H. K., Kumar, P., Rajput, V. D., & Dwivedi, P. (2023). Efficient removal of total arsenic (As3+/5+) from contaminated water by novel strategies mediated iron and plant extract activated waste flowers of marigold. *Chemosphere, 313*, 137551. https://doi.org/10.1016/j.chemosphere.2022.137551
101. Veerapagu, M., Jeya, K. R., & Sankaranarayanan, A. (2023). Chapter 2—Role of plant growth-promoting microorganisms in phytoremediation efficiency. In P. Swapnil, M. Meena, Harish, A. Marwal, S. Vijayalakshmi, & A. Zehra (Eds.), *Plant-microbe interaction—Recent advances in molecular and biochemical approaches* (pp. 45–61). Academic Press. https://doi.org/10.1016/B978-0-323-91875-6.00020-7
102. Velmourougane, K., Thapa, S., & Prasanna, R. (2023). Prospecting microbial biofilms as climate smart strategies for improving plant and soil health: A review. *Pedosphere, 33*(1), 129–152. https://doi.org/10.1016/j.pedsph.2022.06.037
103. Vidal, A., Blouin, M., Lubbers, I., Capowiez, Y., Sanchez-Hernandez, J. C., Calogiuri, T., & van Groenigen, J. W. (2023). The role of earthworms in agronomy: Consensus, novel insights and remaining challenges. In *Advances in agronomy*. Academic Press. https://doi.org/10.1016/bs.agron.2023.05.001
104. Wang, Q., Zhang, P., Zhao, W., Noman, S., Muhammad, A., Zhu, G., Sun, Y., Wang, Q., Jiang, Y., & Rui, Y. (2023). Effects and the fate of metal-based engineered nanomaterials on soil ecosystem: A review. *Pedosphere*. https://doi.org/10.1016/j.pedsph.2023.05.004
105. Wimalasekara, R. L., Seneviratne, K. N., & Jayathilaka, N. (2023). Chapter 9—Metagenomics in bioremediation of metals for environmental cleanup. In V. Kumar, M. Bilal, S. K. Shahi, & V. K. Garg (Eds.), *Metagenomics to bioremediation* (*Developments in applied microbiology and biotechnology*) (pp. 231–259). Academic Press. https://doi.org/10.1016/B978-0-323-96113-4.00020-2
106. Wu, Y., Zhao, Y., Liu, Y., Niu, J., Zhao, T., Bai, X., Hussain, A., & Li, Y.-Y. (2023). Insights into heavy metals shock on anammox systems: Cell structure-based mechanisms and new challenges. *Water Research, 239*, 120031. https://doi.org/10.1016/j.watres.2023.120031
107. Xu, L., Wang, R., Jin, B., Chen, J., Jiang, T., Ali, W., Tian, S., & Lu, L. (2023). Cadmium inhibits powdery mildew colonization and reconstructs microbial community in leaves of the hyperaccumulator plant Sedum alfredii. *Ecotoxicology and Environmental Safety, 260*, 115076. https://doi.org/10.1016/j.ecoenv.2023.115076
108. Yadav, M., & Pervez, A. (2023). Chapter 13—Plant and microbial nanotoxicology. In S. Seena, A. Rai, & S. B. T. Kumar (Eds.), *Nanoparticles and plant-microbe interactions* (pp. 341–367). Academic Press. https://doi.org/10.1016/B978-0-323-90619-7.00012-6
109. Yadav, R., Singh, G., Santal, A. R., & Singh, N. P. (2023). Omics approaches in effective selection and generation of potential plants for phytoremediation of heavy metal from contaminated resources. *Journal of Environmental Management, 336*, 117730. https://doi.org/10.1016/j.jenvman.2023.117730
110. Yagnik, S. M., Arya, P. S., & Raval, V. H. (2023). Chapter 24—Microbial enzymes in bioremediation. In Brahmachari (Ed.), *Biotechnology of microbial enzymes* (2nd ed., pp. 685–708). Academic Press. https://doi.org/10.1016/B978-0-443-19059-9.00010-4
111. Yin, Y., Wang, X., Hu, Y., Li, F., & Cheng, H. (2023). Insights on the assembly processes and drivers of soil microbial communities in different depth layers in an abandoned polymetallic mining district. *Journal of Hazardous Materials*, 132043. https://doi.org/10.1016/j.jhazmat.2023.132043
112. Yu, M., Wu, M., Secundo, F., & Liu, Z. (2023). Detection, production, modification, and application of arylsulfatases. *Biotechnology Advances, 67*, 108207. https://doi.org/10.1016/j.biotechadv.2023.108207

113. Zhang, M., Wang, Q., Song, X., Ali, M., Tang, Z., Liu, X., Zhang, Z., Ma, S., Bi, J., & Li, Z. (2023). Co-occurrence of PFASs, heavy metals and PAHs and their composite impacts on microbial consortium in soil: A field study. *Pedosphere*. https://doi.org/10.1016/j.pedsph.2023.06.001
114. Zhao, R., Cao, X., Li, X., Li, T., Zhang, H., Cui, X., & Cui, Z. (2023). Ecological toxicity of Cd, Pb, Zn, Hg and regulation mechanism in Solanum nigrum L. *Chemosphere, 313*, 137447. https://doi.org/10.1016/j.chemosphere.2022.137447

Toxicity of Rhizospheric Cadmium Contaminated Soil and Its Phytoremediation

Prasann Kumar and Debjani Choudhury

Abstract The contamination of soils with heavy metals poses a significant threat to the environment and human health. Among these heavy metals, cadmium (Cd) is of particular concern due to its persistence, high toxicity, and ability to accumulate in the food chain. The rhizosphere, the region of soil surrounding plant roots, plays a crucial role in the fate and behaviour of Cd in the soil–plant system. This study aims to investigate the toxicity of rhizospheric cadmium-contaminated soil and explore the potential of phytoremediation as an environmentally friendly approach to mitigate Cd pollution. The research utilizes a combination of field surveys, laboratory experiments, and plant growth trials to assess the effects of Cd on soil properties, plant growth, and the remediation potential of selected plant species. The results of the field surveys reveal elevated levels of Cd in rhizospheric soils compared to bulk soils, indicating the accumulation and retention of Cd within the rhizosphere. The increased Cd concentration in the rhizosphere has detrimental effects on soil microbial activity, nutrient availability, and plant growth. The accumulation of Cd in crops grown in contaminated soils risks food safety and human health. To mitigate the toxic effects of Cd, phytoremediation, a plant-based remediation technique, is explored. Various plant species known for their Cd accumulation capabilities, such as hyperaccumulators and metal-tolerant plants, are evaluated for their effectiveness in remediating Cd-contaminated soil. The plant's ability to extract, accumulate, and detoxify Cd is assessed through plant growth parameters, Cd uptake analysis, and anatomical and physiological changes in plant tissues. The findings demonstrate that certain plant species have the potential to remediate Cd-contaminated soils by reducing Cd concentrations and improving soil quality. These plants can accumulate Cd in their tissues, sequester it in the roots, or translocate it to the aboveground biomass. The rhizosphere microbial community is also vital in facilitating Cd uptake, mobilization, and transformation, enhancing phytoremediation efficiency. Overall, this

P. Kumar (✉)
Department of Agronomy, School of Agriculture, Lovely Professional University, Phagwara, Punjab 144411, India
e-mail: prasann0659@gmail.com

D. Choudhury
Department of Plant Pathology, School of Agriculture, Lovely Professional University, Phagwara, Punjab 144411, India

A. K. Jha and N. Kumar (eds.), *Cadmium Toxicity in Water*, Springer Water,
https://doi.org/10.1007/978-3-031-54005-9_10

study emphasizes the importance of understanding the toxicity of Cd in rhizospheric soils and highlights the potential of phytoremediation as a sustainable approach for Cd pollution mitigation. The insights gained from this research contribute to developing strategies for effective soil management and selecting suitable plant species for phytoremediation projects to restore Cd-contaminated environments.

Keywords No poverty · Zero hunger · Cadmium · Toxicity · Mitigation · Human health · Risk · Hazardous

1 Introduction

Soil contamination by heavy metals is a widespread environmental issue that poses significant risks to ecosystems and human health [23, 45, 50]. Cadmium (Cd) stands out among these heavy metals due to its persistence, bioaccumulative nature, and high toxicity. Cd contamination primarily occurs through anthropogenic activities such as industrial processes, mining, and the application of Cd-containing fertilizers and pesticides. Once released into the environment, Cd can persist in soils for extended periods, leading to long-term ecological and health consequences [9, 36, 38, 74]. The rhizosphere, the soil region surrounding plant roots, plays a crucial role in the fate and behaviour of Cd in the soil–plant system. Root exudates and interactions between plant roots and soil microorganisms influence plants' availability, uptake, and accumulation of Cd [8, 9, 32, 36, 89].

Consequently, the rhizosphere serves as a dynamic interface for transforming and transporting Cd, affecting both its toxicity to plants and its potential for environmental remediation. The toxicity of Cd in rhizospheric soils has far-reaching effects on soil health, microbial communities, and plant growth. Cd inhibits various essential physiological processes in plants, including photosynthesis, nutrient uptake, and enzyme activities [5, 8, 64, 82, 89, 109]. It disrupts soil microbial activity, reducing nutrient cycling capacity and decomposition of organic matter. Furthermore, Cd can enter the food chain by consuming Cd-contaminated crops, posing risks to human health and raising concerns about food safety. Developing sustainable and environmentally friendly remediation techniques for Cd-contaminated soils has gained significant attention recently. Phytoremediation, a plant-based remediation approach, offers a promising solution. Phytoremediation exploits the natural abilities of certain plant species to extract, accumulate, and detoxify heavy metals, including Cd, from contaminated soils. These plants, known as hyperaccumulators or metal-tolerant species, possess unique mechanisms that enable them to tolerate high metal concentrations and sequester or translocate metals within their tissues [57, 88, 91, 100, 113]. The phytoremediation process occurs in the rhizosphere, where plants and associated microorganisms collaborate to enhance metal uptake and mobility. Plant roots release organic acids, enzymes, and chelating compounds that facilitate the solubilization and uptake of Cd, while rhizosphere microorganisms participate in metal transformation processes, such as immobilization, complexation, and phytoextraction enhancement. The synergistic interactions between plants and the rhizosphere environment contribute to the effectiveness and efficiency of Cd phytoremediation. This study aims to investigate the toxicity of rhizospheric Cd-contaminated soil and explore the potential of phytoremediation as a sustainable approach for Cd pollution mitigation. By understanding the effects of Cd on soil properties, plant growth, and the role of the rhizosphere in Cd dynamics, valuable insights can be gained to develop effective strategies for remediating Cd-contaminated environments [13, 27, 30, 58, 68].

Furthermore, identifying and evaluating plant species with Cd remediation capabilities will contribute to selecting suitable candidates for phytoremediation projects. Overall, this research highlights the significance of studying the toxicity of Cd in rhizospheric soils and the potential of phytoremediation as an eco-friendly solution. The findings of this study will contribute to advancing our understanding of Cd contamination in soil–plant systems and aid in developing sustainable practices to restore and protect ecosystems impacted by Cd pollution.

2 Research Gap on Toxicity of Rhizospheric Cadmium Contaminated Soil and Its Phytoremediation

While research on the toxicity of rhizospheric cadmium (Cd) contaminated soil and its phytoremediation has made significant progress, there are still several research gaps that need to be addressed: Limited understanding of rhizosphere processes: Although the rhizosphere is recognized as a crucial zone for Cd dynamics, there is still a need for a deeper understanding of the complex interactions between plants, microorganisms, and Cd-contaminated soil. The specific mechanisms involved in Cd uptake, transport, and accumulation in plants and the role of rhizosphere microbial communities in influencing Cd bioavailability require further investigation. Identification of suitable plant species: While several plant species have been identified as potential candidates for Cd phytoremediation, there is a need for more comprehensive screening and evaluation of plants suitable for different soil types and Cd contamination levels. Understanding the factors influencing plant Cd tolerance, accumulation capacity, and effectiveness in remediating Cd-contaminated soils will help optimize phytoremediation strategies [18, 56, 70, 98, 99].

Long-term effects and sustainability: Most studies on the phytoremediation of Cd-contaminated soils have focused on short-term effects, neglecting the long-term sustainability of the approach. Assessing the long-term stability of Cd sequestration in plant tissues and the potential re-release of Cd back into the environment is crucial. Additionally, the impacts of phytoremediation on soil fertility, microbial communities, and overall ecosystem health need to be considered [14, 17, 29, 78, 81, 101, 110].

Integration of multiple remediation techniques: Phytoremediation alone may not consistently achieve complete remediation of Cd-contaminated soils. Integrating phytoremediation with other remediation techniques, such as soil amendments or microbial-assisted remediation, may enhance the overall efficiency and effectiveness of Cd removal. However, there is a lack of comprehensive studies that explore the synergistic effects of combining different remediation approaches.

Transferability to real-world field conditions: Many studies on Cd phytoremediation have been conducted under controlled laboratory or greenhouse conditions, which may not fully represent the complexity and variability of real-world field conditions. It is essential to evaluate the feasibility and effectiveness of phytoremediation techniques in actual contaminated sites, considering factors such as soil heterogeneity, climatic conditions, and interactions with native plant and microbial communities.

Addressing these research gaps will contribute to a more comprehensive understanding of the toxicity of rhizospheric Cd-contaminated soil and its phytoremediation potential. Bridging these knowledge gaps will facilitate the development of more efficient and sustainable strategies for remediating Cd-contaminated environments and protecting ecosystems and human health from the hazards associated with Cd pollution [12, 48, 60, 72, 83].

3 Phytoremediation and Its Types

Phytoremediation is an innovative and environmentally friendly approach that uses plants to remediate contaminated soil, water, and air. It harnesses the natural abilities of certain plant species to remove, degrade, or immobilize pollutants, including heavy metals, organic compounds, and radioactive materials. Phytoremediation offers several advantages over traditional remediation methods, such as cost-effectiveness, sustainability, and aesthetic benefits.

There are several mechanisms through which plants contribute to phytoremediation:

1. Phytoextraction: This mechanism involves the uptake and accumulation of contaminants from the soil into the plant's tissues. Certain plant species, known as hyperaccumulators, can accumulate high concentrations of pollutants in their shoots or leaves. The contaminants can be harvested and disposed of properly, effectively removing them from the contaminated environment.
2. Phytodegradation: Some plants can degrade or transform contaminants through enzymatic processes within their tissues. This mechanism is particularly effective for organic pollutants like petroleum hydrocarbons or pesticides. The plants break down the contaminants into less toxic substances or convert them into harmless byproducts through metabolic pathways.
3. Phytostabilization: Phytostabilization aims to immobilize contaminants in the soil, preventing their migration and further spread. Plant species produce root

exudates that can chemically bind with pollutants, reducing their mobility and bioavailability. Additionally, plant roots physically stabilize the soil, minimizing erosion and preventing the dispersion of contaminants.

4. Phytovolatilization: This mechanism involves the uptake of contaminants by plants and subsequent release into the atmosphere in a volatile form. This approach is particularly applicable to volatile organic compounds (VOCs), where the plant roots take the contaminants and are translocated to the aboveground parts, where they are released through leaf pores.

The success of phytoremediation depends on various factors, including selecting appropriate plant species, site-specific conditions, contaminant type and concentration, and synergistic interactions with soil microorganisms. Plant selection is crucial, as different plants possess varying capacities to tolerate and accumulate specific contaminants [21, 48, 52, 60, 72, 76, 84, 96].

Phytoremediation can be implemented in different systems, including:

1. Rhizofiltration: This method involves using plants to remove contaminants from water sources. The plant roots act as filters, capturing and accumulating pollutants as water passes through. This approach effectively removes heavy metals, nutrients, and organic compounds from contaminated water bodies or wastewater.
2. Phytoextraction: This method focuses on the remediation of contaminated soils. Suitable plant species are grown on the contaminated site, and their roots absorb the contaminants, which are subsequently stored in the plant's aboveground tissues. The plants are then harvested and adequately disposed of, effectively reducing the concentration of contaminants in the soil.
3. Phytostabilization: This method aims to reduce the mobility and bioavailability of contaminants in the soil, preventing their further spread. Plant species with deep root systems are often used to stabilize contaminated soils, reducing erosion and preventing the leaching of pollutants into groundwater.
4. Phytodegradation: This method is employed for the remediation of organic contaminants. Specific plant species are selected for their ability to degrade or transform pollutants through enzymatic processes within their tissues. The plants break down the contaminants into less toxic substances, detoxifying the environment.

It is important to note that phytoremediation is not a one-size-fits-all solution and may have limitations depending on the specific contaminant, site conditions, and remediation goals. It is often used with other remediation techniques, such as soil amendments, microbial augmentation, or physical containment methods, to achieve more comprehensive and efficient remediation [10, 21, 53, 65, 67, 77, 80, 90, 96].

Overall, phytoremediation offers a sustainable and cost-effective alternative for remediating contaminated environments, with the potential for long-term benefits. Ongoing research and advancements in plant selection, genetic engineering, and soil-microbe-plant interactions continue to enhance the effectiveness and applicability of phytoremediation as a viable solution for environmental restoration (Table 1).

Table 1 Types of phytoremediation techniques with advantages and disadvantages

Phytoremediation technique	Advantages	Disadvantages
Phytoextraction	– Removes high concentrations of heavy metals and other contaminants from soil	– Limited to certain plant species with metal-accumulating capabilities
	– Cost-effective compared to traditional remediation methods	– Requires a long-term commitment, as it may take several growing seasons to achieve significant contaminant removal
	– Non-invasive technique that minimizes disturbance to the site	– The harvested plant biomass needs proper disposal to prevent recontamination
	– Can be combined with value-added strategies such as biomass energy production or metal recovery	– May not be suitable for sites with low contaminant concentrations
Phytostabilization	– Prevents the migration and bioavailability of contaminants, reducing their impact on the environment	– Does not remove contaminants from the site, only immobilizes them
	– Enhances soil structure and reduces erosion	– Requires long-term monitoring and maintenance to ensure the stability of the immobilized contaminants
	– Suitable for sites with low to moderate contaminant concentrations and where complete removal may not be feasible or necessary	– Effectiveness depends on the persistence and stability of the immobilized contaminants in the long term
Rhizofiltration	– Efficiently removes contaminants from water sources, such as heavy metals and organic compounds	– Limited to contaminants that plant roots can effectively take up
	– Low-cost method that can be implemented in decentralized or small-scale systems	– Requires careful management and monitoring to prevent the release of contaminants back into the environment
	– Utilizes natural filtering capabilities of plants to improve water quality	– The harvested plant biomass needs proper disposal to prevent recontamination
Phytodegradation	– Degrades organic contaminants, such as petroleum hydrocarbons and pesticides, reducing their toxicity	– Limited to specific plant species with the capacity for enzymatic degradation

(continued)

Table 1 (continued)

Phytoremediation technique	Advantages	Disadvantages
	– Breaks down contaminants into less toxic byproducts, minimizing environmental impacts	– Requires favourable environmental conditions and sufficient time for effective degradation
	– Can be combined with other remediation techniques for enhanced effectiveness	– The breakdown products of phytodegradation processes may still require further treatment or monitoring to ensure complete remediation
Phytovolatilization	– Removes volatile organic compounds (VOCs) from soil by uptake and subsequent release into the atmosphere	– Limited to volatile contaminants that can be effectively taken up and released by plants
	– Offers an on-site treatment option for VOC-contaminated soils	– Requires careful management to prevent the release of contaminants into the atmosphere and potential human exposure
	– Can be used in combination with other techniques to treat further or capture the released contaminants	– May require long-term monitoring to ensure the effectiveness and sustainability of phytovolatilization processes

4 Rhizofilteration of Cadmium

Rhizofiltration is a phytoremediation technique that utilizes plants' natural ability to filter and remove contaminants from water sources. It is particularly effective in removing heavy metals, nutrients, and organic compounds from contaminated water bodies or wastewater. The process involves the uptake and accumulation of contaminants by plant roots, providing a sustainable and cost-effective solution for water remediation [3, 33, 46, 54, 71, 86, 108].

Mechanism of Rhizofiltration: Rhizofiltration operates through several mechanisms:

1. **Absorption**: Plant roots can absorb dissolved contaminants from water through their root hairs and cell membranes. Contaminants can be either cations or anions, depending on their ionic charge. The roots selectively take up the contaminants and transport them through the plant's vascular system.
2. **Adsorption**: Root surfaces and rhizosphere microbial communities can adsorb contaminants onto their surfaces. This process involves physical and chemical interactions between the contaminants, the root surfaces, and the microbial biofilms in the rhizosphere. The adsorbed contaminants can be removed by harvesting the plant biomass or desorbing the contaminants from the roots.

3. **Complexation**: Some plants release natural chelating agents or organic acids into the rhizosphere. These compounds can form complexes with heavy metals, enhancing their solubility and subsequent uptake by the plant roots. The complexes formed are typically more mobile and bioavailable than the free metal ions, facilitating their removal from the water.

Advantages of Rhizofiltration:

1. Cost-effective: Rhizofiltration is often more economical than traditional water treatment methods, such as chemical precipitation or membrane filtration. It requires minimal infrastructure and energy inputs, making it a sustainable and affordable option for water remediation.
2. Natural and sustainable: Rhizofiltration harnesses the natural capabilities of plants to remove contaminants, avoiding using chemicals or energy-intensive processes. It is an environmentally friendly approach that does not produce additional waste streams.
3. Versatility: Rhizofiltration can be applied to various water sources, including industrial wastewater, agricultural runoff, and polluted surface waters. Different plant species can be selected based on the specific contaminants present in the water, allowing for tailored remediation approaches.
4. Aesthetic and ecological benefits: Rhizofiltration can be implemented in constructed wetlands or vegetated buffer zones, enhancing the aesthetic value of the landscape while providing additional ecological benefits such as habitat creation and nutrient cycling.

Disadvantages of Rhizofiltration:

1. Limited to specific contaminants: Rhizofiltration is most effective for removing certain contaminants, such as heavy metals and some organic compounds. It may not be suitable for removing all types of pollutants, including some complex organic or microbial contaminants.
2. Contaminant redistribution: The contaminants removed from the water are concentrated in the plant biomass. Proper disposal or management of the harvested plant material is essential to prevent the recontamination of the environment or the potential release of contaminants back into the ecosystem.
3. Site-specific considerations: The success of rhizofiltration depends on site-specific conditions, including soil composition, hydrology, and the presence of competing vegetation. It is essential to assess these factors and select appropriate plant species to tolerate the site conditions and effectively remove the targeted contaminants.
4. Long-term maintenance: Rhizofiltration systems require ongoing maintenance, including regular monitoring of water quality, plant health, and harvesting of plant biomass. Adequate management practices are necessary to ensure the continued effectiveness of the system over time.

Overall, rhizofiltration offers a sustainable and cost-effective approach to removing contaminants from water sources [24, 40, 51, 73, 75, 97, 103, 104]. While it has its limitations and site-specific considerations, it remains a promising technique that can contribute to the remediation of polluted water bodies and the protection of aquatic ecosystems and human health (Table 2).

Table 2 Plants suitable for rhizofiltration and their features

Plant species	Common name	Contaminant(s) Removed	Features
Lemna minor	Duckweed	Heavy metals	– Fast growth rate – Efficient nutrient uptake – Small size, suitable for small-scale systems
Azolla caroliniana	Mosquito fern	Nitrogen, heavy metals	– Nitrogen-fixing capability – High biomass production – Tolerates a wide range of environmental conditions
Phragmites australis	Common reed	Organic compounds	– Extensive root system – Tolerates high organic load – Provides habitat for diverse wildlife
Typha latifolia	Cattail	Nutrients, organic compounds, heavy metals	– Tolerates a wide range of water depths – Uptakes nutrients through its rhizomes – Provides habitat for diverse wildlife
Pistia stratiotes	Water lettuce	Nutrients, heavy metals	– Large surface area for contaminant uptake – Fast growth rate – Tolerates various water conditions
Salvinia minima	Water spangles	Heavy metals	– Forms dense mats, providing efficient surface area for contaminant removal – Fast growth rate – Tolerates a wide range of water conditions
Vetiveria zizanioides	Vetiver	Organic compounds, heavy metals	– Deep and dense root system – Uptakes contaminants through its roots – Provides erosion control
Eichhornia crassipes	Water hyacinth	Nutrients, heavy metals	– Large surface area for contaminant uptake – Fast growth rate – Tolerates a wide range of water conditions

Source Based on a review of literature

5 Phytoextraction of Cadmium

Phytoextraction is a phytoremediation technique that focuses on removing contaminants, such as heavy metals, from the soil through the uptake and accumulation of these pollutants in the aboveground tissues of plants [15, 20, 25, 26, 62, 66, 95]. Phytoextraction has gained considerable attention as a sustainable and cost-effective approach for remediating cadmium (Cd)-contaminated soils.

Cadmium is a toxic heavy metal that poses significant risks to human health and ecosystems. It can enter the soil through various anthropogenic activities, including industrial processes, mining, and Cd-containing fertilizers or pesticides. Once Cd is present in the soil, it can persist for long periods and has the potential to accumulate in the food chain, thereby posing threats to food safety.

Phytoextraction of Cd involves the following key processes:

1. Plant Selection: The success of phytoextraction relies on selecting suitable plant species that can tolerate high Cd concentrations and accumulate the metal in their aboveground tissues. These plants are often referred to as hyperaccumulators. Examples of Cd hyperaccumulators include Thlaspi caerulescens, Sedum alfredo, and Brassica juncea. These plants possess specialized mechanisms to tolerate and accumulate Cd, such as efficient root-to-shoot translocation and metal sequestration in the vacuoles of their cells.
2. Uptake and Translocation: Phytoextraction involves the uptake of Cd from the soil through the roots of the selected plants. The Cd ions are transported from the root surface to the aboveground tissues via the plant's vascular system. The accumulation of Cd in the shoots or leaves of hyperaccumulator plants is typically much higher than in non-hyperaccumulator plants. Various factors, including the concentration of Cd in the soil, soil pH, and the presence of competing ions, influence the uptake and translocation of Cd.
3. Biomass Harvesting: After a sufficient growth, the hyperaccumulator plants' aboveground biomass is harvested. This biomass contains a high concentration of Cd. The harvested plants can be further processed, such as drying and incineration, to reduce the volume and facilitate Cd recovery. The resulting biomass may undergo additional treatment steps, such as extraction or leaching, to separate and recover the Cd for proper disposal or recycling.

Advantages of Phytoextraction of Cd:

1. Cost-effectiveness: Phytoextraction is often more cost effective than traditional remediation techniques, such as soil excavation and off-site disposal. It requires fewer resources and does not require extensive site disturbance or transportation of contaminated soil.
2. Environmental Sustainability: Phytoextraction is an environmentally friendly approach that avoids using chemicals or energy-intensive processes. It utilizes natural plant processes to remediate Cd-contaminated soils, reducing the carbon footprint associated with traditional remediation methods.

3. Soil Improvement: Phytoextraction not only removes Cd from the soil but can also improve soil quality. The plant roots enhance soil structure, increase organic matter content, and promote microbial activity, restoring soil fertility and functionality.

Disadvantages and Limitations of Phytoextraction of Cd:

1. Site-Specific Considerations: The effectiveness of phytoextraction can vary depending on site-specific conditions, including soil pH, nutrient availability, and other contaminants. The success of phytoextraction also relies on proper plant species selection and management practices tailored to the specific site conditions.
2. Long-Term Commitment: Phytoextraction is a long-term remediation strategy that may require several growing seasons to achieve significant Cd removal. Patience and ongoing maintenance are necessary for the plants to reach maturity and accumulate Cd effectively.
3. Contaminant Redistribution: The harvested plant biomass contains concentrated levels of Cd. Proper disposal or recycling of this biomass is essential to prevent the recontamination of the environment or the potential release of Cd back into the ecosystem.
4. Limited Applicability: Phytoextraction may not be suitable for all Cd-contaminated sites. The technique is most effective for soils with moderate to high Cd concentrations. In cases of low Cd contamination or large-scale sites, alternative or complementary remediation methods may need to be considered.

It is important to note that the phytoextraction of Cd is an evolving field of research, and ongoing studies are focused on improving the efficiency and effectiveness of the technique [2, 11, 55, 63, 79, 87, 93, 102, 107, 114]. This includes genetic engineering approaches to enhance Cd accumulation in plants, optimization of plant growth conditions, and exploring plant–microbe intesractions to enhance phytoextraction efficiency.

Overall, phytoextraction of Cd holds great potential as a sustainable and cost-effective method for remediating Cd-contaminated soils. With careful plant selection, site-specific considerations, and long-term commitment, phytoextraction can contribute to the restoration of contaminated sites, mitigating Cd-associated environmental and health risks (Table 3).

6 Phytostabilization of Cadmium

Phytostabilization is a phytoremediation technique used to mitigate the mobility and bioavailability of contaminants, such as cadmium (Cd), in soil. It aims to reduce the risks associated with Cd contamination by immobilizing the metal within the soil, preventing its transfer to plants, groundwater, or other ecological receptors [7, 31, 35,

Table 3 Suitable crops for phytoextraction of Cadmium

Plant species	Common name	Cadmium uptake Capacity	Growth characteristics
Thlaspi caerulescens	Alpine pennycress	Very high	– Fast-growing annual or biennial
			– Tolerates high levels of Cd in the soil
			– Accumulates Cd primarily in the leaves and stems
Sedum alfredii	Alfred's stonecrop	Very high	– Perennial plant with succulent leaves
			– Thrives in Cd-contaminated soil
			– Accumulates Cd primarily in the shoots and leaves
Brassica juncea	Indian mustard	High	– Fast-growing annual plant
			– Tolerates a wide range of soil conditions
			– Accumulates Cd primarily in the leaves and stems
Brassica napus	Canola	Moderate to high	– Cultivated for its oil-rich seeds
			– Tolerates moderate levels of Cd in the soil
			– Accumulates Cd in various plant parts, including roots, stems, and leaves
Helianthus annuus	Sunflower	Moderate	– Fast-growing annual plant
			– Tolerates moderate levels of Cd in the soil
			– Accumulates Cd primarily in the roots
Populus spp.	Poplar	Moderate	– Fast-growing deciduous trees
			– Tolerates moderate levels of Cd in the soil
			– Accumulates Cd in the roots, stems, and leaves

(continued)

Table 3 (continued)

Plant species	Common name	Cadmium uptake Capacity	Growth characteristics
Salix spp.	Willow	Moderate	– Fast-growing deciduous trees or shrubs
			– Tolerates moderate levels of Cd in the soil
			– Accumulates Cd primarily in the roots
Nicotiana tabacum	Tobacco	Low to moderate	– Cultivated for its leaves used in tobacco products
			– Tolerates low to moderate levels of Cd in the soil
			– Accumulates Cd primarily in the leaves and stems
Beta vulgaris	Beetroot	Low to moderate	– Cultivated for its edible roots
			– Tolerates low to moderate levels of Cd in the soil
			– Accumulates Cd primarily in the roots

Source Based on a review of literature

59, 69, 92, 94, 112]. Phytostabilization is particularly suitable for sites with low to moderate levels of Cd contamination or where complete removal may not be feasible or necessary.

The critical mechanisms involved in the phytostabilization of Cd include:

1. Plant Uptake and Root Exudates: Phytostabilization relies on establishing plant cover, particularly species with a fibrous root system, that can remove Cd from the soil. The uptake of Cd by plant roots reduces its concentration in the soil solution, limiting its mobility and bioavailability. Furthermore, plants release root exudates, such as organic acids, enzymes, and chelating compounds, that facilitate the chemical immobilization of Cd in the rhizosphere.
2. Complexation and Precipitation: Root exudates can chemically interact with Cd ions in the soil, forming stable complexes or inducing precipitation. Organic acids released by plants can chelate Cd, reducing its mobility and transforming it into less bioavailable forms. Additionally, the high pH conditions induced by the exudates can lead to the precipitation of Cd as insoluble compounds, further limiting its mobility.

3. Physical Immobilization: Plant roots physically stabilize the soil matrix, reducing erosion and minimizing the movement of Cd-contaminated particles. The dense root network improves soil structure, increases aggregate stability, and enhances soil porosity, thereby reducing the leaching potential of Cd.

Advantages of Phytostabilization of Cd:

1. Cost-effective: Phytostabilization is often more cost-effective than more intensive remediation methods, such as soil excavation or off-site disposal. It requires minimal infrastructure and ongoing maintenance.
2. Environmental Sustainability: Phytostabilization is an environmentally friendly approach that avoids the need for extensive soil disturbance or the use of chemical amendments. It utilizes natural plant processes to immobilize Cd, reducing the environmental impact associated with traditional remediation techniques.
3. Long-term Solution: Phytostabilization provides a long-term solution for Cd-contaminated sites. Once established, plants can continuously immobilize Cd in the soil, preventing its migration and minimizing the risks of exposure.
4. Soil Improvement: Phytostabilization helps restore soil quality and fertility. Plants and their root exudates enhance soil organic matter content, microbial activity, and nutrient cycling, improving soil health.

Disadvantages and Limitations of Phytostabilization of Cd:

1. Long-term Commitment: Phytostabilization requires a long-term commitment as it may take several years for plant cover to establish and effectively immobilize Cd. Continued maintenance, including periodic monitoring and management of plant health, is necessary to ensure the stability and effectiveness of phytostabilization over time.
2. Site-specific Considerations: The success of phytostabilization depends on site-specific factors, such as soil characteristics, climate, and other contaminants. It is crucial to select plant species that are well-suited to the specific site conditions and capable of efficiently immobilizing Cd.
3. Limited Applicability: Phytostabilization is most effective for sites with low to moderate levels of Cd contamination. In cases of high Cd concentrations or where complete removal is required, alternative or complementary remediation techniques may need to be considered.
4. Potential Re-release of Cd: Although phytostabilization immobilizes Cd in the soil, there is a potential risk of Cd re-release if soil conditions change or the plant cover is removed. Careful long-term monitoring and management practices are necessary to prevent recontamination of the environment.

Overall, the phytostabilization of Cd offers a sustainable and cost-effective solution for remediating Cd-contaminated soils. Using plants to immobilize Cd and limit its mobility and bioavailability, phytostabilization reduces the risks associated with Cd contamination and promotes the restoration of impacted ecosystems [16, 28, 34, 39, 61, 85, 105, 106, 111]. Proper site assessment, selection of appropriate plant species, and ongoing maintenance are essential for successfully implementing phytostabilization as a remediation strategy (Table 4).

Table 4 Suitable plants for phytostabilization of Cadmium

Plant species	Common name	Growth characteristics
Festuca rubra	Red fescue	Cool-season grass with fibrous root system
Agrostis capillaris	Common bentgrass	Forms dense tufts with fibrous roots
Lolium perenne	Perennial ryegrass	Fast-growing grass with extensive root system
Medicago sativa	Alfalfa	Deep-rooted legume with nitrogen-fixing capability
Trifolium pratense	Red clover	Legume with a deep taproot system
Helianthus tuberosus	Jerusalem artichoke	Tall perennial with deep and extensive root system
Sorghum bicolor	Sorghum	Fast-growing crop with fibrous root system
Brassica juncea	Indian mustard	Rapid-growing plant with deep roots
Brassica napus	Canola	Cultivated for its oil-rich seeds
Salix spp.	Willow	Fast-growing trees or shrubs with extensive root system
Populus spp.	Poplar	Fast-growing deciduous trees with deep root system
Robinia pseudoacacia	Black locust	Rapid-growing tree with nitrogen-fixing capability
Spartina alterniflora	Smooth cordgrass	Salt-tolerant perennial with fibrous root system
Phragmites australis	Common reed	Tall grass with the rhizomatous root system
Vetiveria zizanioides	Vetiver	Deep-rooted perennial grass

Note The table provides a selection of plant species suitable for cadmium (Cd) phytostabilisation in contaminated soils. These plants exhibit fibrous root systems, deep-rooted structures, or the ability to produce exudates that aid in Cd immobilization. When implementing phytostabilization, it is essential to consider site-specific conditions and select appropriate plant species based on their adaptability to the specific environment and Cd immobilization capabilities

Source Based on a review of literature

7 Phytodegradation of Cadmium

As a phytoremediation technique, it is essential to note that phytodegradation primarily focuses on the degradation or transformation of organic pollutants rather than heavy metals like cadmium (Cd). Cadmium is a non-biodegradable heavy metal, meaning living organisms cannot break it down. Therefore, the application of phytodegradation for Cd remediation is not applicable [1, 4, 6, 41–44, 47, 49].

Phytodegradation primarily targets organic contaminants, including petroleum hydrocarbons, pesticides, and other organic pollutants. It harnesses the metabolic capabilities of certain plant species to break down or transform these organic compounds enzymatically, rendering them less toxic or converting them into harmless byproducts [19, 22, 37].

The process of phytodegradation involves the following fundamental mechanisms:

1. Uptake and Metabolism: Plants take organic contaminants through their roots and transport them to various plant tissues, including leaves, stems, and roots. Once inside the plant, enzymes present in the plant cells can metabolize or transform the organic compounds through processes such as oxidation, reduction, hydrolysis, or conjugation. This enzymatic metabolism can break complex organic pollutants into more straightforward, less toxic compounds.
2. Rhizodegradation: Rhizodegradation is a specific form of phytodegradation that occurs in the rhizosphere, the zone of soil surrounding plant roots. Plant roots release various enzymes into the rhizosphere, such as peroxidases, dehydrogenases, and cytochrome P450 enzymes. These enzymes can directly degrade or modify organic contaminants in the soil, enhancing their breakdown and degradation.
3. Co-metabolism: Some plants possess enzymes or enzyme systems that metabolize organic contaminants as a byproduct of their normal metabolic processes. Even though the contaminants may not be the preferred substrates for these enzymes, but they can still undergo partial degradation or transformation when exposed. Co-metabolism often occurs in the presence of other carbon sources that support the growth and metabolic activities of the plants.

Advantages of Phytodegradation:

1. Natural and Sustainable: Phytodegradation utilizes natural plant processes and does not rely on the addition of chemicals or energy-intensive treatments. It is an environmentally friendly and sustainable approach to remediate organic contaminants.
2. Cost-effective: Phytodegradation can be a cost-effective remediation method compared to traditional soil excavation or incineration techniques. It can be applied in situ, minimizing the need for soil removal and transportation.
3. Potential for In Situ Application: Phytodegradation can be applied in situ, allowing for the remediation of contaminated sites without extensive excavation or disturbance. This reduces the associated costs, risks, and potential impacts on surrounding ecosystems.

Disadvantages and Limitations of Phytodegradation:

1. Contaminant Specificity: Phytodegradation is limited to certain plant species and contaminants. Different plants possess varying metabolic capacities and may be more effective in degrading specific organic pollutants. Selecting the appropriate plant species capable of metabolizing the target contaminants is crucial for the success of phytodegradation.
2. Site-specific Considerations: The effectiveness of phytodegradation depends on site-specific conditions, including soil properties, contaminant concentrations, and other pollutants or inhibitory substances. Optimal plant selection and management practices should be based on thoroughly understanding the site characteristics and contaminant properties.

3. Time and Implementation Challenges: Phytodegradation is generally slower than other remediation techniques. Achieving significant degradation of organic contaminants may require a longer time frame. Additionally, the successful implementation of phytodegradation requires careful management and monitoring to ensure optimal plant growth and metabolic activity.

Overall, phytodegradation offers a promising approach for the remediation of organic pollutants in contaminated environments. While it does not apply to heavy metals like cadmium, it has proven effective for the degradation of various organic compounds [1, 4, 6, 41, 43, 44, 47, 49]. Ongoing research advances our understanding of plant metabolic pathways, enzyme systems, and plant–microbe interactions, enhancing the potential for phytodegradation as a sustainable and cost-effective remediation strategy (Table 5).

8 Phytovolatilization of Cadmium

Phytovolatilization is a phytoremediation technique that utilizes plants' natural ability to absorb and release volatile contaminants into the atmosphere. It targets explicitly volatile organic compounds (VOCs) and certain elements, such as mercury (Hg) and selenium (Se), that can be transformed into volatile forms. Phytovolatilization offers a sustainable and cost-effective approach to remediation by reducing the concentration of contaminants in soil or water and facilitating their transfer into the atmosphere.

The process of phytovolatilization involves the following fundamental mechanisms:

1. Contaminant Uptake: Plants absorb contaminants, primarily VOCs or volatile elements, through their roots from the soil or water. The contaminants are transported via the plant's vascular system to the aboveground tissues, including the leaves and stems.
2. Translocation and Volatilization: In the aboveground tissues, contaminants can undergo translocation and accumulate in the intercellular spaces or specialized structures, such as glandular trichomes or stomata. From these storage sites, the contaminants can volatilize into the atmosphere through a process known as transpiration or evapotranspiration. Transpiration refers to the loss of water vapour through plant stomata, and during this process, the contaminants are also released into the air.
3. Transformation and Release: In the case of volatile elements, such as Hg and Se, phytovolatilization involves transforming less volatile forms of the element, such as inorganic forms or compounds with lower vapour pressures, into more volatile and readily released forms. The enzymatic processes within the plant often mediate this transformation.

Table 5 Suitable plants for phytodegradation of organic contaminants

Plant species	Common name	Target contaminant(s)	Growth characteristics
Zea mays	Corn	Petroleum hydrocarbons, herbicides	Fast-growing annual crop with extensive root system
Helianthus annuus	Sunflower	Polycyclic aromatic hydrocarbons (PAHs)	Fast-growing annual with a deep taproot system
Salix spp.	Willow	Petroleum hydrocarbons, phenols	Fast-growing trees or shrubs with extensive root system
Populus spp.	Poplar	Chlorinated solvents, petroleum hydrocarbons	Fast-growing deciduous trees with deep root system
Medicago sativa	Alfalfa	Polycyclic aromatic hydrocarbons (PAHs)	Deep-rooted legume with nitrogen-fixing capability
Festuca arundinacea	Tall fescue	Pesticides, polycyclic aromatic hydrocarbons (PAHs)	Perennial grass with fibrous root system
Trifolium repens	White clover	Pesticides, chlorinated solvents	Low-growing perennial with extensive root system
Brassica juncea	Indian mustard	Heavy metals, petroleum hydrocarbons	Rapid-growing plant with deep roots
Phragmites australis	Common reed	Chlorinated solvents, petroleum hydrocarbons	Tall grass with rhizomatous root system
Nicotiana tabacum	Tobacco	Polycyclic aromatic hydrocarbons (PAHs)	Cultivated for its leaves used in tobacco products
Pseudomonas putida	–	Petroleum hydrocarbons, chlorinated solvents	Bacterium with degradative capabilities
Bacillus subtilis	–	Petroleum hydrocarbons, chlorinated solvents	Bacterium with degradative capabilities
Rhodococcus erythropolis	–	Petroleum hydrocarbons, polycyclic aromatic hydrocarbons (PAHs)	Bacterium with degradative capabilities

Note The table provides a selection of plant species and bacterial strains commonly used or studied for their potential in the phytodegradation of organic contaminants. These plants and bacteria possess specific enzymatic systems or metabolic capabilities to degrade or transform organic pollutants. The effectiveness of phytodegradation may vary depending on site-specific conditions and the specific contaminants present. Selecting the appropriate plant species or bacterial strains based on their degradation capabilities and compatibility with the target contaminants is essential for successful phytodegradation [11, 15, 26, 41, 55, 62, 63, 95, 102, 114].
Source Based on a review of literature

Advantages of Phytovolatilization:

1. Cost-effective: Phytovolatilization offers a cost-effective alternative to traditional remediation techniques for VOCs and volatile elements. It requires minimal infrastructure and energy inputs and can be implemented in situ, reducing the need for soil excavation or off-site disposal.
2. Natural and Sustainable: Phytovolatilization relies on natural plant processes and does not require the addition of chemicals or energy-intensive treatments. It is an environmentally friendly approach that utilizes the plants' natural capabilities to remediate volatile contaminants.
3. Versatility: Phytovolatilization can be applied to various contaminants, including VOCs, mercury, selenium, and other volatile elements. Plant species may exhibit varying capacities for volatilizing specific contaminants, allowing for tailored phytovolatilization approaches.

Disadvantages and Limitations of Phytovolatilization:

1. Contaminant Specificity: Phytovolatilization is most effective for VOCs and certain volatile elements. It may not be suitable for non-volatile contaminants or those that do not readily undergo volatilization. The effectiveness of phytovolatilization depends on the specific contaminants present and their volatility.
2. Potential for Air Pollution: Phytovolatilization releases contaminants into the atmosphere, which can contribute to air pollution if not properly managed. The potential risks of releasing volatile contaminants into the air must be considered, and appropriate measures should be taken to minimize the environmental impacts.
3. Contaminant Redistribution: Phytovolatilization can result in the redistribution of contaminants. Volatile contaminants released into the atmosphere may disperse and be transported to other locations, potentially impacting surrounding ecosystems or human populations. Proper monitoring and management strategies are essential to prevent the recontamination of other areas.
4. Limited Applicability: Phytovolatilization may be limited to specific contaminants and environmental conditions. Factors such as soil composition, climate, and contaminant concentrations can influence the effectiveness of phytovolatilization. Considering these factors and selecting appropriate plant species capable of volatilizing the target contaminants is essential.

Overall, phytovolatilization provides a promising approach for the remediation of volatile contaminants. By harnessing the natural volatilization processes of plants, it offers a sustainable and cost-effective method to reduce the concentration of VOCs and volatile elements in soil or water, ultimately minimizing the associated risks to human health and the environment. Proper site assessment, selection of appropriate plant species, and implementation of monitoring and management practices are essential for successful phytovolatilization as a remediation strategy (Table 6).

Table 6 Suitable plants for phytovolatilization

Plant species	Common name	Contaminant(s) targeted	Growth characteristics
Populus deltoides	Eastern cottonwood	Trichloroethylene (TCE), tetrachloroethylene (PCE), benzene	– Fast-growing deciduous tree – Tolerates a wide range of soil and climatic conditions
Brassica juncea	Indian mustard	Volatile organic compounds (VOCs)	– Rapid-growing plant with deep roots – Tolerates various soil types and conditions
Salix spp.	Willow	Trichloroethylene (TCE), tetrachloroethylene (PCE), benzene	– Fast-growing trees or shrubs with extensive root systems – Tolerates wet and dry soil conditions
Vetiveria zizanioides	Vetiver	Volatile organic compounds (VOCs)	– Deep-rooted perennial grass – Tolerates a wide range of soil conditions, including contaminated and wet soils
Zea mays	Corn	Trichloroethylene (TCE), tetrachloroethylene (PCE)	– Fast-growing annual crop with extensive root system – Grows well in various soil types and conditions
Helianthus annuus	Sunflower	Lead (Pb), selenium (Se)	– Fast-growing annual with a deep taproot system – Tolerates a wide range of soil conditions
Bacopa monnieri	Water hyssop	Mercury (Hg)	– Aquatic or semi-aquatic plant – Tolerates waterlogged soil or shallow water conditions
Scirpus validus	Soft stem bulrush	Mercury (Hg)	– Wetland plant with fibrous root system – Tolerates wet and waterlogged soil conditions

(continued)

Table 6 (continued)

Plant species	Common name	Contaminant(s) targeted	Growth characteristics
Brassica juncea	Indian mustard	Selenium (Se)	– Rapid-growing plant with deep roots
			– Tolerates various soil types and conditions

Source Based on a review of literature

9 Biosurfactant-Assisted Phytoremediation of Cadmium

Biosurfactant-assisted phytoremediation is an innovative approach that combines biosurfactants and plants to enhance the remediation of potentially toxic elements (PTEs) in soil. Potentially toxic elements, such as heavy metals, metalloids, and radionuclides, pose significant risks to human health and the environment when present in high concentrations in soil. Traditional phytoremediation techniques alone may have limitations in effectively addressing the remediation of PTEs due to their low bioavailability and limited mobility in the soil matrix. Biosurfactant-assisted phytoremediation aims to overcome these challenges by improving plants' solubility, availability, and uptake of PTEs, leading to enhanced remediation outcomes.

Biosurfactants naturally produce surface-active compounds by microorganisms, including bacteria, fungi, and yeasts. They possess amphiphilic properties, meaning they have both hydrophilic and hydrophobic regions. These properties enable biosurfactants to interact with water and organic contaminants, promoting the solubilization and mobilization of PTEs in the soil and enhancing their bioavailability for plant uptake. Biosurfactants can alter the surface tension, reduce interfacial tension, and increase the wettability of soil, facilitating the movement of PTEs towards the roots of plants.

The biosurfactant-assisted phytoremediation process involves the following fundamental mechanisms:

1. Enhanced Contaminant Solubilization: Biosurfactants interact with PTEs, increasing their solubility in the soil solution. By forming micelles or emulsions, biosurfactants encapsulate PTEs, reducing their association with soil particles and facilitating their release into the soil solution. This improved solubility enhances the availability of PTEs for plant uptake.
2. Improved Plant Uptake: Biosurfactants can enhance the uptake of PTEs by plants. They improve plant cell membranes' permeability, facilitating PTE movement of PTEs from the soil solution into the root cells. Additionally, biosurfactants can promote the translocation of PTEs from the roots to the aboveground tissues of plants, aiding in their accumulation and subsequent removal from the contaminated site.
3. Microbial Interactions: Biosurfactant production by certain microorganisms can stimulate plant growth-promoting activities, such as releasing plant growth

hormones, nutrient mobilization, and enhanced root development. These microbial interactions can further improve phytoremediation efficiency by promoting plant health, increasing root biomass, and facilitating the establishment of beneficial microbial communities in the rhizosphere, which can assist in PTE immobilization or transformation.

Advantages of Biosurfactant-Assisted Phytoremediation:

1. Enhanced Remediation Efficiency: Using biosurfactants enhances the mobilization and bioavailability of PTEs, resulting in improved uptake and removal of contaminants by plants. This leads to higher remediation efficiency compared to traditional phytoremediation techniques alone.
2. Increased Plant Tolerance: Biosurfactants can enhance plant tolerance to PTEs by mitigating the toxic effects of contaminants on plant growth and development. They can reduce the uptake of toxic elements and promote plant stress tolerance mechanisms, allowing plants to withstand better and remediate contaminated environments.
3. Sustainable and Environmentally Friendly: Biosurfactants are naturally derived compounds and offer a sustainable and environmentally friendly approach to enhance phytoremediation. They can be produced from renewable resources and are biodegradable, minimizing the potential for long-term environmental impact.
4. Versatility: Biosurfactants can remediate a wide range of PTEs in different soil types and contaminated sites. They can be applied to various plant species and tailored to specific contaminants and site conditions.

Limitations and Challenges of Biosurfactant-Assisted Phytoremediation:

1. Site-Specific Considerations: The effectiveness of biosurfactant-assisted phytoremediation can vary depending on site-specific factors, including soil properties, contaminant characteristics, and plant–microbe interactions. Selecting suitable biosurfactants and plant species should consider these factors to optimize remediation outcomes.
2. Cost and Availability: Biosurfactant production and application may involve additional costs compared to traditional phytoremediation methods. The availability and affordability of suitable biosurfactants on a large scale can be challenging, especially for widespread application.
3. Long-Term Monitoring and Maintenance: Biosurfactant-assisted phytoremediation requires long-term monitoring and maintenance to ensure sustained effectiveness. Proper management practices, including monitoring plant health, biosurfactant persistence, and contaminant levels, are essential for successful remediation outcomes.

Biosurfactant-assisted phytoremediation shows excellent promise as an effective and sustainable approach for the remediation of PTEs in soil. By harnessing the capabilities of biosurfactants and plants, this technique offers enhanced remediation efficiency, improved plant tolerance, and reduced environmental impact. Ongoing research and development in biosurfactant production, optimization, and field application will further advance the application of biosurfactant-assisted phytoremediation as a valuable tool for contaminated site restoration (Table 7).

10 Plant Growth Regulator and Phytoremediation of Cadmium

Plant growth regulators (PGRs) play a significant role in the phytoremediation of cadmium-contaminated environments. A phytoremediation is a plant-based approach that utilizes the natural abilities of plants to remove, degrade, or immobilize contaminants in soil, water, or air. Cadmium (Cd) is a toxic heavy metal that can harm human health and the environment in high concentrations. PGRs, also known as plant hormones or phytohormones, are naturally occurring chemical compounds that regulate various aspects of plant growth and development. In phytoremediation, PGRs can influence plants' tolerance and remediation capacity towards cadmium contamination through multiple mechanisms.

The role of PGRs in the phytoremediation of cadmium includes:

1. Enhancement of Plant Growth and Biomass: PGRs such as auxins (e.g., indole-3-acetic acid, IAA), cytokinins, and gibberellins can promote root and shoot growth, leading to increased plant biomass. This enhanced growth can facilitate plants' uptake and accumulation of cadmium, improving their ability to remediate contaminated soils.
2. Stimulation of Root Development: PGRs like auxins and ethylene can promote root elongation, branching, and density. Developing a well-developed root system increases the surface area available for cadmium uptake from the soil. Enhanced root growth and architecture can facilitate the exploration of a larger soil volume, leading to improved cadmium extraction from the contaminated site.
3. Activation of Antioxidant Defense Systems: Cadmium induces oxidative stress in plants by generating reactive oxygen species (ROS). PGRs, including abscisic acid (ABA) and salicylic acid (SA), can stimulate the production of antioxidant enzymes (e.g., superoxide dismutase, catalase, peroxidase) and non-enzymatic antioxidants (e.g., glutathione, ascorbic acid) in plants. These antioxidants scavenge ROS, mitigating oxidative damage caused by cadmium and reducing the associated phytotoxicity.
4. Induction of Metal Chelation and Detoxification: PGRs such as phytochelatins (PCs) and metallothioneins (MTs) can be synthesized in response to cadmium stress. PCs are small peptides that chelate and sequester heavy metals, including cadmium, in the vacuoles, minimizing their toxicity in the cytosol. MTs are low-molecular-weight proteins that bind and detoxify heavy metals, including cadmium. PGRs can stimulate the synthesis and accumulation of PCs and MTs, enhancing the cadmium detoxification capacity of plants.
5. Modulation of Cadmium Uptake and Translocation: PGRs can influence the uptake and translocation of cadmium within plants. For instance, auxins and

Table 7 Biosurfactants used in phytoremediation

Biosurfactant	Microorganism source	Properties and applications
Rhamnolipids	*Pseudomonas aeruginosa*	– Excellent surface tension reduction and emulsification abilities
		– Effective in solubilizing hydrophobic contaminants
		– Enhances the bioavailability of hydrophobic compounds
Saponins	*Various plant species*	– Natural plant-derived biosurfactants
		– Surfactant properties aid in solubilization of hydrophobic compounds
		– Found in plant species such as Quillaja Saponaria and Aesculus hippocastanum
Sophorolipids	*Candida bombicola*	– Strong emulsification properties
		– Effective in solubilizing hydrophobic contaminants
		– Produced by yeast species
Surfactin	*Bacillus subtilis*	– Exhibits substantial surface tension reduction and emulsification abilities
		– Enhances solubilization and mobilization of hydrophobic organic pollutants
Bioemulsan	*Acinetobacter species*	– Excellent emulsifying properties
		– Facilitates the solubilization and mobilization of hydrophobic compounds
Lipopeptides	*Bacillus* and *Pseudomonas* spp.	– Surfactant properties aid in solubilization and dispersion of hydrophobic contaminants
		– Produced by different bacterial species
Alasan	*Acinetobacter radioresistens*	– Exhibits surface activity and emulsification properties
		– Shows potential for enhancing the biodegradation of petroleum hydrocarbons
Serrawettins	*Serratia marcescens*	– Surfactant properties and biosurfactant activity
		– Enhances the biodegradation of hydrocarbons and heavy metals

(continued)

Table 7 (continued)

Biosurfactant	Microorganism source	Properties and applications
		– Shows potential for remediating contaminated soil and water
Lichenysin	*Bacillus licheniformis*	– Strong emulsification and surfactant properties
		– Enhances the solubilization and biodegradation of hydrophobic pollutants
		– Shows potential for the remediation of contaminated environments

Note The table overviews some commonly used biosurfactants in phytoremediation. These biosurfactants, produced by various microorganisms and plant species, possess properties that aid in solubilizing and mobilizing hydrophobic contaminants. They have been studied for their effectiveness in enhancing the remediation of contaminated sites. When selecting the appropriate biosurfactant for a phytoremediation project, it is essential to consider specific site conditions, target contaminants, and plant compatibility when

Source Based on a review of literature

cytokinins can enhance cadmium uptake by increasing the expression of transporters involved in metal uptake. Ethylene, on the other hand, can reduce cadmium uptake by inhibiting the expression of metal transporters. PGRs can also affect cadmium translocation from roots to shoots or vice versa, influencing the overall distribution and accumulation of cadmium in different plant tissues.

6. Facilitation of Rhizosphere Microbial Interactions: PGRs can influence the microbial community in the rhizosphere, the soil region influenced by plant roots. They can promote the growth of beneficial rhizobacteria, mycorrhizal fungi, and other soil microorganisms, enhancing plants' remediation capacity. These microbes can aid in the solubilization, immobilization, or transformation of cadmium in the soil, improving overall phytoremediation efficiency.

By modulating plant growth, development, stress responses, and interactions with the soil microbiome, PGRs contribute to the phytoremediation of cadmium-contaminated environments. However, it is essential to note that the effectiveness of PGRs in phytoremediation may vary depending on the plant species, PGR type and concentration, site conditions, and the specific mechanisms involved in cadmium tolerance and detoxification. Proper dosage and timing of PGR application, as well as an understanding of plant-PGR interactions, are essential for maximizing the benefits of PGRs in cadmium phytoremediation strategies (Table 8).

Table 8 Plant growth regulators (PGRs) in phytoremediation of Cadmium

PGR	Function and effects	Examples of PGRs
Auxins	Promote root and shoot growth, enhance cadmium uptake and translocation	Indole-3-acetic acid (IAA), Indole-3-butyric acid (IBA)
Cytokinins	Stimulate cell division, enhance plant growth and biomass, and improve tolerance to cadmium stress	Zeatin, Kinetin, 6-benzylaminopurine (BAP)
Gibberellins	Promote stem and leaf elongation, enhance plant growth and biomass, influence cadmium uptake and translocation	Gibberellic acid (GA3)
Abscisic Acid (ABA)	Regulate stomatal closure, induce antioxidant defense systems, enhance cadmium tolerance	Abscisic acid (ABA)
Ethylene	Influence root development, modulate cadmium uptake, reduce cadmium translocation	Ethylene (ET)
Salicylic Acid (SA)	Activate antioxidant defense systems, enhance cadmium tolerance	Salicylic acid (SA)
Jasmonic Acid (JA)	Regulate stress responses, induce defense mechanisms, and influence cadmium tolerance	Jasmonic acid (JA)
Brassinosteroids	Stimulate cell elongation, enhance plant growth, improve tolerance to cadmium stress	Brassinolide, Epibrassinolide, 24-Epibrassinolide
Polyamines	Enhance antioxidant capacity, regulate metal homeostasis, influence cadmium tolerance	Putrescine, Spermidine, Spermine
Phytochelatins	Chelate and sequester heavy metals, including cadmium, enhance detoxification and reduce cadmium phytotoxicity	Phytochelatins (PCs)
Metallothioneins	Bind and detoxify heavy metals, including cadmium, enhance detoxification and reduce cadmium phytotoxicity	Metallothioneins (MTs)
Mannitol	Act as an osmoprotectant enhances cadmium tolerance and detoxification	Mannitol

Note The table provides an overview of plant growth regulators (PGRs) used in the phytoremediation of cadmium-contaminated environments. These PGRs exhibit various functions and effects contributing to plant growth, stress tolerance, cadmium uptake, detoxification, and translocation. Examples of specific PGRs within each category are mentioned, but it is important to note that the effectiveness of PGRs in phytoremediation may vary depending on the plant species, concentrations, and application methods. The selection and optimization of PGRs for a specific phytoremediation project should be based on site-specific conditions, target contaminants, and the desired outcomes
Source Based on a review of literature

11 Conclusion

In conclusion, the toxicity of rhizospheric cadmium-contaminated soil poses significant risks to the environment and human health. Cadmium, a highly toxic heavy metal, can accumulate in the soil due to various industrial activities, mining operations, and agricultural practices. Its persistence and potential for bioaccumulation make it crucial to address cadmium contamination and mitigate its adverse effects.

Phytoremediation, a sustainable and environmentally friendly approach, offers excellent potential for the remediation of cadmium-contaminated soil. Using plants, phytoremediation exploits the natural abilities of vegetation to uptake, translocate, immobilize, and detoxify contaminants. Different phytoremediation techniques, such as rhizofiltration, phytoextraction, phytostabilization, and phytodegradation, employ specific mechanisms to address cadmium contamination based on site-specific conditions and project objectives.

Rhizospheric phytoremediation techniques, in particular, utilize the interactions between plant roots, the rhizosphere, and associated microorganisms to enhance cadmium removal from the soil. These techniques leverage the plants' root exudates, enhancing microbial activity, increasing metal bioavailability, and stimulating plant–microbe interactions for improved remediation outcomes.

While phytoremediation shows promise, its effectiveness is influenced by plant selection, site conditions, contaminant concentrations, and the specific phytoremediation technique employed. Furthermore, long-term monitoring, maintaining plant health, and optimizing the selection of suitable plant species or bacteria must be addressed for successful implementation.

Nevertheless, the advantages of phytoremediation, such as its cost-effectiveness, sustainability, and ability to restore soil health, make it a valuable tool in addressing rhizospheric cadmium contamination. By harnessing the power of plants, phytoremediation offers a viable and eco-friendly solution to mitigate the risks associated with cadmium-contaminated soil, ultimately contributing to a cleaner and healthier environment.

Continuing research, innovation, and field trials are necessary to advance the understanding and application of phytoremediation techniques for rhizospheric cadmium-contaminated soil. By improving phytoremediation's efficiency, adaptability, and scalability, we can unlock its full potential and contribute to the restoration of contaminated sites, ensuring the protection of ecosystems and human well-being.

12 Prospects for Future

The prospects regarding the toxicity of rhizospheric cadmium-contaminated soil and its phytoremediation hold great potential for advancements in research, technology, and practical applications. Here are some critical prospects in this field:

1. Advanced Plant Selection: Ongoing research efforts can focus on identifying and developing plant species with enhanced cadmium tolerance, accumulation, and remediation capabilities. By understanding different plants' genetic and physiological traits, scientists can select and engineer species that are more efficient in phytoremediation, leading to improved remediation outcomes.
2. Genetic Engineering and Biotechnology: Genetic engineering techniques can be employed to enhance the ability of plants to remediate cadmium-contaminated soil. By introducing genes responsible for metal uptake, translocation, and detoxification, plants can be engineered to exhibit improved phytoremediation properties. Biotechnological approaches, such as plant–microbe interactions and symbiotic bacteria, can also be explored to enhance phytoremediation efficiency further.
3. Innovative Phytoremediation Techniques: Researchers can develop and refine novel phytoremediation techniques tailored to rhizospheric cadmium contamination. These techniques could involve the integration of different plant species, optimising growth conditions, and using soil amendments or additives to enhance cadmium removal and stabilization.
4. Nanotechnology Applications: Nanotechnology offers potential breakthroughs in phytoremediation by utilizing nanoparticles to enhance plant–microbe interactions, improve metal uptake, and facilitate the remediation of cadmium-contaminated soil. Nano-enabled approaches, such as nanostructured materials or nanosensors, can contribute to more efficient and targeted phytoremediation strategies.
5. Microbial Augmentation: Exploring the potential of microbial augmentation can lead to developing effective microbial consortia or bioaugmentation strategies that enhance rhizospheric microbial activities involved in cadmium remediation. This approach can optimize plant–microbe interactions, improve nutrient availability, enhance metal immobilization, and facilitate the transformation of cadmium into less toxic forms.
6. Field-Scale Implementation: The prospects of phytoremediation lie in successfully implementing these techniques at a larger scale. Conducting field trials and demonstrating phytoremediation's feasibility, efficacy, and cost-effectiveness in real-world scenarios will provide essential insights for its more comprehensive application. Scaling up phytoremediation projects and incorporating them into environmental management plans can contribute to sustainable land restoration and pollution mitigation efforts.
7. Integrated Approaches: Integrating phytoremediation with other remediation techniques, such as electrokinetics, soil washing, or microbial-assisted remediation, can result in synergistic effects and improved remediation outcomes. Such integrated approaches can offer comprehensive solutions for the remediation of cadmium-contaminated soil, addressing different aspects of contamination and site-specific conditions.

8. Regulatory Framework and Policy Support: Developing regulatory frameworks and policies that promote and incentivize phytoremediation for cadmium-contaminated soil can facilitate its widespread adoption. Governments and regulatory bodies can encourage research, provide guidelines for implementation, and offer financial incentives to promote phytoremediation as a viable remediation option.

In conclusion, the prospects of rhizospheric cadmium-contaminated soil and its phytoremediation involve advancements in plant selection, genetic engineering, innovative techniques, nanotechnology applications, microbial augmentation, field-scale implementation, integrated approaches, and regulatory support. By embracing these prospects, we can unlock the full potential of phytoremediation and contribute to sustainable and practical solutions for the remediation of cadmium-contaminated soil, ultimately protecting ecosystems, human health, and the environment.

Acknowledgements We would like to express our sincere gratitude to the Department of Agronomy for their support and assistance throughout the writing. The Department's commitment to academic excellence and research has been instrumental in completing this endeavour.

Author's Contribution The authors of this work have made significant contributions to the research project/study. Each author has participated sufficiently in the research project/study, made intellectual contributions, and is responsible for the work's accuracy and integrity. The authors have collaborated closely, ensuring the completion of this work through collective effort, expertise, and dedication.

References

1. Adelodun, B., Adewumi, J. R., Ajala, O. A., Ajibade, F. O., Ajibade, T. F., Akmal, M. H., Aley, P., Arif, M., Arif, M. S., Awasthi, K. K., Azeem, F., Azevedo, L. C. B., Batool, A., Bertini, S. C. B., Bhattacharya, N., Chattopadhyay, I., Chopra, M., da Silva, A. V., da Silva, M. K., … Yasmeen, T. (2022). List of contributors. In A. Kumar, J. Singh, & L. F. R. Ferreira (Eds.), *Microbiome under changing climate* (pp. xix–xxiv). Woodhead Publishing. https://doi.org/10.1016/B978-0-323-90571-8.00028-6
2. Agarwal, P., & Rani, R. (2022). Strategic management of contaminated water bodies: Omics, genome-editing and other recent advances in phytoremediation. *Environmental Technology & Innovation, 27*, 102463. https://doi.org/10.1016/j.eti.2022.102463
3. Aihemaiti, A., Chen, J., Hua, Y., Dong, C., Wei, X., Yan, F., & Zhang, Z. (2022). Effect of ferrous sulfate modified sludge biochar on the mobility, speciation, fractionation and bioaccumulation of vanadium in contaminated soil from a mining area. *Journal of Hazardous Materials, 437*, 129405. https://doi.org/10.1016/j.jhazmat.2022.129405
4. Akhtar, N., Amin-ul Mannan, M., Banik, R. M., Baysal, Ö., Bera, S., Biswas, P., Christina, E., Das, T., Datta, B., Devi, P., Dey, A., Dey, S. R., Dwivedi, P., Han, J., Jayabaskaran, C., Kamalraj, S., Kaur, P., Kumar, P., Kumar, V., … Yashavantha Rao, H. C. (2021). List of contributors. In A. Kumar, J. Singh, & J. Samuel (Eds.), *Volatiles and metabolites of microbes* (pp. xix–xxi). Academic Press. https://doi.org/10.1016/B978-0-12-824523-1.00028-6
5. Alam, S. N., Khalid, Z., Patel, M., Kumari, P., Kumar, A., Singh, B., & Guldhe, A. (2022). Chapter 3—Role of genetic engineering in microbe-assisted phytoremediation of polluted sites (K. Bauddh & Y. Ma (Eds.), *Advances in microbe-assisted phytoremediation of polluted sites* (pp. 63–84). Elsevier. https://doi.org/10.1016/B978-0-12-823443-3.00008-9

6. Aley, P., Singh, J., & Kumar, P. (2022). Chapter 23—Adapting the changing environment: microbial way of life. In A. Kumar, J. Singh, & L. F. R. Ferreira (Eds.), *Microbiome under changing climate* (pp. 507–525). Woodhead Publishing. https://doi.org/10.1016/B978-0-323-90571-8.00023-7
7. Alves, A. R. A., Yin, Q., Oliveira, R. S., Silva, E. F., & Novo, L. A. B. (2022). Plant growth-promoting bacteria in phytoremediation of metal-polluted soils: Current knowledge and future directions. *Science of the Total Environment, 838*, 156435. https://doi.org/10.1016/j.scitotenv.2022.156435
8. Ansari, F., Momina, Ahmad, A., & Rafatullah, M. (2023). Review on bioremediation technologies of polycyclic aromatic hydrocarbons (PAHs) from soil: Mechanisms and future perspective. *International Biodeterioration & Biodegradation, 179*, 105582. https://doi.org/10.1016/j.ibiod.2023.105582
9. Arunraja, D., Romauld, S. I., Devi, P. B., Thiruvengadam, S., & Kumar, V. (2023). Chapter 29—Genetically engineered microbes for bioremediation and phytoremediation of contaminated environment. In V. Kumar, M. Bilal, S. K. Shahi, & V. K. Garg (Eds.), *Metagenomics to bioremediation* (*Developments in applied microbiology and biotechnology*) (pp. 709–721). Academic Press. https://doi.org/10.1016/B978-0-323-96113-4.00039-1
10. Ayub, M. A., Zia ur Rehman, M., Umar, W., Zuhra, N., & Shabaan, M. (2022). Chapter 18—Role of legumes in phytoremediation of heavy metals. In R. S. Meena & S. Kumar (Eds.), *Advances in legumes for sustainable intensification* (pp. 345–360). Academic Press. https://doi.org/10.1016/B978-0-323-85797-0.00011-2
11. Azhar, U., Ahmad, H., Shafqat, H., Babar, M., Shahzad Munir, H. M., Sagir, M., Arif, M., Hassan, A., Rachmadona, N., Rajendran, S., Mubashir, M., & Khoo, K. S. (2022). Remediation techniques for elimination of heavy metal pollutants from soil: A review. *Environmental Research, 214*, 113918. https://doi.org/10.1016/j.envres.2022.113918
12. Bandara, T., Krohn, C., Jin, J., Chathurika, J. B. A. J., Franks, A., Xu, J., Potter, I. D., & Tang, C. (2022). The effects of biochar aging on rhizosphere microbial communities in cadmium-contaminated acid soil. *Chemosphere, 303*, 135153. https://doi.org/10.1016/j.chemosphere.2022.135153
13. Bashir, T., Zafar, S., & Bashir, S. (2022). Chapter 19—Phytoremediation of micropollutants. In M. Z. Hashmi, S. Wang, & Z. Ahmed (Eds.), *Environmental micropollutants* (*Advances in pollution research*) (pp. 365–386). Elsevier. https://doi.org/10.1016/B978-0-323-90555-8.00001-5
14. Bhalla, G., Bhalla, B., Kumar, V., & Sharma, A. (2022). Chapter 16—Bioremediation and phytoremediation of pesticides residues from contaminated water: A novel approach. In M. Hadi Dehghani, R. R. Karri, & I. Anastopoulos (Eds.), *Pesticides remediation technologies from water and wastewater* (pp. 339–363). Elsevier. https://doi.org/10.1016/B978-0-323-90893-1.00016-7
15. Bhanse, P., Kumar, M., Singh, L., Awasthi, M. K., & Qureshi, A. (2022). Role of plant growth-promoting rhizobacteria in boosting the phytoremediation of stressed soils: Opportunities, challenges, and prospects. *Chemosphere, 303*, 134954. https://doi.org/10.1016/j.chemosphere.2022.134954
16. Bhatt, P., Chaudhary, P., Ahmad, S., Bhatt, K., Chandra, D., & Chen, S. (2023). Chapter 2—Recent advances in the application of microbial inoculants in the phytoremediation of xenobiotic compounds. In D. Chandra & P. Bhatt (Eds.), *Unravelling plant-microbe synergy* (*Developments in applied microbiology and biotechnology*) (pp. 37–48). Academic Press. https://doi.org/10.1016/B978-0-323-99896-3.00013-8
17. Borah, P., Rene, E. R., Rangan, L., & Mitra, S. (2023). Phytoremediation of nickel and zinc using Jatropha curcas and Pongamia pinnata from the soils contaminated by municipal solid wastes and paper mill wastes. *Environmental Research, 219*, 115055. https://doi.org/10.1016/j.envres.2022.115055
18. Breton-Deval, L., Guevara-García, A., Juarez, K., Lara, P., Rubio-Noguez, D., & Tovar-Sanchez, E. (2022). Chapter 13—Role of rhizosphere microbiome during phytoremediation

of heavy metals. In S. Das & H. R. Dash (Eds.), *Microbial biodegradation and bioremediation* (2nd ed., pp. 263–291). Elsevier. https://doi.org/10.1016/B978-0-323-85455-9.00016-3
19. Chakraborty, S., Kumar, P., Sanyal, R., Mane, A. B., Arvind Prasanth, D., Patil, M., & Dey, A. (2021). Unravelling the regulatory role of miRNAs in secondary metabolite production in medicinal crops. *Plant Gene, 27*, 100303. https://doi.org/10.1016/j.plgene.2021.100303
20. Chen, L., Beiyuan, J., Hu, W., Zhang, Z., Duan, C., Cui, Q., Zhu, X., He, H., Huang, X., & Fang, L. (2022). Phytoremediation of potentially toxic elements (PTEs) contaminated soils using alfalfa (Medicago sativa L.): A comprehensive review. *Chemosphere, 293*, 133577. https://doi.org/10.1016/j.chemosphere.2022.133577
21. Daraz, U., Li, Y., Ahmad, I., Iqbal, R., & Ditta, A. (2023). Remediation technologies for acid mine drainage: Recent trends and future perspectives. *Chemosphere, 311*, 137089. https://doi.org/10.1016/j.chemosphere.2022.137089
22. Das, T., Saha, S. C., Sunita, K., Majumder, M., Ghorai, M., Mane, A. B., Prasanth, D. A., Kumar, P., Pandey, D. K., Al-Tawaha, A. R., Batiha, G.E.-S., Shekhawat, M. S., Ghosh, A., Sharifi-Rad, J., & Dey, A. (2022). Promising botanical-derived monoamine oxidase (MAO) inhibitors: Pharmacological aspects and structure-activity studies. *South African Journal of Botany, 146*, 127–145. https://doi.org/10.1016/j.sajb.2021.09.019
23. Devi, A., Ferreira, L. F. R., Saratale, G. D., Mulla, S. I., More, N., & Bharagava, R. N. (2022). Chapter 1—Microbe-assisted phytoremediation of environmental contaminants. In K. Bauddh & Y. Ma (Eds.), *Advances in microbe-assisted phytoremediation of polluted sites* (pp. 3–26). Elsevier. https://doi.org/10.1016/B978-0-12-823443-3.00001-6
24. Dixit, S., Tracy, P., Vishnoi, N., Swain, A. A., Bauddh, K., & Kumar, M. (2022). Chapter 9—Phytoremediation of heavy metal contaminated soil in association with arbuscular mycorrhizal fungi. In K. Bauddh & Y. Ma (Eds.), *Advances in microbe-assisted phytoremediation of polluted sites* (pp. 207–230). Elsevier. https://doi.org/10.1016/B978-0-12-823443-3.00016-8
25. Etesami, H., & Srivastava, A. K. (2022). Chapter 23—Bacterial induced alleviation of cadmium and arsenic toxicity stress in plants: Mechanisms and future prospects. In R. C. Dubey & P. Kumar (Eds.), *Rhizosphere engineering* (pp. 445–469). Academic Press. https://doi.org/10.1016/B978-0-323-89973-4.00013-2
26. Farooqi, Z. U. R., Hussain, M. M., Ayub, M. A., Qadir, A. A., & Ilic, P. (2022). Chapter 2—Potentially toxic elements and phytoremediation: Opportunities and challenges. In R. A. Bhat, F. M. P. Tonelli, G. H. Dar, & K. Hakeem (Eds.), *Phytoremediation* (pp. 19–36). Academic Press. https://doi.org/10.1016/B978-0-323-89874-4.00020-0
27. Gangola, S., Joshi, S., Bhandari, G., Bhatt, P., Kumar, S., Bhandari, N. S., & Mittal, A. (2023). Chapter 3—Remediation of heavy metals by rhizospheric bacteria and their mechanism of detoxification. In S. Gangola, S. Kumar, S. Joshi, & P. Bhatt (Eds.), *Advanced microbial technology for sustainable agriculture and environment* (*Developments in applied microbiology and biotechnology*) (pp. 31–46). Academic Press. https://doi.org/10.1016/B978-0-323-95090-9.00005-4
28. Goud, E. L., Singh, J., & Kumar, P. (2022). Chapter 19—Climate change and their impact on global food production. In A. Kumar, J. Singh, & L. F. R. Ferreira (Eds.), *Microbiome under changing climate* (pp. 415–436). Woodhead Publishing. https://doi.org/10.1016/B978-0-323-90571-8.00019-5
29. Gudeta, K., Kumar, V., Bhagat, A., Julka, J. M., Bhat, S. A., Ameen, F., Qadri, H., Singh, S., & Amarowicz, R. (2023). Ecological adaptation of earthworms for coping with plant polyphenols, heavy metals, and microplastics in the soil: A review. *Heliyon, 9*(3), e14572. https://doi.org/10.1016/j.heliyon.2023.e14572
30. Gupta, A., Majumdar, A., & Srivastava, S. (2022). Chapter 9—Approaches for assisted phytoremediation of arsenic contaminated sites. In V. Pandey (Ed.), *Assisted phytoremediation* (pp. 221–242). Elsevier. https://doi.org/10.1016/B978-0-12-822893-7.00007-0
31. Islam, N. F., Patowary, R., & Sarma, H. (2022). Chapter 15—Biosurfactant-assisted phytoremediation for a sustainable future. In V. Pandey (Ed.), *Assisted phytoremediation* (pp. 399–414). Elsevier. https://doi.org/10.1016/B978-0-12-822893-7.00003-3

32. Jha, S. S., & Songachan, L. S. (2022). Mycorrhizoremediation: Understanding the science behind it and it's future prospects. *Materials Today: Proceedings, 51*, 2431–2436. https://doi.org/10.1016/j.matpr.2021.11.605
33. Kafle, A., Timilsina, A., Gautam, A., Adhikari, K., Bhattarai, A., & Aryal, N. (2022). Phytoremediation: Mechanisms, plant selection and enhancement by natural and synthetic agents. *Environmental Advances, 8*, 100203. https://doi.org/10.1016/j.envadv.2022.100203
34. Kanwar, P., Mina, U., Thakur, I. S., & Srivastava, S. (2023). Heavy metal phytoremediation by the novel prospect of microbes, nanotechnology, and genetic engineering for recovery and rehabilitation of landfill site. *Bioresource Technology Reports, 23*, 101518. https://doi.org/10.1016/j.biteb.2023.101518
35. Kaur, H., Tashima, Singh, S., & Kumar, P. (2023). Reconditioning of plant metabolism by arbuscular mycorrhizal networks in cadmium contaminated soils: Recent perspectives. *Microbiological Research, 268*, 127293. https://doi.org/10.1016/j.micres.2022.127293
36. Kaur, M., Sharma, S., & Sodhi, H. S. (2022). Chapter 16—Phytoremediation of pesticides. In R. A. Bhat, F. M. P. Tonelli, G. H. Dar, & K. Hakeem (Eds.), *Phytoremediation* (pp. 323–350). Academic Press. https://doi.org/10.1016/B978-0-323-89874-4.00026-1
37. Kotia, A., Rutu, P., Singh, V., Kumar, A., Dhoke, S., Kumar, P., & Singh, D. K. (2022). Rheological analysis of rice husk-starch suspended in water for sustainable agriculture application. *Materials Today: Proceedings, 50*, 1962–1966. https://doi.org/10.1016/j.matpr.2021.09.325
38. Kristanti, R. A., & Hadibarata, T. (2023). Phytoremediation of contaminated water using aquatic plants, its mechanism and enhancement. *Current Opinion in Environmental Science & Health, 32*, 100451. https://doi.org/10.1016/j.coesh.2023.100451
39. Kumar, A., Naroju, S. P., Langthasa, M., Behera, I., Tyagi, S., Sharma, N., & Gautam, R. K. (2023). Bioremediation potential of green wastes and plant growth promoting rhizobacteria and its enhancement by their combination: A review. *Environmental Advances, 12*, 100379. https://doi.org/10.1016/j.envadv.2023.100379
40. Kumar, M., Bolan, N., Jasemizad, T., Padhye, L. P., Sridharan, S., Singh, L., Bolan, S., O'Connor, J., Zhao, H., Shaheen, S. M., Song, H., Siddique, K. H. M., Wang, H., Kirkham, M. B., & Rinklebe, J. (2022). Mobilization of contaminants: Potential for soil remediation and unintended consequences. *Science of the Total Environment, 839*, 156373. https://doi.org/10.1016/j.scitotenv.2022.156373
41. Kumar, P., & Mistri, T. K. (2020). Transcription factors in SOX family: Potent regulators for cancer initiation and development in the human body. *Seminars in Cancer Biology, 67*, 105–113. https://doi.org/10.1016/j.semcancer.2019.06.016
42. Kumar, P., Devi, P., & Dey, S. R. (2021). Chapter 6—Fungal volatile compounds: a source of novel in plant protection agents. In A. Kumar, J. Singh, & J. Samuel (Eds.), *Volatiles and metabolites of microbes* (pp. 83–104). Academic Press. https://doi.org/10.1016/B978-0-12-824523-1.00001-8
43. Kumar, P., Kumar, T., Singh, S., Tuteja, N., Prasad, R., & Singh, J. (2020). Potassium: A key modulator for cell homeostasis. *Journal of Biotechnology, 324*, 198–210.
44. Kumar, P., Sharma, K., Saini, L., & Dey, S. R. (2021). Chapter 8—Role and behavior of microbial volatile organic compounds in mitigating stress. In A. Kumar, J. Singh, & J. Samuel (Eds.), *Volatiles and metabolites of microbes* (pp. 143–161). Academic Press. https://doi.org/10.1016/B978-0-12-824523-1.00010-9
45. Kumar, S., Thakur, N., Singh, A. K., Gudade, B. A., Ghimire, D., & Das, S. (2022). 26—Microbes-assisted phytoremediation of contaminated environment: Global status, progress, challenges, and future prospects. In V. Kumar, M. P. Shah, & S. K. Shahi (Eds.), *Phytoremediation technology for the removal of heavy metals and other contaminants from soil and water* (pp. 555–570). Elsevier. https://doi.org/10.1016/B978-0-323-85763-5.00007-6
46. Kumar, V., Umrao, P. D., & Kaistha, S. D. (2022). Chapter 12—Beneficial plant microbiome assisted chromium phytoremediation. In K. Bauddh & Y. Ma (Eds.), *Advances in microbe-assisted phytoremediation of polluted sites* (pp. 301–346). Elsevier. https://doi.org/10.1016/B978-0-12-823443-3.00018-1

47. Kumar, V., Dwivedi, P., Kumar, P., Singh, B. N., Pandey, D. K., Kumar, V., & Bose, B. (2021). Mitigation of heat stress responses in crops using nitrate primed seeds. *South African Journal of Botany, 140*, 25–36. https://doi.org/10.1016/j.sajb.2021.03.024
48. Kumari, K., Cherian, S., & Bauddh, K. (2022). Chapter 2—Microbial augmented phytoremediation with improved ecosystems services. In K. Bauddh & Y. Ma (Eds.), *Advances in microbe-assisted phytoremediation of polluted sites* (pp. 27–62). Elsevier. https://doi.org/10.1016/B978-0-12-823443-3.00017-X
49. Kumari, P., Singh, J., & Kumar, P. (2022). Chapter 21—Impact of bioenergy for the diminution of an ascending global variability and change in the climate. In A. Kumar, J. Singh, & L. F. R. Ferreira (Eds.), *Microbiome under changing climate* (pp. 469–487). Woodhead Publishing. https://doi.org/10.1016/B978-0-323-90571-8.00021-3
50. Kurniawan, S. B., Ramli, N. N., Said, N. S. M., Alias, J., Imron, M. F., Abdullah, S. R. S., Othman, A. R., Purwanti, I. F., & Hasan, H. A. (2022). Practical limitations of bioaugmentation in treating heavy metal contaminated soil and role of plant growth promoting bacteria in phytoremediation as a promising alternative approach. *Heliyon, 8*(4), e08995. https://doi.org/10.1016/j.heliyon.2022.e08995
51. Li, Q., Yin, J., Wu, L., Li, S., & Chen, L. (2023). Effects of biochar and zero valent iron on the bioavailability and potential toxicity of heavy metals in contaminated soil at the field scale. *Science of the Total Environment*, 165386. https://doi.org/10.1016/j.scitotenv.2023.165386
52. Li, X., Xiao, J., Gai, X., Du, Z., Salam, M. M. A., & Chen, G. (2023). Facilitated remediation of heavy metals contaminated land using Quercus spp. with different strategies: Variations in amendments and experiment periods. *Science of the Total Environment, 876*, 163245. https://doi.org/10.1016/j.scitotenv.2023.163245
53. Li, X., Li, B., Zheng, Y., Luo, L., Qin, X., Yang, Y., & Xu, J. (2022). Physiological and rhizospheric response characteristics to cadmium of a newly identified cadmium accumulator Coreopsis grandiflora Hogg. (Asteraceae). *Ecotoxicology and Environmental Safety, 241*, 113739. https://doi.org/10.1016/j.ecoenv.2022.113739
54. Li, X., Tian, L., Li, B., Chen, H., Zhao, G., Qin, X., Liu, Y., Yang, Y., & Xu, J. (2022). Polyaspartic acid enhances the Cd phytoextraction efficiency of Bidens pilosa by remolding the rhizospheric environment and reprogramming plant metabolism. *Chemosphere, 307*, 136068. https://doi.org/10.1016/j.chemosphere.2022.136068
55. Li, X., Li, Y., Zhu, X., Gui, X., Ma, C., Peng, W., Li, Y., Zhang, Y., Huang, W., Hua, D., Jia, S., & Wu, M. (2022). Evaluation of the cadmium phytoextraction potential of tobacco (Nicotiana tabacum) and rhizosphere micro-characteristics under different cadmium levels. *Chemosphere, 286*, 131714. https://doi.org/10.1016/j.chemosphere.2021.131714
56. Liu, H., Huang, H., Xie, Y., Liu, Y., Shangguan, Y., & Xu, H. (2023). Integrated biochemical and transcriptomic analysis reveals the effects of Burkholderia sp. SRB-1 on cadmium accumulating in Chrysopogon zizanioides L. under Cd stress. *Journal of Environmental Management, 337*, 117723. https://doi.org/10.1016/j.jenvman.2023.117723
57. Liu, Y., Cui, W., Li, W., Xu, S., Sun, Y., Xu, G., & Wang, F. (2023). Effects of microplastics on cadmium accumulation by rice and arbuscular mycorrhizal fungal communities in cadmium-contaminated soil. *Journal of Hazardous Materials, 442*, 130102. https://doi.org/10.1016/j.jhazmat.2022.130102
58. Maiti, S. K., Ghosh, D., & Raj, D. (2022). 16—Phytoremediation of fly ash: bioaccumulation and translocation of metals in natural colonizing vegetation on fly ash lagoons. In K. K. Kar (Ed.), *Handbook of fly ash* (pp. 501–523). Butterworth-Heinemann. https://doi.org/10.1016/B978-0-12-817686-3.00011-6
59. Majumdar, A., Upadhyay, M. K., Ojha, M., Afsal, F., Giri, B., Srivastava, S., & Bose, S. (2022). Enhanced phytoremediation of Metal(loid)s via spiked ZVI nanoparticles: An urban clean-up strategy with ornamental plants. *Chemosphere, 288*, 132588. https://doi.org/10.1016/j.chemosphere.2021.132588
60. Mansoor, S., Khan, N. F., Farooq, I., Kaur, N., Manhas, S., Raina, S., & Khan, I. F. (2022). Chapter 4—Phytoremediation at molecular level. In R. A. Bhat, F. M. P. Tonelli, G. H. Dar, &

K. Hakeem (Eds.), *Phytoremediation* (pp. 65–90). Academic Press. https://doi.org/10.1016/B978-0-323-89874-4.00011-X

61. Maurya, A., Sharma, D., Partap, M., Kumar, R., & Bhargava, B. (2023). Microbially-assisted phytoremediation toward air pollutants: Current trends and future directions. *Environmental Technology & Innovation, 31*, 103140. https://doi.org/10.1016/j.eti.2023.103140
62. Miranda, R. de S., Boechat, C. L., Bomfim, M. R., Santos, J. A. G., Coelho, D. G., Assunção, S. J. R., Cardoso, K. M., & Cardoso, E. B. (2022). 3—Phytoremediation: A sustainable green approach for environmental cleanup. In V. Kumar, M. P. Shah, & S. K. Shahi (Eds.), *Phytoremediation technology for the removal of heavy metals and other contaminants from soil and water* (pp. 49–75). Elsevier. https://doi.org/10.1016/B978-0-323-85763-5.00017-9
63. Munir, R., Jan, M., Muhammad, S., Afzal, M., Jan, N., Yasin, M. U., Munir, I., Iqbal, A., Yang, S., Zhou, W., & Gan, Y. (2023). Detrimental effects of Cd and temperature on rice and functions of microbial community in paddy soils. *Environmental Pollution, 324*, 121371. https://doi.org/10.1016/j.envpol.2023.121371
64. Nicolas, G., Alessandro, B., Denis, Z., Silvia, S., & Enrico, D. (2022). Helicrysum italicum (roth) G. Don, a promising species for the phytostabilization of polluted mine sites: A case study in the Montevecchio mine (Sardinia, Italy). *Journal of Geochemical Exploration, 242*, 107088. https://doi.org/10.1016/j.gexplo.2022.107088
65. Njoku, K. L., & Nwani, S. O. (2022). Phytoremediation of heavy metals contaminated soil samples obtained from mechanic workshop and dumpsite using Amaranthus spinosus. *Scientific African, 17*, e01278. https://doi.org/10.1016/j.sciaf.2022.e01278
66. Ogundola, A. F., Adebayo, E. A., & Ajao, S. O. (2022). 2—Phytoremediation: The ultimate technique for reinstating soil contaminated with heavy metals and other pollutants. In V. Kumar, M. P. Shah, & S. K. Shahi (Eds.), *Phytoremediation technology for the removal of heavy metals and other contaminants from soil and water* (pp. 19–49). Elsevier. https://doi.org/10.1016/B978-0-323-85763-5.00012-X
67. Ojuederie, O. B., Amoo, A. E., Owonubi, S. J., & Ayangbenro, A. S. (2022). Chapter 6—Nanoparticles-assisted phytoremediation: Advances and applications. In V. Pandey (Ed.), *Assisted phytoremediation* (pp. 155–178). Elsevier. https://doi.org/10.1016/B978-0-12-822893-7.00011-2
68. Oubohssaine, M., Dahmani, I., Sbabou, L., Bruneel, O., & Aurag, J. (2022). The rhizosphere of Sulla spinosissima growing in abandoned mining soils is a reservoir of heavy metals tolerant plant growth-promoting rhizobacteria. *Biocatalysis and Agricultural Biotechnology, 39*, 102236. https://doi.org/10.1016/j.bcab.2021.102236
69. Pan, J., Cao, S., Xu, G., Rehman, M., Li, X., Luo, D., Wang, C., Fang, W., Xiao, H., Liao, C., & Chen, P. (2023). Comprehensive analysis reveals the underlying mechanism of arbuscular mycorrhizal fungi in kenaf cadmium stress alleviation. *Chemosphere, 314*, 137566. https://doi.org/10.1016/j.chemosphere.2022.137566
70. Pandey, V. C., Gajić, G., Sharma, P., & Roy, M. (2022). 5—Adaptive phytoremediation practices for sustaining ecosystem services. In V. C. Pandey, G. Gajić, P. Sharma, & M. Roy (Eds.), *Adaptive phytoremediation practices* (pp. 181–225). Elsevier. https://doi.org/10.1016/B978-0-12-823831-8.00008-6
71. Pandey, V. C., Mahajan, P., Saikia, P., & Praveen, A. (2022). Chapter 3—Grass fiber crops in phytoremediation. In V. C. Pandey, P. Mahajan, P. Saikia, & A. Praveen (Eds.), *Fiber crop-based phytoremediation* (pp. 57–87). Elsevier. https://doi.org/10.1016/B978-0-12-823993-3.00007-3
72. Pandey, V. C., Mahajan, P., Saikia, P., & Praveen, A. (2022). Chapter 6—Transgenic fiber crops in phytoremediation. In V. C. Pandey, P. Mahajan, P. Saikia, & A. Praveen (Eds.), *Fiber crop-based phytoremediation* (pp. 157–180). Elsevier. https://doi.org/10.1016/B978-0-12-823993-3.00003-6
73. Pandita, D. (2022). Chapter 23—Plant growth promoting bacteria (PGPB): applications and challenges in bioremediation of metal and metalloid contaminated soils. In T. Aftab & K. Hakeem (Eds.), *Metals metalloids soil plant water systems* (pp. 485–500). Academic Press. https://doi.org/10.1016/B978-0-323-91675-2.00002-0

74. Parihar, A., & Malaviya, P. (2023). Textile wastewater phytoremediation using Spirodela polyrhiza (L.) Schleid. assisted by novel bacterial consortium in a two-step remediation system. *Environmental Research, 221*, 115307. https://doi.org/10.1016/j.envres.2023.115307
75. Peera Sheikh Kulsum, P. G., Khanam, R., Das, S., Nayak, A. K., Tack, F. M. G., Meers, E., Vithanage, M., Shahid, M., Kumar, A., Chakraborty, S., Bhattacharya, T., & Biswas, J. K. (2023). A state-of-the-art review on cadmium uptake, toxicity, and tolerance in rice: From physiological response to remediation process. *Environmental Research, 220*, 115098. https://doi.org/10.1016/j.envres.2022.115098
76. Qiao, Y., Hou, D., Lin, Z., Wei, S., Chen, J., Li, J., Zhao, J., Xu, K., Lu, L., & Tian, S. (2023). Sulfur fertilization and water management ensure phytoremediation coupled with argo-production by mediating rhizosphere microbiota in the Oryza sativa L.-Sedum alfredii Hance rotation system. *Journal of Hazardous Materials, 457*, 131686. https://doi.org/10.1016/j.jhazmat.2023.131686
77. Rabani, M. S., Hameed, I., Mir, T. A., Wani, B. A., Gupta, M. K., Habib, A., Jan, M., Hussain, H., Tripathi, S., Pathak, A., Ahad, M. B., & Gupta, C. (2022). Chapter 5—Microbial-assisted phytoremediation. In R. A. Bhat, F. M. P. Tonelli, G. H. Dar, & K. Hakeem (Eds.), *Phytoremediation* (pp. 91–114). Academic Press. https://doi.org/10.1016/B978-0-323-89874-4.00006-6
78. Rachna, Gupta, S. K., Singh, B., Nema, A. K., & Pandey, V. C. (2022). Chapter 3—Integrated phytoremediation approaches for abatement of aquatic pollution and element recovery. In V. C. Pandey (Ed.), *Algae and aquatic macrophytes in cities* (pp. 39–64). Elsevier. https://doi.org/10.1016/B978-0-12-824270-4.00001-8
79. Rahman, S. ur, Khalid, M., Hui, N., Rehman, A., Kayani, S.-I., Fu, X., Zheng, H., Shao, J., Khan, A. A., Ali, M., Taheri, A., Liu, H., Yan, X., Hu, X., Qin, W., Peng, B., Li, M., Xinghao, Y., Zhang, Y., & Tang, K. (2023). Piriformospora indica alter root-associated microbiome structure to enhance Artemisia annua L. tolerance to arsenic. *Journal of Hazardous Materials, 457*, 131752. https://doi.org/10.1016/j.jhazmat.2023.131752
80. Rana, V., Bandyopadhyay, S., & Maiti, S. K. (2022). 8—Potential and prospects of weed plants in phytoremediation and eco-restoration of heavy metals polluted sites. In V. Kumar, M. P. Shah, & S. K. Shahi (Eds.), *Phytoremediation technology for the removal of heavy metals and other contaminants from soil and water* (pp. 187–205). Elsevier. https://doi.org/10.1016/B978-0-323-85763-5.00015-5
81. Rane, N. R., Tapase, S., Kanojia, A., Watharkar, A., Salama, E.-S., Jang, M., Kumar Yadav, K., Amin, M. A., Cabral-Pinto, M. M. S., Jadhav, J. P., & Jeon, B.-H. (2022). Molecular insights into plant–microbe interactions for sustainable remediation of contaminated environment. *Bioresource Technology, 344*, 126246. https://doi.org/10.1016/j.biortech.2021.126246
82. Riaz, U., Athar, T., Mustafa, U., & Iqbal, R. (2022). Chapter 23—Economic feasibility of phytoremediation. In R. A. Bhat, F. M. P. Tonelli, G. H. Dar, & K. Hakeem (Eds.), *Phytoremediation* (pp. 481–502). Academic Press. https://doi.org/10.1016/B978-0-323-89874-4.00025-X
83. Ruley, J. A., Amoding, A., Tumuhairwe, J. B., & Basamba, T. A. (2022). Chapter 14—Rhizoremediation of petroleum hydrocarbon–contaminated soils: A systematic review of mutualism between phytoremediation species and soil living microorganisms. In R. A. Bhat, F. M. P. Tonelli, G. H. Dar, & K. Hakeem (Eds.), *Phytoremediation* (pp. 263–296). Academic Press. https://doi.org/10.1016/B978-0-323-89874-4.00008-X
84. Sahu, S., Rajbonshi, M. P., Gujre, N., Gupta, M. K., Shelke, R. G., Ghose, A., Rangan, L., Pakshirajan, K., & Mitra, S. (2022). Bacterial strains found in the soils of a municipal solid waste dumping site facilitated phosphate solubilization along with cadmium remediation. *Chemosphere, 287*, 132320. https://doi.org/10.1016/j.chemosphere.2021.132320
85. Sarker, A., Al Masud, M. A., Deepo, D. M., Das, K., Nandi, R., Ansary, M. W. R., Islam, A. R. M. T., & Islam, T. (2023). Biological and green remediation of heavy metal contaminated water and soils: A state-of-the-art review. *Chemosphere, 332*, 138861. https://doi.org/10.1016/j.chemosphere.2023.138861

86. Sarma, H., Narayan, M., Peralta-Videa, J. R., & Lam, S. S. (2022). Exploring the significance of nanomaterials and organic amendments—Prospect for phytoremediation of contaminated agroecosystem. *Environmental Pollution, 308*, 119601. https://doi.org/10.1016/j.envpol.2022.119601
87. Sevak, P., & Pushkar, B. (2023). Bacterial responses towards arsenic toxicity and in-depth analysis of molecular mechanism along with possible on-field application. *Journal of Environmental Chemical Engineering, 11*(4), 110187. https://doi.org/10.1016/j.jece.2023.110187
88. Sharma, B., Chaudhary, T., & Shukla, P. (2022). Chapter 4—Combinatorial genetic engineering approaches in phytoremediation of pollutants. In P. Sharma, A. Pandey, Y. W. Tong, & H. H. Ngo (Eds.), *Current developments in biotechnology and bioengineering* (pp. 55–71). Elsevier. https://doi.org/10.1016/B978-0-323-99907-6.00001-3
89. Sharma, P., Bano, A., Singh, S. P., & Tong, Y. W. (2023). Chapter 4—Metal-tolerant microbial inoculants for improved phytoextraction. In V. K. Sharma, A. Kumar, M. R. Z. Passarini, S. Parmar, & V. K. Singh (Eds.), *Microbial inoculants* (*Developments in applied microbiology and biotechnology*) (pp. 75–98). Academic Press. https://doi.org/10.1016/B978-0-323-99043-1.00005-0
90. Sharma, P., Chaturvedi, P., Chandra, R., & Kumar, S. (2022). Identification of heavy metals tolerant Brevundimonas sp. from rhizospheric zone of Saccharum munja L. and their efficacy in in-situ phytoremediation. *Chemosphere, 295*, 133823. https://doi.org/10.1016/j.chemosphere.2022.133823
91. Sharma, P., Rai, S., Gautam, K., & Sharma, S. (2023). Chapter 11—Phytoremediation strategies of plants: Challenges and opportunities. In A. Husen (Ed.), *Plants and their interaction to environmental pollution* (pp. 211–229). Elsevier. https://doi.org/10.1016/B978-0-323-99978-6.00012-1
92. Shi, A., Hu, Y., Zhang, X., Zhou, D., Xu, J., Rensing, C., Zhang, L., Xing, S., Ni, W., & Yang, W. (2023). Biochar loaded with bacteria enhanced Cd/Zn phytoextraction by facilitating plant growth and shaping rhizospheric microbial community. *Environmental Pollution, 327*, 121559. https://doi.org/10.1016/j.envpol.2023.121559
93. Silambarasan, S., Logeswari, P., Vangnai, A. S., Kamaraj, B., & Cornejo, P. (2022). Plant growth-promoting actinobacterial inoculant assisted phytoremediation increases cadmium uptake in Sorghum bicolor under drought and heat stresses. *Environmental Pollution, 307*, 119489. https://doi.org/10.1016/j.envpol.2022.119489
94. Singh, B. S. M., Singh, D., & Dhal, N. K. (2022). Enhanced phytoremediation strategy for sustainable management of heavy metals and radionuclides. *Case Studies in Chemical and Environmental Engineering, 5*, 100176. https://doi.org/10.1016/j.cscee.2021.100176
95. Singh, K., Tripathi, S., & Chandra, R. (2023). Bacterial assisted phytoremediation of heavy metals and organic pollutants by Cannabis sativa as accumulator plants growing on distillery sludge for ecorestoration of polluted site. *Journal of Environmental Management, 332*, 117294. https://doi.org/10.1016/j.jenvman.2023.117294
96. Singh, M., Nanda, K., Singh, V., & Singh, S. P. (2022). Chapter 12—Metal polluted soil detoxification using phytoremediation technology. In P. Sharma, A. Pandey, Y. W. Tong, & H. H. Ngo (Eds.), *Current developments in biotechnology and bioengineering* (pp. 243–260). Elsevier. https://doi.org/10.1016/B978-0-323-99907-6.00011-6
97. Singh, V. K., Singh, R., Rajput, V. D., & Singh, V. K. (2023). Halophytes for the sustainable remediation of heavy metal-contaminated sites: Recent developments and future perspectives. *Chemosphere, 313*, 137524. https://doi.org/10.1016/j.chemosphere.2022.137524
98. Song, L., Xu, X., Zheng, Y., Hong, W., Li, X., Ai, Y., Wang, Y., Zhang, Z., Chen, H., Huang, Y., Zhang, J., & Zhou, J. (2023). Dynamic mechanisms of cadmium accumulation and detoxification by Lolium perenne grown in soil inoculated with the cadmium-tolerant bacterium strain Cdq4-2. *Science of the Total Environment, 873*, 162314. https://doi.org/10.1016/j.scitotenv.2023.162314
99. Sonowal, S., Nava, A. R., Joshi, S. J., Borah, S. N., Islam, N. F., Pandit, S., Prasad, R., & Sarma, H. (2022). Biosurfactant-assisted phytoremediation of potentially toxic elements in soil: Green

technology for meeting the United Nations Sustainable Development Goals. *Pedosphere, 32*(1), 198–210. https://doi.org/10.1016/S1002-0160(21)60067-X
100. Swain, A. A., Dwivedi, N., Bauddh, K., & Kumar, M. (2022). Chapter 15—Role of microorganism in phytoremediation of mine spoiled soils. In K. Bauddh & Y. Ma (Eds.), *Advances in microbe-assisted phytoremediation of polluted sites* (pp. 379–400). Elsevier. https://doi.org/10.1016/B978-0-12-823443-3.00009-0
101. Takkar, S., Shandilya, C., Agrahari, R., Chaurasia, A., Vishwakarma, K., Mohapatra, S., Varma, A., & Mishra, A. (2022). 17—Green technology: Phytoremediation for pesticide pollution. In V. Kumar, M. P. Shah, & S. K. Shahi (Eds.), *Phytoremediation technology for the removal of heavy metals and other contaminants from soil and water* (pp. 353–375). Elsevier. https://doi.org/10.1016/B978-0-323-85763-5.00008-8
102. Tang, S., Xu, Y., Zeng, K., Liang, X., Shi, X., Liu, K., Ma, J., Yu, F., & Li, Y. (2023). Comparative study on plant growth-promoting bacterial inoculation by irrigation and spraying for promoting Bidens pilosa L. phytoremediation of cadmium-contaminated soil. *Ecotoxicology and Environmental Safety, 254*, 114764. https://doi.org/10.1016/j.ecoenv.2023.114764
103. Timalsina, H., Gyawali, T., Ghimire, S., & Paudel, S. R. (2022). Potential application of enhanced phytoremediation for heavy metals treatment in Nepal. *Chemosphere, 306*, 135581. https://doi.org/10.1016/j.chemosphere.2022.135581
104. Turan, M., Argin, S., Bolouri, P., Arjumend, T., Ersoy, N., Yıldırım, E., Güneş, A., Ekinci, M., & Birinci, D. (2022). Chapter 11—Bioremediation of contaminated soil with plant growth rhizobium bacteria, In V. Pandey (Ed.), *Assisted phytoremediation* (pp. 265–284). Elsevier. https://doi.org/10.1016/B978-0-12-822893-7.00013-6
105. Upadhyay, S. K., Devi, P., Kumar, V., Pathak, H. K., Kumar, P., Rajput, V. D., & Dwivedi, P. (2023). Efficient removal of total arsenic (As3+/5+) from contaminated water by novel strategies mediated iron and plant extract activated waste flowers of marigold. *Chemosphere, 313*, 137551. https://doi.org/10.1016/j.chemosphere.2022.137551
106. Veerapagu, M., Jeya, K. R., & Sankaranarayanan, A. (2023). Chapter 2—Role of plant growth-promoting microorganisms in phytoremediation efficiency. In P. Swapnil, M. Meena, Harish, A. Marwal, S. Vijayalakshmi, & A. Zehra (Eds.), *Plant-microbe interaction—Recent advances in molecular and biochemical approaches* (pp. 45–61). Academic Press. https://doi.org/10.1016/B978-0-323-91875-6.00020-7
107. Venkataraman, S., & Hefferon, K. (2023). Chapter 7—Phytoremediation strategies using biotechnology. In S. Venkataraman & K. Hefferon (Eds.), *Agricultural biotechnology* (pp. 137–169). Academic Press. https://doi.org/10.1016/B978-0-323-88476-1.00007-2
108. Wang, X., Jain, A., Chen, B., Wang, Y., Jin, Q., Yugandhar, P., Xu, Y., Sun, S., & Hu, F. (2022). Differential efficacy of water lily cultivars in phytoremediation of eutrophic water contaminated with phosphorus and nitrogen. *Plant Physiology and Biochemistry, 171*, 139–146. https://doi.org/10.1016/j.plaphy.2021.12.001
109. Xiao, Y., Chen, L., Teng, K., Ma, J., Xiang, S., Jiang, L., Liu, G., Yang, B., & Fang, J. (2023). Potential roles of the rhizospheric bacterial community in assisting Miscanthus floridulus in remediating multi-metal(loid)s contaminated soils. *Environmental Research, 227*, 115749. https://doi.org/10.1016/j.envres.2023.115749
110. Yadav, R., Singh, S., Kumar, A., & Singh, A. N. (2022). Chapter 15—Phytoremediation: A wonderful cost-effective tool. In S. Kathi, S. Devipriya, & K. Thamaraiselvi (Eds.), *Cost effective technologies for solid waste and wastewater treatment* (*Advances in environmental pollution research*) (pp. 179–208). Elsevier. https://doi.org/10.1016/B978-0-12-822933-0.00008-5
111. Yadav, R., Singh, G., Santal, A. R., & Singh, N. P. (2023). Omics approaches in effective selection and generation of potential plants for phytoremediation of heavy metal from contaminated resources. *Journal of Environmental Management, 336*, 117730. https://doi.org/10.1016/j.jenvman.2023.117730
112. Yang, J., Huang, Y., Zhao, G., Li, B., Qin, X., Xu, J., & Li, X. (2022). Phytoremediation potential evaluation of three rhubarb species and comparative analysis of their rhizosphere

characteristics in a Cd- and Pb-contaminated soil. *Chemosphere, 296*, 134045. https://doi.org/10.1016/j.chemosphere.2022.134045

113. Zeng, X., Yang, Y., Zhang, Q., Zeng, C., Deng, X., Yuan, H., Gong, X., Zou, D., & Zeng, Q. (2023). Field-scale differences in rhizosphere micro-characteristics of Cichorium intybus, Ixeris polycephala, sunflower, and Sedum alfredii in the phytoremediation of Cd-contaminated soil. *Ecotoxicology and Environmental Safety, 262*, 115137. https://doi.org/10.1016/j.ecoenv.2023.115137
114. Zhang, X., Zeng, B., Li, H., Huang, J., Jiang, L., Zhang, X., Tan, Z., Wu, Z., Qin, X., Feng, C., & Huang, Z. (2022). Soil heavy metals and phytoremediation by Populus deltoides alter the structure and function of bacterial community in mine ecosystems. *Applied Soil Ecology, 172*, 104359. https://doi.org/10.1016/j.apsoil.2021.104359

Root Iron Plaque Formation and Cadmium Accumulation in Paddy Rice: A Literature-Based Study

Abu Bakkar Siddique, Md. Rafiqul Islam, Mohammad Shahid, Md. Maruf Billah, Ravi Naidu, and Mohammad Mahmudur Rahman

Abstract Cadmium (Cd) is considered as one of the most hazardous elements and its existence in crop field is of great environmental concern due to its detrimental impacts, persistence, and intense level of bioaccumulation. In the developing nations, rice is counted as the most essential food crop and over half of the global population relies on rice for the survival. The consumption of rice is a significant dietary source of Cd and may pose human health risk. Therefore, to ensure food safety and reducing Cd toxicity, a comprehensive review incorporating innovative and sustainable, ecofriendly approach are needed. The origin of Cd in paddy soils and effect of dominant soil components like soil pH, soil redox potential on Cd availability, Cd toxicity in rice and genotypic variation in Cd uptake are highlighted in this review. This review also focused on the application of various organic and inorganic amendment to reduce Cd accumulation in rice plant. Moreover, this review provides an insight of Fe plaque formation and its adsorptive action to decrease Cd addition in rice. Furthermore, the present review discusses the Cd transport mechanism in rice and future research thrusts and guidelines to moderate Cd uptake in rice.

A. B. Siddique · Md. M. Billah · R. Naidu · M. M. Rahman (✉)
Global Centre for Environmental Remediation, College of Engineering, Science and Environment, The University of Newcastle, Callaghan, NSW 2308, Australia
e-mail: mahmud.rahman@newcastle.edu.au

crcCARE, The University of Newcastle, Callaghan, NSW 2308, Australia

A. B. Siddique
Department of Agriculture, Noakhali Science and Technology University (NSTU), Noakhali 3814, Bangladesh

Md. R. Islam
Department of Soil Science, Bangladesh Agricultural University (BAU), Mymensingh 2202, Bangladesh

M. Shahid
ICAR—National Rice Research Institute, Cuttack, Odisha 753006, India

Md. M. Billah
Agrotechnology Discipline, Khulna University, Khulna 9208, Bangladesh

A. K. Jha and N. Kumar (eds.), *Cadmium Toxicity in Water*, Springer Water,
https://doi.org/10.1007/978-3-031-54005-9_11

Keywords Cadmium · Cd accumulation and uptake · Organic amendment · Inorganic amendment · Rice genotypes

1 Introduction

Cadmium (Cd) is identified as a potent nephrotoxin component and feasibly detrimental contaminant in the soil–plant systems, and it is very difficult to remove due to its persistence and non-degradation behaviour [1, 2]. Cd is positioned seventh (7th) among the leading 20 contaminated metals and is categorised as a chronic human carcinogen (class I) [3, 4]. Cd is readily discharged into the agro-ecosystems via different anthropogenic activities including metal mining, smelting activity, metallurgical works, industrial waste, municipal waste, sewage irrigation, use of agrochemicals such as phosphatic fertilizers, pesticides, insecticides etc. [5, 6, 7]. Once in the soil system, Cd is immediately stored in crops leading to food chain contamination [8].

Cadmium doesn't have any identified role in plant metabolism due to its non-essential forms in physiological functions of plants [9]. Cd is easily accumulated into the consumable portion of crops through root uptake when produced in soil contaminated with Cd [10, 11]. The bioavailability of Cd in the soil as well as the genetic makeup of the plants affect how much Cd is delivered and distributed from soil to the aboveground edible parts of crop [12]. When Cd concertation in crop plants exceeds the marginal level, the overall development of crop plant is adversely pretended by Cd which results in lower production, poor quality and weakening food security [9, 10]. The toxicity action of Cd affects the crop yield extremely and causes severe human health hazards including anaemia, cancer, cerebrovascular infection, cardiac failure, renal dysfunction, proteinuria, hypertension, emphysema, lung damage, osteoporosis etc. [13].

In the developing nations, rice is counted as the most essential food crop and majority of the population relies on rice based diets for the survival [14]. Globally, almost 90% of the earth's rice is consumed in Asian countries with a continued rise in demand [14]. Besides, per capita intake of rice is also increasing even in developed countries and areas like the European Union and the United States partly due to immigration from Asian countries [14]. Rice consumption, all over the world by the large population is considered as the principal route of Cd exposure [15]. Elevated level of Cd in rice poses serious threats to overall agricultural production and has received extensive public attention related to human health [16]. The first outbreak of chronic Cd intoxication through Cd contaminated rice consumption is the well documented Itai-Itai disease, found in the Jinzu River Basin of Toyama Territory, Japan in the 1950s [17]. Rice based diet is documented as the significant avenue of Cd exposure by human population in different geological locations, including India, China, Bangladesh, Pakistan, Japan, and Sri Lanka [18, 19]. The research report exposed that rice plants are more efficient in absorbing Cd and storing it in grain, which is 3–4 times greater than in other cereal crops that poses serious problems to produce quality food and human health as well [20, 11]. Therefore, to identify efficient management practices is imperative to prevent Cd translocation and uptake in rice grain to lower human exposure to Cd as well as related health risks.

Cadmium translocation and uptake in plants is influenced and controlled by several factors. In case of rice cultivation, the key Cd transport process involved in Cd accretion in edible portions from roots are: (i) uptake by roots, (ii) xylem loading (transports from roots to shoots), (iii) redirection transport of Cd at node through transfer and (iv) finally phloem mediated Cd transfer from leaf blades intervascular to grains [15, 21]. Roots are the first organ in the soil and performs as a potential obstacle to combat with Cd and its sequential deposition in aerial parts of plants via series of the process [22, 21]. The translocation of Cd from root to shoot is considered as the fundamental procedure in overall Cd deposition in rice grain [23, 24]. Thus, blocking Cd entry into roots of crop plant is the promising option to encounter Cd toxicity in rice [25]. It was well established that Fe is generally deposited on the surface of roots and act as obstacle to inhibit the accumulation of toxic elements into roots [26, 27]. Contrasting other wetland plants, rice plants have developed widespread porous aerenchymatous tissues that allows infiltration of O_2 from shoot to root due to respiration activity in roots [28]. To adapt to waterlogged conditions, root aerenchyma discharged O_2 in the root zone area that aided in the creation of Fe plaque through oxidation of ferrous (Fe^{2+}) to ferric (Fe^{3+}) iron [29, 30]. On the outer layer of the roots, Fe plaques deposits are located in the form of Fe oxides or hydroxides like ferrihydrite ($5Fe_2O_3{\cdot}9H_2O$), goethite (α-FeOOH) and lepidocrocite (γ-FeOOH) [31]. Fe plaque possess a abundant efficient groups with numerous adsorption sites leading to sequestration of Cd by surface assimilation or co-precipitation and hence hinders the bioavailability of Cd in root zone, its uptake as well as translocation into the plants [31, 19]. Sebastian and Prasad [25] showed that Fe plaque was accumulated across the root surfaces after exogenous Fe accumulation which prevents Fe deficiency and Cd accretion in rice plants. It is exposed that Fe application increased plaque formation capabilities and decreasing Cd gathering in upper parts of rice plants through increasing Cd sequestration in Fe plaque [29].

Rice cultivar and soil pH are the two influential factors controlling Cd translocation and uptake in rice plants. Basically, a considerable variation existed among the different rice cultivars in relation to Cd accumulation and uptake even under low soil Cd situations [32, 33]. Duan et al. [32] detected a 10–32-fold deviation in Cd intensity in rice grain through field investigation over two years. Several studies [34, 35, 36] explored negative association between translocation and uptake of Cd in rice grain and soil pH. It has been stated that bioavailability and solubility of Cd enhanced with reduction in soil pH due to high protonation, leading to a decrease in the negative charge on the soil particles to bind Cd and thus increased more bioavailable Cd for plant uptake [34, 37, 38]. In this context, the conversion of Cd from stable fractions to readily exchangeable fractions enhance Cd solubility and phytoaccumulation [36, 5, 15]. On the contrary, the increase of soil pH, reduces the spreading and solubility of Cd through increased complexation by adsorptive sites on the soil particles [36, 39].

2 Sources and Contamination of Cd in Paddy Soils

The contamination of Cd to the soil at variable quantity is mainly derived from the different natural and anthropogenic activities [40, 41, 7]. It has been observed that Cd does not exist as a pure metal in the soil conditions rather, it is normally coexisted with the ores of zinc sulphide and is also obtained as impurity in the ores of copper and lead [42, 43]. According to the research report, the key natural sources of Cd enhancement in soils are fossil fuels, forest fires, volcanic eruption, wind-blown dust etc. [5, 7]. Geological weathering of the parent materials is also important contributor towards Cd deposition in soil [44, 5, 6]. Moreover, the soil contains a superior concentration of Cd when originated from the weathering of mafic and ultramafic rocks [45]. The soils derived from the breakdown of black shale are rich in Cd and the intensity of Cd in these soils was measured up to 200 mg kg^{-1} [46]. The average Cd intensity of 1.3 mg kg^{-1} in sedimentary rocks was higher than igneous rocks (0.16 mg kg^{-1}) [47]. The Cd intensity was varied from 0.37 to 21 mg kg^{-1} in carbonaceous mudstone and siltstone while in carbonate rocks it was varied from 0.22 to 3.6 mg kg^{-1} [48]. However, it was estimated that different natural sources contributed about 10% of total Cd released into the environment [49]. On the other hand, anthropogenic input of Cd to the paddy soil environments is accounted for 90% of total emission, which is mainly due to metal mining and smelting activity, industrial and municipal waste, vehicular emissions, metallurgical works and utilization of agrochemicals like chemical fertilizers (e.g. phosphatic fertilizers), pesticide, sewage irrigation, waste water irrigation, etc. [36, 6, 40, 50, 49, 7]. The Cd composites enriched with nitrogen ($CdNO_3$) are also exist in the land due to decline of NO_3^- with Cd [51, 52]. Research on the vehicular emission of Cd to environment revealed that crop plant uptake more Cd near traffic zone where the potential source of Cd is the motor oil containing zinc (Zn) extracts and rubber tires vulcanization using Zn compounds [51]. The production of Ni–Cd batteries, coal incineration as well as electrodeposition activities accelerate the Cd deposition in the atmosphere resulting into gathering of Cd in plants and soil [53]. It was observed that use of manure in the agricultural fields added 3.6 g Cd ha^{-1} $year^{-1}$ whilst manure dispensed at a level of N equal to 0.25 t N ha^{-1} $year^{-1}$ [54]. The usage of sewage-sludge in the land accelerate in increasing the background Cd concentration of soil [55]. Nicholson et al. [54] observed that if sludge having Cd of 3.4 mg kg^{-1} dry weight used at the rate of N equal to 0.25 t N ha^{-1} $year^{-1}$ contributed up to 19 g Cd ha^{-1} $year^{-1}$. Among the various anthropogenic inputs, the usage of phosphate fertilizers and wastewater irrigation from industries are primary route of excessive Cd input in paddy fields [42]. In Europe, it was estimated that the Cd intensity in phosphatic fertilizer ranges from 0.1 to 120 mg Cd kg^{-1} [56] while in China it was 15.3 mg kg^{-1} which is much lower compared to phosphate fertilizers produced (16–45 mg Cd kg^{-1}) in North America [42].

2.1 *Effect of Soil pH on Cd Bioavailability and Uptake by Rice*

Among the soil properties, soil pH is considered as a prominent component influencing the translocation and uptake of Cd in rice plants because of its strong effect on the solubility, bioavailability and speciation of Cd in soils [57]. It was well established in numerous studies that the Cd accretion in rice grain exhibited a significant and adverse association to soil pH which implied that the mobility and accretion of Cd in rice is decreased with increasing soil pH [35, 58, 59, 57]. It was noted that in acidic soils (pH < 7), the movement and solubility of heavy metal is high as compared to the alkaline soil (pH > 8) [60]. At low acidic soil, the bioavailability and solubility of Cd was improved because of increase in protonation at low soil pH which results into inhibiting the sorption site of the soil particles to bind Cd [34, 37, 38]. In this case, the desorption and bioaccumulation are promoted through conversion of Cd from Fe–Mn oxide and carbonate bound forms (fixed and unavailable form) to mobile and bioavailable forms (exchangeable form) [36, 5, 15]. Whereas Cd immobilization is increased in alkaline soils because of binding Cd with carbonates and coprecipitation with hydroxide ion (OH^-) and other anions [57]. Moreover, the Cd adsorption capacity by Fe-oxides substantially enhanced when soil pH increased from 5.5 to 6.5 following in a substantial reduction of exchangeable Cd for plant uptake [1]. Similarly, Chen et al. [61] also reported that $CaCl_2$ extractable Cd was decreased by 2.6–3.0-fold at soil pH 6.5 raised from an initial value of 5.5. Christensen [62] exposed that the Cd adsorption capacity of soils was observed to rise by 3 unit for 1 unit increase in pH between 4.0 and 7.7. The fluctuation of soil pH was also uneven in rice soils [63]. At the initial stage of cultivation, soil solution pH was measured as neutral pH and changed to alkaline conditions at the time of mature stage [64]. The pH of acidic soils was found to increase under reduced condition due to utilization of proton while the pH of alkaline soils was declined as result a of CO_2 accumulation due to mineralization of organic matter in paddy soils [65]. The Cd addition in rice plant is enhanced through increase in respiration activity is soils [51]. The movement and addition of Cd is diminished with increase in soil pH by promoting Cd complexation into reactive sites of the soil particles [36, 39]. The deposition of $Cd(OH)^+$ is augmented due to hydrolysis reaction under high soil pH which ultimately increased the Cd sorption to the soil surfaces thereby reducing Cd mobility and availability for plant accumulation [66]. So, with a view to preventing the discharge of bioavailable Cd in soil, soil acidification must be avoided [67].

2.2 *Soil Redox Potential and Cd Uptake by Rice*

Soil redox potential (Eh) is an assess of the oxidation and reduction condition of soils and is defined as the tendency of electron transfer from electron donors (reduced agents) to electron acceptors (oxidized agents) [68]. Generally, oxidized soil conditions prevail at high Eh whereas reduced soil conditions dominate at soil low Eh [69, 70]. The soil pH is accelerated by reduction reactions through consumption of protons while the soil pH is declined at oxidized conditions due to generation of protons [71]. The redox status of soil is greatly varied, and it can be categorised as oxidizing (Eh > 400), weakly reducing (Eh 200 to 400), moderately reducing (Eh − 100 to 200) and strongly reducing (Eh < − 100) [72]. Moreover, the variation of soil Eh in relation to water managements in rice fields can be the dominating issues influencing both the Cd bioavailability and solubility in soils [35, 73]. On the contrary, in flooding situations, higher soil pH and lower Eh were observed whereas in dry and well aerated soil conditions lower soil pH and higher Eh had been recorded [36]. In flooded rice fields, microbial reduction is occurred which utilizing oxidized compounds such as NO_3^-, SO_4^{2-} Mn^{4-}, Fe^{3+} as well as disintegration substances of soil organic matter [15]. These compounds accept electron during reduction processes followed by the discharge of NO_2^-, S^{2-}, Mn^{2+}, Fe^{2+} and organic acids results into declining of soil Eh [74, 36, 15]. Subsequently, the solvency of Cd in reduced soil condition is declined because of adsorption of Cd to Fe–Mn oxyhydroxides as well as production insoluble CdS or $CdCO_3$ [75, 76]. Therefore, it is recommended that before panicle initiation flooding condition should be practiced in order to limit Cd availability and addition in rice plants [77]. The Cd absorption in rice plant is reduced by 16-fold when rice filed was flooded for a duration of 8 days before the heading stage [75]. However, in aerobic soils, Eh is increased as a result oxidative reactions of soils [5]. Under oxidized conditions, CdS is dissolved to form $CdSO_4$ which increases the solubility of Cd taken by rice plants [67]. On the contrary, Chloride absorption in water is another aspect that ameliorate Cd accumulation as it forms an exchangeable $CdCl_2$ complex [78]. Chlorine mediated salinity also decreases Cd adsorption on soil and increases solution Cd that accelerates to intensify Cd solubility as well as sorption to plants root [79, 80].

2.3 *Effect of Cd on Rice Growth and Accumulation*

Cd not being an essential element for plants, does not involve in beneficial physiological role, however, it has an adverse effect on every aspect of plants' growth when it accumulates. Cd in high concentrations reported to hamper cell and the growth of plants like maize, wheat, alfalfa, lentil, chickpea, soybean and spinach [81, 82, 83, 84, 85]. It has been widely reported that, in case of rice cultivation, Cd stress hinders the growth and development seriously [86, 87, 88]. Toxic levels of Cd specially in rice plants limit germination, restrict plant growth and development, disrupt seedling metabolic procedures, and impair farming efficiency [89, 90]. A high level (between 0.2 and 1.0 mM) of Cd exposure to rice seeds may lead to decrease in rate of germination, shoot elongation, water content and biomass yield [86]. Additionally, Cd toxicity also leads to chlorosis, wilting, necrosis, and leaf rolling [91, 92]. It was explored that Cd toxicity diminished the growth and development of rice plant as estimated with regard to root and shoot length, root and leaf area and numbers of roots and leaves per plant [11, 93]. Under Cd toxicity, rice plant showed a reduction in gas exchange and photosynthesis properties such as chlorophyll, net photosynthetic rate, carotenoids, transpiration rate, stomatal conductance etc. [94, 88]. Under Cd stress in rice, plant height, biomass, and photosynthetic capacity were all decreased while oxidative stress was elevated [95]. The decrease in photosynthesis perhaps the reason, because of the lowered biomass yield of rice under Cd toxicity [81, 95]. Cd toxicity decreased the chlorophyll content, seedling height and fresh weight of rice plants [96]. It has been stated by Kanu et al. [97] that in soils 1, 2, and 3 mg kg^{-1} of Cd caused 34, 43, and 63% decrease in grain yield of a rice variety with subsequent decrease in protein and amylose contents. Moreover, Siddique et al. [98] observed 12.2 to 33.3% decline in grain and straw yield of rice at different Cd exposure level. According to reports, rice plants exposed to Cd stress exhibited substantially decreased tillering ability, leaf area, plant height, generation of grain yield and dry matter [99]. A recent study by Feng et al. [100] showed that soil Cd treatments of 1 and 5 mg kg^{-1} enhanced leaf Cd intensity and addition but decreased net photosynthetic rate, plant height, and biomass production with a more rigorous response in the case of the 5 mg kg^{-1} treatment. Toxicity symptoms like short shoot, immature roots, growth retardation and green-yellow banded leaves were examined in rice seedlings under Cd treatment (5 mg L^{-1}) subsequent to 35 days, and the biomass, carotenoid contents, and chlorophyll were lessened [101]. Acute Cd stress (150 μM) restricted the growth of roots due to the conspicuous generation of superoxide which leads to widespread cell death in root tips, whereas mild Cd stress (15 μM) evoke morphogenesis responses like root elongation and radial root development, consequently produce thicker and shorter roots [102]. Cd poisoning reduces the amount of N, P, Ca, and Mg in plant shoots and roots [85] while Cd increase in leaves restricts the mechanism of stomatal opening [84, 85]. It was found that, Cd lowered gas exchange parameters, causing photosynthetic pigments and chloroplast structure to deteriorate [103].

2.4 Genotypic Variation of Rice Plant in Cd Accumulation

The research reports exposed that various rice genotypes demonstrate a substantial distinction in Cd uptake and accumulation. [104]. The transportation and uptake of Cd in rice plants varied markedly between plant species as well as cultivars of the same species [105, 79, 51]. Cd uptake and its distribution into different parts of rice genotypes is attributed to dominant heritable traits [106, 107]. In a research, rice (49 varieties) produced on Cd polluted soil, a significant variation in grain's Cd content of roughly 13–23 times has been observed [108, 109] whereas Liu et al. [105] reported the values from 0.14 to 1.43 mg kg^{-1} in the Cd concentration of rice. The Cd concentration in brown rice varied between 0.96 and 0.99 mg kg^{-1} in 38 different rice genotypes [110]. Accordingly Ye et al. [111] examined the impact of rice genotypes on Cd levels in several rice plant parts and observed that indica rice varieties had greater Cd levels in different organs than hybrid and japonica cultivars. Römkens et al. [112] also showed that for indica cultivars, only pure soils were appropriate as under field conditions these had better Cd absorption even in low Cd soils. Song et al. [11] performed a pot experiment in a glasshouse and found that 20 different rice cultivars had significantly different Cd contents at the equal Cd levels (0.3 or 0.6 mg kg^{-1}) in soil. Similarly, the level of Cd in both grains and rachises of early rice variety was noticeably lower than that of late rice variety grown at the same area [113]. It has been reported that, in hydroponics experiment, the Cd tolerance and Cd accumulation varied significantly among 146 rice lines while the same results was recorded in case of 17 rice lines in a pot culture experiment [33]. The Cd intensity of grain in hybrid rice varieties was greater than that of conventional ones [114]. The ability of the roots to oxidise, acidify, and secrete organic acids were associated to the alteration in Cd accretion and uptake in many rice cultivars [111]. Moreover, Wang et al. [115] performed a hydroponic experiment with low and high grain-Cd accruing rice varieties and observed that low grain-Cd accruing cultivars were more tolerant to Cd stress and had higher capability in debarring Cd carriage to soluble organelles because of distinctive character of cell wall for binding additional Cd. In a hydroponic experiment with 10 μM L^{-1} Cd, Barman et al. [116] investigated the Cd accretion in four rice genotypes and explored that Netiya and Maharaj amassed 0.033 and 0.029 mg kg^{-1} Cd in grains, correspondingly, though I.E.T4786 and Jamini had no Cd accretion in grains. Siddique et al. [98] reported that application of 1 and 3 mg kg^{-1} Cd in two soils with a pH of 4.6 and 6.6 resulted in Cd uptake of 0.29 to 1.92 mg kg^{-1} grain in Langi cultivar and 0.17 to 1.53 mg kg^{-1} grain in Quest cultivar of rice. The effect of Cd (0.14 mg kg^{-1}) exposure on Cd uptake of brown rice in 12 genotypes was studied and it was found to be in the sort of 0.020 to 0.222 mg kg^{-1} [117, 118]. In case of brown rice, CLY4418 had the utmost Cd absorption, while TYJ1 and NJ9108 had the least. Furthermore, Yuan et al. [119] discovered 0.16 to 0.96 and 0.08 to 0.58 mg kg^{-1} Cd accretion in late and early growing rice cultivars, respectively, from a polluted paddy soil (0.13–0.84 mg Cd kg^{-1} soil). Additionally, they noted that the concentration of amino acids dropped by 9.2% for early rice cultivars while 11.1% for late ones. The rate of Cd incorporation by the shoots of rice plants in a field trial with soils contaminated with 0.42, 0.85, and 2.80 mg Cd kg^{-1} soil was 6.17, 4.52, and 3.89 times higher in the four indica cultivars than japonica cultivars, correspondingly [120, 121, 122]. It has been reported that hybrid low-Cd accruing cultivar (XY), conventional low Cd accruing cultivar (WY), hybrid high-Cd accruing cultivar (XZ) and traditional high Cd accruing cultivar (WZ), have 18.09, 10.75, 25.73 and 21.26 μg pot^{-1} Cd accumulation, respectively [123, 124]. In a hydroponic culture with 1.0 mg L^{-1} Cd, [125] reported that 284 and 421 mg kg^{-1} Cd uptake in roots and 41 and 75 mg kg^{-1} Cd uptake in shoots in indica and japonica rice cultivars, respectively. Therefore, it is best to avoid using cultivars that gather high volumes of Cd in agricultural production.

3 Approaches to Diminish Cd in Rice

Since Cd consumption from rice represents such a significant route of exposure, particularly for persons intaking substantial amounts of contaminated rice regularly, alleviation approaches to lessen Cd spread in rice are immediately required. Various approaches may be used to moderate the addition of Cd in rice by breeding rice cultivars with repel Cd [126, 127, 128], identification of existing rice cultivars which accumulate low Cd or genotype selection [116, 29], remediation of Cd polluted soils to restrict Cd accumulation by rice plants through paddy soil amendments [61], phytoremediation [90, 129], microbes assisted remediation [130, 131, 132, 133, 134] and the development of effective water management strategies [73, 135, 77, 136], which provide important opportunities to abate the Cd accumulation in rice. In this article, we chiefly focus on inorganic minerals amendments in paddy soils to lower Cd addition and uptake by rice plant. In addition, biochar amendments in lessening Cd in rice is also discussed.

3.1 Amendment of Si

Silicon (Si) in the soil is the second most profuse element absorbed up by plant as monosilicic acid and finally polymerized to form silica gel [137, 138]. It is well established that Si lessens the metals toxicity and improves tolerance of plants to metal stress [137, 139]. The usage of Si immobilized Cd in paddy soils which may be ascribed due to the pH rise in soils. The impacts of silicon fertiliser on soil and Cd convenience and uptake by rice plants cultivated in acid soil sustained under two distinct moisture environment were examined by Babu and Nagabovanalli [137]. The study found that Si fertilizer enhanced root biomass in rice plant as grown in Cd-treated plants and considerably decreased the intensities of Cd in grain, husk, straw, and root [137, 140].

A research that looked into the effectiveness of split Si application at various growth stages of rice cultivated in Cd-spiked alkaline soil found that Si increased rice growth (up to 98%) and photosynthesis rate (84%) and reduced Cd concentration in rice grain [139]. Another study explored the role of Si (1.5 mM $NaSiO_3 \cdot 9H_2O$) and B (16.23 μM H_3BO_3) in declining Cd accrual in rice plant grown under glasshouse hydroponic experiment with Cd level of 100 μM $CdCl_2$. The results showed that application and usages of either Si or B extensively inhibited Cd deposition and Cd-induced oxidative stress by enhancing antioxidative approaches [141]. Another study reported that rice biomass can be boosted up to 15.5% by utilizing silicate-rich soil amendments and fertilisers and decreased total leaf Cd by 8.5 to 21.9% in addition to reducing the uptake of total Cd and soluble forms of Cd in rice leaves [142].

3.2 Amendment of Se

Selenium is an essential micronutrient and moderate doses of Se application can improve plant growth, enhance Se levels in grain and also effective in diminishing harmful Cd transport and uptake in rice grain [143, 144]. A recent study found that joint use of Se along with organics in high Cd-contaminated soils decreased Cd bioaccumulation in rice and the decline was influenced by rice genotypes, types and doses of organic amendments used [145]. The study also reported that the rice grain yield and photosynthesis was significantly mounted because of combined usage of Se and organic amendment for all tested rice cultivars [145]. On the other hand, the role of Se on soil Cd availability and Cd uptake was investigated by another study in rice plants which showed that addition of inorganic Se compounds (both selenite and selenate) pointedly enhanced rice grain Se by 4.25-fold and decreased Cd by 36.5% [146]. Pore-water analysis results indicated that Cd level in the soil solution was decreased by 11.2–13.0% and recommended that Se amendment in Cd-polluted soil help decreased Cd concentration and increased Se levels in rice grain by raising soil pH, decreasing Cd availability, and preventing Cd transport in rice plants [146].

In a study, rice was cultivated in two Cd-contaminated soils (Jiangxi soil, pH: 5.25 and Hunan soil, pH: 7.26) under glasshouse conditions at two different conditions including aerobic and flooded [135]. The results indicated that rice cultivation using flooding condition of the soil reduced grain Cd in Jiangxi soil but not in Hunan soil and the accumulation of Se declined the intensity of Cd in rice grain in Hunan soil but not in Jiangxi soil [135]. The study also explored that use of Se as selenite (0.5 mg kg^{-1}) reduced the Cd contents in rice grain by 45.2% and 67.7% under aerobic and flooding situations, correspondingly [135].

A strong synergistic impact of both Se and Si in reducing Cd noxiousness in rice plant was observed by a current study [147]. The study exhibited that the interaction of Se–Si diminished transportation and addition of Cd in rice shoots (up to 73.2%) from roots [147]. The study also observed that glutathione and phytochelatin levels in rice plants were enhanced by the interaction of Se–Si [147]. Another study explained that utilization of Se along with different treatments such as use of Cd safe varieties, water management, lime application implicitly condensed the Cd content in husked rice by 72.7% when rice cultivated in slight—moderate Cd-polluted soils [148].

3.3 Application of Metal Nanoparticles

Engineered nanoparticles like metal and metal-oxide nanoparticles (MNPs or MONPs) enhanced the tolerance of metal toxicity in various plants including rice plants. However, limited studies examined the reduction of Cd accumulation by

MNPs or MONPs and more investigations are required to determine the effectiveness and probable mechanism in reducing Cd, especially using sophisticated practices like synchrotron-based x-ray fluorescence (XRF) and x-ray absorption spectroscopy (XAS). A research revealed that ZnO NPs, which is used as well-known nano-fertilizer, significantly lowered Cd contents (26%) in rice shoots, when rice was grown under glasshouse conditions [149]. The ZnO NPs substantially lessened the absorption of Cd and improved the concentrations of Zn in rice plants [150]. The study stated that foliar application of ZnO NPs (100 mg/L) reduced Cd in shoot and roots of rice plants by 30% and 31%, respectively. However, while ZnO NPs was applied along with biochar, the Cd reduction was somewhat increased (39% in shoot and 38% in root) compared to ZnO NPs alone [150]. Another study suggested that, Iron oxide nanoparticles (Fe NPs) were also effective as amendment in reducing Cd in plants which reported that the mixture of Fe NPs and biochar significantly decreased the Cd uptake in rice shoots (up to 42% vs. 31%) and roots (39% vs. 26%) compared to single Fe NPs (30 mg/L foliar application) as well as increased the rice growth and photosynthesis capacity [95, 151].

3.4 *Liming Treatment*

In a study revealed that Cd absorption in rice grain was decreased by 70–80 with lime ($CaCO_3$) amendment in paddy soil without altering grain productivity [61]. This research evaluated the impact of lime in combination of low Cd cultivars rice on Cd addition in rice grain, which enhanced the soil pH from 5.5 to 6.5 [61]. Finally, the study recommended that increasing soil pH up to 6.5 through liming along with low Cd cultivars was applicable in lowering Cd uptake in rice grain [61]. The possible role of Ca in rice under Cd stress was examined by Kanu et al. [152] who observed that Cd stress declined the photosynthesis pigments substantially as well as hampered the enzymatic antioxidant activities. The application and usages of Ca declined the Cd intensities in leaves, stems, roots and grains and enhanced the proline, soluble protein, and soluble sugar contents in rice plants under Cd stress situation [152].

3.5 *Biochar Amendment*

Several studies explored that the application of biochar in paddy soil might be an efficient approach to reduce Cd addition in rice. A study found that use of moderate rate of sewage sludge biochar (1.5–3 t/ha) decreased the Cd absorption in rice grain significantly when produced in Cd-polluted paddy soil (0.82 mg kg^{-1}) and enhanced grain yield [153]. When biochar was mixed with zero valent iron as amendment and applied to paddy soils, the bioavailability of Cd decreased up to 93% due to the increased amount of indistinct Fe oxides in paddy soils and the interaction of Fe and biochar. However, even biochar amendments alone lessen the addition of Cd in rice plant parts (roots, straw, husks and grain) substantially [154]. However, contrary report is also available where biochar application enhanced the Cd uptake in rice cultivated in Cd-polluted soils [155]. The authors showed that when 5% biochar was applied to Cd-contaminated soils collected near mine site and rice was grown, Cd level in soil decreased by 34% whereas grain Cd was increased by 451% in brown rice [155]. Rizwan et al. [95] showed that biochar application raised the soil pH and decreased the bioavailability of Cd based on their study when rice was grown under continuous flooding condition and biochar was applied as soil amendment under Cd stress. Nevertheless, plant growth including plant height, biomass and photosynthesis were decreased due to biochar application and increased the oxidative stress [95].

3.6 *Iron Plaque Formation and Cd Uptake by Rice*

It was well documented that like other aquatic plants, rice plants can develop extensive porous aerenchyma tissues on its root systems [28, 156]. This aerenchyma tissue is composed of higher intracellular spaces which perform central action in O_2 transportation from the aerial parts to roots for respiration activity [157]. For physiological adaptation in anaerobic conditions, the aerenchymatous tissues of rice roots diffuses a part of the O_2 into the root zone. This phenomenon is defined as the radial oxygen loss (ROL) [158] which considerably differed among the rice varieties depending on the waterlogged conditions and/or availability of O_2 in soil environment [159, 160]. ROL is closely linked to the oxidation of soil substances in the rhizosphere and responsible for deposition of toxic metals in the root zone and on the root surfaces [158, 161]. Some research indicated that rice genotypes with high ROL is attributed to less Cd translocation and accumulation [162] and As [163] in above ground plants' part and as a result of O_2 discharge from the roots through ROL, ferrous iron (Fe^{2+}) is oxidised to form ferric iron (Fe^{3+}) in the root zone and deposited as a brown or dark brown layer on the root surfaces is termed as Fe plaque [158]. Fe plaques possess a abundant functional groups with many adsorption sites leading to sequestration of Cd by adsorption or co-precipitation and therefore restricts the bioavailability of Cd in root zone and translocation and collection of Cd in plants [31, 19]. Fe plaque deposition in the roots of plants raised in aquatic medium and thus the development of Fe plaque through radial oxygen loss in the rhizosphere and behaviour of Fe plaque in heavy metal sequestration was reported earlier [164]. Fe formation is affected by several agents like soil redox potentials, radial oxygen loss (ROL), moisture condition, rice cultivar etc. [158, 165]. The Fe plaque formation on root exteriors is widely seen in rice and ROL from roots is considered as the important cause affecting Fe plaque formation [162]. However, Cheng et al. [158] exposed a profound positive affiliation between ROL and Fe plaque formation while negative relationship is observed between metal translocation in root/plaque and distribution in root and shoot tissues.

According to several studies, the unstructured and crystalline Fe oxides and hydroxides found in plant Fe plaques have an efficient capability for sharing and adsorbing heavy metals [166, 167, 19]. Fe plaque deposition is enhanced through addition of rapeseed seed cake and this effectively restricted Cd translocation and uptake in rice seedlings [168]. Selenium and lime treatment improved the ability of Fe plaque to inhibit Cd uptake by rice seedlings thus significantly decreased Cd deposition in the root and shoot [169]. A pot culture study was conducted in two paddy soils with Cd stress (0.96 denotes as soil L and 3.76 mg Cd/kg as soil H) and it was shown that the contents and quantity of Cd in Fe plaque were considerably higher than those in rice plants, and a substantial difference is observed at the various growth stages [170]. They also observed that distribution of Cd in Fe plaque is accounted for 17.0–49.4% in soil L and 24.5–47.4% in soil H subsequently shoot and root tissues, throughout the entire cultivation period of rice except for heading stages. Fe plaque was produced around the root after addition of Fe and Fe concentration in rice plants improved that prevents the rice plants from Fe deficiency and toxicity induced by Cd [25]. This is substantiated by the higher biomass yield, chlorophyll contents and quantum efficacy of photosynthesis among the plants with higher iron plaque. According to Liu et al. [171], rice plants (particularly high ROL-ability cultivars) treated with Fe-rich amendments (red mud) tended to have a high potential to enhance Fe plaque on root exterior as well as in the rhizosphere and decline Cd uptake and translocation from root to grain. A hydroponic experiment was performed by Xu and Yu [172] and found that N_2 flushing increased root Fe plaque and enhanced Cd and Zn desolation in the iron plaques of rice seedlings. It has been shown that the development of iron plaque on rice roots serves as a blockade to soil Cd toxicity and act as a buffer or reservoir that could lower Cd uptake into rice roots, possibly because of the adsorption and immobilization of the metals on the plaques [173, 174]. The high adsorption ability of the iron hydroxides efficient groups in iron plaque have been observed to limit the uptake of metals by plants when it is present on the root surfaces [175].

The arrangement of the soil and soil amendments influence the development of iron plaque and its capacity to seize contaminant metals. According to a report, increasing the amount of Fe amendments causes an increase in both iron plaque formation and iron plaque adsorbed Cd, whereas rising the Cd level in the soil causes a drop in the ratio of iron plaque adsorbed Cd [176]. Biochar amendment in a typically multi-metal poisoned soil significantly changed the iron plaque formation in rice plants and metal mobility indicating that it has a potential role in decreasing Cd and Zn addition in rice root [177]. The basic purpose for this is to decrease in the pore water concentrations of Cd and Zn and strengthening the iron plaque blocking capacity for Cd and Pb by biochar amendment. Recent study had shown that Fe application increased plaque formation capabilities and reducing Cd uptake in upper parts of rice plant through increasing Cd sequestration in Fe plaque in both hydroponic and soil culture [29, 30]. However, Fe plaque formation capabilities utilizing both hydroponic and soil culture with different rice cultivar and protective functions of Fe plaque in regulating Cd accretion and translocation in rice are précised in Table 1.

Table 1 Fe plaque formation: a new promising technology to reduce Cd accumulation in rice plant based on recent studies

Type of experiment	Rice variety	Rice type	Fe amendment doses	Duration of Fe plaque formation	Fe plaque formation capacity	Key findings	References
Hydroponic	Huanghuazhan	Conventional late rice cultivar	160 mg L^{-1} as $FeSO_4.7H_2O$ along with 1 and 6 mg L^{-1} of P as NaH_2PO_4	–	–	The addition of P enhanced the formation of Fe plaque but prevent the retention capacity of Cd. A decrease in Cd in Fe plaque by 35–62% was observed. In addition, an increase of Cd (72–113 mg kg^{-1}) in rice roots was also noticed. The study concluded that P alone was adequate for the formation of Fe plaque and subsequent decrease of root Fe plaque, which finally reduced the Cd in Fe plaque	Jiaofeng et al. [178]
Soil	H You 518	*indica* hybrid rice	Biochar (BC)-Fe as nanoFe_3O_4 were mixed with soil (0.05 - 1.6%)	119 days	–	The concentration of Cd in rice was considerably decreased by 49, 36 and 47% due to the addition of Fe-BC. The study reported that Fe-BC addition promoted Fe plaque formation and DCB extracted Fe content in rice roots	[19, 8]

(continued)

Table 1 (continued)

Type of experiment	Rice variety	Rice type	Fe amendment doses	Duration of Fe plaque formation	Fe plaque formation capacity	Key findings	References
Solution	–	–	50 mg L^{-1} Fe_2O_3 NPs	10 days	The level of Fe in Fe plaque was 32 g kg^{-1} with Fe_2O_3 NPs treatment compared to control	The concentration of Cd in Fe plaque was considerably higher with Fe_2O_3 NPs treatment than the control. The study finding showed that FeNP accumulation in rice roots which favored the immobilization of Cd	Huang et al. [179]
Hydroponic	Vandana, Opus, Reizic, Quest and Koshihikari	Japonica rice	0, 50 and 100 mg L^{-1} as $FeSO_4.7H_2O$	3 days	Ranges from 20,076 ± 202 to 45,600 ± 351 mg kg^{-1} Fe depending on the variety of rice	Higher iron plaque formation restricts the Cd accrual and translocation in rice. Quest variety showed the highest Fe plaque development capacity and less sensitive Cd stress compared to other rice variety	[98]

(continued)

Table 1 (continued)

Type of experiment	Rice variety	Rice type	Fe amendment doses	Duration of Fe plaque formation	Fe plaque formation capacity	Key findings	References
Soil	Quest and Langi	Japonica rice	0, 1.0 and 2.0 g kg^{-1} as $FeSO_4 \cdot 7H_2O$	7 weeks	Fe plaque formation capacity was 8 and tenfold higher under 1.0 and 2.0 g kg^{-1} compared to control treatment	Biomass production was promoted by 11–20% in Quest variety and 26–43% in Langi variety. 13% higher Fe plaque production was observed in Quest cultivar compared to Langi cultivar. Cd absorption was lowered by 35% and 47% in shoots and 36% and 58% in roots at 1.0 and 2.0 g kg^{-1} Fe supply individually	[29]
Hydroponic	Xiangwanxian 13	*indica* late rice	40 and 320 mg L^{-1} as EDTA-Fe, biochar (BC)-Fe and $FeSO_4.7H_2O$	6 days	1.19–2.61 g kg^{-1} Fe in EDTA-Fe, 0.33–1.04 g kg^{-1} Fe in BC-Fe and 1.0–3.48 g kg^{-1} Fe in $FeSO_4$-Fe	The formation of Fe plaque by BC-Fe was much higher than that formed by EDTA-Fe and $FeSO_4 \cdot 7H_2O$. The translocation factors were decreased by 85% and 80% from root to stem and stem to leaf, respectively	[2, 180]

(continued)

Table 1 (continued)

Type of experiment	Rice variety	Rice type	Fe amendment doses	Duration of Fe plaque formation	Fe plaque formation capacity	Key findings	References
Hydroponic	Changliangyou and Yuzhenxiang	Indica rice	0 and 50 mg L^{-1} as $FeSO_4 \cdot 7H_2O$	3 days	Ranges from 6000 to 14,000 mg kg^{-1} Fe depending on organic amendment and rice cultivar	Fe^{2+} application decreased Cd accumulation by 8–59% in shoots and 6–56% in roots. Under Fe^{2+} application, Cd concentration in rice plant were remarkedly lower than Fe deficient conditions signifying that Fe plaque demonstrated a vital role to limit Cd accretion and distribution in rice plant as relatively higher quantity of Fe plaque formation was observed by Fe^{2+} application	[181]

(continued)

Table 1 (continued)

Type of experiment	Rice variety	Rice type	Fe amendment doses	Duration of Fe plaque formation	Fe plaque formation capacity	Key findings	References
Solution culture	Wuyunjing21	Japonica rice	1.0 mmol L^{-1} Fe^{2+}	12 days	8000–34,000 mg kg^{-1} Fe	Total Cd absorption in rice plant was decreased by 23.3% because of the enriched Fe plaque deposition with Fe addition due to the development of Fe plaque formation by added Se. The study indicated that addition of Se along with Fe can reduce the accumulation of Cd in rice	[143]
Hydroponic	Huahangsimiao	Mutant variety	0.2 mM Fe^{2+}	7 days	41,000 mg kg^{-1} Fe	Fe plaque production was appeared on the root surfaces as intricate network structures which diminished the adsorption of Cd in roots by 54.8% compared Cd only treatment. In comparison to Cd control, Cd absorption in shoots, roots translocation from root to shoot was declined by 36%, 34.1% and 20.1% respectively in presence of Fe formation	[26]

4 Cadmium Transport Pathway and Its Transporters in Rice

Rice plant root uptake Cd from soil by root acquisition as the form of Cd^{2+} and after going through numerous procedures, it eventually accumulated into rice grains [15, 23] and after absorption, Cd is transported to shoots from root through xylem promptly [182, 24]. The xylem regulated Cd transportation extensively induced the accumulation of Cd in various plants including rice [183, 184, 185]. Substantial quantity of Cd is held in the xylem sap and shoot tissue after Cd addition to roots for a period of 1 h, and this action involved in movement of Cd from root to shoot via xylem which is the key contributing factor for Cd accretion in shoot [185, 186]. In contrast, the panicle neck, phloem act as a dominant pathway for Cd accumulation into grains [187]. It was also observed that the impact of phloem in Cd translocation in rice grain was 91–100% [187]. In the phloem, Cd predominantly confines to protein molecules and marginally to thiol-compounds and Cd concentration in phloem sap may be the principal processes regulating of its grain deposition [188]. Non-invasive imaging of real time responses of living plants utilizing a positron-emitting ^{107}Cd tracer imaging system (PETIS) elucidated the specific behaviour of Cd in rice plants after acquisition [189]. The findings indicated that Cd is instantly translocated into shoots from roots via culms and Cd tended to detain in nodes. It has been observed that, after 7 h of Cd exposure, Cd arrived at the panicles and accumulated constantly in lieu of leaf blades. The results exposed that for switching Cd transport from xylem to phloem nodes are significant tissues and are essential for the journey of Cd from the soil to the rice grains [189]. It is explored that in contrast to Cd captivated from roots, Cd remobilization leaf lamina is also likely to promote Cd uptake in rice grain [190]. It was stated that considerable extent of Cd is stored in the leaf blades before heading and redistributed and translocated into grains throughout the reproductive stage [190].

It was exposed that Cd uptake by rice roots from the rhizosphere is a dose supported practice and shows saturable kinetic attributes [189, 191]. On the contrary, [189] determined the kinetics of Cd uptake in rice root and demonstrated that Cd uptake rate by japonica rice was relational to the absorbable Cd absorption in the culture solution within the tested range varied from 0.05 to 100 nM, indicating a direct association to uptake rate and Cd absorption in certain range. Due to non-essential forms, Cd can readily enter plant body utilizing same uptake mechanisms with critical plant nutrients, like Ca, Fe and Zn [192, 193]. As for example, it is believed that Cd is competed for same binding sites with Fe and Mn [194, 195].

In rice plants, several transporters associated to Cd absorption and tolerance have been reported. Transporters of natural resistance associated macrophage protein (NRAMP) family are also known to facilitate Cd transport. It is reported that *AtNRAMP1*, *AtNRAMP3* and *AtNRAMP4* facilitate the uptake of Cd in yeasts and overexpression of *AtNRAMP3* genes leads to an increase in Cd^{2+} sensitivity and Cd^{2+} accumulation [196]. There are seven NRAMP genes in rice [197]. *OsNRAMP5* expression is confined to exodermis and epidermis of roots, exterior part of cortical tissues as well as in surrounding tissues of the xylem and liable for Fe, Mn and Cd transportation into root cells of rice plants from the external solution [194, 195]. Silencing and knockout of *OsNRAMP5* leads to significant reduction of Cd deposition both in straw and grains to insignificant levels [198], it also decreased the growth of rice plant, especially when cultivated at low Mn concentrations [195]. The down-regulation of *OsNRAMP5* is a better option to reduce Cd sequestration in roots [198]. *OsNRAMP5* not only facilitates Cd and Mn uptake but also involved in consecutive uptake of Fe under Fe deficiency [195]. *OsNRAMP1*, iron transporter protein has been found to be specified in the plasma membrane, indicating its role in cellular Cd uptake and Cd transport within rice plant [199]. The higher *OsNRAMP1* expression was examined in roots when 1 μM Cd applied under Fe deficient conditions leads to elevated level of Cd uptake in rice [199].

In *Arabidopsis thaliana*, *AtHMA3* mediates the influx of Cd into the vacuolar compartment and thereby carries out an intracellular sequestration of the toxic metal [200, 201]. The vacuolar membrane-specific localization of *OsHMA3* specifies that the protein is responsible for sequestration of Cd from the root cytoplasm into the vacuole in a parallel manner to *AtHMA3* Similar to *AtHMA3*, *OsHMA3* is believed to be responsible for sequestering Cd from the root cytoplasm into the vacuole based on its vacuolar membrane-specific localization [202, 201]. The cytoplasm of root cells of rice plants having high Cd accumulating capabilities exhibited more available Cd for xylem loading in the absence of *OsHMA3*-regulated Cd transfer into the vacuoles. This discrepancy yielded in higher accumulation in shoots of rice plants [202].

After root absorption, a portion of Cd existed as Cd-phytochelatin complexes are segregated in the vacuoles under the action of *OsHMA3* and the remaining portions are distributed to xylem facilitated by *OsHMA2* transporter. The transporter *OsHMA2* in rice is observed mainly in root pericycle, and accountable for xylem loading of Cd at the vegetative phase [203] while at the reproductive phase, the *OsHMA2* are also expressed in the phloem of expanded and disperse vascular bundles of the nodes and play a crucial role in preferential transportation of Cd through phloem to the emerging tissues [204]. There are two core possible corridors of Cd transportation in rice grain while ripening: (a) Cd is taken up straight away through the xylem to the emerging grain and (b) Cd is taken up to vigorously transpiring parts of the panicle, and then relatively rapidly remobilized via phloem to the grain [190] and these two pathways have been established in literature [24, 205]. Moreover, Uraguchi et al. [206] have identified a transporter gene for phloem Cd movement in rice. This rice gene known as *OsLCT1*, is the homolog of wheat Low-affinity Cation Transporter1 [207] and shows a Cd-efflux activity on the plasma membrane [206]. A strong *OsLCT1* expression was observed in leaf blades and nodes of rice plants during reproductive phase and particularly in node I, the upper most node, *OsLCT1* is highly articulated and participates in Cd transport to grain [206]. Suppression of *OsLCT1* expression can effectively reduce grain Cd level and the adjustment of *OsLCT1* supports the development of "low-Cd rice" [206].

5 Conclusion

The study demonstrated that Cd contamination in soils is mainly originated from both anthropogenic and natural activities. Anthropogenic activities like usage of low grade phosphatic fertilizers, pesticides, biosolids and mining activities are the primary reasons of Cd disgrace in paddy soils. Cd contamination in rice has been emerged extensively, in Asian countries like India, China, Bangladesh, Japan and Sri Lanka where exposure of Cd from rice-based diets have been reported. Two crucial soil parameters such as soil pH and Eh are mainly dominating Cd uptake by paddy rice. As flooded conditions are preferred for rice cultivation, so wet and dry conditions of paddy fields controlled the Cd chemistry and accumulation in rice grain. The study also exposed that Cd has detrimental effects on the growth of paddy plants, which also affect the Cd accumulation in paddy rice. The uptake behaviour of Cd is reliant on rice genotype. Various strategies including inorganic minerals such as Fe, Mn, Se and Si are proven to decrease the Cd uptake in paddy rice. However, blend of inorganic minerals along with organic amendments also key to minimize Cd contamination in paddy soils. Iron plaque formation occurs when Fe was added as amendment with paddy soils which lessen the Cd addition in rice. However, in many Asian countries, Fe is dominant in irrigation water and the accrual of Cd in rice grain is less and Cd compete for the similar transporters (OsNRAMP1, OsNRAMP5 and OsIRT1) with essential plant nutrients like Fe, Mn and Zn.

Acknowledgements The first author gratefully acknowledges the University of Newcastle as well as Australian Government Research Training Program for its stipend and fee offset scholarship.

References

1. Chen, H., Wang, P., Gu, Y., Kretzschmar, R., Kopittke, P. M., & Zhao, F. J. (2020). The within-field spatial variation in rice grain Cd concentration is determined by soil redox status and pH during grain filling. *Environmental Pollution, 261*, 114151.
2. Zhang, D., Ding, A., Li, T., Wu, X., Liu, Y., & Naidu, R. (2021) Immobilization of Cd and Pb in a contaminated acidic soil amended with hydroxyapatite, bentonite, and biochar. *Journal of Soils and Sediments.*
3. Dong, X. B., Huang, W., Bian, Y. B., Feng, X., Ibrahim, S. A., Shi, D. F., Qiao, X., & Liu, Y. (2019). Remediation and mechanisms of Cadmium biosorption by a Cadmium-binding protein from Lentinula edodes. *Journal of Agriculture and Food Chemistry, 67*, 11373–11379.
4. Xu, Y., Feng, J., & Li, H. (2021). Water management increased rhizosphere redox potential and decreased Cd uptake in a low-Cd rice cultivar but decreased redox potential and increased Cd uptake in a high-Cd rice cultivar under intercropping. *Science of the Total Environment, 751*, 141701.
5. Hussain, B., Umer, M. J., Li, J., Ma, Y., Abbas, Y., Ashraf, M. N., Tahir, N., Ullah, A., Gogoi, N., & Farooq, M. (2021). Strategies for reducing cadmium accumulation in rice grains. *Journal of Cleaner Production, 286.*
6. Nazia, T., Ullah, A., Tahir, A., Rashid, H. U., ur Rehman, T., Danish, S., Hussain, B., & Akca, H. (2021). Strategies for reducing Cd concentration in paddy soil for rice safety. *Journal of Cleaner Production, 316.*

7. Zou, M., Zhou, S., Zhou, Y., Jia, Z., Guo, T., & Wang, J. (2021). Cadmium pollution of soil-rice ecosystems in rice cultivation dominated regions in China: A review. *Environmental Pollution, 280*, 116965.
8. Zhang, L., Zhang, C., Du, B., Lu, B., Zhou, D., Zhou, J., & Zhou, J. (2020). Effects of node restriction on cadmium accumulation in eight Chinese wheat (Triticum turgidum) cultivars. *Science of the Total Environment, 725*, 138358.
9. Khan, M. A., Khan, S., Khan, A., & Alam, M. (2017). Soil contamination with cadmium, consequences and remediation using organic amendments. *Science of the Total Environment*, 601–602, 1591–1605.
10. Shanmugaraj, B. M., Malla, A., & Ramalingam, S. (2019). Chapter 1—Cadmium stress and toxicity in plants: An overview. In M. Hasanuzzaman, M. N. V. Prasad, & M. Fujita (Eds.), *Cadmium toxicity and tolerance in plants* (pp. 1–17). Academic Press.
11. Song, W.-E., Chen, S.-B., Liu, J.-F., Chen, L., Song, N.-N., Li, N., & Liu, B. (2015). Variation of Cd concentration in various rice cultivars and derivation of cadmium toxicity thresholds for paddy soil by species-sensitivity distribution. *Journal of Integrative Agriculture, 14*, 1845–1854.
12. Grant, C. A., Clarke, J. M., Duguid, S., & Chaney, R. L. (2008). Selection and breeding of plant cultivars to minimize cadmium accumulation. *Science of the Total Environment, 390*, 301–310.
13. IARC. (2012). Cadmium and Cadmium compounds. International Agency for Research on Cancer (IARC) monograph on the evaluation of carcinogenic risks to humans. Review of Human Carcinogens: Arsenic, Metals, Fibre and Dusts. 100c.
14. Ricepedia. (2021). *An online counterpart of Rice Almanac, an encyclopedia on rice.* http://ricepedia.org/rice-as-food/the-global-staple-rice-consumers. Accessed March 26, 2021.
15. Li, H., Luo, N., Li, Y. W., Cai, Q. Y., Li, H. Y., Mo, C. H., & Wong, M. H. (2017). Cadmium in rice: Transport mechanisms, influencing factors, and minimizing measures. *Environmental Pollution, 224*, 622–630.
16. Khaliq, M. A., James, B., Chen, Y. H., Ahmed Saqib, H. S., Li, H. H., Jayasuriya, P., & Guo, W. (2019). Uptake, translocation, and accumulation of Cd and its interaction with mineral nutrients (Fe, Zn, Ni, Ca, Mg) in upland rice. *Chemosphere, 215*, 916–924.
17. Aoshima, K. (2016). Itai-itai disease: Renal tubular osteomalacia induced by environmental exposure to cadmium—Historical review and perspectives. *Soil Science and Plant Nutrition, 62*, 319–326.
18. Mao, C., Song, Y., Chen, L., Ji, J., Li, J., Yuan, X., Yang, Z., Ayoko, G. A., Frost, R. L., & Theiss, F. (2019). Human health risks of heavy metals in paddy rice based on transfer characteristics of heavy metals from soil to rice. *CATENA, 175*, 339–348.
19. Zhang, J. Y., Zhou, H., Gu, J. F., Huang, F., Yang, W. J., Wang, S. L., Yuan, T. Y., & Liao, B. H. (2020). Effects of nano-Fe_3O_4-modified biochar on iron plaque formation and Cd accumulation in rice (Oryza sativa L.). *Environmental Pollution, 260*, 113970.
20. Meharg, A. A., Norton, G., Deacon, C., Williams, P., Adomako, E. E., Price, A., Zhu, Y., Li, G., Zhao, F. J., McGrath, S., Villada, A., Sommella, A., De Silva, P. M., Brammer, H., Dasgupta, T., & Islam, M. R. (2013). Variation in rice cadmium related to human exposure. *Environmental Science and Technology, 47*, 5613–5618.
21. Uraguchi, S., & Fujiwara, T. (2013). Rice breaks ground for cadmium-free cereals. *Current Opinion in Plant Biology, 16*, 328–334.
22. Lux, A., Martinka, M., Vaculik, M., & White, P. J. (2011). Root responses to cadmium in the rhizosphere: A review. *Journal of Experimental Botany, 62*, 21–37.
23. Qi, X., Tam, N. F., Li, W. C., & Ye, Z. (2020). The role of root apoplastic barriers in cadmium translocation and accumulation in cultivars of rice (Oryza sativa L.) with different Cd-accumulating characteristics. *Environmental Pollution, 264*, 114736.
24. Uraguchi, S., Mori, S., Kuramata, M., Kawasaki, A., Arao, T., & Ishikawa, S. (2009). Root-to-shoot Cd translocation via the xylem is the major process determining shoot and grain cadmium accumulation in rice. *Journal of Experimental Botany, 60*, 2677–2688.

25. Sebastian, A., & Prasad, M. N. V. (2016). Iron plaque decreases cadmium accumulation in Oryza sativa L. and serves as a source of iron. *Plant Biology (Stuttgart, Germany), 18*, 1008–1015.
26. Fu, Y., Yang, X., & Shen, H. (2018). Root iron plaque alleviates cadmium toxicity to rice (Oryza sativa) seedlings. *Ecotoxicology and Environmental Safety, 161*, 534–541.
27. Zhang, Q., Chen, H., Xu, C., Zhu, H., & Zhu, Q. (2019). Heavy metal uptake in rice is regulated by pH-dependent iron plaque formation and the expression of the metal transporter genes. *Environmental and Experimental Botany, 162*, 392–398.
28. Wang, S., & Reid, M. C. (2020). Kinetics of nitrous oxide mass transfer from porewater into root aerenchyma of wetland plants. *Journal of Environmental Quality, 49*, 1717–1729.
29. Siddique, A. B., Rahman, M. M., Islam, M. R., & Naidu, R. (2021). Varietal variation and formation of iron plaques on cadmium accumulation in rice seedling. *Environmental Advances, 5*, 100075.
30. Siddique, A. B., Rahman, M. M., Islam, M. R., Shehzad, M. T., Nath, B., & Naidu, R. (2021). Influence of iron plaque on accumulation and translocation of Cadmium by rice seedlings. *Sustainability, 13*, 10307.
31. Li, S., Chen, S., Wang, M., Lei, X., Zheng, H., Sun, X., Wang, L., & Han, Y. (2020). Iron fractions responsible for the variation of Cd bioavailability in paddy soil under variable pe + pH conditions. *Chemosphere, 251*, 126355.
32. Duan, G., Shao, G., Tang, Z., Chen, H., Wang, B., Tang, Z., Yang, Y., Liu, Y., & Zhao, F. J. (2017). Genotypic and environmental variations in grain Cadmium and Arsenic concentrations among a panel of high yielding rice cultivars. *Rice (NY), 10*, 9.
33. Hongjiang, Z., Xizhou, Z., Tingxuan, L., & Fu, H. (2013). Variation of cadmium uptake, translocation among rice lines and detecting for potential cadmium-safe cultivars. *Environmental Earth Sciences, 71*, 277–286.
34. Gu, Q., Yang, Z., Yu, T., Yang, Q., Hou, Q., & Zhang, Q. (2018). From soil to rice—A typical study of transfer and bioaccumulation of heavy metals in China. *Acta Agriculturae Scandinavica, Section B—Soil & Plant Science, 68*, 631–642.
35. Honma, T., Ohba, H., Kaneko-Kadokura, A., Makino, T., Nakamura, K., & Katou, H. (2016). Optimal soil Eh, pH, and water management for simultaneously minimizing arsenic and Cadmium concentrations in rice grains. *Environmental Science and Technology, 50*, 4178–4185.
36. Hussain, B., Ashraf, M. N., Shafeeq Ur, R., Abbas, A., Li, J., & Farooq, M. (2021). Cadmium stress in paddy fields: Effects of soil conditions and remediation strategies. *Science of the Total Environment, 754*, 142188.
37. Rajkumar, M., Sandhya, S., Prasad, M. N., & Freitas, H. (2012). Perspectives of plant-associated microbes in heavy metal phytoremediation. *Biotechnology Advances, 30*, 1562–1574.
38. Zhao, F.-J., & Wang, P. (2020). Arsenic and cadmium accumulation in rice and mitigation strategies. *Plant and Soil, 446*, 1–21.
39. Wang, P., Chen, H., Kopittke, P. M., & Zhao, F. J. (2019). Cadmium contamination in agricultural soils of China and the impact on food safety. *Environmental Pollution, 249*, 1038–1048.
40. Qin, G., Niu, Z., Yu, J., Li, Z., Ma, J., & Xiang, P. (2021). Soil heavy metal pollution and food safety in China: Effects, sources and removing technology. *Chemosphere, 267*, 129205.
41. Vareda, J. P., Valente, A. J. M., & Duraes, L. (2019). Assessment of heavy metal pollution from anthropogenic activities and remediation strategies: A review. *Journal of Environmental Management, 246*, 101–118.
42. Hu, Y., Cheng, H., & Tao, S. (2016). The challenges and solutions for Cadmium-contaminated rice in China: A critical review. *Environment International, 92–93*, 515–532.
43. Khanam, R., Kumar, A., Nayak, A. K., Shahid, M., Tripathi, R., Vijayakumar, S., Bhaduri, D., Kumar, U., Mohanty, S., Panneerselvam, P., Chatterjee, D., Satapathy, B. S., & Pathak, H. (2020). Metal(loid)s (As, Hg, Se, Pb and Cd) in paddy soil: Bioavailability and potential risk to human health. *Science of the Total Environment, 699*, 134330.

44. Bashir, S., Ali, U., Shaaban, M., Gulshan, A. B., Iqbal, J., Khan, S., Husain, A., Ahmed, N., Mehmood, S., Kamran, M., & Hu, H. (2020). Role of sepiolite for cadmium (Cd) polluted soil restoration and spinach growth in wastewater irrigated agricultural soil. *Journal of Environmental Management, 258*, 110020.
45. Shah, M. T., Begum, S., & Khan, S. (2009). Pedo and biogeochemical studies of mafic and ultramfic rocks in the Mingora and Kabal areas, Swat, Pakistan. *Environmental Earth Sciences, 60*, 1091–1102.
46. He, Z. L., Yang, X. E., & Stoffella, P. J. (2005). Trace elements in agroecosystems and impacts on the environment. *Journal of Trace Elements in Medicine and Biology, 19*, 125–140.
47. Kubier, A., Wilkin, R. T., & Pichler, T. (2019). Cadmium in soils and groundwater: A review. *Applied Geochemistry, 108*, 1–16.
48. Liu, Y., Xiao, T., Ning, Z., Li, H., Tang, J., & Zhou, G. (2013). High cadmium concentration in soil in the Three Gorges region: Geogenic source and potential bioavailability. *Applied Geochemistry, 37*, 149–156.
49. Zhao, Y., Deng, Q., Lin, Q., Zeng, C., & Zhong, C. (2020). Cadmium source identification in soils and high-risk regions predicted by geographical detector method. *Environmental Pollution, 263*, 114338.
50. Roberts, T. L. (2014). Cadmium and phosphorous fertilizers: The issues and the science. *Procedia Engineering, 83*, 52–59.
51. Sebastian, A., & Prasad, M. N. V. (2013). Cadmium minimization in rice. A review. *Agronomy for Sustainable Development, 34*, 155–173.
52. Yang, Y., Xiong, J., Tao, L., Cao, Z., Tang, W., Zhang, J., Yu, X., Fu, G., Zhang, X., & Lu, Y. (2020). Regulatory mechanisms of nitrogen (N) on cadmium (Cd) uptake and accumulation in plants: A review. *Science of the Total Environment, 708*, 135186.
53. Schoeters, G., Den Hond, E., Zuurbier, M., Naginiene, R., van den Hazel, P., Stilianakis, N., Ronchetti, R., & Koppe, J. G. (2006). Cadmium and children: Exposure and health effects. *Acta Paediatrica, 95*, 50–54.
54. Nicholson, F. A., Smith, S. R., Alloway, B. J., Carlton-Smith, C., & Chambers, B. J. (2003). An inventory of heavy metals inputs to agricultural soils in England and Wales. *Science of the Total Environment, 311*, 205–219.
55. Chu, L., & He, W. (2021). Toxic metals in soil due to the land application of sewage sludge in China: Spatiotemporal variations and influencing factors. *Science of the Total Environment, 757*, 143813.
56. Nziguheba, G., & Smolders, E. (2008). Inputs of trace elements in agricultural soils via phosphate fertilizers in European countries. *Science of the Total Environment, 390*, 53–57.
57. Wang, Y., Su, Y., & Lu, S. (2020) Predicting accumulation of Cd in rice (Oryza sativa L.) and soil threshold concentration of Cd for rice safe production. *Science of the Total Environment, 738*, 139805.
58. Rafiq, M. T., Aziz, R., Yang, X., Xiao, W., Rafiq, M. K., Ali, B., & Li, T. (2014). Cadmium phytoavailability to rice (Oryza sativa L.) grown in representative Chinese soils. A model to improve soil environmental quality guidelines for food safety. *Ecotoxicology and Environmental Safety, 103*, 101–107.
59. Shahid, M., Dumat, C., Khalid, S., Niazi, N. K., & Antunes, P. M. C. (2017). Cadmium bioavailability, uptake, toxicity and detoxification in soil-plant system. In P. de Voogt (Ed.), *Reviews of environmental contamination and toxicology* (Vol. 241, pp. 73–137). Springer International Publishing.
60. Shahid, M., Dumat, C., Aslam, M., & Pinelli, E. (2015). Assessment of lead speciation by organic ligands using speciation models. *Chemical Speciation & Bioavailability, 24*, 248–252.
61. Chen, H., Zhang, W., Yang, X., Wang, P., McGrath, S. P., & Zhao, F.-J. (2018). Effective methods to reduce cadmium accumulation in rice grain. *Chemosphere, 207*, 699–707.
62. Christensen, T. H. (1984). Cadmium soil sorption at low concentrations: I. Effect of time, cadmium load, pH, and calcium. *Water, Air, and Soil Pollution, 21*, 105–114.
63. Yu, H. Y., Liu, C., Zhu, J., Li, F., Deng, D. M., Wang, Q., & Liu, C. (2016). Cadmium availability in rice paddy fields from a mining area: The effects of soil properties highlighting iron fractions and pH value. *Environmental Pollution, 209*, 38–45.

64. Ding, C., Du, S., Ma, Y., Li, X., Zhang, T., & Wang, X. (2019). Changes in the pH of paddy soils after flooding and drainage: Modeling and validation. *Geoderma, 337*, 511–513.
65. Kögel-Knabner, I., Amelung, W., Cao, Z., Fiedler, S., Frenzel, P., Jahn, R., Kalbitz, K., Kölbl, A., & Schloter, M. (2010). Biogeochemistry of paddy soils. *Geoderma, 157*, 1–14.
66. Wang, A. S., Angle, J. S., Chaney, R. L., Delorme, T. A., & Reeves, R. D. (2006). Soil pH Effects on Uptake of Cd and Zn by Thlaspi caerulescens. *Plant and Soil, 281*, 325–337.
67. Sebastian, A., & Prasad, M. N. (2013). Cadmium accumulation retard activity of functional components of photo assimilation and growth of rice cultivars amended with vermicompost. *International Journal of Phytoremediation, 15*, 965–978.
68. Zhang, Z., & Furman, A. (2021). Soil redox dynamics under dynamic hydrologic regimes—A review. *Science of the Total Environment, 763*, 143026.
69. Frohne, T., Diaz-Bone, R. A., Du Laing, G., & Rinklebe, J. (2014). Impact of systematic change of redox potential on the leaching of Ba, Cr, Sr, and V from a riverine soil into water. *Journal of Soils and Sediments, 15*, 623–633.
70. Shaheen, S. M., & Rinklebe, J. (2014). Geochemical fractions of chromium, copper, and zinc and their vertical distribution in floodplain soil profiles along the Central Elbe River, Germany. *Geoderma, 228–229*, 142–159.
71. Shahid, M., Shamshad, S., Rafiq, M., Khalid, S., Bibi, I., Niazi, N. K., Dumat, C., & Rashid, M. I. (2017). Chromium speciation, bioavailability, uptake, toxicity and detoxification in soil-plant system: A review. *Chemosphere, 178*, 513–533.
72. Shen, B., Wang, X., Zhang, Y., Zhang, M., Wang, K., Xie, P., & Ji, H. (2020). The optimum pH and Eh for simultaneously minimizing bioavailable cadmium and arsenic contents in soils under the organic fertilizer application. *Science of the Total Environment, 711*, 135229.
73. Pan, Y., Bonten, L. T. C., Koopmans, G. F., Song, J., Luo, Y., Temminghoff, E. J. M., & Comans, R. N. J. (2016). Solubility of trace metals in two contaminated paddy soils exposed to alternating flooding and drainage. *Geoderma, 261*, 59–69.
74. de Livera, J., McLaughlin, M. J., Hettiarachchi, G. M., Kirby, J. K., & Beak, D. G. (2011). Cadmium solubility in paddy soils: Effects of soil oxidation, metal sulfides and competitive ions. *Science of the Total Environment, 409*, 1489–1497.
75. Arao, T., Kawasaki, A., Baba, K., Mori, S., & Matsumoto, S. (2009). Effects of water management on cadmium and arsenic accumulation and dimethylarsinic acid concentrations in Japanese rice. *Environmental Science & Technology, 43*, 9361–9367.
76. Sarwar, N., Saifullah, Malhi, S. S., Zia, M. H., Naeem, A., Bibi, S., & Farid, G. (2010). Role of mineral nutrition in minimizing cadmium accumulation by plants. *Journal of the Science of Food and Agriculture, 90*, 925–937.
77. Wan, Y., Huang, Q., Camara, A. Y., Wang, Q., & Li, H. (2019). Water management impacts on the solubility of Cd, Pb, As, and Cr and their uptake by rice in two contaminated paddy soils. *Chemosphere, 228*, 360–369.
78. Smolders, E., & McLaughlin, M. J. (1996). Chloride increases Cadmium uptake in Swiss Chard in a resin-buffered nutrient solution. *Soil Science Society of America Journal, 60*, 1443–1447.
79. Liu, W., Zhou, Q., An, J., Sun, Y., & Liu, R. (2010). Variations in cadmium accumulation among Chinese cabbage cultivars and screening for Cd-safe cultivars. *Journal of Hazardous Materials, 173*, 737–743.
80. Lopez-Chuken, U. J., Young, S. D., & Sanchez-Gonzalez, M. N. (2010). The use of chloro-complexation to enhance cadmium uptake by Zea mays and Brassica juncea: Testing a "free ion activity model" and implications for phytoremediation. *International Journal of Phytoremediation, 12*, 680–696.
81. Abbas, T., Rizwan, M., Ali, S., Adrees, M., Mahmood, A., Zia-ur-Rehman, M., Ibrahim, M., Arshad, M., & Qayyum, M. F. (2018). Biochar application increased the growth and yield and reduced cadmium in drought stressed wheat grown in an aged contaminated soil. *Ecotoxicology and Environmental Safety, 148*, 825–833.
82. Rehman, M. Z., Rizwan, M., Ghafoor, A., Naeem, A., Ali, S., Sabir, M., & Qayyum, M. F. (2015). Effect of inorganic amendments for in situ stabilization of cadmium in contaminated

soils and its phyto-availability to wheat and rice under rotation. *Environmental Science and Pollution Research International, 22*, 16897–16906.
83. Rizwan, M., Meunier, J.-D., Miche, H., & Keller, C. (2012). Effect of silicon on reducing cadmium toxicity in durum wheat (Triticum turgidum L. cv. Claudio W.) grown in a soil with aged contamination. *Journal of Hazardous Materials, 209–210*, 326–334.
84. Younis, U., Malik, S. A., Rizwan, M., Qayyum, M. F., Ok, Y. S., Shah, M. H. R., Rehman, R. A., & Ahmad, N. (2016). Biochar enhances the cadmium tolerance in spinach (Spinacia oleracea) through modification of Cd uptake and physiological and biochemical attributes. *Environmental Science and Pollution Research, 23*, 21385–21394.
85. Zhang, F., Liu, M., Li, Y., Che, Y., & Xiao, Y. (2019). Effects of arbuscular mycorrhizal fungi, biochar and cadmium on the yield and element uptake of Medicago sativa. *Science of the Total Environment, 655*, 1150–1158.
86. Ahsan, N., Lee, S.-H., Lee, D.-G., Lee, H., Lee, S. W., Bahk, J. D., & Lee, B.-H. (2007). Physiological and protein profiles alternation of germinating rice seedlings exposed to acute cadmium toxicity. *Comptes Rendus Biologies, 330*, 735–746.
87. Li, B., Wang, X., Qi, X., Huang, L., & Ye, Z. (2012). Identification of rice cultivars with low brown rice mixed cadmium and lead contents and their interactions with the micronutrients iron, zinc, nickel and manganese. *Journal of Environmental Sciences, 24*, 1790–1798.
88. Wang, Y., Jiang, X., Li, K., Wu, M., Zhang, R., Zhang, L., & Chen, G. (2014). Photosynthetic responses of Oryza sativa L. seedlings to cadmium stress: Physiological, biochemical and ultrastructural analyses. *BioMetals, 27*, 389–401.
89. Guilherme, M. d. F. S., Oliveira, H. M., & Silva, E. D. (2015). Cadmium toxicity on seed germination and seedling growth of wheat *Triticum aestivum.* Acta Scientiarum. *Biological Sciences, 37.*
90. Raza, A., Habib, M., Kakavand, S. N., Zahid, Z., Zahra, N., Sharif, R., & Hasanuzzaman, M. (2020). Phytoremediation of Cadmium: Physiological, biochemical, and molecular mechanisms. *Biology (Basel), 9.*
91. Bari, M. A., Akther, M. S., Reza, M. A., & Kabir, A. H. (2019). Cadmium tolerance is associated with the root-driven coordination of cadmium sequestration, iron regulation, and ROS scavenging in rice. *Plant Physiology and Biochemistry, 136*, 22–33.
92. Kabir, A. H. (2016). Biochemical and molecular changes in rice seedlings (Oryza sativa L.) to cope with chromium stress. *Plant Biology, 18*, 710–719.
93. Yu, H., Wang, J., Fang, W., Yuan, J., & Yang, Z. (2006). Cadmium accumulation in different rice cultivars and screening for pollution-safe cultivars of rice. *Science of the Total Environment, 370*, 302–309.
94. Rascio, N., Dalla Vecchia, F., La Rocca, N., Barbato, R., Pagliano, C., Raviolo, M., Gonnelli, C., & Gabbrielli, R. (2008). Metal accumulation and damage in rice (cv. Vialone nano) seedlings exposed to cadmium. *Environmental and Experimental Botany, 62*, 267–278.
95. Rizwan, M., Ali, S., Abbas, T., Adrees, M., Zia-ur-Rehman, M., Ibrahim, M., Abbas, F., Qayyum, M. F., & Nawaz, R. (2018). Residual effects of biochar on growth, photosynthesis and cadmium uptake in rice (Oryza sativa L.) under Cd stress with different water conditions. *Journal of Environmental Management, 206*, 676–683.
96. Wu, M., Wang, P. Y., Sun, L. G., Zhang, J. J., Yu, J., Wang, Y. W., & Chen, G. X. (2014). Alleviation of cadmium toxicity by cerium in rice seedlings is related to improved photosynthesis, elevated antioxidant enzymes and decreased oxidative stress. *Plant Growth Regulation, 74*, 251–260.
97. Kanu, A. S., Ashraf, U., Mo, Z., Fuseini, I., Mansaray, L. R., Duan, M., Pan, S., & Tang, X. (2017). Cadmium uptake and distribution in fragrant rice genotypes and related consequences on yield and grain quality traits. *Journal of Chemistry, 2017*, 1405878.
98. Siddique, A. B., Rahman, M. M., Islam, M. R., Mondal, D., & Naidu, R. (2021). Response of iron and Cadmium on yield and yield components of rice and translocation in grain: Health risk estimation. *Frontiers in Environmental Science, 9.*
99. Liu, J., Cai, G., Qian, M., Wang, D., Xu, J., Yang, J., & Zhu, Q. (2007). Effect of Cd on the growth, dry matter accumulation and grain yield of different rice cultivars. *Journal of the Science of Food and Agriculture, 87*, 1088–1095.

100. Feng, X., Liu, W. X., Sehar, S., Zheng, W. T., Zhang, G. P., & Wu, F. B. (2018). Application of sulfur fertilizer reduces cadmium accumulation and toxicity in tobacco seedlings (Nicotiana tabacum). *Plant Growth Regulation, 85*, 165–170.
101. Guo, L., Chen, A., He, N., Yang, D., & Liu, M. (2017). Exogenous silicon alleviates cadmium toxicity in rice seedlings in relation to Cd distribution and ultrastructure changes. *Journal of Soils and Sediments, 18*, 1691–1700.
102. Tamás, L., Bočová, B., Huttová, J., Mistrík, I., & Zelinová, V. (2017). Depletion of extracellular calcium increases cadmium toxicity in barley root tip via enhanced Cd uptake-mediated superoxide generation and cell death. *Acta Physiologiae Plantarum, 39.*
103. Noor, W., Umar, S., Mir, M. Y., Shah, D., Majeed, G., Hafeez, S., Yaqoob, S., Gulzar, A., & Kamali, A. (2018). Effect of cadmium on growth, photosynthesis and nitrogen metabolism of crop plants. *IBM Journal of Research and Development, 18*, 100–106.
104. Yu, C., Sun, C., Shen, C., Wang, S., Liu, F., Liu, Y., Chen, Y., Li, C., Qian, Q., & Aryal, B. (2015). The auxin transporter, Os AUX 1, is involved in primary root and root hair elongation and in Cd stress responses in rice (Oryza sativa L.). *The Plant Journal, 83*, 818–830.
105. Liu, J., Qian, M., Cai, G., Yang, J., & Zhu, Q. (2007). Uptake and translocation of Cd in different rice cultivars and the relation with Cd accumulation in rice grain. *Journal of Hazardous Materials, 143*, 443–447.
106. Ishikawa, S., Abe, T., Kuramata, M., Yamaguchi, M., Ando, T., Yamamoto, T., & Yano, M. (2010). A major quantitative trait locus for increasing cadmium-specific concentration in rice grain is located on the short arm of chromosome 7. *Journal of Experimental Botany, 61*, 923–934.
107. Ueno, D., Koyama, E., Yamaji, N., & Ma, J. F. (2011). Physiological, genetic, and molecular characterization of a high-Cd-accumulating rice cultivar, Jarjan. *Journal of Experimental Botany, 62*, 2265–2272.
108. Arao, T., & Ae, N. (2003). Genotypic variations in cadmium levels of rice grain. *Soil Science and Plant Nutrition, 49*, 473–479.
109. Ueno, D., Koyama, E., Kono, I., Ando, T., Yano, M., & Ma, J. F. (2009). Identification of a novel major quantitative trait locus controlling distribution of Cd between roots and shoots in rice. *Plant and Cell Physiology, 50*, 2223–2233.
110. Junyu, H., Cheng, Z., Yanfang, R., Yuping, Y., & Dean, J. (2006). Genotypic variation in grain cadmium concentration of lowland rice. *Journal of Plant Nutrition and Soil Science, 169*, 711–716.
111. Ye, X., Ma, Y., & Sun, B. (2012). Influence of soil type and genotype on Cd bioavailability and uptake by rice and implications for food safety. *Journal of Environmental Sciences, 24*, 1647–1654.
112. Römkens, P. F. A. M., Brus, D. J., Guo, H. Y., Chu, C. L., Chiang, C. M., & Koopmans, G. F. (2011). Impact of model uncertainty on soil quality standards for cadmium in rice paddy fields. *Science of the Total Environment, 409*, 3098–3105.
113. Liu, Y., Zhang, C., Zhao, Y., Sun, S., & Liu, Z. (2017). Effects of growing seasons and genotypes on the accumulation of cadmium and mineral nutrients in rice grown in cadmium contaminated soil. *Science of the Total Environment, 579*, 1282–1288.
114. Cao, F., Wang, R., Cheng, W., Zeng, F., Ahmed, I. M., Hu, X., Zhang, G., & Wu, F. (2014). Genotypic and environmental variation in cadmium, chromium, lead and copper in rice and approaches for reducing the accumulation. *Science of the Total Environment, 496*, 275–281.
115. Wang, F., Wang, M., Liu, Z., Shi, Y., Han, T., Ye, Y., Gong, N., Sun, J., & Zhu, C. (2015). Different responses of low grain-Cd-accumulating and high grain-Cd-accumulating rice cultivars to Cd stress. *Plant Physiology and Biochemistry, 96*, 261–269.
116. Barman, F., Majumdar, S., Arzoo, S. H., & Kundu, R. (2020). Genotypic variation among 20 rice cultivars/landraces in response to cadmium stress grown locally in West Bengal, India. *Plant Physiology and Biochemistry, 148*, 193–206.
117. Huang, G., Ding, C., Guo, F., Li, X., Zhou, Z., Zhang, T., & Wang, X. (2017). The role of node restriction on Cadmium accumulation in the brown rice of 12 Chinese rice (Oryza sativa L.) cultivars. *Journal of Agriculture and Food Chemistry, 65*, 10157–10164.

118. Huang, Q.-N., An, H., Yang, Y.-J., Liang, Y., & Shao, G.-S. (2017). Effects of Mn-Cd antagonistic interaction on Cd accumulation and major agronomic traits in rice genotypes by different Mn forms. *Plant Growth Regulation, 82*, 317–331.
119. Yuan, K., Wang, C., Zhang, C., Huang, Y., Wang, P., & Liu, Z. (2020). Rice grains alleviate cadmium toxicity by expending glutamate and increasing manganese in the cadmium contaminated farmland. *Environmental Pollution, 262*, 114236.
120. Chen, H., Yang, Y., Ye, Y., Tao, L., Fu, X., Liu, B., & Wu, Y. (2019). Differences in cadmium accumulation between indica and japonica rice cultivars in the reproductive stage. *Ecotoxicology and Environmental Safety, 186*, 109795.
121. Chen, L., Wu, W., Han, F., Li, J., Ye, W., Fu, H., Yan, Y., Ma, Y., & Wang, Q. (2019). Agronomic management and rice varieties controlling Cd bioaccumulation in rice. *International Journal of Environmental Research and Public Health, 16*.
122. Chen, X. W., Wu, L., Luo, N., Mo, C. H., Wong, M. H., & Li, H. (2019). Arbuscular mycorrhizal fungi and the associated bacterial community influence the uptake of cadmium in rice. *Geoderma, 337*, 749–757.
123. Li, X., Peng, P., Long, J., Dong, X., Jiang, K., & Hou, H. (2020) Plant-induced insoluble Cd mobilization and Cd redistribution among different rice cultivars. *Journal of Cleaner Production, 256*.
124. Li, Y., Guo, P., Liu, Y., Su, H., Zhang, Y., Deng, J., & Wu, Y. (2020). Effects of sulfur on the toxicity of cadmium to Folsomia candida in red earth and paddy soil in southern Fujian. *Journal of Hazardous Materials, 387*, 121683.
125. Huang, L., Li, W. C., Tam, N. F. Y., & Ye, Z. (2019). Effects of root morphology and anatomy on cadmium uptake and translocation in rice (Oryza sativa L.). *Journal of Environmental Sciences (China), 75*, 296–306.
126. Ishikawa, S. (2020). Mechanisms of cadmium accumulation in rice grains and molecular breeding for its reduction. *Soil Science and Plant Nutrition, 66*, 28–33.
127. Liu, C., Ding, S., Zhang, A., Hong, K., Jiang, H., Yang, S., Ruan, B., Zhang, B., Dong, G., Guo, L., Zeng, D., Qian, Q., & Gao, Z. (2020). Development of nutritious rice with high zinc/selenium and low cadmium in grains through QTL pyramiding. *Journal of Integrative Plant Biology, 62*, 349–359.
128. Songmei, L., Jie, J., Yang, L., Jun, M., Shouling, X., Yuanyuan, T., Youfa, L., Qingyao, S., & Jianzhong, H. (2019). Characterization and evaluation of OsLCT1 and OsNramp5 mutants generated through CRISPR/Cas9-mediated mutagenesis for breeding low Cd rice. *Rice Science, 26*, 88–97.
129. Yang, X., Qin, J., Li, J., Lai, Z., & Li, H. (2021). Upland rice intercropping with Solanum nigrum inoculated with arbuscular mycorrhizal fungi reduces grain Cd while promoting phytoremediation of Cd-contaminated soil. *Journal of Hazardous Materials, 406*, 124325.
130. Gong, L., Wang, J., Abbas, T., Zhang, Q., Cai, M., Tahir, M., Wu, D., & Di, H. (2021). Immobilization of exchangeable Cd in soil using mixed amendment and its effect on soil microbial communities under paddy upland rotation system. *Chemosphere, 262*, 127828.
131. Halim, M. A., Rahman, M. M., Megharaj, M., & Naidu, R. (2020). Cadmium immobilization in the rhizosphere and plant cellular detoxification: Role of plant-growth-promoting rhizobacteria as a sustainable solution. *Journal of Agricultural and Food Chemistry, 68*, 13497–13529.
132. Lin, X., Mou, R., Cao, Z., Xu, P., Wu, X., Zhu, Z., & Chen, M. (2016). Characterization of cadmium-resistant bacteria and their potential for reducing accumulation of cadmium in rice grains. *Science of the Total Environment, 569*, 97–104.
133. Pramanik, K., Mitra, S., Sarkar, A., & Maiti, T. K. (2018). Alleviation of phytotoxic effects of cadmium on rice seedlings by cadmium resistant PGPR strain Enterobacter aerogenes MCC 3092. *Journal of hazardous materials, 351*, 317–329.
134. Wang, M., Chen, S., Chen, L., & Wang, D. (2020). Microbial mechanisms responsible for the variation of soil Cd availability under different pe + pH environments. *Ecotoxicology and Environmental Safety, 206*, 111057.

135. Wan, Y., Camara, A. Y., Yu, Y., Wang, Q., Guo, T., Zhu, L., & Li, H. (2018). Cadmium dynamics in soil pore water and uptake by rice: Influences of soil-applied selenite with different water managements. *Environmental Pollution, 240*, 523–533.
136. Wang, J., Li, D., Lu, Q., Zhang, Y., Xu, H., Wang, X., & Li, Y. (2020). Effect of water-driven changes in rice rhizosphere on Cd lability in three soils with different pH. *Journal of Environmental Sciences (China), 87*, 82–92.
137. Babu, T., & Nagabovanalli, P. (2017). Effect of silicon amendment on soil-cadmium availability and uptake in rice grown in different moisture regimes. *Journal of Plant Nutrition, 40*, 2440–2457.
138. Sohail, M. I., ur Rehman, M. Z., Rizwan, M., Yousaf, B., Ali, S., ul Haq, M. A., Anayat, A., & Waris, A. A. (2020). Efficiency of various silicon rich amendments on growth and cadmium accumulation in field grown cereals and health risk assessment. *Chemosphere, 244*, 125481.
139. ur Rehman, M. Z., Rizwan, M., Rauf, A., Ayub, M. A., Ali, S., Qayyum, M. F., Waris, A. A., Naeem, A., & Sanaullah, M. (2019). Split application of silicon in cadmium (Cd) spiked alkaline soil plays a vital role in decreasing Cd accumulation in rice (Oryza sativa L.) grains. *Chemosphere, 226*, 454–462.
140. Ma, C., Ci, K., Zhu, J., Sun, Z., Liu, Z., Li, X., Zhu, Y., Tang, C., Wang, P., & Liu, Z. (2021). Impacts of exogenous mineral silicon on cadmium migration and transformation in the soil-rice system and on soil health. *Science of the Total Environment, 759*, 143501.
141. Chen, D., Chen, D., Xue, R., Long, J., Lin, X., Lin, Y., Jia, L., Zeng, R., & Song, Y. (2019). Effects of boron, silicon and their interactions on cadmium accumulation and toxicity in rice plants. *Journal of Hazardous Materials, 367*, 447–455.
142. Ji, X., Liu, S., Juan, H., Bocharnikova, E. A., & Matichenkov, V. V. (2017). Effect of silicon fertilizers on cadmium in rice (Oryza sativa) tissue at tillering stage. *Environmental Science and Pollution Research International, 24*, 10740–10748.
143. Huang, G., Ding, C., Li, Y., Zhang, T., & Wang, X. (2020). Selenium enhances iron plaque formation by elevating the radial oxygen loss of roots to reduce cadmium accumulation in rice (Oryza sativa L.). *Journal of Hazardous Materials, 398*, 122860.
144. Wan, Y., Yu, Y., Wang, Q., Qiao, Y., & Li, H. (2016). Cadmium uptake dynamics and translocation in rice seedling: Influence of different forms of selenium. *Ecotoxicology and Environmental Safety, 133*, 127–134.
145. Liu, N., Jiang, Z., Li, X., Liu, H., Li, N., & Wei, S. (2020). Mitigation of rice cadmium (Cd) accumulation by joint application of organic amendments and selenium (Se) in high-Cd-contaminated soils. *Chemosphere, 241*, 125106.
146. Huang, Q., Xu, Y., Liu, Y., Qin, X., Huang, R., & Liang, X. (2018). Selenium application alters soil cadmium bioavailability and reduces its accumulation in rice grown in Cd-contaminated soil. *Environmental Science and Pollution Research, 25*, 31175–31182.
147. Huang, H., Li, M., Rizwan, M., Dai, Z., Yuan, Y., Hossain, M. M., Cao, M., Xiong, S., & Tu, S. (2021). Synergistic effect of silicon and selenium on the alleviation of cadmium toxicity in rice plants. *Journal of Hazardous Materials, 401*, 123393.
148. Xue, T., Liao, X., Wang, L., Gong, X., Zhao, F., Ai, J., & Zhang, Y. (2020). Effects of adding selenium on different remediation measures of paddy fields with slight–moderate cadmium contamination. *Environmental geochemistry and health, 42*, 377–388.
149. Ma, X., Sharifan, H., Dou, F., & Sun, W. (2020). Simultaneous reduction of arsenic (As) and cadmium (Cd) accumulation in rice by zinc oxide nanoparticles. *Chemical Engineering Journal, 384*, 123802.
150. Ali, S., Rizwan, M., Noureen, S., Anwar, S., Ali, B., Naveed, M., Abd_Allah, E. F., Alqarawi, A. A., & Ahmad, P. (2019). Combined use of biochar and zinc oxide nanoparticle foliar spray improved the plant growth and decreased the cadmium accumulation in rice (Oryza sativa L.) plant. *Environmental Science and Pollution Research, 26*, 11288–11299.
151. Rizwan, M., Noureen, S., Ali, S., Anwar, S., ur Rehman, M. Z., Qayyum, M. F., & Hussain, A. (2019). Influence of biochar amendment and foliar application of iron oxide nanoparticles on growth, photosynthesis, and cadmium accumulation in rice biomass. *Journal of Soils and Sediments, 19*, 3749–3759.

152. Kanu, A. S., Ashraf, U., Mo, Z., Sabir, S.-u.-R., Baggie, I., Charley, C. S., & Tang, X. (2019). Calcium amendment improved the performance of fragrant rice and reduced metal uptake under cadmium toxicity. *Environmental Science and Pollution Research, 26*, 24748–24757.
153. Zhang, Y., Chen, T., Liao, Y., Reid, B. J., Chi, H., Hou, Y., & Cai, C. (2016). Modest amendment of sewage sludge biochar to reduce the accumulation of cadmium into rice(Oryza sativa L.): A field study. *Environmental Pollution, 216*, 819–825.
154. Qiao, J.-T., Liu, T.-X., Wang, X.-Q., Li, F.-B., Lv, Y.-H., Cui, J.-H., Zeng, X.-D., Yuan, Y.-Z., & Liu, C.-P. (2018). Simultaneous alleviation of cadmium and arsenic accumulation in rice by applying zero-valent iron and biochar to contaminated paddy soils. *Chemosphere, 195*, 260–271.
155. Li, H., Yu, Y., Chen, Y., Li, Y., Wang, M., & Wang, G. (2019). Biochar reduced soil extractable Cd but increased its accumulation in rice (Oryza sativa L.) cultivated on contaminated soils. *Journal of Soils and Sediments, 19*, 862–871.
156. Wang, X., Yao, H., Wong, M. H., & Ye, Z. (2013). Dynamic changes in radial oxygen loss and iron plaque formation and their effects on Cd and As accumulation in rice (Oryza sativa L.). *Environmental Geochemistry and Health, 35*, 779–788.
157. Gao, C., Wang, M., Ding, L., Chen, Y., Lu, Z., Hu, J., & Guo, S. (2020). High water uptake ability was associated with root aerenchyma formation in rice: Evidence from local ammonium supply under osmotic stress conditions. *Plant Physiology and Biochemistry, 150*, 171–179.
158. Cheng, H., Wang, M., Wong, M. H., & Ye, Z. (2014). Does radial oxygen loss and iron plaque formation on roots alter Cd and Pb uptake and distribution in rice plant tissues? *Plant and Soil, 375*, 137–148.
159. Colmer, T. D., Cox, M. C., & Voesenek, L. A. (2006). Root aeration in rice (Oryza sativa): Evaluation of oxygen, carbon dioxide, and ethylene as possible regulators of root acclimatizations. *New Phytologist, 170*, 767–777.
160. Mei, X., Li, Q., Wang, H., Fang, H., Chen, H., Chen, X., Yang, Y., Rizwan, M., & Ye, Z. (2020). Effects of cultivars, water regimes, and growth stages on cadmium accumulation in rice with different radial oxygen loss. *Plant and Soil, 453*, 529–543.
161. Otte, M. L., Rozema, J., Koster, L., Haarsma, M. S., & Broekman, R. A. (1989). Iron plaque on roots of Aster tripolium L.: interaction with zinc uptake. *New Phytologist, 111*, 309–317.@@
162. Wang, M. Y., Chen, A. K., Wong, M. H., Qiu, R. L., Cheng, H., & Ye, Z. H. (2011). Cadmium accumulation in and tolerance of rice (Oryza sativa L.) varieties with different rates of radial oxygen loss. *Environmental Pollution, 159*, 1730–1736.
163. Mei, X. Q., Ye, Z. H., & Wong, M. H. (2009). The relationship of root porosity and radial oxygen loss on arsenic tolerance and uptake in rice grains and straw. *Environmental Pollution, 157*, 2550–2557.
164. Tripathi, R. D., Tripathi, P., Dwivedi, S., Kumar, A., Mishra, A., Chauhan, P. S., Norton, G. J., & Nautiyal, C. S. (2014). Roles for root iron plaque in sequestration and uptake of heavy metals and metalloids in aquatic and wetland plants. *Metallomics, 6*, 1789–1800.
165. Wu, C., Zou, Q., Xue, S.-G., Pan, W.-S., Huang, L., Hartley, W., Mo, J.-Y., & Wong, M.-H. (2016). The effect of silicon on iron plaque formation and arsenic accumulation in rice genotypes with different radial oxygen loss (ROL). *Environmental Pollution, 212*, 27–33.
166. Khan, N., Seshadri, B., Bolan, N., Saint, C. P., Kirkham, M. B., Chowdhury, S., Yamaguchi, N., Lee, D. Y., Li, G., Kunhikrishnan, A., Qi, F., Karunanithi, R., Qiu, R., Zhu, Y. G., & Syu, C. H. (2016). Chapter One—Root iron plaque on wetland plants as a dynamic pool of nutrients and contaminants. In D. L. Sparks (Ed.), *Advances in agronomy* (pp. 1–96). Academic Press.
167. Li, X.-C., Yang, Z.-Z., Zhang, C., Wei, J.-J., Zhang, H.-Q., Li, Z.-H., Ma, C., Wang, M.-S., Chen, J.-Q., & Hu, J.-W. (2019). Effects of different crystalline iron oxides on immobilization and bioavailability of Cd in contaminated sediment. *Chemical Engineering Journal, 373*, 307–317.
168. Yang, W. T., Zhou, H., Gu, J. F., Zeng, Q. R., & Liao, B. H. (2017). Influence of rapeseed cake on iron plaque formation and Cd uptake by rice (Oryza sativa L.) seedlings exposed to excess Cd. *Bulletin of Environmental Contamination and Toxicology, 99*, 601–606.

169. Huang, G., Ding, C., Guo, F., Li, X., Zhang, T., & Wang, X. (2017). Underlying mechanisms and effects of hydrated lime and selenium application on cadmium uptake by rice (Oryza sativa L.) seedlings. *Environmental Science and Pollution Research, 24*, 18926–18935.
170. Zhou, H., Zhu, W., Yang, W. T., Gu, J. F., Gao, Z. X., Chen, L. W., Du, W. Q., Zhang, P., Peng, P. Q., & Liao, B. H. (2018). Cadmium uptake, accumulation, and remobilization in iron plaque and rice tissues at different growth stages. *Ecotoxicology and Environmental Safety, 152*, 91–97.
171. Liu, Z., Yang, J., Wan, X., Peng, Y., Liu, J., Wang, X., & Zeng, M. (2016). How red mud-induced enhancement of iron plaque formation reduces Cadmium accumulation in rice with different radial oxygen loss. *Polish Journal of Environmental Studies, 25*, 1603–1613.
172. Xu, B., & Yu, S. (2013). Root iron plaque formation and characteristics under N_2 flushing and its effects on translocation of Zn and Cd in paddy rice seedlings (Oryza sativa). *Annals of Botany, 111*, 1189–1195.
173. Liu, J., Cao, C., Wong, M., Zhang, Z., & Chai, Y. (2010). Variations between rice cultivars in iron and manganese plaque on roots and the relation with plant cadmium uptake. *Journal of Environmental Sciences, 22*, 1067–1072.
174. Liu, H.-J., Zhang, J. L., & Zhang, F. S. (2007) Role of iron plaque in Cd uptake by and translocation within rice (Oryza sativa L.) seedlings grown in solution culture. *Environmental and Experimental Botany, 59*, 314–320.
175. Liu, J., Leng, X., Wang, M., Zhu, Z., & Dai, Q. (2011). Iron plaque formation on roots of different rice cultivars and the relation with lead uptake. *Ecotoxicology and Environmental Safety, 74*, 1304–1309.
176. Liu, H., Zhang, J., Christie, P., & Zhang, F. (2008). Influence of iron plaque on uptake and accumulation of Cd by rice (Oryza sativa L.) seedlings grown in soil. *Science of the Total Environment, 394*, 361–368.
177. Zheng, R. L., Cai, C., Liang, J. H., Huang, Q., Chen, Z., Huang, Y. Z., Arp, H. P., & Sun, G. X. (2012). The effects of biochars from rice residue on the formation of iron plaque and the accumulation of Cd, Zn, Pb, As in rice (Oryza sativa L.) seedlings. *Chemosphere, 89*, 856–862.
178. Jiaofeng, G., Yang, H., Peng, Z., Bohan, L. & Hang, Z. (2022). Increasing phosphorus inhibits the retention and prevention of cadmium by iron plaque and promotes cadmium accumulation in rice plants. *Chemosphere, 307*, 135642.
179. Huang, G., Pan, D., Wang, M., Zhong, S., Huang, Y., Li, F., Li, X. & Xing, B. (2022). Regulation of iron and cadmium uptake in rice roots by iron (III) oxide nanoparticles: insights from iron plaque formation, gene expression, and nanoparticle accumulation. *Environmental Science: Nano, 9*(11), 4093–4103.
180. Zhang, J. Y., Zhou, H., Zeng, P., Wang, S. L., Yang, W. J., Huang, F., Huo, Y., Yu, S. N., Gu, J. F., & Liao, B. H. (2021). Nano-Fe_3O_4-modified biochar promotes the formation of iron plaque and cadmium immobilization in rice root. *Chemosphere, 276*, 130212.
181. Liu, N., Lou, X., Li, X., Shuai, Z., Liu, H., Jiang, Z., & Wei, S. (2021). Rhizosphere dissolved organic matter and iron plaque modified by organic amendments and its relations to cadmium bioavailability and accumulation in rice. *Science of the Total Environment, 792*, 148216.
182. Tan, L., Zhu, Y., Fan, T., Peng, C., Wang, J., Sun, L., & Chen, C. (2019). OsZIP7 functions in xylem loading in roots and inter-vascular transfer in nodes to deliver Zn/Cd to grain in rice. *Biochemical and Biophysical Research Communications, 512*, 112–118.
183. Hart, J. J., Welch, R. M., Norvell, W. A., & Kochian, L. V. (2006). Characterization of cadmium uptake, translocation and storage in near-isogenic lines of durum wheat that differ in grain cadmium concentration. *New Phytologist, 172*, 261–271.
184. Mori, S., Uraguchi, S., Ishikawa, S., & Arao, T. (2009). Xylem loading process is a critical factor in determining Cd accumulation in the shoots of Solanum melongena and Solanum torvum. *Environmental and Experimental Botany, 67*, 127–132.
185. Uraguchi, S., & Fujiwara, T. (2012). Cadmium transport and tolerance in rice: Perspectives for reducing grain cadmium accumulation. *Rice (N Y), 5*, 5.

186. Zhong, S., Li, X., Li, F., Huang, Y., Liu, T., Yin, H., Qiao, J., Chen, G., & Huang, F. (2021). Cadmium uptake and transport processes in rice revealed by stable isotope fractionation and Cd-related gene expression. *Science of the Total Environment, 806*, 150633.
187. Kensuke, T., Shu, F., Toru, F., Tadakatsu, Y., & Hiroaki, H. (2007). Quantitative estimation of the contribution of the phloem in cadmium transport to grains in rice plants (Oryza sativa L.). *Soil Science & Plant Nutrition, 53*, 72–77.
188. Mariyo, K., Satoru, I., Kazumi, I., Koichi, C., Hiroaki, H., Shuichi, Y., & Tadakatsu, Y. (2010). Possible chemical forms of cadmium and varietal differences in cadmium concentrations in the phloem sap of rice plants (Oryza sativa L.). *Soil Science & Plant Nutrition, 56*, 839–847.
189. Fujimaki, S., Suzui, N., Ishioka, N. S., Kawachi, N., Ito, S., Chino, M., & Nakamura, S. (2010). Tracing cadmium from culture to spikelet: Noninvasive imaging and quantitative characterization of absorption, transport, and accumulation of cadmium in an intact rice plant. *Plant Physiology, 152*, 1796–1806.
190. Rodda, M. S., Li, G., & Reid, R. J. (2011). The timing of grain Cd accumulation in rice plants: The relative importance of remobilisation within the plant and root Cd uptake post-flowering. *Plant and Soil, 347*, 105–114.
191. Ishikawa, S., Suzui, N., Ito-Tanabata, S., Ishii, S., Igura, M., Abe, T., Kuramata, M., Kawachi, N., & Fujimaki, S. (2011). Real-time imaging and analysis of differences in cadmium dynamics in rice cultivars (Oryza sativa) using positron-emitting Cd-107 tracer. *Bmc Plant Biology, 11*.
192. Li, L., Zhang, Y., Ippolito, J. A., Xing, W., Qiu, K., & Wang, Y. (2020). Cadmium foliar application affects wheat Cd, Cu Pb and Zn accumulation. *Environmental Pollution, 262*, 114329.
193. Lu, L. L., Tian, S. K., Yang, X. E., Li, T. Q., & He, Z. L. (2009). Cadmium uptake and xylem loading are active processes in the hyperaccumulator Sedum alfredii. *Journal of Plant Physiology, 166*, 579–587.
194. Ishimaru, Y., Takahashi, R., Bashir, K., Shimo, H., Senoura, T., Sugimoto, K., Ono, K., Yano, M., Ishikawa, S., Arao, T., Nakanishi, H., & Nishizawa, N. K. (2012). Characterizing the role of rice NRAMP5 in manganese, iron and cadmium transport. *Science Report, 2*, 286.
195. Sasaki, A., Yamaji, N., Yokosho, K., & Ma, J. F. (2012). Nramp5 is a major transporter responsible for manganese and cadmium uptake in rice. *The Plant Cell, 24*, 2155–2167.
196. Thomine, S., Wang, R., Ward, J. M., Crawford, N. M., & Schroeder, J. I. (2000). Cadmium and iron transport by members of a plant metal transporter family in *Arabidopsis* with homology to *Nramp* genes. *Proceedings of the National Academy of Sciences, 97*, 4991–4996.
197. Narayanan, N. N., Vasconcelos, M. W., & Grusak, M. A. (2007). Expression profiling of Oryza sativa metal homeostasis genes in different rice cultivars using a cDNA macroarray. *Plant Physiology and Biochemistry, 45*, 277–286.
198. Slamet-Loedin, I. H., Johnson-Beebout, S. E., Impa, S., & Tsakirpaloglou, N. (2015). Enriching rice with Zn and Fe while minimizing Cd risk. *Frontiers in Plant Science, 6*.
199. Takahashi, R., Ishimaru, Y., Senoura, T., Shimo, H., Ishikawa, S., Arao, T., Nakanishi, H., & Nishizawa, N. K. (2011). The OsNRAMP1 iron transporter is involved in Cd accumulation in rice. *Journal of Experimental Botany, 62*, 4843–4850.
200. Gravot, A., Lieutaud, A., Verret, F., Auroy, P., Vavasseur, A., & Richaud, P. (2004). AtHMA3, a plant P1B-ATPase, functions as a Cd/Pb transporter in yeast. *FEBS Letters, 561*, 22–28.
201. Morel, M., Crouzet, J., Gravot, A., Auroy, P., Leonhardt, N., Vavasseur, A., & Richaud, P. (2009). AtHMA3, a P1B-ATPase allowing Cd/Zn/Co/Pb vacuolar storage in Arabidopsis. *Plant Physiology, 149*, 894–904.
202. Miyadate, H., Adachi, S., Hiraizumi, A., Tezuka, K., Nakazawa, N., Kawamoto, T., Katou, K., Kodama, I., Sakurai, K., Takahashi, H., Satoh-Nagasawa, N., Watanabe, A., Fujimura, T., & Akagi, H. (2011). OsHMA3, a P1B-type of ATPase affects root-to-shoot cadmium translocation in rice by mediating efflux into vacuoles. *New Phytologist, 189*, 190–199.
203. Takahashi, R., Ishimaru, Y., Shimo, H., Ogo, Y., Senoura, T., Nishizawa, N. K., & Nakanishi, H. (2012). The OsHMA2 transporter is involved in root-to-shoot translocation of Zn and Cd in rice. *Plant Cell and Environment, 35*, 1948–1957.

204. Yamaji, N., Xia, J., Mitani-Ueno, N., Yokosho, K., & Feng Ma, J. (2013). Preferential delivery of zinc to developing tissues in rice is mediated by P-type heavy metal ATPase OsHMA2. *Plant Physiology, 162*, 927–939.
205. Yoneyama, T., Gosho, T., Kato, M., Goto, S., & Hayashi, H. (2010). Xylem and phloem transport of Cd, Zn and Fe into the grains of rice plants (Oryza sativaL.) grown in continuously flooded Cd-contaminated soil. *Soil Science and Plant Nutrition, 56*, 445–453.
206. Uraguchi, S., Kamiya, T., Sakamoto, T., Kasai, K., Sato, Y., Nagamura, Y., Yoshida, A., Kyozuka, J., Ishikawa, S., & Fujiwara, T. (2011). Low-affinity cation transporter (OsLCT1) regulates cadmium transport into rice grains. *Proceedings of the National Academy of Sciences USA, 108*, 20959–20964.
207. Clemens, S., Antosiewicz, D. M., Ward, J. M., Schachtman, D. P., & Schroeder, J. I. (1998). The plant cDNA LCT1 mediates the uptake of calcium and cadmium in yeast. *Proceedings of the National Academy of Sciences of the United States of America, 95*, 12043–12048.

Antioxidant Defence: A Key Frontier in Conferring Cadmium Tolerance in Plants

Ankita Biswas and Suparna Pal

Abstract Under favourable circumstances, reactive oxygen species (ROS) is frequently produced at safe levels and controlled by multifarious networks. The delicate equilibrium between ROS ($O_2^{\bullet-}$, H_2O_2, $^1O_2$, $^{\bullet}OH$) origination and ROS scavenging is constantly disordered by environmental stress factors like drought, heat, salinity and heavy metal. The heavy metal ions including Cd extensively disrupt redox homeostasis making ROS extremely deadly and impair cellular homeostasis. This in turn indicates scanty plant health and crop output. To ensure the existence and adapt to such Cd mediated stress conditions, plants have amalgamated antioxidant machinery having two well equipped arms, enzymatic superoxide dismutase (SOD), catalase (CAT), ascorbate peroxidase (APX), glutathione reductase (GR), glutathione peroxidase (GPX), monodehydroascorbate reductase (MDHAR), dehydroascorbate reductase (DHAR) and non-enzymatic ascorbic acid (AsA), glutathione (GSH), carotenoids, α-Tocopherol, phenol, proline. These two arms scavenge ROS in a cooperative fashion. In this study, we pivot on Cd stress mediated ROS generation, ROS chemistry and localization and their scavenging mechanism resolved by both arms of the antioxidative defence systems, emphasizing their promising role in Cd stress tolerance. Such inclusive knowledge on antioxidant defence will permit us to evolve genetically engineered Cd stress-tolerant plants.

Keywords Antioxidant defence · Asada-Halliwell pathway · Cd stress tolerance · Oxidative stress · Reactive oxygen species · Stress signalling

1 Introduction

Plants, being sessile type, are frequently confronted with copious environmental stresses, like salinity, drought, UV-B radiations, water, temperatures and other miscellanies of soil and airborne pathogens, industrial wastage, heavy metals and metalloids

A. Biswas · S. Pal (✉)
Department of Botany, Lady Brabourne College, P-1/2, Suhrawardy Avenue, Beniapukur, Kolkata, West Bengal 700017, India
e-mail: suparnap486@gmail.com

A. K. Jha and N. Kumar (eds.), *Cadmium Toxicity in Water*, Springer Water,
https://doi.org/10.1007/978-3-031-54005-9_12

[38]. Escalated anthropogenic activity, rapid industrialization, and modern farming practices in recent years have all been devoted to contamination of the environment with heavy metal elements, which is one of the major issues experienced by plants, especially in developing nations. On the basis of atomic number and density higher than 20 and 5 g/cm^3, the 53 d-block elements have been assigned as "heavy metals" [79]. For their basic metabolic functioning, plant cells need mineral elements, including C, O, H, N, Mg, S, P and K (macro elements) and B, Cu, Fe, Zn, Mn, Mo, Ni, Co, Cl, and Br (micro elements) [38]. These macro and micro elements are necessary for various physiological and biological activities in plants including photosynthesis, protein modification, intra-compartmental redox reactions, carbohydrate metabolism, nitrogen fixation and nucleic acid synthesis and growth. While some heavy metals (Cu, Zn, Co and Mo) are employed as microelements, others, like As, Cd, Cr, Pb, Hg have manifested severe harmful effects on plants including inadequate biomass production, poor yield, leaf chlorosis, abated photosynthesis, water and nutrient imbalance [133]. According to ATSDR [11], As, Pb, Hg, Cd, and Cr were ranked first, second, third, seventh, and seventeenth in the priority index of hazardous substances for toxicological profile, which is chiefly based on heavy metal's frequency, toxicity, and potential for human exposure at national priorities list (NPL).

Cadmium (Cd; atomic number 48, atomic weight 114.41, electron configuration $[Kr]4d^{10}5s^2$) is a chemical element that belongs to group XII of the periodic chart. This delicate, silvery-white metal is chemically close to zinc and mercury in its physical and chemical characteristics. The density of Cd is 8.64 g/cm^3, the electronegativity of this metal is 1.5 with crystal ionic radius of 0.97, the ionization potential of 8.993, the boiling point and melting point are 765 °C and 320.9 °C respectively at 100 kPa [112]. In addition, most Cd's compounds favour an oxidation state of + 2, which rarely exists in + 1 state. It is an inflammable metal that is insoluble in water. Typically, elemental Cd reacts with other elements such as chlorine (cadmium chloride), oxygen (cadmium oxide), or sulphur (cadmium sulphide, cadmium sulphate) [108]. With a biological half-life of 10–30 years, Cd is one of the most dangerous heavy metals [105].

Exposure of Cd in agricultural soil occurs through natural origin including weathering of minerals, geologic remains and volcanic rocks. Whereas anthropogenic origin such as sewage sludge, liming, wastewater irrigation, chemical fertilizer (phosphate fertilizer), coal mines, battery and petroleum industries, vehicular emission, aerosol emission adds to Cd load of the agricultural lands [50, 93, 155].

Being highly movable, Cd is easily transferred from the root cortical tissue to xylem tissue by constructing complexes with phytochelatins and finally stock up into aboveground parts of plants [113]. Moreover, plants have not evolved any distinct transporter for Cd, preferably its uptake into plant's body is allowed by the Fe, Mn, and Zn transporters including NRAMP (Natural Resistance Associated Macrophage Protein), ZIP (ZRT-IRT similar protein), YSL (Yellow-Strip-Like), CAX (Cation/proton exchanger), ABC transporter (ATP-binding cassette transporter), P-type ATPase and LCT transporter [87, 89, 135]. Additionally, soil bioavailable Cd concentration depends on various soil physiochemical ingredients notably pH, organic matter, nitrogen source, soil microorganisms, and presence of other minerals [23, 61]. Cd can freely cumulate in crop plants and invade the food chain, endorsing a negative footprint on both environment and people's wellness [127].

Within plant cells, Cd absorption could greatly encourage the production of reactive oxygen species (ROS), like superoxide radicals ($O_2^{\bullet-}$), hydrogen peroxide (H_2O_2), and the hydroxyl radical ($^{\bullet}OH$) and leading to the large-scale propagation of oxidative stress, disrupting preliminary defence strategies as well as accumulating cytotoxic molecule, methylglyoxal in plant tissue provoking cellular death [16, 62]. According to [52] and Singh et al. [133], this oxidative stress sequels to membrane damage and hinders the downstream cascade such as perturbation of DNA and protein structures, poor growth and development, lower photosynthetic efficiency, chlorosis etc. [129]. Also, Cd is reported to interact with the sulfhydryl group of proteins, which disrupts the redox equilibrium and critically affects plant's metabolic pathway [103]. The activities of abundant enzymatic antioxidants namely SOD, CAT, APX, GPX, GR, guaiacol peroxidase (GPOD) and polyphenol oxidase (PPO) as well as many non-enzymatic antioxidants namely phenol, flavonoid, carotenoid, anthocyanin, melatonin, AsA, GSH are dramatically amplified to counteract ROS [16,151]. By rendering thiol compounds (phytochelatins), which chelate Cd in root vacuoles, plants are competent to display tolerance to the plants. Cd tolerance is also governed by several compatible solutes like proline, glycine betaine and trehalose [27].

2 Global Status of Cd Contamination in Agricultural Lands

India, UK, and Japan have comparatively less Cd in soils than in China [80]. Although Awashthi [13] suggested the standard range for the Cd content in Indian soil has been set to be 3–6 mg/kg soil, a fresh study by Panwar and Ahmed [111] exhibited the presence of Cd in soil at greater concentration. Soil Cd content in different constituents of Earth's surface is given in Table 1.

Table 1 Soil Cd content of different constituents of Earth's surface

S. No.	Study area	Maximum Cd content in soil (mg/kg)	Reference
1	Jinding, China	531	Wen et al. [145]
2	Taoyuan, Egypt	378	Abd El-Salam and Abu-Zuid [2]
3	Bolnisi, Georgia	121.5	Avkopashvili et al. [12]
4	Haridwar, India	83.6	Deepali and Gangwar [32]
5	Krishna Vihar, India	47.7	Panwar and Ahmed [111]
6	Staffordshire, UK	16	Vangronsveld et al. [140]

3 Toxic Effects of Non-essential Heavy Metal: Cadmium

3.1 Effects of Cd on Plants

The most noticeable symptom of Cd exposure is reduction of root length, dry and fresh biomass which is linked to the shrinkage of mitotic index in root meristems. It accelerates root browning, roots tend to become mucilaginous, and lateral root development is compromised, which prompts the hardening and twisting nature of roots [40, 53]. Leaf chlorosis, desiccation, necrotic lesions, and stunting in leaf growth are very prominent signs of Cd stress in shoots [136].

Cd acutely affects physiological activities of plants by disorganizing complex II of photosynthetic apparatus as well as two photosystems (PSI and PSII), which leads to alteration of the chloroplast ultrastructure [35]. Cd impedes enzymes involved in the C3 cycle, downregulating photosynthetic rate. Cd can positively substitute Mg^{2+} in chlorophyll pigments and Calcium (Ca^{2+}) ions in oxygen-evolving complexes and accordingly diminish chlorophyll, carotenoid content [96]. Moreover, in contrast to mesophyll cells, depletion in photosynthetic pigment is more pronounced in stomatal guard cells. Even at minimal dosages, Cd can urge water stress, poor gas exchange, wilting, and it can suppress the uptake of vital macro- and microelements. It has been exemplified that Cd frequently competes with essential nutrients including Ca, Mg, P, and K [52].

Under the influence of Cd, suppression of electron transport is noted by blocking cysteine groups in enzymes, and competitively binds to Ca^{2+} binding motifs in water-splitting complex of photosystem II, which induces elicitation of toxic free radicals leading to an outburst of ROS and initiating apoptosis in plants [43, 47]. It has been extensively documented that overproduced ROS triggers genotoxic effects. The consequence of genotoxicity includes DNA damage (8-dihydro-8-oxoguanine) from nucleotide lesions, base excision, and structural abnormalities [114]. As a result, these aberrations have genotoxic consequences, which check the cell cycle progression and alter the plant's capacity for overall growth and development [1]. Table 2 summarizes Cd induced ROS formation in plants.

3.2 Effects of Cd on Humans

The foremost route of Cd exposure in human body is through food, cigarette smoking and inhalation of polluted dust, food, or water [71]. It poses severe health risks even at very low concentrations in the body as it cannot be metabolically converted into less hazardous species and has an extremely inferior elimination mechanism due to the lack of efficient chelating agents [121]. According to The International Agency for Research on Cancer, Cd has been categorized as a group I human carcinogen [88]. It portrays the prime causative agent of DNA damage and mutation that may aid in the growth of cancer [29].

Table 2 Cd inflicted ROS generation in plants grown under different experimental conditions

Plant species	Cd treatment	ROS	References
Gossypium hirsutum	500 μM $CdCl_2$ (hydroponics)	H_2O_2, $O_2^{\bullet-}$	Daud et al. [31]
Nicotiana tabacum	100, 500 μm $CdCl_2$ (hydroponics)	H_2O_2	Iannone et al. [63]
Triticum aestivum	200 mg/Kg Cd (soil)	H_2O_2	Iqbal et al. [66]
Vigna radiata	1 mM, and 1.5 mM $CdCl_2$ (hydroponics)	H_2O_2	Nahar et al. [100]
Sorghum bicolor	50, 100, 200 μM (hydroponics)	$O_2^{\bullet-}$ and H_2O_2	Zhan et al. [152]
Zea mays	5 mg/kg Cd (soil)	$O_2^{\bullet-}$	Zhao et al. [154]
Vicia faba	50 μM $CdCl_2$ (hydroponics)	$O_2^{\bullet-}$ and H_2O_2	Nouairi et al. [106]
Zea mays	0.5 mM $CdSO_4$ (hydroponics)	$O_2^{\bullet-}$ and H_2O_2	Alharby et al. [7]
Brassica napus	1.66 mg/Kg Cd (soil)	$O_2^{\bullet-}$ and H_2O_2	Lei et al. [84]
Oryza sativa	10 μM $CdCl_2$ (hydroponics)	H_2O_2 and $O_2^{\bullet-}$	Barman et al. [17], Biswas and Pal [22]
Zea mays	500 μM $CdCl_2$ (soil)	H_2O_2	Saleem et al. [124]
Mentha piperita	1.6 mg/L $CdCl_2$ (hydroponics)	H_2O_2, $O_2^{\bullet-}$	Wang et al. [142]
Solanum lycopersicum	100 μM Cd $(NO_3)_2$ (hydroponics)	H_2O_2, and $O_2^{\bullet-}$	Yan et al. [147]
Nicotiana tabacum	100 μmol L^{-1} $CdCl_2$ (hydroponics)	$O_2^{\bullet-}$ and H_2O_2	Zhang et al. [153]

Cd poisoning uprises oxidative stress, which disorders the body's antioxidant defence system and generates ROS, weakens the immune system [137]. ROS also damages biological macromolecules such as carbohydrates, proteins, and enzymes and facilitates the lipid peroxidation of cell membranes. Additionally, it leads to conformational alteration of the Na/K^+ ATPase, as a consequence the membrane potential drops, ionic homeostasis disturbs and affects neurotransmitter signaling cascade [107]. Ca^{2+} channels, can occasionally get obstructed owing to Cd toxicity (Cd^{2+}), modify intracellular Ca^{2+} influx levels which lower acetylcholinesterase activity in central nervous system [141]. Cd poisoning chiefly targets superoxide dismutase and glutathione activity, which reflects in neurotoxicity and behavioural aberrations [25]. Chronic exposures to Cd can compromise the respiratory system, and influence anosmia, respiratory stress, emphysema, bronchitis, chronic rhinitis, obstructive pulmonary disorder, destruction of olfactory epithelium and induce lung malfunction. In the kidney, Cd can efficiently bind via metallothionein and accumulate in renal tissue. Cd discourages activation of vitamin D and dwindles phosphate absorption in bones [137].

Cd exposure resulted in infertility, abnormal embryonic development, and structural and functional problems with the male or female reproductive system [37]. Male-specific consequences include reduced sperm motility, testicular damage, organ dysfunction, vacuolization of the seminiferous tubules, and prostate cancer, whereas female-specific consequences are alteration of progesterone synthesis, placental damage, haemorrhagic deformities in the ovaries, and ovarian cancer [10, 94]. Cd has a multitude of destructive consequences on biological systems, leading to disease development that is chronic and fatal.

4 Cd Stress Activates ROS Dependent Signaling Cascades in Plants

Literature is replete with several reports of ROS signal transduction pathways and stress responsiveness in plants. Similar to other stress related responses, plants acknowledge Cd stress through typical signaling pathways which involve sensing of Cd stress signal, transferring that signal to downstream components and initiating appropriate defence response to counteract negative impacts of stress in plants [38]. Cd accumulation in plants expedites the activation of the ROS-mediated signal transduction network [15, 35, 48]. Numerous lines of evidence have approved the dual function of ROS in signaling (like cell growth, and differentiation process) and toxicity (imbalance of redox environment) in plant bodies under Cd stress [67, 81, 83]. Ample studies already suggested that Cd exposure undoubtedly triggered ROS signaling cascade in various plant species such as fenugreek [150], rice [118], tomato [9, 77], and maize [89]. Cd, as a plant stressor, provokes Ca^{2+} deficiency by replacing as well as activating miR398 expression which downregulates the antioxidative enzymes (Cu/Zn SOD, CAT) and promotes ROS accumulation, NADPH oxidases activity [147]. Cd also governs ROS generation by alternative processes, including protein phosphatases, Ca^{2+} channels, Fenton or Haber–Weiss reaction and cGMP, in plant cells [138]. The accumulated ROS are delicately sensed and decoded by specific ROS sensors like EX1, EX2 and OXI1 proteins that trigger downstream signaling hub, particularly MAPKs protein kinases cascades (MAPKKK-MAPKK-MAPK) by phosphorylation [67, 69]. The MAPK kinase dependent response is further transduced to diverse stress specific transcriptional regulators and regulates stress responsive gene expression [59]. These findings implicate that plant's response to Cd persuaded ROS by activating coordinated intracellular networks which included MAP mediated phosphorylation cascades [62].

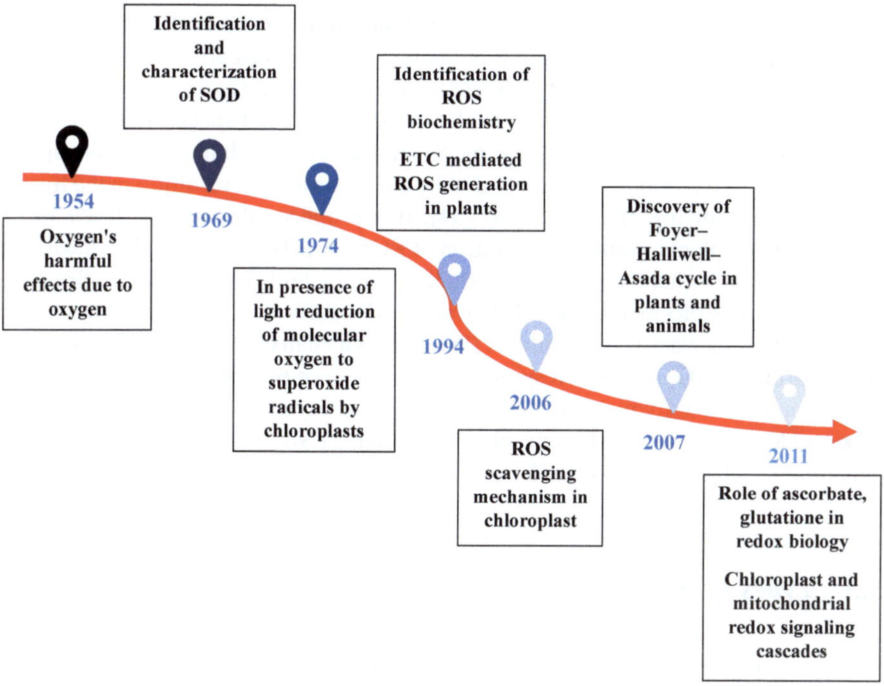

Fig. 1 Graphical representation of ROS evolution

5 Chemistry and Cellular Location of ROS in Plants

5.1 Timeline of ROS

Worldwide, the research on ROS has been carried out since ancient times. We would like to represent a graphical illustration of major pioneer landmark in the field of ROS research. Our modern knowledge and understanding of ROS biology is the product of scientists' noteworthy contributions. It is prominent that end of the twentieth century imprints the most significant milestones of this discipline (Fig. 1).

5.2 Chemical Nature of ROS

The ground state of atmospheric O_2 (triplet oxygen, $^3O_2$) is referred to as free nonreactive molecule confining two identically numbered unpaired parallel spin electrons. But accessory energy from electron transport chains (ETC), UV-B irradiation, and ionising radiation assist $^3O_2$ to overcome the spin constraint, which sparks ROS production [55]. Mitochondria, chloroplasts, peroxisomes, and plasma membranes are the primary cellular compartments to generate ROS in plant cells. By absorbing high light irradiance, the chlorophyll pigments of chloroplast are stimulated to their triplet excited state and rapidly converted $^3O_2$ into excited O_2 molecule (a form of ROS, half-life of 3.1–3.9 μs) if this triplet chlorophyll is not successfully quenched [34]. Additionally, ETC pathway or nicotinamide adenine dinucleotide phosphate (NADPH) activity oxidase can freely donate electrons to $^3O_2$, resulting in the formation of $O_2^{\bullet-}$ (half-life of 1–1000 μs) [54]. It is proposed that mostly 1–2% of O_2 expenditure by plants results in generation of ROS [30]. This $O_2^{\bullet-}$ combines with H^+ producing $HO_2^{\bullet-}$ (half-life of < 1 s), which is significantly more reactive, firm, and enterable across bio-membranes. Likewise, SOD isoforms, NADPH oxidases, and heme-containing class III peroxidases (POD) activity can efficiently dismutase $O_2^{\bullet-}$/$HO_2^{\bullet-}$ to elicit H_2O_2 [19]. Correspondingly, in presence of sunlight, $^\bullet OH$ is produced by Fenton reaction, hydroperoxides activity. Heme oxygenases, cytochrome P450s, superoxide reductases, and some photosystem II (PSII) proteins also evoke $^\bullet OH$ generation (half-life of 1 ns) [55].

5.3 *Types of ROS*

Based on properties and chemical reactivities, cellular ROS are comprised of both free radicals and nonradicals. Among the free radicals, superoxide ($O_2^{\bullet-}$), Hydroxyl ($^\bullet OH$), alkoxyl ($RO^\bullet$), and peroxyl radical ($ROO^\bullet$) and nonradicals, hydrogen peroxide (H_2O_2), singlet oxygen ($^1O_2$), and ozone (O_3) are most prevalent examples and produced independently in living cells [95]. Some other nonradical ROS are also found in plant cell compartments, especially hydroperoxides (ROOH), and excited carbonyls (RO*) [72]. Further, reactive oxygen intermediates (ROI) are categorized as reactive oxygen molecules triggered by partial O_2 reduction. Various forms of acids like hypobromous acid (HOBr), hypoiodous acid (HOI), and hypochlorous acid (HOCl) and radicals like carbonate radical ($CO_3^{\bullet-}$) and semiquinone ($SQ^{\bullet-}$) are additional forms of ROS in plant systems [42, 144].

5.4 *Localization of ROS*

ROS are naturally spawned by both stressed and non stressed plant cells [20]. The key intracellular sites of ROS production in plant cells are chloroplasts, mitochondria, peroxisomes, endoplasmic reticulum, plasma membrane, apoplast, and cell wall [78, 104] (Fig. 2). In the context of ETC and high energy exposure, chloroplasts act as a central hub of ROS generation (30–100 times greater than mitochondrial ROS) [36, 132]. Following Cd stress, chlorophyll is frequently overexcited and contributes to

$^1O_2$ generation by substituting O_2 at PSII reaction centre. It is revealed that chloroplast and peroxisomes are involved in light mediated ROS generation, whereas mitochondria have been shown to generate ROS in dark conditions [146]. Within mitochondria, electron leakage from both complex I and III of the ETC results in $O_2^{\bullet-}$ evolution, which is subsequently converted into H_2O_2 by Mn-SOD and Cu/Zn-SOD. In peroxisome, the primary catalyst for the formation of ROS (H_2O_2) is glycolate oxidase (GOX), xanthine oxidase (XOD), polyamine oxidase, flavin oxidase, copper amine oxidase, sulfite oxidase and sarcosine oxidase [28]. Quinine reductase, class III POX, germin and amine-like oxalate oxidases, lipoxygenases (LOX), and NADPH oxidase, which conveys electrons from NADPH on the cytosolic side to O_2 leading $O_2^{\bullet-}$ formation in the apoplast [97]. The sources of ROS accumulation within plant cell wall includes amine oxidases, POX and LOX activity [73]. Besides, Quinone reductase and NADPH oxidase both potentially hike ROS levels in the plasma membrane. In addition to these primary sites, the endoplasmic reticulum and glyoxysomes can also induce H_2O_2 production by cytochrome (Cyt) P450 and GOX activity respectively [55]. Interestingly, through the sub-cellular network, higher ROS levels in one organelle may also transfer to lower ROS levels in the nucleus or other organelles and cause stress-specific acclimatization [98].

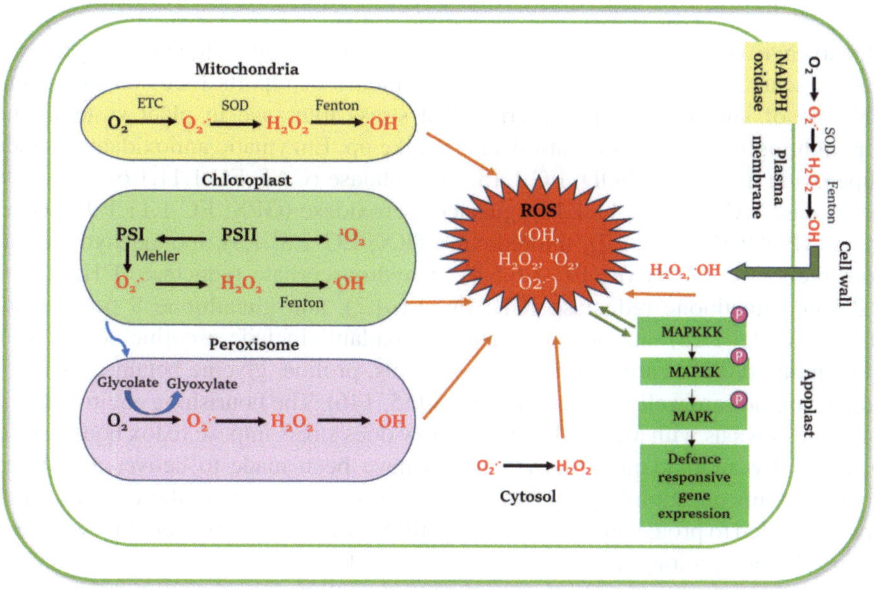

Fig. 2 Origin of ROS in plant cell compartments

6 Tolerance Mechanism of Plant Against Cd-Inflicted Oxidative Tress

As lofty ROS concentration is detrimental to the plant system, it is inevitable to balance ROS at non-toxic levels. Plants have evolved myriads of mechanisms to withstand and avoid the adverse consequences of Cd mediated toxicity. Therefore, tolerance towards Cd induced oxidative stress can be achieved by two preferred pathways including the antioxidative defence pathway and apoptosis pathway (programme cell death) which avert the accumulation of damaged cells within plant body [101]. Higher plants possess very well-stocked antioxidant defence complexes i.e., enzymatic i.e., SOD, CAT, APX GR, and non-enzymatic i.e., AsA, phenylpropanoids, carotenoid, GSH to synchronize compatible levels of ROS [38]. According to Singh et al. [133], *Lycopersicum esculentum* succeeded in Cd tolerance by activating apoptotic pathway to solve daunting problem of excessive ROS accumulation.

6.1 Antioxidants as a Countermeasure for Cd Tolerance in Plants: ROS Scavengers

Due to environmental stress including Cd, plant growth and yield in recent years are indeed scary. In the wake of this situation, plenty of reports focus on the potential role of antioxidants in conferring Cd stress tolerance in plant systems and highlighting the activities of antioxidants spike up. Enzymatic antioxidants include superoxide dismutase (SOD; EC 1.15.1.1), catalase (CAT; EC 1.11.1.6), ascorbate peroxidase (APX; EC 1.11.1.11), guaiacol peroxidase (GPX; EC 1.11.1.9), peroxidases (POD; EC 1.11.1.7), thioredoxin (TRX; EC 1.8. 1.9), monodehydroascorbate reductase (MDHAR; EC 1.6.5.4), dehydroascorbate reductase (DHAR; EC 1.8.5.1), glutathione reductase (GR; EC 1.6.4.2), and glutathione S transferases (GST; EC 2.5.1.18) and nonenzymatic antioxidants include ascorbic acid (AsA), glutathione (GSH), carotenoids, α-Tocopherols, proline, glycine betaine, alkaloids and flavonoids are well familiar with stress [55, 146]. The flourishing environmental stress bestows us with a crucial question, how does stress impose redox homeostasis balanced by antioxidants? Many ventures have been made to deliver a befitting answer to this question (Fig. 3). As a safeguard in plant system, these antioxidants are stimulated to protect plant system from ROS by transforming them to lower toxic forms, thereby prolonging cell redox symmetry (Table 3).

6.1.1 Enzymatic Antioxidants

According to aforesaid facts, an inspection of the antioxidative defence strategy employed by plants subjected to Cd notified that they are expressed in Cd stressed plants vehemently. SOD possesses metalloenzyme family and serves as front line of

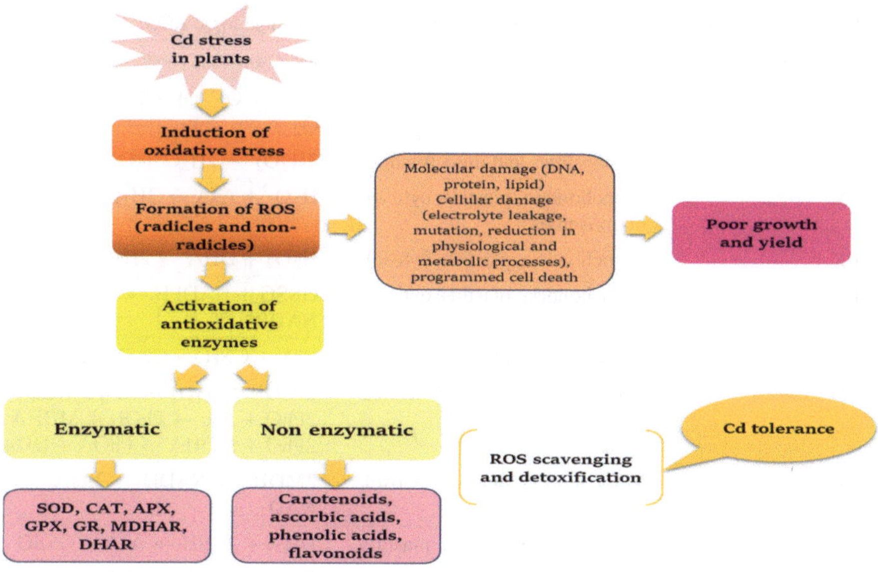

Fig. 3 Schematic representation showing Cd induced oxidative stress response in plants and activation of antioxidative defence system for balancing redox homeostasis

defence for ROS detoxification (converting $O_2^{\bullet-}$ to H_2O_2, narrowing the chances of $^{\bullet}OH$ generation). In light of metal cofactors, they are classified into three major forms: Cu/Zn-SODs (found in mitochondria, chloroplasts, peroxisomes, and cytoplasm of eukaryotic cells), Fe-SODs (found in peroxisomes and chloroplasts) and Mn-SODs (exclusive to mitochondria) [33]. Tetrameric heme is a component of CAT enzymes (converting H_2O_2 to H_2O and O_2), which exists in either mono-functional (dismutating enzymes of higher plants like rice, maize, Arabidopsis) or bi-functional form (dismutating and peroxidative enzymes of prokaryotes and lower eukaryotes). Moreover, when CAT is depleted in photosynthesizing chloroplast, plant body compiles some metabolites including ascorbate, GSH, thylakoid-bound ascorbate peroxidase (tAPX) and soluble stromal ascorbate peroxidase (sAPX) [43]. Powerful reductases such as DHAR, MDHAR, reduced ferredoxin and NADPH are engaged in ascorbate reduction [45]. In this route, APX enzymes (belonging to heme peroxidases) cordially reduce H_2O_2 to form monodehydroascorbate radicals by using ascorbate that consequently diverse to reduced AsA and DHA by MDAR using NADPH or reduced ferredoxin [99]. In sight of DHAR, DHA is efficiently recycled to AsA along with alteration of GSH (reduced form) into GSSG (oxidised form) (Fig. 4). Due to high pKa value of GSH at normal physiological pH, it restricts its affinity towards ROS, therefore acclimatizing high GSH/GSSG ratio via Asada-Halliwell cycle in plants [43]. POD predominately oxidises PhOH to induce phenoxyl radical (PhO or Q_A) which obtains electrons from H_2O_2 and forms H_2O within apoplast and vacuole [134]. PPO (EC 1.14.18.1) which is located in chloroplast nurtures ROS scavenging through interaction with POD. In presence of O_2, it oxidises PhOA to

Table 3 Reaction mechanisms of the main antioxidants that scavenge toxic ROS

Antioxidative enzymes	Localization	Reaction catalyzed/Function
SOD	Mitochondria, plastids, cytosol	$O_2^{\bullet-} + O_2^{\bullet-} + 2H^+ \rightarrow 2H_2O_2 + O_2$
CAT	Peroxisomes and mitochondria	$2H_2O_2 \rightarrow 2H_2O + O_2$
APX	Mitochondria, plastids, cytosol, peroxisomes	$H_2O_2 + AsA \rightarrow 2H_2O + DHA$
GPX	Mitochondria, plastids, cytosol	$H_2O_2 + DHA \rightarrow 2H_2O + GSSG$
GR	Mitochondria, plastids, cytosol	$GSSG + NADPH \rightarrow 2GSH + NADP^+$
GST	Mitochondria, plastids, cytosol	$R\text{-}X + GSH \rightarrow GS\text{-}R + H\text{-}X$
POD	Apoplast, cell wall and vacuole	$2PhOH + H_2O_2 \rightarrow 2PhO + 2H_2O$ $2PhO + Asc \rightarrow PhOH + MDHA$ $2PhO + MDHA \rightarrow PhOA + DHA$
MDHAR	Mitochondria, plastids, cytosol	$2MDHA + NADH \rightarrow 2AsA + NAD$
DHAR	Cytosol, mitochondria, plastids	$DHA + 2GSH \rightarrow AsA + GSSG$
AsA	Cytosol, mitochondria, plastids, peroxisome, apoplast and vacuole	Detoxify H_2O_2 via APX
GSH	Cytosol, mitochondria, plastids, peroxisome, apoplast and vacuole	Detoxify co-substrate for enzymes like GR, GST and POD
α-Tocopherol	Mainly in membranes	Detoxify products of LPO
Flavonoids	Vacuole	Directly scavenges H_2O_2, $^1O_2$,·OH
Phenolics compound	Cell wall	Directly scavenges $^1O_2$,·OH, $ROO^{\bullet}$, $ONOO^-$
Alkaloids	Vacuole	Scavenges H_2O_2, $^1O_2$,·OH, $O_2^{\bullet-}$
Carotenoids	Chloroplast	Quenches surplus energy from photosystem

H_2O and Q_A. Correspondingly, several thiol-based peroxidases (Prxs- members of heme free enzymes) are spotted in plants [24]. Prxs are partly antioxidants that utilize cysteinyl thiolate-based cycle and ardently reduce H_2O_2 in mitochondria, chloroplast and cytosol. Amidst six types of Prx family (Prx A-E), Prx A-D are unique to higher plants including *Oryza sativa, Arabidopsis thaliana, Populus trichocarpa* etc. [43]. Similarly, Trx has now been recognized as antioxidative defence gene. They facilitate redox stability by reducing the 2-CysPrx group in chloroplast and mitochondria [55].

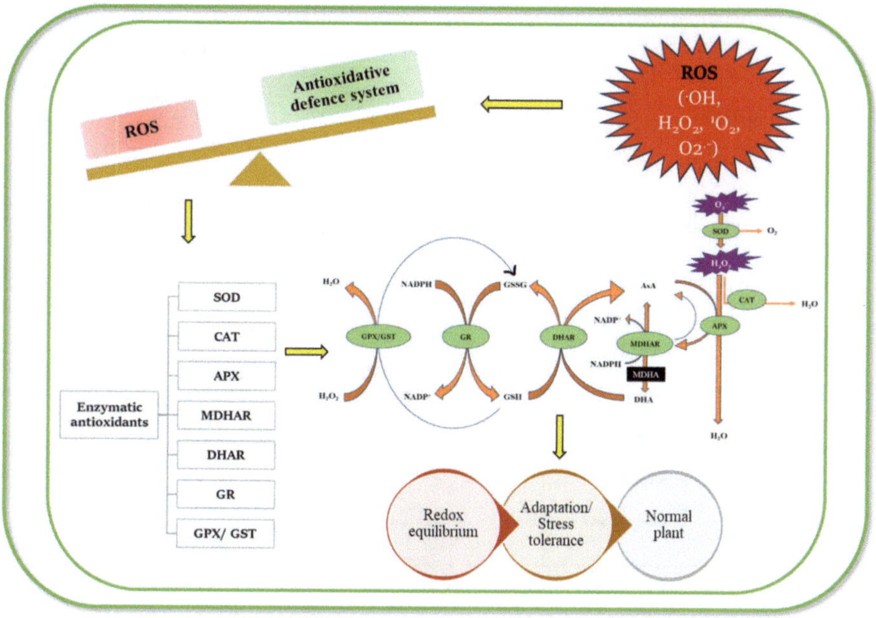

Fig. 4 Mechanism of enzymatic antioxidants against Cd stress

6.1.2 Non-enzymatic Antioxidants

The non enzymatic antioxidants are proclaimed to be pivotal players in minimizing stress induced oxidative damage. The non-enzymatic antioxidants such as AsA, GSH, α-Tocopherol, phenolic acids, carotenoids, flavonoids, and proline impede the downstream cascade of ROS generated during Cd stress [3, 110, 119] (Fig. 5). An omnipresent thiol peptide, GSH assists in breakdown of H_2O_2 through GPX catalyzed reaction. As a robust reductant for DHAR in Asada-Halliwell pathway, GSH scavenges H_2O_2, decays H_2O_2 and lipid peroxides [109]. Ascorbate contributes tocopherol and xanthophyll formation and correlates with the quenching of surplus energy [120]. Carotenoids, being a light harvesting accessory pigment, suppress oxidative damage and assure safe dissipation of superfluous energy in the form of heat generated during intense light irradiation. By scavenging $^1Chl^*$, $^3Chl^*$, and $^1O_2$ in thylakoid membrane of chloroplasts, carotenoids as well hamper over-stimulation of PS-II complex, yielding thylakoid membrane stability [120, 157]. GABA (γ-aminobutyric acid), a familiar non protein amino acid confers stress tolerance by monitoring free radicals' accumulation in plant cells. Being an osmolyte, GABA encourages pile of other osmolytes principally proline under stress conditions promoting osmotic balance in plants [68]. By proficiently scavenging abundant ROS ($^1O_2$, $OH^{\bullet -}$), and sustaining photosynthesis, α-Tocopherol (lipophilic antioxidant) acts as safeguard of chloroplast [158]. Low molecular weight non enzymatic antioxidants i.e., flavonoids (especially flavones and flavonols) play a fundamental

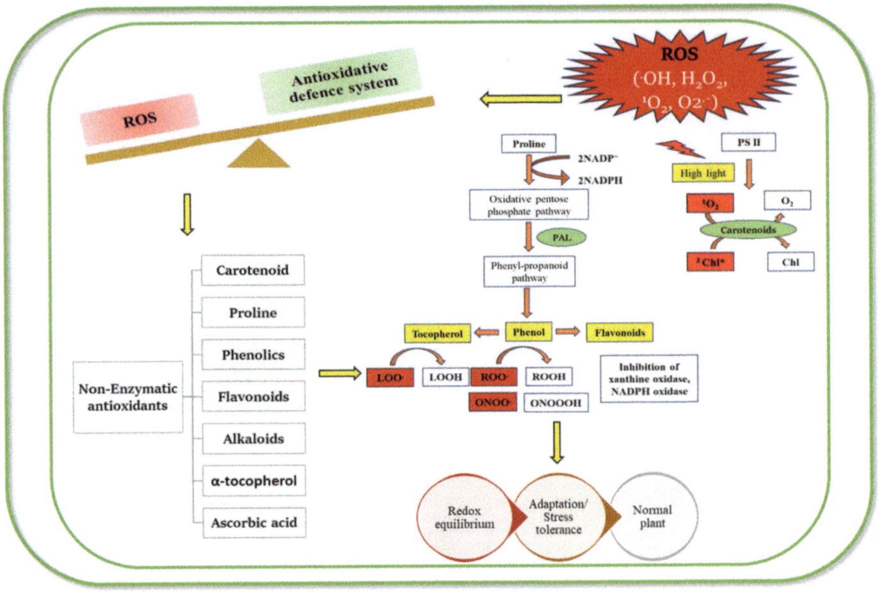

Fig. 5 Mechanism of non-enzymatic antioxidants against Cd stress

role to scavenge harmful free radicals and restore stress induced membranes [130]. Parallelly, another antioxidant like phenolic acids upgrades stress tolerance by acting as a chelator as well as scavenger of free radicals (O_2^-, $OH^{\bullet -}$, ROO^-, $ONOO^-$) [55]. Alkaloids also manifest antioxidant capacity by evading H_2O_2 accumulation and scavenging ROS [91]. Likewise, ornithine and citrulline are identified as potent non enzymatic antioxidants of plant systems [55].

7 Antioxidant Response Against Cd Toxicity: A Literature Review

Several reports comprehend the expression and modulation of Cd induced antioxidative defence genes. It was exhibited that rice seedlings under Cd stress showed a sharp increment of APX, GR, GPX and SOD activities compared to control seedlings, however, DHAR, GST and CAT activities declines. The stress tolerance was also correlated with production of non enzymatic antioxidants like phenolic and DHA content with reduction of AsA content resulting in ROS detoxification [117]. Coping with metal load, magnification of POX (60%) and SOD (40%) activities and proline content were detected in mustard seeds under Cd treatment (5 and 10 mg/Kg soil) [131]. In maize, the overexpression of *ZmWRKY4* gene (positive modulator of stress response) in nucleus of maize mesophyll cells was proclaimed to boost antioxidant enzymes such as SOD (up to 1.7-fold) and APX (up to 3.2-fold) activity in presence of Cd (200, 400 and 800 μM $CdCl_2$), improving stress tolerance [58]. Data by Alaraidh et al. [6] disclosed that upregulation of gene expression and transcription pattern (mRNA synthesis) of antioxidative enzyme encoding genes such as CAT (0.71-fold), APX (1.06-fold) and POX (0.76-fold) in fenugreek when exposed for 7 days of Cd treatment (2.5–10 mg/L). Kaya et al. [75] found that treatment of Cd (0.1 mM $CdCl_2$) to *Capsicum annuum* manipulated ROS accumulation by increasing enzymatic antioxidants mainly SOD, CAT, POD, LOX, GSH and proline compared with that under control. Induction of antioxidants was entirely analogous to the mitigation of oxidative damage. A recent study revealed that transgenic *Arabidopsis thaliana* tolerated high Cd levels (500 μM) by overexpressing glutaredoxin (CaGrx, LOC101493651) derived from chickpea leaves. This tolerance was accomplished by enhancing growth, enzyme activity (GR, GPX, GST, APX, SOD and CAT), and physiological characteristics [82]. Moreover, it was reported Cd tolerance in transgenic tobacco plants offered by overstimulation of *ZmVTE4* gene which encodes α-Tocopherol-methyl-transferase led to elevation of non enzymatic antioxidants chiefly α-Tocopherol levels [156]. In Menthol mint (*Mentha arvensis* L.), exposure of 30 days old seedlings at 50 μM $CdCl_2$ was found to trigger all antioxidant enzymes like SOD, CAT, POX and GR with respect to control plants [149]. In *Brassica napus* seedlings, Cd at a concentration (25 μM) sustained redox equilibrium and mitigated Cd incited oxidative damage through induction of antioxidative enzymes i.e., SOD, POD and APX [125]. El-Esawi et al. [41] reported that Cd (150 and 300 μM $CdCl_2$) stressed soybean significantly boosted various antioxidative enzymes (SOD, CAT, APX, POD) as well as upregulated antioxidative defence gene expression namely *APX* (1.75 and 2.8-fold), *CAT* (1.5 and 2.25-fold), *Fe-SOD* (1.65 and 2.6-fold), *POD* (1.5 and 2.3-fold) as compared to control plants and downregulated DPPH activity (13.20% and 28.30%) respectively thereby diminishing ROS levels and cellular toxicity. As well, plant also alleviated oxidative damage by elevating non enzymatic antioxidant content like proline content by 2.69 and 3.98-fold, respectively, compared to control plants. $CdCl_2$ (0.3 and 0.5 mM) treated mung bean resulted in an augment of antioxidative enzymes (CAT, GPX, APX and SOD) than control set and considerably attenuated Cd-induced ROS generation [57]. Exposure of 1.0- and 2.0-mM Cd in rice increased enzymatic antioxidants including APX activity by about (12%, 39%), MDHAR activity (38%, 15%), GR activity (7%, 23%), SOD activity (9%, 13%), and GPX activity (35% and 38%) as well as non enzymatic antioxidants including DHA content by (106%, 131%), GSH content (41%, 21%) owing to fight against ROS, scavenged overproduced H_2O_2 led to stress tolerance [21]. It was demonstrated by Li et al. [85] that all antioxidative defence enzymes like SOD (10.26%), CAT (78.70%) and POD (67.74%) were enhanced by application of Cd (100 μmol/L $CdCl_2$) in cucumber seedlings than the control plant. In response to Cd (50, 100, and 200 μmol/L), low molecular weight enzymatic antioxidants such as SOD (112%, 195%, and 317%), CAT (106%,158% and 221%) and POD (59%, 110% and 168%) and non enzymatic antioxidants such as proline content (56%, 86%, and 103%) respectively were reported to raise in *Triticum aestivum* for minimizing ROS levels, controlling Cd induced toxicity at cellular levels [139]. To endure Cd

(50 and 100 μM) stress, *Oryza sativa* evolved well organized antioxidative defence system which encompassed enzymatic antioxidants like SOD, CAT, APX and POD as well as non enzymatic antioxidants including flavonoids, phenolics, anthocyanin and ascorbic acid in comparison with control reflecting immediate scavenging mechanism of toxic free radicals [5]. Yeboah et al. [148] proclaimed the enrichment of enzymatic enzymes (SOD, CAT) to balance redox homeostasis under Cd at a concentration of 100 mg/L in *Ricinus communis*. Cd (25 and 50 mg kg^{-1}) treated *Pisum sativum* exhibited ROS detoxifying antioxidative defence system mainly enzymatic i.e., CAT (up to 475%), SOD (up to 47.25%), and APX (up to 125%) and non-enzymatic i.e., proline (up to 47.74%) and total phenolics content (up to 66.67%) to quench excess ROS, shield cellular machinery [26]. A similar trend for improved antioxidative activity (enzymatic and non enzymatic) was also recorded in *Brassica oleracea* [90], *Triticum aestivum* [64], *Hordeum vulgare* [102], *Solanum lycopersicum* [14] on account of Cd stress. The recent experiments in Cd induced activation of antioxidative defence enzymes have been tabulated in Table 4.

8 Recent Approaches to Enhance Antioxidants in Plants

Established data and facts have clearly shown that to minimize oxidative stress induced crop yield retardation, there is a dire need to enhance antioxidative defence apparatus in plant body. It is anticipated to be the best solution for plants to survive in growing competition for stress including Cd stress. Although, the potentiality of antioxidant defence varies between plant genotypes and species, between different stress environment and their respective frequency [46]. In rice, treatment of mineral elements has been shown to enhance enzymatic and non-enzymatic antioxidants with concomitant improvement of growth, yield under Cd stress [65]. Studies in *Mentha arvensis* subjected to gibberellic acid (GA) addition under Cd stress (50 μM $CdCl_2$) pointed to the fact that GA induced upregulation of antioxidative enzymes like SOD, CAT, GR and POX. Mahmud et al. [92] also claimed that application of citric acid elevated CAT, MDHAR, DHAR and GR activities in Cd treated (0.5 and 1.0 mM $CdCl_2$) *Brassica juncea* seedlings. Another recent study in sunflower has demonstrated that introduction of phytohormone-producing, plant-growth-promoting rhizobacteria *Acinetobacter bouvetii* can efficiently improve enzymatic and non enzymatic antioxidants, including CAT, APX, SOD, and POD, flavonoids, phenolics, proline, GSH under chromium (Cr) stress [115]. In addition, methyl jasmonate treated wheat plants showed improved antioxidants, declined H_2O_2 content and provided resistance against heat stress [44]. In rice, seed priming with brassinosteroids (EBL) has been found to upregulate expression of antioxidative enzymes (SOD, POD) in response to Cr stress [18]. Treatment of GABA results in upregulated gene expression of antioxidative enzymes (SOD, CAT, POD) as well as primary enzymes of Asada-Halliwell cycle and mitigate ROS accumulation in white clover [116]. Melatonin application in water stressed sites has shown elevated antioxidative defence and repressed ROS and RNS levels in plants [159]. Again,

Table 4 Antioxidant defence in plants under different concentrations of Cd stress

Plant species	Cd treatment	Enzymatic activity altered	Reference
Triticum aestivum	3 mg/Kg Cd (soil)	SOD (4.80–9.51%) and POD (12.4–23.7%), CAT (22.6–75.5%) increased	Guo et al. [51]
Mentha arvensis	50 μM $CdCl_2$ solution (soil)	SOD (50.14%), CAT (11.46%), POX (58.31%), GR (70.13%), carotenoid (76.92%) increased	Zaid et al. [149]
Brassica oleracea	5 mg/L Cd (hydroponics)	SOD, CAT and POD, proline (52.17%) increased MDHAR, DHAR, APX, GR, GST and GSH decreased	Shah et al. [128]
Sorghum bicolor	5, 25, 50, 100 μM $CdCl_2$ (hydroponics)	SOD (12%, 22%, 28%, 9%), CAT (7%, 8%, 12%, 13%), POD (12%, 11%, 15%, 19%) increased	Jawad Hassan et al. [70]
Oryza sativa	10 μM $CdCl_2$ (hydroponics)	SOD (32.23%), CAT (86.96%), APX (23.26%), GPX (13.47%), GR (72.0%) increased	Paul et al. [113]
Mentha piperita	1.6 mg/L $CdCl_2$ (hydroponics)	SOD, CAT, POD phenolics, PPO gene, flavonoid content increased	Wang et al. [142]
Solanum lycopersicum	100 μM Cd $(NO_3)_2$ (hydroponics)	APX, GR, GPOD and DHAR increased in root SOD, CAT, APX decreased in shoot	Yan et al. [147]
Nicotiana tabacum	100 μmol/L $CdCl_2$ (hydroponics)	DHAR, MDHAR, GPX, GR, POD and GST increased SOD, CAT, APX, TPX, and proteins of Trx-Prx pathway decreased	Zhang et al. [153]
Solanum lycopersicum	2 mg/Kg Cd (soil)	SOD, CAT, APX, proline content increased	Sarwar et al. [126]
Pisum sativum	300 μM $CdCl_2$ (hydroponics)	CAT (43%) increased SOD (25%), APX (60%) decreased	Sakouhi et al. [123]
Withania somnifera	100 and 300 μM Cd (hydroponics)	Phenolic content, flavonoids increased	Goncharuk and Zagoskina [49]
Vigna radiata	25 μM $CdSO_4$ (hydroponics)	SOD (38%), CAT (16%), and POX (82%) in plumule increased	Kaur et al. [74]
Cicer arietinum	200 μM $CdCl_2$ (hydroponics)	APX (1.7-fold), MDHAR (1.8-fold), DHAR (1.6-fold), GR (6-fold), and GPX (1.6-fold) increased	Sakouhi et al. [122]
Capsicum annuum	50 and 100 μM $CdCl_2$ (soil)	SOD, POD, CAT, APX, and their gene expression as well as the content of phenolic, flavonoid, anthocyanin and ascorbic acid increased	Alshegaihi et al. [8]
Lactuca sativa	250 μM $CdCl_2$ (soil)	POD (63%), APX (39%) increased	Waris et al. [143]

(continued)

Table 4 (continued)

Plant species	Cd treatment	Enzymatic activity altered	Reference
Oryza sativa	5.0 μM $CdSO_4$ (hydroponics)	GSH increased (153.4%) SOD (71.69%), POD (7.68%) decreased	Huang et al. [60]
Borago officinalis	81, 162, 243 and 324 μM $Cd(NO_3)_2{\cdot}4H_2O$	Proline, anthocyanin, phenolic compounds, and flavonoid increased	Aboutalebi et al. [4]
Ficus parvifolia	2 and 4 mmol/L $CdCl_2$ (soil)	SOD, CAT, APX and POD increased	Li et al. [86]

in mangrove species, sodium nitroprusside (SNP) performs a vital role in supervising activities of antioxidant enzymes (CAT, APX, MDHAR, DHAR) and recovering from salt stress [56]. However, approaches to enhancing antioxidative defence systems in plants under Cd stress are still limited. Further research is indispensable to shed light on recent approaches that will impart crucial avenues for improved stress responsiveness in developing countries.

9 Future Roadmap of Antioxidants: Nano-antioxidants

Modern agriculture is steadily advancing into precise agriculture with the help of new-age technologies and inventions. Application of nano-antioxidant can be pondered as future roadmap to fortify the plants from any kind of stress while shrinking ROS accumulation in plants. Although, nano-antioxidant is an emerging practice of nanotechnology in order to aptly scavenge ROS or free radical induced disease in humans. In this regard, antioxidants have been covalently combined with nanoparticles for attenuating oxidative damage by improving stability, bioaccessibility, greater sensitivity, less cytotoxicity and target specific delivery of antioxidants [76]. A study has been carried out by Eftekhari et al. [39] to implicate that addition of nano antioxidant can succour to reduce environmental induced toxicity in humans by improving solubility, permeability, and ROS scavenging properties. We should undoubtedly consider that nano-antioxidants have met with huge success in biomedical fields. On that account, advanced research concerning the formulation of nano-antioxidant is a prerequisite to conferring Cd tolerance and retaining redox equilibrium in plant body.

References

1. Abbas, T., Rizwan, M., Ali, S., Zia-ur-Rehman, M., Qayyum, M. F., Abbas, F., Hannan, F., Rinklebe, J., & Ok, Y. S. (2017). Effect of biochar on cadmium bioavailability and uptake in

wheat (Triticum aestivum L.) grown in a soil with aged contamination. *Ecotoxicology and Environmental Safety, 140*, 37–47.
2. Abd El-Salam, M. M., & Abu-Zuid, G. I. (2015). Impact of landfill leachate on the groundwater quality: A case study in Egypt. *Journal of Advanced Research, 6*(4), 579–586.
3. Abdulfatah, H. F. (2022). Non-enzymatic antioxidants in stressed plants: A review. *Concern, 16*(2), 25–37.
4. Aboutalebi, S., Zare, N., & Sheikhzadeh, P. (2023). Physio-biochemical response of *Borago officinalis* L. roots to cadmium toxicity. *Russian Journal of Plant Physiology, 70*(3), 37.
5. Afzal, J., Saleem, M. H., Batool, F., Elyamine, A. M., Rana, M. S., Shaheen, A., El-Esawi, M. A., Tariq Javed, M., Ali, Q., Arslan Ashraf, M., & Hussain, G. S. (2020). Role of ferrous sulfate ($FeSO_4$) in resistance to cadmium stress in two rice (Oryza sativa L.) genotypes. *Biomolecules, 10*(12), 1693.
6. Alaraidh, I. A., Alsahli, A. A., & Razik, E. A. (2018). Alteration of antioxidant gene expression in response to heavy metal stress in Trigonella foenum-graecum L. *South African Journal of Botany, 115*, 90–93.
7. Alharby, H. F., Al-Zahrani, H. S., Hakeem, K. R., Alsamadany, H., Desoky, E. S. M., & Rady, M. M. (2021). Silymarin-enriched biostimulant foliar application minimizes the toxicity of cadmium in maize by suppressing oxidative stress and elevating antioxidant gene expression. *Biomolecules, 11*(3), 465.
8. Alshegaihi, R. M., Mfarrej, M. F. B., Saleem, M. H., Parveen, A., Ahmad, K. S., Ali, B., Abeed, A. H., Alshehri, D., Alghamdi, S. A., Alghanem, S. M., & Lone, J. A. (2023). Effective citric acid and EDTA treatments in cadmium stress tolerance in pepper (Capsicum annuum L.) seedlings by regulating specific gene expression. *South African Journal of Botany, 159*, 367–380.
9. Alves, L. R., Monteiro, C. C., Carvalho, R. F., Ribeiro, P. C., Tezotto, T., Azevedo, R. A., & Gratão, P. L. (2017). Cadmium stress related to root-to-shoot communication depends on ethylene and auxin in tomato plants. *Environmental and Experimental Botany, 134*, 102–115.
10. Amegah, A. K., Sewor, C., & Jaakkola, J. J. (2021). Cadmium exposure and risk of adverse pregnancy and birth outcomes: A systematic review and dose–response meta-analysis of cohort and cohort-based case–control studies. *Journal of Exposure Science & Environmental Epidemiology, 31*(2), 299–317.
11. ATSDR, U. (2015). Toxicologial profiles, toxic substances portal.
12. Avkopashvili, G., Avkopashvili, M., Gongadze, A., Tsulukidze, M., & Shengelia, E. (2017). Determination of Cu, Zn and Cd in soil, water and food products in the vicinity of RMG gold and copper mine, Kazreti, Georgia. *Annals of Agrarian Science, 15*(2), 269–272.
13. Awashthi, S. K. (2000). Prevention of food adulteration act no 37 of 1954. Central and state rules as amended for, 3.
14. Badawy, I. H., Hmed, A. A., Sofy, M. R., & Al-Mokadem, A. Z. (2022). Alleviation of cadmium and nickel toxicity and phyto-stimulation of tomato plant l. by endophytic micrococcus luteus and enterobacter cloacae. *Plants, 11*(15), 2018.
15. Bali, A. S., Sidhu, G. P. S., Kumar, V., & Bhardwaj, R. (2019). Mitigating cadmium toxicity in plants by phytohormones. In *Cadmium toxicity and tolerance in plants* (pp. 375–396). Academic Press.
16. Banerjee, A., Singh, A., & Roychoudhury, A. (2019). Spermidine application reduces fluoride uptake and ameliorates physiological injuries in a susceptible rice cultivar by activating diverse regulators of the defense machinery. *Environmental Science and Pollution Research, 26*, 36598–36614.
17. Barman, F., Majumdar, S., Arzoo, S. H., & Kundu, R. (2020). Genotypic variation among 20 rice cultivars/landraces in response to cadmium stress grown locally in West Bengal, India. *Plant Physiology and Biochemistry, 148*, 193–206.
18. Basit, F., Chen, M., Ahmed, T., Shahid, M., Noman, M., Liu, J., An, J., Hashem, A., Fahad Al-Arjani, A. B., Alqarawi, A. A., & Alsayed, M. F. S. (2021). Seed priming with brassinosteroids alleviates chromium stress in rice cultivars via improving ROS metabolism and antioxidant defense response at biochemical and molecular levels. *Antioxidants, 10*(7), 1089.

19. Berwal, M., & Ram, C. (2018). Superoxide dismutase: A stable biochemical marker for abiotic stress tolerance in higher plants. In *Abiotic and biotic stress in plants* (pp. 1–10).
20. Bhattacharjee, S., & Bhattacharjee, S. (2019). ROS and oxidative stress: Origin and implication. In *Reactive oxygen species in plant biology* (pp. 1–31).
21. Bhuyan, M. B., Parvin, K., Mohsin, S. M., Mahmud, J. A., Hasanuzzaman, M., & Fujita, M. (2020). Modulation of cadmium tolerance in rice: Insight into vanillic acid-induced upregulation of antioxidant defense and glyoxalase systems. *Plants, 9*(2), 188.
22. Biswas, A., & Pal, S. (2022). Comparative screening of cadmium stress responses among three rice cultivars. *Environmental and Experimental Biology, 20*(3), 179–192.
23. Biswas, A., Pal, S., & Paul, S. (2023). Recent status of cadmium-contaminated paddy field and its impact on food safety: A field study over two years in North 24 Parganas, West Bengal, India. *Journal of Crop Science and Biotechnology*, 1–17.
24. Boeckx, T., Winters, A. L., Webb, K. J., & Kingston-Smith, A. H. (2015). Polyphenol oxidase in leaves: Is there any significance to the chloroplastic localization? *Journal of Experimental Botany, 66*(12), 3571–3579.
25. Briffa, J., Sinagra, E., & Blundell, R. (2020). Heavy metal pollution in the environment and their toxicological effects on humans. *Heliyon, 6*(9).
26. Chaturvedi, R., Favas, P. J., Pratas, J., Varun, M., & Paul, M. S. (2021). Harnessing *Pisum sativum–Glomus mosseae* symbiosis for phytoremediation of soil contaminated with lead, cadmium, and arsenic. *International Journal of Phytoremediation, 23*(3), 279–290.
27. Cha-um, S., Rai, V., & Takabe, T. (2019). Proline, glycinebetaine, and trehalose uptake and inter-organ transport in plants under stress. In *Osmoprotectant-mediated abiotic stress tolerance in plants: Recent advances and future perspectives* (pp. 201–223).
28. Corpas, F. J., González-Gordo, S., & Palma, J. M. (2020). Plant peroxisomes: A factory of reactive species. *Frontiers in Plant Science, 11*, 853.
29. Cui, Z. G., Ahmed, K., Zaidi, S. F., & Muhammad, J. S. (2021). Ins and outs of cadmium-induced carcinogenesis: Mechanism and prevention. *Cancer Treatment and Research Communications, 27*, 100372.
30. Das, K., & Roychoudhury, A. (2014). Reactive oxygen species (ROS) and response of antioxidants as ROS-scavengers during environmental stress in plants. *Frontiers in Environmental Science, 2*, 53.
31. Daud, M. K., Quiling, H., Lei, M., Ali, B., & Zhu, S. J. (2015). Ultrastructural, metabolic and proteomic changes in leaves of upland cotton in response to cadmium stress. *Chemosphere, 120*, 309–320.
32. Deepali, K. K., & Gangwar, K. (2010). Metals concentration in textile and tannery effluents, associated soils and ground water. *New York Science Journal, 3*(4), 82–89.
33. del Río, L. A., Corpas, F. J., López-Huertas, E., & Palma, J. M. (2018). Plant superoxide dismutases: Function under abiotic stress conditions. In *Antioxidants and antioxidant enzymes in higher plants* (pp. 1–26).
34. Dmitrieva, V. A., Tyutereva, E. V., & Voitsekhovskaja, O. V. (2020). Singlet oxygen in plants: Generation, detection, and signaling roles. *International Journal of Molecular Sciences, 21*(9), 3237.
35. Dobrikova, A. G., & Apostolova, E. L. (2019). Damage and protection of the photosynthetic apparatus under cadmium stress. In *Cadmium toxicity and tolerance in plants* (pp. 275–298). Academic Press.
36. Dogra, V., & Kim, C. (2020). Singlet oxygen metabolism: From genesis to signaling. *Frontiers in Plant Science,* 1640.
37. Dutta, S., Gorain, B., Choudhury, H., Roychoudhury, S., & Sengupta, P. (2022). Environmental and occupational exposure of metals and female reproductive health. *Environmental Science and Pollution Research, 29*(41), 62067–62092.
38. Dutta, S., Mitra, M., Agarwal, P., Mahapatra, K., De, S., Sett, U., & Roy, S. (2018). Oxidative and genotoxic damages in plants in response to heavy metal stress and maintenance of genome stability. *Plant Signaling & Behavior, 13*(8), e1460048.

39. Eftekhari, A., Dizaj, S. M., Chodari, L., Sunar, S., Hasanzadeh, A., Ahmadian, E., & Hasanzadeh, M. (2018). The promising future of nano-antioxidant therapy against environmental pollutants induced-toxicities. *Biomedicine & Pharmacotherapy, 103*, 1018–1027.
40. El Rasafi, T., Oukarroum, A., Haddioui, A., Song, H., Kwon, E. E., Bolan, N., Tack, F. M., Sebastian, A., Prasad, M. N. V., & Rinklebe, J. (2022). Cadmium stress in plants: A critical review of the effects, mechanisms, and tolerance strategies. *Critical Reviews in Environmental Science and Technology, 52*(5), 675–726.
41. El-Esawi, M. A., Elkelish, A., Soliman, M., Elansary, H. O., Zaid, A., & Wani, S. H. (2020). Serratia marcescens BM1 enhances cadmium stress tolerance and phytoremediation potential of soybean through modulation of osmolytes, leaf gas exchange, antioxidant machinery, and stress-responsive genes expression. *Antioxidants, 9*(1), 43.
42. Farnese, F. S., Menezes-Silva, P. E., Gusman, G. S., & Oliveira, J. A. (2016). When bad guys become good ones: The key role of reactive oxygen species and nitric oxide in the plant responses to abiotic stress. *Frontiers in Plant Science, 7*, 471.
43. Farooq, M. A., Niazi, A. K., Akhtar, J., Farooq, M., Souri, Z., Karimi, N., & Rengel, Z. (2019). Acquiring control: The evolution of ROS-Induced oxidative stress and redox signaling pathways in plant stress responses. *Plant Physiology and Biochemistry, 141*, 353–369.
44. Fatma, M., Iqbal, N., Sehar, Z., Alyemeni, M. N., Kaushik, P., Khan, N. A., & Ahmad, P. (2021). Methyl jasmonate protects the PS II system by maintaining the stability of chloroplast D1 protein and accelerating enzymatic antioxidants in heat-stressed wheat plants. *Antioxidants, 10*(8), 1216.
45. Foyer, C. H. (2018). Reactive oxygen species, oxidative signaling and the regulation of photosynthesis. *Environmental and Experimental Botany, 154*, 134–142.
46. Fujita, M., & Hasanuzzaman, M. (2022). Approaches to enhancing antioxidant defense in plants. *Antioxidants (Basel)., 11*(5), 925.
47. Genchi, G., Sinicropi, M. S., Lauria, G., Carocci, A., & Catalano, A. (2020). The effects of cadmium toxicity. *International Journal of Environmental Research and Public Health, 17*(11), 3782.
48. Ghosh, R., & Roy, S. (2019). Cadmium toxicity in plants: Unveiling the physicochemical and molecular aspects. In *Cadmium tolerance in plants* (pp. 223–246). Academic Press.
49. Goncharuk, E. A., & Zagoskina, N. V. (2023). Heavy metals, their phytotoxicity, and the role of phenolic antioxidants in plant stress responses with focus on cadmium. *Molecules, 28*(9), 3921.
50. Gope, M., Masto, R. E., George, J., Hoque, R. R., & Balachandran, S. (2017). Bioavailability and health risk of some potentially toxic elements (Cd, Cu, Pb and Zn) in street dust of Asansol, India. *Ecotoxicology and Environmental Safety, 138*, 231–241.
51. Guo, J., Qin, S., Rengel, Z., Gao, W., Nie, Z., Liu, H., Li, C., & Zhao, P. (2019). Cadmium stress increases antioxidant enzyme activities and decreases endogenous hormone concentrations more in Cd-tolerant than Cd-sensitive wheat varieties. *Ecotoxicology and Environmental Safety, 172*, 380–387.
52. Haider, F. U., Liqun, C., Coulter, J. A., Cheema, S. A., Wu, J., Zhang, R., Wenjun, M., & Farooq, M. (2021). Cadmium toxicity in plants: Impacts and remediation strategies. *Ecotoxicology and Environmental Safety, 211*, 111887.
53. Hameed, A., Rasool, S., Azooz, M. M., Hossain, M. A., Ahanger, M. A., & Ahmad, P. (2016). Heavy metal stress: plant responses and signaling. *Plant Metal Interaction*, 557–583.
54. Hasanuzzaman, M., Alam, M. M., Nahar, K., Mohsin, S. M., Bhuyan, M. B., Parvin, K., Hawrylak-Nowak, B., & Fujita, M. (2019). Silicon-induced antioxidant defense and methylglyoxal detoxification works coordinately in alleviating nickel toxicity in Oryza sativa L. *Ecotoxicology, 28*, 261–276.
55. Hasanuzzaman, M., Bhuyan, M. B., Zulfiqar, F., Raza, A., Mohsin, S. M., Mahmud, J. A., Fujita, M., & Fotopoulos, V. (2020). Reactive oxygen species and antioxidant defense in plants under abiotic stress: Revisiting the crucial role of a universal defense regulator. *Antioxidants, 9*(8), 681.

56. Hasanuzzaman, M., Inafuku, M., Nahar, K., Fujita, M., & Oku, H. (2021). Nitric oxide regulates plant growth, physiology, antioxidant defense, and ion homeostasis to confer salt tolerance in the mangrove species, Kandelia obovata. *Antioxidants, 10*(4), 611.
57. Hassan, M., Israr, M., Mansoor, S., Hussain, S. A., Basheer, F., Azizullah, A., & Ur Rehman, S. (2021). Acclimation of cadmium-induced genotoxicity and oxidative stress in mung bean seedlings by priming effect of phytohormones and proline. *PLoS ONE, 16*(9), e0257924.
58. Hong, C., Cheng, D., Zhang, G., Zhu, D., Chen, Y., & Tan, M. (2017). The role of ZmWRKY4 in regulating maize antioxidant defense under cadmium stress. *Biochemical and Biophysical Research Communications, 482*(4), 1504–1510.
59. Huang, H., Ullah, F., Zhou, D. X., Yi, M., & Zhao, Y. (2019). Mechanisms of ROS regulation of plant development and stress responses. *Frontiers in Plant Science, 10*, 800.
60. Huang, J., Jing, H. K., Zhang, Y., Chen, S. Y., Wang, H. Y., Cao, Y., Zhang, Z., Lu, Y. H., Zheng, Q. S., Shen, R. F., & Zhu, X. F. (2023). Melatonin reduces cadmium accumulation via mediating the nitric oxide accumulation and increasing the cell wall fixation capacity of cadmium in rice. *Journal of Hazardous Materials, 445*, 130529.
61. Hussain, B., Ashraf, M. N., Abbas, A., Li, J., & Farooq, M. (2021). Cadmium stress in paddy fields: Effects of soil conditions and remediation strategies. *Science of The Total Environment, 754*, 142188.
62. Huybrechts, M., Cuypers, A., Deckers, J., Iven, V., Vandionant, S., Jozefczak, M., & Hendrix, S. (2020). Cadmium and plant development: An agony from seed to seed. *International Journal of Molecular Sciences, 20*(16), 3971.
63. Iannone, M. F., Groppa, M. D., & Benavides, M. P. (2015). Cadmium induces different biochemical responses in wild type and catalase-deficient tobacco plants. *Environmental and Experimental Botany, 109*, 201–211.
64. Ilyas, N., Akhtar, N., Yasmin, H., Sahreen, S., Hasnain, Z., Kaushik, P., Ahmad, A., & Ahmad, P. (2022). Efficacy of citric acid chelate and Bacillus sp. in amelioration of cadmium and chromium toxicity in wheat. *Chemosphere, 290*, 133342.
65. Imran, M., Hussain, S., He, L., Ashraf, M. F., Ihtisham, M., Warraich, E. A., & Tang, X. (2021). Molybdenum-induced regulation of antioxidant defense-mitigated cadmium stress in aromatic rice and improved crop growth, yield, and quality traits. *Antioxidants, 10*(6), 838.
66. Iqbal, M., Hussain, I., Liaqat, H., Ashraf, M. A., Rasheed, R., & Rehman, A. U. (2015). Exogenously applied selenium reduces oxidative stress and induces heat tolerance in spring wheat. *Plant Physiology and Biochemistry, 94*, 95–103.
67. Jain, S., Muneer, S., Guerriero, G., Liu, S., Vishwakarma, K., Chauhan, D. K., Dubey, N. K., Tripathi, D. K., & Sharma, S. (2018). Tracing the role of plant proteins in the response to metal toxicity: A comprehensive review. *Plant Signaling & Behavior, 13*(9), 1507401.
68. Jalil, S. U., & Ansari, M. I. (2020). Physiological role of Gamma-aminobutyric acid in salt stress tolerance. In *Salt and drought stress tolerance in plants: Signaling networks and adaptive mechanisms* (pp. 337–350).
69. Jalmi, S. K., Bhagat, P. K., Verma, D., Noryang, S., Tayyeba, S., Singh, K., Sharma, D., & Sinha, A. K. (2018). Traversing the links between heavy metal stress and plant signaling. *Frontiers in Plant Science, 9*, 12.
70. Jawad Hassan, M., Ali Raza, M., Ur Rehman, S., Ansar, M., Gitari, H., Khan, I., Wajid, M., Ahmed, M., Abbas Shah, G., Peng, Y., & Li, Z. (2020). Effect of cadmium toxicity on growth, oxidative damage, antioxidant defense system and cadmium accumulation in two sorghum cultivars. *Plants, 9*(11), 1575.
71. Jean, J., Sirot, V., Hulin, M., Le Calvez, E., Zinck, J., Noël, L., Vasseur, P., Nesslany, F., Gorecki, S., Guérin, T., & Rivière, G. (2018). Dietary exposure to cadmium and health risk assessment in children—Results of the French infant total diet study. *Food and Chemical Toxicology, 115*, 358–364.
72. Kapoor, D., Sharma, R., Handa, N., Kaur, H., Rattan, A., Yadav, P., Gautam, V., Kaur, R., & Bhardwaj, R. (2015). Redox homeostasis in plants under abiotic stress: Role of electron carriers, energy metabolism mediators and proteinaceous thiols. *Frontiers in Environmental Science, 3*, 13.

73. Kar, R. K. (2015). ROS signaling: Relevance with site of production and metabolism of ROS. In *Reactive oxygen species and oxidative damage in plants under stress* (pp. 115–125).
74. Kaur, H., Nazir, F., Hussain, S. J., Kaur, R., Rajurkar, A. B., Kumari, S., Siddiqui, M. H., Mahajan, M., Khatoon, S., & Khan, M. I. R. (2023). *Gibberellic acid alleviates cadmium-induced seed germination inhibition through modulation of carbohydrate metabolism and antioxidant capacity in mung bean seedlings. Sustainability, 15*(4), 3790.
75. Kaya, C., Akram, N. A., Sürücü, A., & Ashraf, M. (2019). Alleviating effect of nitric oxide on oxidative stress and antioxidant defence system in pepper (Capsicum annuum L.) plants exposed to cadmium and lead toxicity applied separately or in combination. *Scientia Horticulturae, 255*, 52–60.
76. Khalil, I., Yehye, W. A., Etxeberria, A. E., Alhadi, A. A., Dezfooli, S. M., Julkapli, N. B. M., Basirun, W. J., & Seyfoddin, A. (2019). Nanoantioxidants: Recent trends in antioxidant delivery applications. *Antioxidants, 9*(1), 24.
77. Khan, M. Y., Prakash, V., Yadav, V., Chauhan, D. K., Prasad, S. M., Ramawat, N., Singh, V. P., Tripathi, D. K., & Sharma, S. (2019). Regulation of cadmium toxicity in roots of tomato by indole acetic acid with special emphasis on reactive oxygen species production and their scavenging. *Plant Physiology and Biochemistry, 142*, 193–201.
78. Kohli, S. K., Khanna, K., Bhardwaj, R., Abd_Allah, E. F., Ahmad, P., & Corpas, F. J. (2019). Assessment of subcellular ROS and NO metabolism in higher plants: Multifunctional signaling molecules. *Antioxidants, 8*(12), 641.
79. Koller, M., & Saleh, H. M. (2018). Introductory chapter: Introducing heavy metals. *Heavy Metals, 1*, 3–11.
80. Kubier, A., Wilkin, R. T., & Pichler, T. (2019). Cadmium in soils and groundwater: A review. *Applied Geochemistry, 108*, 104388.
81. Kumar, A., & Prasad, M. N. V. (2018). Plant-lead interactions: Transport, toxicity, tolerance, and detoxification mechanisms. *Ecotoxicology and Environmental Safety, 166*, 401–418.
82. Kumar, K., Raina, S. K., & Sultan, S. M. (2020). Arabidopsis MAPK signaling pathways and their cross talks in abiotic stress response. *Journal of Plant Biochemistry and Biotechnology, 29*(4), 700–714.
83. Kundu, P., Gill, R., Ahlawat, S., Anjum, N. A., Sharma, K. K., Ansari, A. A., Hasanuzzaman, M., Ramakrishna, A., Chauhan, N., Tuteja, N., & Gill, S. S. (2018). Targeting the redox regulatory mechanisms for abiotic stress tolerance in crops. In *Biochemical, physiological and molecular avenues for combating abiotic stress tolerance in plants* (pp. 151–220).
84. Lei, M., Li, Z., Zhang, B., Wang, X., Tie, B., Ayaz, T., & Lu, X. (2022). Mechanisms of stress alleviation after lime and biochar applications for Brassica napus L. In *Cadmium-contaminated soil*. Adsorption Science & Technology.
85. Li, H., Chen, H., Chen, L., & Wang, C. (2022). The role of hydrogen sulfide in plant roots during development and in response to abiotic stress. *International Journal of Molecular Sciences, 23*(3), 1024.
86. Li, Y., Cheng, X., Feng, C., & Huang, X. (2023). Interaction of lead and cadmium reduced cadmium toxicity in Ficus parvifolia seedlings. *Toxics, 11*(3), 271.
87. Li, Z., Liang, Y., Hu, H., Shaheen, S. M., Zhong, H., Tack, F. M., Wu, M., Li, Y. F., Gao, Y., Rinklebe, J., & Zhao, J. (2021). Speciation, transportation, and pathways of cadmium in soil-rice systems: A review on the environmental implications and remediation approaches for food safety. *Environment International, 156*, 106749.
88. Liang, Y., Yi, L., Deng, P., Wang, L., Yue, Y., Wang, H., Tian, L., Xie, J., Chen, M., Luo, Y., & Yu, Z. (2021). Rapamycin antagonizes cadmium-induced breast cancer cell proliferation and metastasis through directly modulating ACSS2. *Ecotoxicology and Environmental Safety, 224*, 112626.
89. Liu, X. S., Feng, S. J., Zhang, B. Q., Wang, M. Q., Cao, H. W., Rono, J. K., Chen, X., & Yang, Z. M. (2019). OsZIP1 functions as a metal efflux transporter limiting excess zinc, copper and cadmium accumulation in rice. *BMC Plant Biology, 19*, 1–16.
90. Ma, J., Saleem, M. H., Alsafran, M., Al Jabri, H., Rizwan, M., Nawaz, M., Ali, S., & Usman, K. (2022). Response of cauliflower (*Brassica oleracea* L.) to nitric oxide application under cadmium stress. *Ecotoxicology and Environmental Safety, 243*, 113969.

91. Macáková, K., Afonso, R., Saso, L., & Mladěnka, P. (2019). The influence of alkaloids on oxidative stress and related signaling pathways. *Free Radical Biology and Medicine, 134*, 429–444.
92. Mahmud, J. A., Hasanuzzaman, M., Nahar, K., Bhuyan, M. H. M., & Fujita, M. (2018). Insights into citric acid-induced cadmium tolerance and phytoremediation in *Brassica juncea* L.: Coordinated functions of metal chelation, antioxidant defense and glyoxalase systems. *Ecotoxicology and Environmental Safety, 147*, 990–1001.
93. Marrugo-Negrete, J., Pinedo-Hernández, J., & Díez, S. (2017). Assessment of heavy metal pollution, spatial distribution and origin in agricultural soils along the Sinú River Basin, Colombia. *Environmental Research, 154*, 380–388.
94. Massányi, P., Massányi, M., Madeddu, R., Stawarz, R., & Lukáč, N. (2020). Effects of cadmium, lead, and mercury on the structure and function of reproductive organs. *Toxics, 8*(4), 94.
95. Maurya, A. K. (2020). Oxidative stress in crop plants. In *Agronomic crops: Volume 3: Stress responses and tolerance* (pp. 349–380).
96. Mazur, R., Sadowska, M., Kowalewska, Ł., Abratowska, A., Kalaji, H. M., Mostowska, A., Garstka, M., & Krasnodębska-Ostręga, B. (2016). Overlapping toxic effect of long term thallium exposure on white mustard (*Sinapis alba* L.) photosynthetic activity. *BMC Plant Biology.*
97. Mittler, R. (2017). ROS are good. *Trends in plant science, 22*(1), 11–19.
98. Mittler, R., Zandalinas, S. I., Fichman, Y., & Van Breusegem, F. (2022). Reactive oxygen species signalling in plant stress responses. *Nature Reviews Molecular Cell Biology, 23*(10), 663–679.
99. Moursi, Y. S., Dawood, M. F., Sallam, A., Thabet, S. G., & Alqudah, A. M. (2021). Antioxidant enzymes and their genetic mechanism in alleviating drought stress in plants. In *Organic solutes, oxidative stress, and antioxidant enzymes under abiotic stressors* (pp. 233–262).
100. Nahar, K., Hasanuzzaman, M., Rahman, A., Alam, M. M., Mahmud, J. A., Suzuki, T., & Fujita, M. (2016). Polyamines confer salt tolerance in mung bean (*Vigna radiata* L.) by reducing sodium uptake, improving nutrient homeostasis, antioxidant defense, and methylglyoxal detoxification systems. *Frontiers in Plant Science, 7*, 1104.
101. Nair, A. R., DeGheselle, O., Smeets, K., Van Kerkhove, E., & Cuypers, A. (2013). Cadmium-induced pathologies: Where is the oxidative balance lost (or not)? *International Journal of Molecular Sciences, 14*(3), 6116–6143.
102. Nazir, M. M., Noman, M., Ahmed, T., Ali, S., Ulhassan, Z., Zeng, F., & Zhang, G. (2022). Exogenous calcium oxide nanoparticles alleviate cadmium toxicity by reducing Cd uptake and enhancing antioxidative capacity in barley seedlings. *Journal of Hazardous Materials, 438*, 129498.
103. Nemmiche, S. (2017). Oxidative signaling response to cadmium exposure. *Toxicological Sciences, 156*(1), 4–10.
104. Noctor, G., & Foyer, C. H. (2016). Intracellular redox compartmentation and ROS-related communication in regulation and signaling. *Plant Physiology, 171*(3), 1581–1592.
105. Nordberg, G. F., Nordberg, M., & Costa, M. (2022). Toxicology of metals: Overview, definitions, concepts, and trends. In *Handbook on the toxicology of metals* (pp. 1–14).
106. Nouairi, I., Jalali, K., Essid, S., Zribi, K., & Mhadhbi, H. (2019). Alleviation of cadmium-induced genotoxicity and cytotoxicity by calcium chloride in faba bean (*Vicia faba* L. var. minor) roots. *Physiology and Molecular Biology of Plants, 25*, 921–931.
107. Olaniyan, O. T., Ojewale, A. O., Eweoya, O. O., Adedoyin, A. A., Adesanya, O. A., Adeoye, A. O., & Okeniran, O. S. (2021). Modulatory role of vitamin E on proton pump (ATPase) activity of cadmium chloride-induced testicular damage in Wistar rats. *BioMed Research International, 2021.*
108. Olvera, D. A. C., & Avelizapa, N. G. R. (2022). Environmental pollution and current bioremediation strategies for cadmium containing residues. In *Biotechnology for treatment of residual wastes containing metals* (pp. 157–171).

109. Pandey, P., Singh, J., Achary, V. M. M., & Reddy, M. K. (2015). Redox homeostasis via gene families of ascorbate-glutathione pathway. *Frontiers in Environmental Science, 3*, 25.
110. Pandey, S., Fartyal, D., Agarwal, A., Shukla, T., James, D., Kaul, T., Negi, Y. K., Arora, S., & Reddy, M. K. (2017). Abiotic stress tolerance in plants: Myriad roles of ascorbate peroxidase. *Frontiers in Plant Science, 8*, 581.
111. Panwar, R. M., & Ahmed, S. (2018). Assessment of contamination of soil and groundwater due to e-waste handling. *Current Science*, 166–173.
112. Pattanayak, C., & Behera, B. (2022). Toxic effects of cadmium on Clarias batrachus—A review. *International Journal of All Research Education and Scientific Methods, 10*, 1812–1819.
113. Paul, S., Guha, T., Dey, S., Paul, S., & Kundu, R. (2022). Amelioration of cadmium toxicity by enhancing nitrogen assimilation and photosynthetic activity by two different nitrogen supplements in rice (*Oryza sativa* L.) cv. Lalat. *Plant Stress, 4*, 100082.
114. Praveen, A., Pandey, C., Ehasanullah, K. H. A. N., Panthri, M., & Gupta, M. (2020). Silicon-mediated genotoxic alterations in Brassica juncea under arsenic stress: A comparative study of biochemical and molecular markers. *Pedosphere, 30*(4), 517–527.
115. Qadir, M., Hussain, A., Hamayun, M., Shah, M., Iqbal, A., Irshad, M., Ahmad, A., Lodhi, M. A., & Lee, I. J. (2021). Phytohormones producing Acinetobacter bouvetii P1 mitigates chromate stress in sunflower by provoking host antioxidant response. *Antioxidants, 10*(12), 1868.
116. Qi, H., Kang, D., Zeng, W., Jawad Hassan, M., Peng, Y., Zhang, X., Zhang, Y., Feng, G., & Li, Z. (2021). Alterations of endogenous hormones, antioxidant metabolism, and aquaporin gene expression in relation to γ-aminobutyric acid-regulated thermotolerance in white clover. *Antioxidants, 10*(7), 1099.
117. Rahman, A., Mostofa, M. G., Nahar, K., Hasanuzzaman, M., & Fujita, M. (2016). Exogenous calcium alleviates cadmium-induced oxidative stress in rice (*Oryza sativa* L.) seedlings by regulating the antioxidant defense and glyoxalase systems: Calcium-induced cadmium stress tolerance in rice. *Brazilian Journal of Botany, 39*, 393–407.
118. Rahman, A., Nahar, K., Hasanuzzaman, M., & Fujita, M. (2016). Manganese-induced cadmium stress tolerance in rice seedlings: Coordinated action of antioxidant defense, glyoxalase system and nutrient homeostasis. *Comptes Rendus Biologies, 339*(11–12), 462–474.
119. Ramadoss, B. R., Subramanian, U., Alagarsamy, M., & Gangola, M. P. (2021). Non-enzymatic antioxidants' significant role in abiotic stress tolerance in crop plants. In *Organic solutes, oxidative stress, and antioxidant enzymes under abiotic stressors* (pp. 365–392).
120. Sachdev, S., Ansari, S. A., Ansari, M. I., Fujita, M., & Hasanuzzaman, M. (2021). Abiotic stress and reactive oxygen species: Generation, signaling, and defense mechanisms. *Antioxidants, 10*(2), 277.
121. Saini, S., & Dhania, G. (2020). Cadmium as an environmental pollutant: ecotoxicological effects, health hazards, and bioremediation approaches for its detoxification from contaminated sites. In *Bioremediation of industrial waste for environmental safety: Volume II: Biological agents and methods for industrial waste management* (pp. 357–387).
122. Sakouhi, L., Kharbech, O., Ben Massoud, M., Munemasa, S., Murata, Y., & Chaoui, A. (2022). Exogenous oxalic acid protects germinating chickpea seeds against cadmium injury. *Journal of Soil Science and Plant Nutrition*, 1–13.
123. Sakouhi, L., Mahjoubi, Y., Labben, A., Kharbech, O., Chaoui, A., & Djebali, W. (2023). Effects of cadmium–selenium interaction on glyoxalase and antioxidant systems of *Pisum sativum* germinating seeds. *Journal of Plant Growth Regulation, 42*(5), 3084–3099.
124. Saleem, M. H., Parveen, A., Khan, S. U., Hussain, I., Wang, X., Alshaya, H., El-Sheikh, M. A., & Ali, S. (2022). Silicon fertigation regimes attenuates cadmium toxicity and phytoremediation potential in two maize (Zea mays L.) cultivars by minimizing its uptake and oxidative stress. *Sustainability, 14*(3), 1462.
125. Sami, A., Shah, F. A., Abdullah, M., Zhou, X., Yan, Y., Zhu, Z., & Zhou, K. (2020). Melatonin mitigates cadmium and aluminium toxicity through modulation of antioxidant potential in *Brassica napus* L. *Plant Biology, 22*(4), 679–690.

126. Sarwar, M. J., Zahir, Z. A., Asghar, H. N., Shabaan, M., & Ayyub, M. (2023). Co-application of organic amendments and Cd-tolerant rhizobacteria for suppression of cadmium uptake and regulation of antioxidants in tomato. *Chemosphere, 327*, 138478.
127. Shaari, N. E. M., Tajudin, M. T. F. M., Khandaker, M. M., Majrashi, A., Alenazi, M. M., Abdullahi, U. A., & Mohd, K. S. (2022). Cadmium toxicity symptoms and uptake mechanism in plants: A review. *Brazilian Journal of Biology, 84*, e252143.
128. Shah, A. A., Khan, W. U., Yasin, N. A., Akram, W., Ahmad, A., Abbas, M., Ali, A., & Safdar, M. N. (2020). Butanolide alleviated cadmium stress by improving plant growth, photosynthetic parameters and antioxidant defense system of Brassica oleracea. *Chemosphere, 261*, 27728.
129. Shahid, M., Dumat, C., Khalid, S., Niazi, N. K., & Antunes, P. M. (2017). Cadmium bioavailability, uptake, toxicity and detoxification in soil-plant system. *Reviews of Environmental Contamination and Toxicology, 241*, 73–137.
130. Sharma, K., Mahato, N., & Lee, Y. R. (2019). Extraction, characterization and biological activity of citrus flavonoids. *Reviews in Chemical Engineering, 35*(2), 265–284.
131. Sheetal, K. R., Singh, S. D., Anand, A., & Prasad, S. (2016). Heavy metal accumulation and effects on growth, biomass and physiological processes in mustard. *Indian Journal of Plant Physiology, 21*, 219–223.
132. Singh, A., Kumar, A., Yadav, S., & Singh, I. K. (2019). Reactive oxygen species-mediated signaling during abiotic stress. *Plant Gene, 18*, 100173.
133. Singh, S., Parihar, P., Singh, R., Singh, V. P., & Prasad, S. M. (2016). Heavy metal tolerance in plants: Role of transcriptomics, proteomics, metabolomics, and ionomics. *Frontiers in Plant Science, 6*, 1143.
134. Smirnoff, N., & Arnaud, D. (2019). Hydrogen peroxide metabolism and functions in plants. *New Phytologist, 221*(3), 1197–1214.
135. Song, Y., Jin, L., & Wang, X. (2017). Cadmium absorption and transportation pathways in plants. *International Journal of Phytoremediation, 19*(2), 133–141.
136. Subašić, M., Šamec, D., Selović, A., & Karalija, E. (2022). Phytoremediation of cadmium polluted soils: Current status and approaches for enhancing. *Soil Systems, 6*(1), 3.
137. Suhani, I., Sahab, S., Srivastava, V., & Singh, R. P. (2021). Impact of cadmium pollution on food safety and human health. *Current Opinion in Toxicology, 27*, 1–7.
138. Tang, Z., Wang, H. Q., Chen, J., Chang, J. D., & Zhao, F. J. (2023). Molecular mechanisms underlying the toxicity and detoxification of trace metals and metalloids in plants. *Journal of Integrative Plant Biology, 65*(2), 570–593.
139. Ur Rahman, S. H., Qi, X., Zhang, Z., Ashraf, M. N., Du, Z., Zhong, Y. L., Mehmood, F., Ur Rahman, S., & Shehzad, M. (2020). The effect of silicon foliar and root application on growth, physiology, and antioxidant enzyme activity of wheat plants under cadmium toxicity. *Applied Ecology & Environmental Research, 18*(2).
140. Vangronsveld, J., Herzig, R., Weyens, N., Boulet, J., Adriaensen, K., Ruttens, A., Thewys, T., Vassilev, A., Meers, E., Nehnevajova, E., & van Der Lelie, D. (2009). Phytoremediation of contaminated soils and groundwater: Lessons from the field. *Environmental Science and Pollution Research, 16*, 765–794.
141. Wang, B., & Du, Y. (2013). Cadmium and its neurotoxic effects. In *Oxidative medicine and cellular longevity* (Vol. 2013).
142. Wang, B., Yuan, X., Zhu, Y., Wang, Y., Li, D., He, J., & Xiao, Y. (2023). Low-level cadmium exposure induced hormesis in peppermint young plant by constantly activating antioxidant activity based on physiological and transcriptomic analyses. *Frontiers in Plant Science, 14*, 1088285.
143. Waris, Z., Noreen, Z., Shah, A. A., Usman, S., Shah, A. N., Rizwan, M., Casini, R., & Elansary, H. O. (2023). Efficacy of γ-aminobutyric acid (GABA) on physio-biochemical attributes of lettuce (*Lactuca sativa* L.) under cadmium toxicity. *Journal of Plant Growth Regulation*, 1–17.
144. Waszczak, C., Carmody, M., & Kangasjärvi, J. (2018). Reactive oxygen species in plant signaling. *Annual Review of Plant Biology, 69*, 209–236.

145. Wen, H., Zhang, Y., Cloquet, C., Zhu, C., Fan, H., & Luo, C. (2015). Tracing sources of pollution in soils from the Jinding Pb–Zn mining district in China using cadmium and lead isotopes. *Applied Geochemistry, 52*, 147–154.
146. Xie, X., He, Z., Chen, N., Tang, Z., Wang, Q., & Cai, Y. (2019). The roles of environmental factors in regulation of oxidative stress in plant. *BioMed Research International.*
147. Yan, G., Hua, Y., Jin, H., Huang, Q., Zhou, G., Xu, Y., He, Y., & Zhu, Z. (2023). Sly-miR398 participates in cadmium stress acclimation by regulating antioxidant system and cadmium transport in tomato (*Solanum lycopersicum*). *International Journal of Molecular Sciences, 24*(3), 1953.
148. Yeboah, A., Lu, J., Ting, Y., Karikari, B., Gu, S., Xie, Y., Liu, H., & Yin, X. (2021). Genome-wide association study identifies loci, beneficial alleles, and candidate genes for cadmium tolerance in castor (*Ricinus communis* L.). *Industrial Crops and Products, 171*, 113842.
149. Zaid, A., Mohammad, F., & Fariduddin, Q. (2020). Plant growth regulators improve growth, photosynthesis, mineral nutrient and antioxidant system under cadmium stress in menthol mint (*Mentha arvensis* L.). *Physiology and Molecular Biology of Plants, 26*, 25–39.
150. Zayneb, C., Bassem, K., Zeineb, K., Grubb, C. D., Noureddine, D., Hafedh, M., & Amine, E. (2015). Physiological responses of fenugreek seedlings and plants treated with cadmium. *Environmental Science and Pollution Research, 22*, 10679–10689.
151. Zeng, Q., Ling, Q., Hu, F., Wu, J., Yang, Z., Qi, Y., & Li, Q. (2017). Genotypic differences in growth and antioxidant enzyme activities under cadmium stress in sugarcane. *Bulletin of Environmental Contamination and Toxicology, 99*, 607–613.
152. Zhan, Y. H., Zhang, C. H., Zheng, Q. X., Huang, Z. A., & Yu, C. L. (2017). Cadmium stress inhibits the growth of primary roots by interfering auxin homeostasis in Sorghum bicolor seedlings. *Journal of Plant Biology, 60*, 593–603.
153. Zhang, H., Yao, T., Wang, Y., Wang, J., Song, J., Cui, C., Ji, G., Cao, J., Muhammad, S., Ao, H., & Zhang, H. (2023). Trx CDSP32-overexpressing tobacco plants improves cadmium tolerance by modulating antioxidant mechanism. *Plant Physiology and Biochemistry, 194*, 524–532.
154. Zhao, L. J., Xie, J. F., Zhang, H., Wang, Z. T., Jiang, H. J., & Gao, S. L. (2018). Enzymatic activity and chlorophyll fluorescence imaging of maize seedlings (*Zea mays* L.) after exposure to low doses of chlorsulfuron and cadmium. *Journal of Integrative Agriculture, 17*(4), 826–836.
155. Zhu, C., Tian, H., Cheng, K., Liu, K., Wang, K., Hua, S., Gao, J., & Zhou, J. (2016). Potentials of whole process control of heavy metals emissions from coal-fired power plants in China. *Journal of Cleaner Production, 114*, 343–351.
156. Zhu, Q., Zhang, J., Yu, H., Li, L., Chen, X., Jiang, M., & Tan, M. (2019). Maize cd-tolerant ZmVTE4 encoding γ-tocopherol-methyl-transferase alleviated cd-toxicity through its product α-tocopherol. *Environmental and Experimental Botany, 158*, 171–179.
157. Zulfiqar, S., Sharif, S., Saeed, M., & Tahir, A. (2021). Role of carotenoids in photosynthesis. In *Carotenoids: Structure and function in the human body* (pp. 147–187).
158. Shehzad, J., & Mustafa, G. (2023). Mechanism of reactive oxygen species regulation in plants. In *Reactive Oxygen Species: Prospects in Plant Metabolism* (pp. 17–41).
159. Moustafa-Farag, M., Mahmoud, A., Arnao, M.B., Sheteiwy, M.S., Dafea, M., Soltan, M., Elkelish, A., Hasanuzzaman, M., & Ai, S. (2020). Melatonin-induced water stress tolerance in plants: Recent advances. *Antioxidants*, 9(9), 809.

Effect and Responses of Cadmium in Plants

Rohtas Kumar, Vikas Kumar, Vikas Tandon, Sushil Kumar, and Roohi

Abstract Cadmium (Cd) is one of the non-essential, extremely harmful contaminants that are generating major environmental and agricultural issues globally. From germination to yield, plants are harmed by Cd toxicity, but the magnitude of the harm depends on the dosage and the period of exposure. Due to prolonged Cd exposure, it interference with enzymatic and photosynthetic activities, as well as membrane degradation whereas, seed germination and plant growth are reduced. At higher levels, Higher levels of cadmium exposure disrupt plant nutrition and water relationships and result in oxidative damage. Additionally, Cd-induced structural alterations in the photosynthetic system reduce crop yield. Cd can be removed from and stabilized in soil by phytoremediation, whereas Cd can also be absorbed into the bodies of microorganisms. Increased Cd absorption in hyper accumulator plants to remove and change the poisonous form of Cd into harmless forms. Cd can be washed out, immobilized, and stabilised in the soil using chemical treatments during chemical remediation. Due to its eco-friendly characteristics, bioremediation of polluted environments is regarded as effective and credible. Furthermore, the application of nutrients antagonistic to Cd reduce uptake and toxicity in crops. The pH of the soil may be raised by the organic additions, and they may also bind to its functional groups, create complexations, and turn exchangeable forms into residual ones. Adopting certain agricultural practises is also found to be beneficial in lowering Cd uptake and accumulation in plants and harvesting high-quality food from Cd-contaminated soils.

Keywords Cadmium · Toxicity · Anti-oxidant enzyme · Lipid peroxidation

---

R. Kumar · V. Kumar (✉) · V. Tandon · S. Kumar
Department of Soil Science, CCS Haryana Agricultural University, Hisar 125001, India
e-mail: vikasoils52@yahoo.com

Roohi
CCS HAU, Regional Research Station, Karnal 132001, India

A. K. Jha and N. Kumar (eds.), *Cadmium Toxicity in Water*, Springer Water,
https://doi.org/10.1007/978-3-031-54005-9_13

1 Introduction

Cadmium is non-essential toxic element generally present in soil in traces but due to different anthropogenic activities, its high concentration in cultivated soils affect crop growth and productivity. Excessive Cd doses are carcinogenic to humans. It is ranked 7th in the list of top 20 toxic metals and classified under group-1 of carcinogens. Therange of Cd dietary intake (66.5–116 μg Cd kg^{-1} body weight per month) is much higher than maximum permissible limit (25 μg Cd kg^{-1} body weight per month) reported by FAO/WHO. Different anthropogenic activity includes, industrial discharge, mining activities, pollution from industries, whereas, agriculture-based source includes use of chemical inputs includes fertilizers, pesticides and insecticides resulting in increased Cd related hazards in environment [1]. High levels of Cd are harmful to plant, animals and humans due to its high mobility, toxicity and long-term persistence in the environment [2]. It is toxic to plants even at very low concentrations and interfere at various morphological, physiological, biochemical and molecular levels [3]. Its adverse impact on plants can be observed as inhibition of seed germination, root elongation, lower plant height and decreased number of leaves per plant [4]. Plants also adopt different defense mechanism in order to avoid Cd toxicity either through avoiding Cd uptake or through accumulation, storage and immobilization in different plant tissues. Once such toxic metal enters into the plant system, defense mechanism of plant activates through calcium calmodulin pathway, ROS overproduction and synthesis of different phytohormones [5]. This review paper is intended to present the effect of Cd on plant health and relative response of plants by adopting different signaling pathways and roles of different phytohormones and microorganism to avoid Cd toxicity are discussed in detail.

This chapter encompassing the overview of different morphological, physiological and molecular changes in plants occurring due to prolonged exposure to Cd stress and signaling pathway as response adopted by plants to coup up with Cd toxicity. Moreover, the mechanisms of toxicity and detoxification strategies employed by plants to confront Cd stress is emphasized. The roles of rhizosphere microorganisms, phytohormones and Phytochelatins in Cd tolerance and resistance strategies are discussed.

2 Sources of Cd in Soil

Cadmium (Cd) can originate from various sources and be deposited into soils through natural processes or human activities (Fig. 1). Natural source of Cd includes weathering of rocks, volcanic eruption and forest fire whereas, anthropogenic source of Cd in soil includes phosphatic fertilizers, industrial waste, mining and electroplating activities and battery waste [6]. Concentration of Cd is very low in normal soils but its solubility increases by root exudates (organic acids) and its concentration is found to be high in acid soils [7]. Cadmium in soil either occur as naturally through weathering of rocks such as sedimentary rock, basaltic rock, granite, shale and clays consist about 15, 0.006–0.6, 0.003–0.18, 0–11 and 0.2–8.4 ppm of Cd [8]. Anthropogenically through mining or refining of Cd rich ores contributes about 13,000 to 30,000 t Cd yr^{-1}, whereas, Cd containing phosphatic fertilizers contribute about 2–200 mg kg^{-1}, in agricultural soil, Cd concentration varies from 0.01 to 0.7 mg kg^{-1} [9–11]. The presence and intensity of cadmium (Cd) contamination in soils can differ based on specific local factors, such as the distance from pollution sources, the historical use of the land, and the practices employed for soil management. Effective soil management techniques, remediation efforts, and compliance with regulations concerning materials containing Cd can all contribute to reducing Cd contamination and safeguarding soil quality.

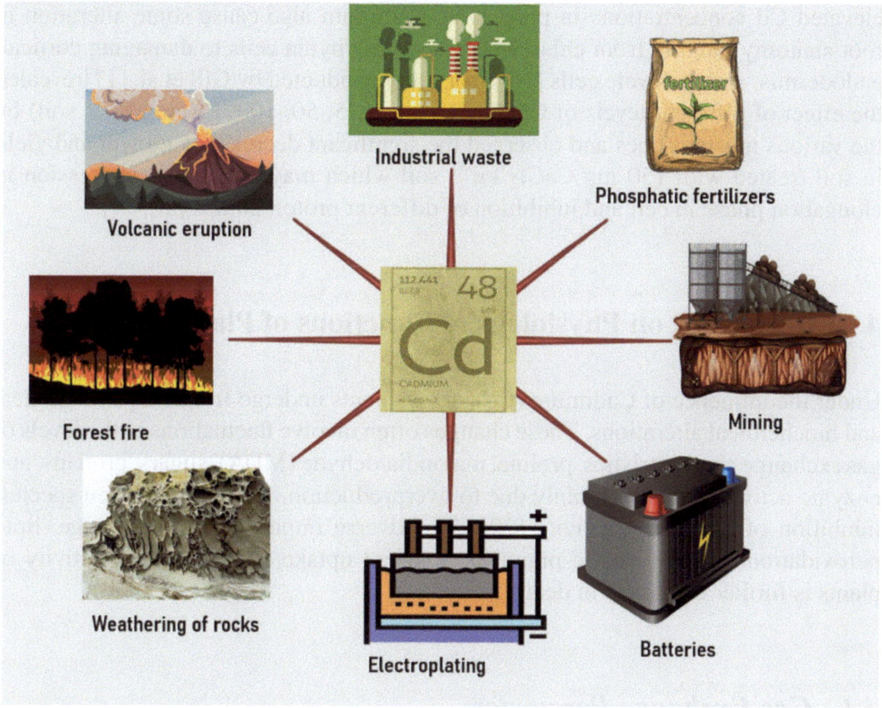

Fig. 1 Natural and anthropogenic sources of Cd in soil

3 Effect of Cd Toxicity on Growth and Development of Plants

Higher concentrations of cadmium (Cd) in plants can lead to various damages and adverse effects on their growth and development (Table 1). The damages due to higher Cd concentration in soil can be observed first on roots by damaging root nucleoli and decrease absorption of nitrogen inhibiting nitrate reductase activity [12]. Due to inhibition of root Fe^{3+} reductase induced by Cd toxicity resulting in Fe^{2+} deficiency and ultimately affect photosynthesis [13]. Visible symptoms of Cd toxicity include chlorosis, leaf roll and retarted plant growth [14]. High Cd concentrations in soil also influence the germination of seedling of crops. Plant also face several oxidation damages to plant cells due production of reactive oxygen species (ROS) due to elevated Cd concentrations in plant [15]. Cadmium also cause some alteration in root anatomy ranging from enlargement of parenchyma cells to damaging cortical, endodermis, and pericycle cells [16]. In a study conducted by Gill et al. [17]revealed the effect of different levels of $CdCl_2$ exposure (25, 50, 100, 150 mg kg^{-1} soil) on the various mustard lines and observed the significant decrease in growth and yield in soil treated with 150 mg $CaCl_2$ kg^{-1} soil which may be due to suppression in elongation phase in cell and inhibition of different proton pumps [18, 19].

4 Effect of Cd on Physiological Functions of Plants

Under the influence of Cadmium (Cd) stress, plants undergo frequent physiological and biochemical alterations. These changes often involve fluctuations in the levels of gas exchange characteristics, proline, malondialdehyde (MDA), sugars, proteins, and enzyme activity which is mainly due to overproduction of reactive oxygen species, inhibition of enzymatic activity [20–23]. Adverse impact on gas exchange, lipid peroxidation, photosynthetic pigments, nutrient uptake and enzymatic activity of plants is further discussed in detail.

4.1 Gas Exchange Parameters

Cadmium (Cd) has been observed to have significant effects on gas exchange parameters in plants. The presence of Cd can lead to alterations in various aspects of gas exchange, including stomatal conductance, photosynthesis, transpiration, and respiration. A study conducted by Younis et al. [24] on fenugreek grown in soil contaminated with Cd at concentrations of 25, 50, and 100 mg kg^{-1}, reductions in relative water content, stomatal conductance, transpiration rate and electrolyte leakage were reported. In studies conducted on carrot (*Daucus carota*), maize (*Zea mays*), and spinach (*Spinacia oleracea*), it was observed that higher concentrations of Cadmium

Table 1 Effect of elevated concentrations of Cd on growth and development of different

Plant species	Cd levels	Effects	Reference
Brassica compestris	0, 25, 50 and 100 mg $CdCl_2$ kg^{-1} soil	Decreased plant height, fresh mass, dry mass and leaf area	Irfan et al. [47]
Triticum aestivum	0, 2, 4, 6, 8 and 10 mg l^{-1}	Inhibited root and shoot elongation and seed germination	Amirjani et al. [48]
Saccharum officinarum	0, 2, 5 $CdCl_2$ mM	Severe damage to leaves and inhibited plant growth	Fornazier [49]
Momordica charantia	2 and 5 mg kg^{-1}	Decreased plant height and lower fruit per plant	Zafar-Ul-hye et al. [50]
Zea maize	12.5, 25, 50, 100, 200, 400 and 800 μmol Cd l^{-1}	Decreased root biomass	Ling et al. [35]
Sorghum bicolor	40, 80, 160, 320, 640 mg kg^{-1}	Shoot and root growth was decreased	An [51]
Virolasurinamensis	0, 15, 30, 45 and 60 mg l^{-1}	Lower root and shoot length	De Maria et al. [52]
Hordeum vulgare	10^{-5} M $CaCl_2$	Wilting and leaf curling and less biomass and growth	Lentini et al. [53]
Ageratum conyzoides	0, 3, 6, 12 and 30 mg kg^{-1}	Lower biomass and yield	Zhu et al. [54]
Cucumis sativus	40, 80, 160, 320, 640 mg kg^{-1}	Decreased shoot and root growth	An [51]
Raphanus Sativus	20, 40, 60, 80, 100 and 150 mg kg^{-1}	Root elongation was highly affected	Chen et al. [55]
Solanum lycopersicum	0, 0.1, 1, 5 and 10 μmol l^{-1}	Reduction in plant height and chlorophyll content	Dong et al. [56]
Allium cepa	0, 1, 10 and 100 μM	Increased chrosomal aberration and decreased plant and fruit size	Zhou et al. [57]
Lactuca sativa	1, 10, 20, 50 and 100 mg l^{-1}	Decreased root growth	Wang et al. [58]

(Cd) resulted in a reduction in stomatal conductance and transpiration rate. Sharma et al. [25] reported decreased transpiration rate and stomatal conductance in carrot exposed to increased Cd concentrations. Similarly, Zhang et al. [26] found a decrease in transpiration rate and stomatal conductance in Cd-exposed maize. Younis et al. [27] observed the same effects in Cd-exposed spinach. These findings suggest that Cd exposure can lead to reduced transpiration and hinder the opening and functioning of stomata in these plant species. Andrade et al. [28] reported the lowest values of water potential (– 0.46 Mpa), net CO_2 assimilation (1.6 μmol m^{-2} s^{-1}), stomatal conductance (13.0 mmol m^{-2} s^{-1}) and transpiration (0.48 mol m^{-2} s^{-1}) at concentration of 60 mg l^{-1} of Cd. Similar results of decreasing stomatal conductance in barley exposed to Cd was observed by Vassilev et al. [29].

4.2 Chlorophyll

Cadmium (Cd) toxicity plays a role in the reduction of chlorophyll content in plants. High levels of Cd can interfere with the biosynthesis of chlorophyll, which is essential for photosynthesis. It can disrupt the enzymes and processes involved in chlorophyll production, leading to a decline in its content within plant tissues and its visual symptoms includes leaf chlorosis and necrosis [30]. Cadmium induced depleted chlorophyll content in plants have reported by several researchers [31–34]. Concentration of chlorophyll a and chlorophyll b decreased significantly in leaves when higher level (800 μmol l^{-1}) of Cd supplied to plants [35]. Under Cd stress, the decrease in chlorophyll content caused by reduced biosynthesis leads to a significant decline in photosynthetic activity in plants [36–39].

4.3 Lipid Peroxidation

Cd stress in plants can trigger lipid peroxidation which results oxidative degradation of membrane lipids and production resulting in the production of reactive oxygen species (ROS) and the formation of lipid peroxidation products such as malondialdehyde (MDA) [40]. Lipid peroxidation can be measured by assessment of thiobarbituric acid reactive substances (TBARS) [41]. MDA concentration was found to be increased in hydroponically grown chrysanthemum under Cd stress (4 mg l^{-1}) [42]. Similar results were reported in *Lactuca sativa, Prunus dulcis, Ricinus communis, Spinacia oleracea* [43, 44]. In another study TBARS concentration was found to be increased in Brassicaceae and Fabaceae family [17, 45, 46].

4.4 Enzynatic Activity

Cd stress has been found to have significant effects on enzymatic activity in plants. Specifically, Cd toxicity can lead to alterations in the activity of various enzymes involved in crucial metabolic processes. Modification was generally seen in Antioxidative, Detoxyfying, photosynthetic, respiratory enzymes as well as enzymes involved in nitrogen metabolism. In a hydroponic study on Sorghum bicolor exposed to different levels of Cd from 50 to 2000 or 5000 μM for 28 days and found significant increase in catalase (CAT) while peroxidase (POD) and glutathione s transferase (GST) decreased whereas, no significant change was observed in ascorbate peroxidase (APX) [59]. In a study where Chrysanthemum coranarium grown hydroponically under Cd concentration of 4 mg l^{-1}there was no significant variation in superoxide dismutase (SOD) and POD was observed in shoot and roots [42]. Similar results were reported by Guo et al. [60], Zayneb et al. [61].

5 Cadmium Uptake and Translocation in Plants

As Cd is non-essential toxic metal present in soil in traces and its uptake through roots mediated through various processes. Being divalent cation, uptake of Cd^{2+} is similar as that of other divalent cations (Ca^{2+}, Mg^{2+}, Fe^{2+}, Zn^{2+} and Cu^{2+}) [62]. Cadmium uptake also influence by soil pH, organic matter content, concentration of Cd in soil rhizosphere [63]. Recent studies suggest that higher chloride concentration in soil help to mobilize inherent Cd present in soil and increase its uptake in wheat [64]. Transport of Cd from soil to root uptake is carried out through diffusion directly from soil solution to plant roots (Passive transport) as well as through active transport to cross the plasma membrane of root cells, which involves the participation of non-specific membrane transport proteins such as Zinc Transporter (ZIP) and Iron Transporter (IRT), as well as ATPase for transport of Cd [65]. Various other transporters were found to be involved in Cd translocation, P-type ATPase, ABC transporter, LCT transporter, NRAMP family, CAX family, CE family [66]. When Cd ion reaches in the plant cell, both symplastic and apoplastic pathways are involved to transport it to xylem [67]. It also has been reported that Cd is chelated with phosphorus in apoplast and with sulphurin symplast thereafter absorbed Cd moves to cytosol and then pericycle, finally reaches the xylem to be translocated in different plant parts [68].

6 Response of Plants Against Cadmium Toxicity

Cadmium (Cd) detoxification in plants refers to the mechanisms by which plants mitigate the toxic effects of Cd and minimize its accumulation in their tissues (Figs. 2). Plants have evolved several strategies to detoxify Cd, allowing them to tolerate and survive in Cd-contaminated environments [69]. Different pathways adopted by plants under cadmium stress includes phytohormones production, calcium calmodulin pathway, reactive oxygen overproduction [70]. Plants use chelation and complexation as primary mechanism of Cd detoxification, such as phytochelatins (PCs), metallothioneins (MTs), organic acids, and amino acids. These ligands bind to Cd ions, forming stable complexes that reduce the bioavailability and toxicity of Cd [71]. Plants can sequester Cd into vacuoles, which are membrane-bound organelles within plant cells. By compartmentalizing Cd in vacuoles, plants minimize its interaction with cellular components and reduce its impact on vital cellular processes [72]. Cd-induced oxidative stress is a common consequence of Cd toxicity in plants. To counteract this, plants activate their antioxidant defense system, which includes enzymes such as superoxide dismutase (SOD), catalase (CAT), peroxidases, and glutathione reductase. These enzymes help neutralize reactive oxygen species (ROS) generated by Cd stress, protecting plant cells from oxidative damage [73]. Plants employ transporters to efflux Cd ions from their cells, limiting its intracellular accumulation. Certain ATP-binding cassette (ABC) transporters and heavy metal ATPases (HMAs) are involved in the active efflux of Cd ions across the plasma membrane or tonoplast (vacuolar membrane) [74].

During Cd exposure, the intermediates of the tricarboxylic acid cycle (TCA) are utilized for the synthesis of metal chelators, such as siderophores and organic acids. Cd can enter the cell walls through transporters for ions like Zn, Fe, Cu, and Ca [75]. The entry of Cd triggers the production of reactive oxygen species (ROS) like H_2O_2, O_2^-, and OH^-, as well as nitric oxide (NO) [76]. This, in turn, leads to the upregulation of antioxidant enzyme activity, including ascorbate peroxidase (APX), guaiacol peroxidase (POD), superoxide dismutase (SOD), and glutathione reductase (GR), as well as the synthesis of antioxidants [77]. Excessive Cd exposure results in protein S-nitrosylation and the production of metallothioneins. The key

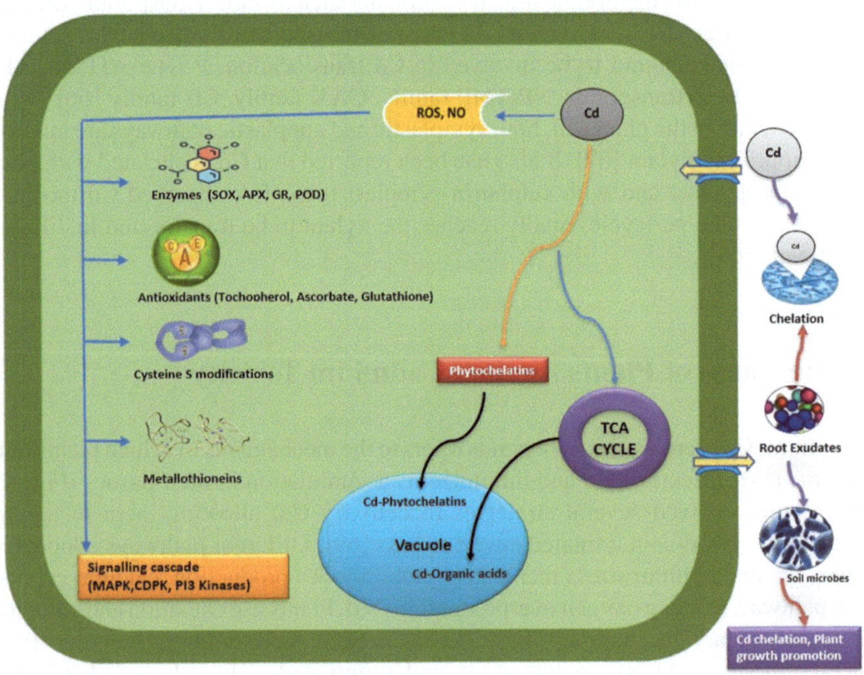

Fig. 2 Schematic presentation of steps involved in Cd detoxification in plants

signaling pathways involved in Cd stress response include mitogen-activated protein kinase (MAPK), calcium-dependent protein kinase (CDPK), and phosphoinositide 3-kinases (PI3 kinases). Cd stress also activates auxin-responsive promoters (e.g., DR5-GUS), auxin efflux carriers (e.g., PINs), and genes associated with pathogen response (e.g., PR genes). In response to Cd stress, plants synthesize phytochelatins as part of their defense system [74]. Phytochelatins aid in the sequestration of Cd into vacuoles, helping to minimize its toxicity within the plant [78].

The mechanism of cadmium stress and detoxification in plants involves several processes presented in Fig. 2.Tricarboxylic acid cycle (TCA) intermediates are utilized for the synthesis of metal chelators like siderophores and organic acids, which aid in binding and sequestering cadmium during exposure. Cadmium enters the plant cells through transporter proteins that also facilitate the transport of zinc, iron, copper, and calcium ions. Upon entering, cadmium induces the overproduction of reactive oxygen species (ROS) such as H_2O_2, O_2, and OH^-, as well as nitric oxide (NO). In response, the plant activates antioxidant enzymes like ascorbate peroxidase (APX), guaiacol peroxidase (POD), superoxide dismutase (SOD), and glutathione reductase (GR), leading to an increase in their activity. The plant also synthesizes additional antioxidants to counteract the oxidative stress caused by cadmium. Excessive cadmium exposure results in protein S-nitrosylation and the production of metallothioneins. Several signaling pathways are involved in the plant's response to cadmium stress, including mitogen-activated protein kinase (MAPK), calcium-dependent protein kinase (CDPK), and phosphoinositide 3-kinases (PI3 kinases). Cadmium stress also activates auxin-responsive promoters (DR5-GUS), auxin efflux carriers (PINs), and pathogen-related genes (PR genes). Phytochelatins, synthesized in response to cadmium stress, act as a defense system by sequestering cadmium in vacuoles.

7 Ca Calmodulin Pathway

The Ca calmodulin pathway plays a significant role in mitigating Cd toxicity in plants by activating defense mechanisms and regulating cellular processes. Exposure to Cd disrupts calcium homeostasis in plant cells, leading to an influx of calcium ions (Ca^{2+}). This influx can occur through various transporters, including those involved in the transport of zinc (Zn^{2+}), iron (Fe^{2+}), copper (Cu^{2+}), and calcium itself [79]. The increased cytosolic calcium levels in response to Cd stress lead to the activation of Ca calmodulin. Calcium ions bind to calmodulin, inducing a conformational change and activating its function as a signaling molecule [80]. Activated Ca calmodulin interacts with various target proteins, including enzymes, kinases, and transcription factors. This interaction triggers a cascade of downstream signaling events, resulting in the activation of defense responses [81]. The Ca calmodulin pathway can upregulate antioxidant systems in plants exposed to Cd. This includes the activation of antioxidant enzymes such as ascorbate peroxidase (APX), guaiacol peroxidase (POD), superoxide dismutase (SOD), and glutathione reductase (GR). These enzymes help scavenge and neutralize reactive oxygen species (ROS) generated during Cd stress, reducing oxidative damage [9]. The Ca calmodulin pathway influences gene expression patterns in response to Cd stress. It can activate transcription factors and regulatory proteins that control the expression of genes involved in Cd detoxification, metal chelation, and stress tolerance. This modulation of gene expression enables plants to enhance their defense mechanisms against Cd toxicity [82]. Phytochelatins are metal-binding peptides synthesized in response to Cd stress. The Ca calmodulin pathway can regulate the synthesis of phytochelatins, which act as chelators to sequester Cd ions and prevent their accumulation in vital cellular compartments [83]. The Ca calmodulin pathway interacts with other signaling pathways, such as MAPK (mitogen-activated protein kinase) and ROS signaling, to coordinate the plant's response to Cd stress. This cross-talk ensures a comprehensive and integrated defense response [84]. By activating antioxidant systems, modulating gene expression, and regulating phytochelatin synthesis, the Ca calmodulin pathway helps plants mitigate the toxic effects of Cd. It enables plants to enhance their tolerance to Cd stress and maintain cellular homeostasis, ultimately promoting their survival and growth in Cd-contaminated environments.

7.1 Influence of Plant Hormones

Plant hormones play a significant role in modulating the response of plants to Cd stress. Abscisic Acid (ABA) is known to be involved in regulating various physiological processes during Cd stress [72]. It helps in stomatal closure, reducing water loss and limiting Cd uptake through the roots. ABA also regulates antioxidant defense systems, enhancing the plant's ability to cope with oxidative stress induced by Cd. In soyabean crop, concentration of ABA was initially increased to highest during first 24 h but after 140 h ABA concentration decreased gradually [85]. Ethylene is involved in regulating plant growth and development under Cd stress. It promotes root growth and lateral root formation, facilitating Cd uptake and accumulation in roots. Ethylene also regulates the expression of genes involved in metal transport and detoxification processes. Salicylic Acid (SA) is a signaling molecule involved in plant defense responses against various stresses, including Cd stress. It plays a role in activating antioxidant defense systems, inducing the synthesis of phytochelatins, and regulating gene expression related to Cd detoxification. In different studies it has been observed that increase in SA concentration in plants under Cd stress [3, 86]. Jasmonic Acid (JA) is known for its role in plant defense against insect herbivory and pathogens, but it also plays a role in Cd stress response. It regulates the production of reactive oxygen species (ROS), activates defense genes, and enhances antioxidant capacity in plants exposed to Cd. It has been reported in several studies that concentration of JA was found to be increased in leaves of plants under Cd stress [86, 87]. Gibberellins (GA) has been reported to alleviate the inhibitory effects of Cd on plant growth. It promotes root elongation and enhances nutrient uptake, potentially reducing Cd uptake by roots. GA also regulates gene expression related to detoxification and antioxidative defense mechanisms. Concentration of gibberellic acid found to be decreased in leaves of plants under Cd stress reported under several studies [88–90]. Soaking barley seeds in GB3 reduces the adverse impact of Cd on coleoptile growth [91]. Cytokinins are involved in regulating cell division, growth, and stress responses in plants. Under Cd stress, cytokinins can modulate root development, promote lateral root formation, and enhance the plant's ability to tolerate Cd-induced oxidative stress. Many studies reported decrease in cytokinin concentration in plants under Cd stress [90, 92]. Auxins are crucial for plant growth and development, and they also play a role in Cd stress response. They regulate root system architecture, control nutrient uptake, and modulate the expression of genes involved in Cd transport and detoxification. In a study reported modification in expression of IAA biosynthetic genes and transporter genes under Cd stress and concentration of IAA was also found decreased in leaves [93].

7.2 Phytochelatin

Phytochelatins serve an important role in mitigating the toxicity of Cd. They act as natural chelators, binding to Cd ions and forming stable complexes, thereby reducing the availability and potential harm of Cd in plant cells [84]. By sequestering Cd, phytochelatins help prevent its accumulation and distribution to sensitive plant tissues, such as the roots and shoots. This detoxification mechanism assists in maintaining essential physiological processes and minimizing the negative impact of Cd on plant growth and development [94, 95]. The concentration of phytochelatin in rice crop was 23 and 47% under 1 and 2 mM Cd concentration. Studies have indicated that the addition of exogenous GSH leads to enhanced Cd accumulation in the roots of *B. campestris* plants exposed to 20 mM Cd, while simultaneously reducing its translocation to the shoots [95].

7.3 Proline Content

The presence of proline in plants plays a crucial role in alleviating the stress caused by Cd. When plants are exposed to Cd toxicity, they often encounter oxidative stress and cellular damage [39]. Proline, an amino acid, acts as an essential osmoprotectant and antioxidant to counteract such stress. Increased levels of proline aid plants in combating oxidative stress induced by Cd, as it functions as a scavenger for free radicals and promotes the stabilization of cellular structures. Additionally, proline contributes to the maintenance of cellular water balance and assists in regulating osmotic processes, thereby helping plants cope with Cd-induced osmotic stress [96]. Furthermore, proline actively participates in the regulation of gene expression associated with stress responses. It influences the expression of genes involved in responding to stress and facilitates the activation of defense mechanisms against Cd toxicity [97]. It has been reported in several studies that plants under high Cd exposure resulted in increased proline concentration in plants [96, 98]. The accumulation of proline in plants is acknowledged as a strategy to mitigate the effects of Cd stress by regulating osmotic potential and stabilizing membrane structures [99, 100].

8 Microorganism

Cd toxicity is mitigated by microorganism in many ways, either through production of chelating agents like siderophores and organic acids which help to immobilize Cd and reduce its bioavailability and potential toxicity. Various endophytic and rizosphere microbes plays a critical role in mitigating Cd stress through transformation of mobile Cd ion into immobile form, accumulation of Cd ion in body to avoid uptake of Cd through plant roots [101]. In heavy metal contaminated soils, the predominance of genera *bacillus*, *pseudomonas* and *athrobacter* is observed which includes firmicutes, proteobacteria and actinobacteria [102]. Certain strains of rhizospheric bacteria that are tolerant to heavy metals and promote plant growth, known as heavy metal tolerant plant growth promoting (HMR-PGP) strains, play a crucial role in regulating plant growth and modifying the physico-chemical characteristics of the soil. These strains have the ability to enhance the bioavailability of heavy metals in the soil, leading to the rapid detoxification and removal of these toxic elements [103]. It has been reported that Cd-tolerant endophytic bacteria increased the bio accumulation of Cd by *Brachiaria mutica* [104]. In a study it has been reported that the presence of an endophytic fungus (*Neotyphodium coenophialum*) has a positive impact on the growth of tall fescue (*Lolium arundinaceum*) under Cd-stress conditions. The endophytic fungus enhances the transport of Cd from the plant's roots to its shoots, resulting in increased Cd accumulation within the plant. This interaction between the fungus and the plant promotes the overall growth and development of the tall fescue in the presence of Cd stress [105]. After the assessment of the environmental adaptability of Cd, Pb, and Cu resistant bacteria found in the rhizospheric soil of *Boehmeria nivea* plants growing near mine refineries. Their findings revealed that rhizosphere bacteria belonging to the genera *Cupriavidus*, *Pseudomonas*, *Bacillus*, and *Acinetobacter* exhibited a remarkable tolerance to high concentrations of Cd, Pb, and Cu. These bacteria demonstrated their ability to thrive and survive in the presence of these heavy metals, indicating their adaptability to the contaminated environment [106]. Application of oxalic acid and citric acid on Brassica acid in Cd contaminated soil lowers the phyto availability of Cd [107]. Recent research indicates that investigations into the plant microbiome derived from polluted soils have the potential to enhance current phytoremediation techniques for the remediation of heavy metals [108].

9 Immune System Developed by Plants to Mitigate Cd Toxicity

Plants have receptor proteins which recognize and bind to Cd ions and after recognition triggers immune signaling pathways, resulted in activation of defense responses [109]. Specific genes responsible for defense against Cd stress, such genes encode for various defense proteins, enzymes, and signaling molecules that help alleviate Cd toxicity [110]. The triggered plant immune system in Cd polluted environment respond by increasing the production of antioxidants, such as catalase (CAT), superoxide dismutase (SOD) and glutathione peroxidase (GPX), to neutralize ROS [111]. Cd stress can trigger systemic responses in plants, directing to activation of induced systemic resistance (ISR). This response enhances the plant's overall resistance against Cd and other stresses by priming the immune system and activating defense mechanisms [112]. Activation of signaling network is the first step in Cd stress condition which resulted in release of root exudates consisting of several metabolites in plants such as, *Medicago polymorpha* and *Malva sylvestris* [113], *Sedum alfredii* [114].

Certain genes are responsible for regulation and plant tolerance to Cd stress in plants. In a study conducted on *Brassica napus*, genes IRT1, HMA2 and HMA4 found activated at different Cd concentrations varied from 0.03 to 0.3 mg l^{-1} and observed the role of these genes in transportation and accumulation of Cd [68]. Similar results were obtained by Yu et al. [115], Zhang et al. [116]. In Zea mays, gene ZmWRKY4 was found responsible for upregulating antioxidant defense system at Cd concentrations [117]. Increase in pectin methylesterase activity resulted in increase in ion binding sites to counteract Cd stress.

10 Conclusion

Cadmium (Cd) toxicity in plants induces a range of physiological and molecular alterations. These include the activation of metal chelation mechanisms, increased production of antioxidants, synthesis of phytochelatins, and modulation of gene expression. Plant–microbe interactions, particularly with Cd-tolerant rhizospheric bacteria and endophytic fungi, play a significant role in Cd detoxification and promoting plant growth. However, understanding on various causes of Cd toxicity and its tolerant and resistance mechanisms employed by plants and its ecosystem is yet incomplete. Hence, the research gap can be merged through following studies that can mitigate Cd contamination, safeguard plant health, and promote sustainable agricultural practices are:

- Identification and utilization of different strains of soil microbes which are efficient in mitigating Cd toxicity in plants.
- Identification of specific genes responsible for heritable changes in gene expression to halt Cd toxicity in plants.

- Correlation of different environmental parameters and soil characteristics which are responsible for Cd toxicity and development of management strategies accordingly.

References

1. Palansooriya, K. N., Shaheen, S. M., Chen, S. S., Tsang, D. C. W., Hashimoto, Y., Hou, D., Bolan, N. S., Rinklebe, J., & Ok, Y. S. (2020). Soil amendments for immobilization of potentially toxic elements in contaminated soils: A critical review. *Environment International, 134*. https://doi.org/10.1016/j.envint.2019.105046
2. Hussain, A., Ali, S., Rizwan, M., ur Rehman, M. Z., Qayyum, M. F., Wang, H., & Rinklebe, J. (2019). Responses of wheat (*Triticum aestivum*) plants grown in a Cd contaminated soil to the application of iron oxide nanoparticles. *Ecotoxicology and Environmental Safety, 173*, 156–164.https://doi.org/10.1016/j.ecoenv.2019.01.118
3. Rizwan, M., Ali, S., Adrees, M., Ibrahim, M., Tsang, D. C. W., Zia-ur-Rehman, M., Zahir, Z. A., Rinklebe, J., Tack, F. M. G., & Ok, Y. S. (2017). A critical review on effects, tolerance mechanisms and management of cadmium in vegetables. *Chemosphere, 182*, 90–105. https://doi.org/10.1016/j.chemosphere.2017.05.013
4. Bae, J., Benoit, D. L., & Watson, A. K. (2016). Effect of heavy metals on seed germination and seedling growth of common ragweed and roadside ground cover legumes. *Environmental Pollution, 213*, 112–118. https://doi.org/10.1016/j.envpol.2015.11.041
5. Bali, A. S., Sidhu, G. P. S., Kumar, V., & Bhardwaj, R. (2018). Mitigating cadmium toxicity in plants by phytohormones. In *Cadmium toxicity and tolerance in plants: From physiology to remediation* (pp. 375–396). Elsevier. https://doi.org/10.1016/B978-0-12-814864-8.00015-2
6. El Rasafi, T., Oukarroum, A., Haddioui, A., Song, H., Kwon, E. E., Bolan, N., Tack, F. M. G., Sebastian, A., Prasad, M. N. V., & Rinklebe, J. (2022). Cadmium stress in plants: A critical review of the effects, mechanisms, and tolerance strategies. *Critical Reviews in Environmental Science and Technology, 52*(5), 675–726. https://doi.org/10.1080/10643389.2020.1835435
7. Mengel, K., Kirkby, E. A., Kosegarten, H., & Appel, T. (2001). *Principles of plant nutrition* (5th ed.). Springer.
8. Cannon, H. L., Connally, G. G., Epstein, J. B., Parker, J. G., Thornton, I., & Wixson, G. (1978). Geological sources of most trace elements. *Geochem Environment, 3*, 17–31.
9. Gallego, S. M., Pena, L. B., Barcia, R. A., Azpilicueta, C. E., Iannone, M. F., Rosales, E. P., Zawoznik, M. S., Groppa, M. D., & Benavides, M. P. (2012). Unravelling cadmium toxicity and tolerance in plants: Insight into regulatory mechanisms. *Environmental and Experimental Botany, 83*, 33–46. https://doi.org/10.1016/j.envexpbot.2012.04.006
10. Pierre, W. H., & Norman, A. G. (1954). Soil and fertilizer phosphorus in crop nutrition. *Soil Science, 77*(2), 171. https://doi.org/10.1097/00010694-195402000-00025
11. Szolnoki, Z., Farsang, A., & Puskás, I. (2013). Cumulative impacts of human activities on urban garden soils: Origin and accumulation of metals. *Environmental Pollution, 177*, 106–115. https://doi.org/10.1016/j.envpol.2013.02.007
12. Mathys, W. (1975). Enzymes of heavy-metal-resistant and non-resistant populations of *Silene cucubalus* and their interaction with some heavy metals *in vitro* and *in vivo*. *Physiologia Plantarum, 33*(2), 161–165. https://doi.org/10.1111/j.1399-3054.1975.tb03785.x
13. Alcántara, E., Romera, F. J., Cañete, M., & De La Guardia, M. D. (1994). Effects of heavy metals on both induction and function of root fe(lll) reductase in fe-deficient cucumber (*Cucumis sativus* L.) plants. *Journal of Experimental Botany, 45*(12), 1893–1898. https://doi.org/10.1093/jxb/45.12.1893
14. Menon, P., Joshi, N., & Joshi, A. (2016). Effect of heavy metals on seed germination of *Trigonella foenum-graceum* L. *International Journal of Life-Sciences Scientific Research, 2*(4), 488–493. https://doi.org/10.21276/ijlssr.2016.2.4.27

15. Cho, U. H., & Seo, N. H. (2005). Oxidative stress in Arabidopsis thaliana exposed to cadmium is due to hydrogen peroxide accumulation. *Plant Science, 168*(1), 113–120. https://doi.org/10.1016/j.plantsci.2004.07.021
16. Mongkhonsin, B., Nakbanpote, W., Meesungnoen, O., & Prasad, M. N. V. (2018). Adaptive and tolerance mechanisms in herbaceous plants exposed to cadmium. In *Cadmium toxicity and tolerance in plants: From physiology to remediation.* Elsevier Inc. https://doi.org/10.1016/B978-0-12-814864-8.00004-8
17. Gill, S. S., Khan, N. A., & Tuteja, N. (2011). Differential cadmium stress tolerance in five Indian mustard (*Brassica juncea* L) cultivars: An evaluation of the role of antioxidant machinery. *Plant Signaling and Behavior, 6*(2), 293–300. https://doi.org/10.4161/psb.6.2.15049
18. John, R., Ahmad, P., Gadgil, K., & Sharma, S. (2009). Heavy metal toxicity: Effect on plant growth, biochemical parameters and metal accumulation by *Brassica juncea* L. *International Journal of Plant Production, 3*(3), 65–75.
19. Khan, Z. I., Ashraf, M., Ahmad, K., & Akram, N. A. (2011). A study on the transfer of cadmium from soil to pasture under semi-arid conditions in Sargodha, Pakistan. *Biological Trace Element Research, 142*(2), 143–147. https://doi.org/10.1007/s12011-010-8755-5
20. Anjum, S. A., Ashraf, U., Khan, I., Tanveer, M., Ali, M., Hussain, I., & Wang, L. C. (2016). Chromium and aluminum phytotoxicity in maize: Morpho-physiological responses and metal uptake. *Clean—Soil, Air, Water, 44*(8), 1075–1084. https://doi.org/10.1002/clen.201500532
21. Hasanuzzaman, M., Nahar, K., & Fujita, M. (2018). *Plants under metal and metalloid stress: Responses, tolerance and remediation* (pp. 1–424).https://doi.org/10.1007/978-981-13-2242-6
22. Marzban, L., Akhzari, D., Ariapour, A., Mohammadparast, B., & Pessarakli, M. (2017). Effects of cadmium stress on seedlings of various rangeland plant species (*Avena fatua* L., *Lathyrus sativus* L., and *Lolium temulentum* L.): Growth, physiological traits, and cadmium accumulation. *Journal of Plant Nutrition, 40*(15), 2127–2137. https://doi.org/10.1080/01904167.2016.1269347
23. Redondo-Gómez, S., Mateos-Naranjo, E., & Andrades-Moreno, L. (2010). Accumulation and tolerance characteristics of cadmium in a halophytic Cd-hyperaccumulator, Arthrocnemum macrostachyum. *Journal of Hazardous Materials, 184*(1–3), 299–307. https://doi.org/10.1016/j.jhazmat.2010.08.036
24. Younis, U., Malik, S. A., Qayyum, M. F., Raza Shah, M. H., Shahzad, A. N., & Mahmood, S. (2015). Biochar affects growth and biochemical activities of fenugreek (*Trigonella corniculata*) in cadmium polluted soil. *Journal of Applied Botany and Food Quality, 88*(February), 29–33. https://doi.org/10.5073/JABFQ.2015.088.006
25. Sharma, R. K., Agrawal, M., & Agrawal, S. B. (2010). Physiological and biochemical responses resulting from cadmium and zinc accumulation in carrot plants. *Journal of Plant Nutrition, 33*(7), 1066–1079. https://doi.org/10.1080/01904161003729774
26. Zhang, L., Zhang, H., Guo, W., Tian, Y., Chen, Z., & Wei, X. (2012). Photosynthetic responses of energy plant maize under cadmium contamination stress. *Advanced Materials Research, 356–360*, 283–286. https://doi.org/10.4028/www.scientific.net/AMR.356-360.283
27. Younis, U., Malik, S. A., Rizwan, M., Qayyum, M. F., Ok, Y. S., Shah, M. H. R., Rehman, R. A., & Ahmad, N. (2016). Biochar enhances the cadmium tolerance in spinach (*Spinacia oleracea*) through modification of Cd uptake and physiological and biochemical attributes. *Environmental Science and Pollution Research, 23*(21), 21385–21394. https://doi.org/10.1007/s11356-016-7344-3
28. Andrade, W. V., de Oliveira Neto, C. F., dos Santos Filho, B. G., do Amarante, C. B., Cruz, E. D., Okumura, R. S., Correa Barbosa, A. V., Palheta de Sousa, D. J., Silva Teixeira, J. S., & de Santana Botelho, A. (2019). Effect of cadmium on young plants of Virola surinamensis. *AoB PLANTS, 11*(3), 1–11.https://doi.org/10.1093/aobpla/plz022
29. Vassilev, A., Lidon, F. C., Matos, M. D. C., Ramalho, J. C., & Yordanov, I. (2002). Photosynthetic performance and content of some nutrients in cadmium- and copper-treated barley plants. *Journal of Plant Nutrition, 25*(11), 2343–2360. https://doi.org/10.1081/PLN-120014699

30. Molins, H., Michelet, L., Lanquar, V., Agorio, A., Giraudat, J., Roach, T., Krieger-Liszkay, A., & Thomine, S. (2013). Mutants impaired in vacuolar metal mobilization identify chloroplasts as a target for cadmium hypersensitivity in *Arabidopsis thaliana. Plant, Cell and Environment, 36*(4), 804–817. https://doi.org/10.1111/pce.12016
31. Sharavanan, P. S., Ravimycin, T., Anbazhagan, C., & Selvaraju, M. (2007). Toxic effect of cadmium on seed germination seedling growth and biochemical changes of cowpea [Vigna unguiculata (L.) Walp]. *Plant Archives, 7*(1), 219–221.
32. Yadav, M. (2013). Chlorophyll content of gram (*Cicer arietinum* L.) Under cadmium and mercury treatments. *International Journal of Chemical Sciences, 11*(3), 1486–1492.
33. Zengin, F. K., & Munzuroglu, O. (2005). Effects of some heavy metals on content of chlorophyll, proline and some antioxidant chemicals in bean (*Phaseolus vulgaris* L.) seedlings. *Acta Biologica Cracoviensia Series Botanica, 47*(2), 157–164.
34. Zhang, J. B., & Huang, W. N. (2007). Effects of cadmium stress on photosynthetic functions of strawberry. *The Journal of Applied Ecology, 18*(7), 1673–1676.
35. Ling, T., Gao, Q., Du, H., Zhao, Q., & Ren, J. (2017). Growing, physiological responses and Cd uptake of Corn (*Zea mays* L.) under different Cd supply. *Chemical Speciation and Bioavailability, 29*(1), 216–221. https://doi.org/10.1080/09542299.2017.1400924
36. Elloumi, N., Zouari, M., Chaari, L., Jomni, C., Ben Rouina, B., & Ben Abdallah, F. (2014). Ecophysiological responses of almond (*Prunus dulcis*) seedlings to cadmium stress. *Biologia (Poland), 69*(5), 604–609. https://doi.org/10.2478/s11756-014-0348-x
37. Nikolić, N., Zorić, L., Cvetković, I., Pajević, S., Borišev, M., Orlović, S., & Pilipović, A. (2017). Assessment of cadmium tolerance and phytoextraction ability in young populus deltoides L. And populus × euramericana plants through morpho-anatomical and physiological responses to growth in cadmium enriched soil. *IForest, 10*(3), 635–644. https://doi.org/10.3832/ifor2165-010
38. Silva, J. R. R., Fernandes, A. R., Silva Junior, M. L., Santos, C. R. C., & Lobato, A. K. S. (2018). Tolerance mechanisms in Cassia alata exposed to cadmium toxicity—Potential use for phytoremediation. *Photosynthetica, 56*(2), 495–504. https://doi.org/10.1007/s11099-017-0698-z
39. Zouari, M., Ben Ahmed, C., Elloumi, N., Bellassoued, K., Delmail, D., Labrousse, P., Ben Abdallah, F., & Ben Rouina, B. (2016). Impact of proline application on cadmium accumulation, mineral nutrition and enzymatic antioxidant defense system of Olea europaea L. cv Chemlali exposed to cadmium stress. *Ecotoxicology and Environmental Safety, 128*, 195–205. https://doi.org/10.1016/j.ecoenv.2016.02.024
40. Mukherjee, M. (2017). Lead and cadmium toxicity on seedling growth and metabolism of *Trigonellafoenum-graecum* L. *International Journal of Science and Research (IJSR), 6*(7), 1685–1689. https://doi.org/10.21275/art20175574
41. Kalai, T., Khamassi, K., Teixeira da Silva, J. A., Gouia, H., & Bettaieb Ben-Kaab, L. (2014). Cadmium and copper stress affect seedling growth and enzymatic activities in germinating barley seeds. *Archives of Agronomy and Soil Science, 60*(6), 765–783. https://doi.org/10.1080/03650340.2013.838001
42. Meng, Y., Zhang, L., Wang, L., Zhou, C., Shangguan, Y., & Yang, Y. (2019). Antioxidative enzymes activity and thiol metabolism in three leafy vegetables under Cd stress. *Ecotoxicology and Environmental Safety, 173*(December 2018), 214–224. https://doi.org/10.1016/j.ecoenv.2019.02.026
43. Bauddh, K., & Singh, R. P. (2012). Growth, tolerance efficiency and phytoremediation potential of *Ricinus communis* (L.) and *Brassica juncea* (L.) in salinity and drought affected cadmium contaminated soil. *Ecotoxicology and Environmental Safety, 85*, 13–22. https://doi.org/10.1016/j.ecoenv.2012.08.019
44. Nada, E., Ferjani, B. A., Ali, R., Bechir, B. R., Imed, M., & Makki, B. (2007). Cadmium-induced growth inhibition and alteration of biochemical parameters in almond seedlings grown in solution culture. *Acta Physiologiae Plantarum, 29*(1), 57–62. https://doi.org/10.1007/s11738-006-0009-y

45. Anjum, N. A., Umar, S., & Iqbal, M. (2014). Assessment of cadmium accumulation, toxicity, and tolerance in Brassicaceae and Fabaceae plants—Implications for phytoremediation. *Environmental Science and Pollution Research, 21*(17), 10286–10293. https://doi.org/10.1007/s11356-014-2889-5
46. Mobin, M., & Khan, N. A. (2007). Photosynthetic activity, pigment composition and antioxidative response of two mustard (*Brassica juncea*) cultivars differing in photosynthetic capacity subjected to cadmium stress. *Journal of Plant Physiology, 164*(5), 601–610. https://doi.org/10.1016/j.jplph.2006.03.003
47. Irfan, M., Ahmad, A., & Hayat, S. (2014). Effect of cadmium on the growth and antioxidant enzymes in two varieties of *Brassica juncea. Saudi Journal of Biological Sciences, 21*(2), 125–131. https://doi.org/10.1016/j.sjbs.2013.08.001
48. Amirjani, M. R., Mohammad, I., & Amirjani, R. (2012). Effects of cadmium on wheat growth and some physiological factors. *International Journal of Forest, Soil and Erosion, 2*(1), 50–58. www.ijfse.com
49. Fornazier, R. F., Ferreira, R. R., Pereira, G. J. G., Molina, S. M. G., John Smith, R., Lea, P. J., & Azevedo, R. A. (2002). Cadmium stress in sugar cane callus cultures: Effect on antioxidant enzymes. *Plant Cell, Tissue and Organ Culture, 71*, 125–131.
50. Zafar-Ul-hye, M., Naeem, M., Danish, S., Fahad, S., Datta, R., Abbas, M., Rahi, A. A., Brtnicky, M., Holátko, J., Tarar, Z. H., & Nasir, M. (2020). Alleviation of cadmium adverse effects by improving nutrients uptake in bitter gourd through cadmium tolerant rhizobacteria. *Environments—MDPI, 7*(8), 1–16. https://doi.org/10.3390/environments7080054
51. An, Y. J. (2004). Soil ecotoxicity assessment using cadmium sensitive plants. *Environmental Pollution, 127*(1), 21–26. https://doi.org/10.1016/S0269-7491(03)00263-X
52. De Maria, S., Puschenreiter, M., & Rivelli, A. R. (2013). Cadmium accumulation and physiological response of sunflower plants to Cd during the vegetative growing cycle. *Plant, Soil and Environment, 59*(6), 254–261. https://doi.org/10.17221/788/2012-pse
53. Lentini, M., De Lillo, A., Paradisone, V., Liberti, D., Landi, S., & Esposito, S. (2018). Early responses to cadmium exposure in barley plants: Effects on biometric and physiological parameters. *Acta Physiologiae Plantarum, 40*(10).https://doi.org/10.1007/s11738-018-2752-2
54. Zhu, G., Xiao, H., Guo, Q., Zhang, Z., Zhao, J., & Yang, D. (2018). Effects of cadmium stress on growth and amino acid metabolism in two Compositae plants. *Ecotoxicology and Environmental Safety, 158*(April), 300–308. https://doi.org/10.1016/j.ecoenv.2018.04.045
55. Chen, Y. X., He, Y. F., Luo, Y. M., Yu, Y. L., Lin, Q., & Wong, M. H. (2003). Physiological mechanism of plant roots exposed to cadmium. *Chemosphere, 50*(6), 789–793. https://doi.org/10.1016/S0045-6535(02)00220-5
56. Dong, J., Wu, F. B., & Zhang, G. P. (2005). Effect of cadmium on growth and photosynthesis of tomato seedlings. *Journal of Zhejiang University: Science, 6B*(10), 974–980. https://doi.org/10.1631/jzus.2005.B0974
57. Zhou, Z. S., Song, J. B., & Yang, Z. M. (2012). Genome-wide identification of *Brassica napus* microRNAs and their targets in response to cadmium. *Journal of Experimental Botany, 63*(12), 4597–4613. https://doi.org/10.1093/jxb/err313
58. Wang, M., Chen, L., Chen, S., & Ma, Y. (2012). Alleviation of cadmium-induced root growth inhibition in crop seedlings by nanoparticles. *Ecotoxicology and Environmental Safety, 79*, 48–54. https://doi.org/10.1016/j.ecoenv.2011.11.044
59. Soudek, P., Petrová, Š, Vaňková, R., Song, J., & Vaněk, T. (2014). Accumulation of heavy metals using Sorghum sp. *Chemosphere, 104*, 15–24. https://doi.org/10.1016/j.chemosphere.2013.09.079
60. Guo, J., Qin, S., Rengel, Z., Gao, W., Nie, Z., Liu, H., Li, C., & Zhao, P. (2019). Cadmium stress increases antioxidant enzyme activities and decreases endogenous hormone concentrations more in Cd-tolerant than Cd-sensitive wheat varieties. *Ecotoxicology and Environmental Safety, 172*(63), 380–387. https://doi.org/10.1016/j.ecoenv.2019.01.069

61. Zayneb, C., Bassem, K., Zeineb, K., Grubb, C. D., Noureddine, D., Hafedh, M., & Amine, E. (2015). Physiological responses of fenugreek seedlings and plants treated with cadmium. *Environmental Science and Pollution Research, 22*(14), 10679–10689. https://doi.org/10.1007/s11356-015-4270-8
62. Ismael, M. A., Elyamine, A. M., Moussa, M. G., Cai, M., Zhao, X., & Hu, C. (2019). Cadmium in plants: Uptake, toxicity, and its interactions with selenium fertilizers. *Metallomics, 11*(2), 255–277. https://doi.org/10.1039/c8mt00247a
63. Qadir, S., Qureshi, M. I., Javed, S., & Abdin, M. Z. (2004). Genotypic variation in phytoremediation potential of *Brassica juncea* cultivars exposed to Cd stress. *Plant Science, 167*(5), 1171–1181. https://doi.org/10.1016/j.plantsci.2004.06.018
64. Adeniji, BA, Budimir-Hussey, M. T., & Macfie, S. (2010). Production of organic acids and adsorption of Cd on roots of durum wheat (*Triticum turgidum* L. var. durum). *Acta Physiologiae Plantarum, 32*, 1063–1072.
65. Sebastian, A., & Prasad, M. N. V. (2019). Photosynthetic light reactions in *Oryza sativa* L. under Cd stress: Influence of iron, calcium, and zinc supplements. *Eurobiotech Journal, 3*(4), 175–181. https://doi.org/10.2478/ebtj-2019-0021
66. Song, Y., Jin, L., & Wang, X. (2017). Cadmium absorption and transportation pathways in plants. *International Journal of Phytoremediation, 19*(2), 133–141. https://doi.org/10.1080/15226514.2016.1207598
67. Thakur, S., Singh, L., Wahid, Z. A., Siddiqui, M. F., Atnaw, S. M., & Din, M. F. M. (2016). Plant-driven removal of heavy metals from soil: Uptake, translocation, tolerance mechanism, challenges, and future perspectives. *Environmental Monitoring and Assessment, 188*(4). https://doi.org/10.1007/s10661-016-5211-9
68. Wu, Z., Zhao, X., Sun, X., Tan, Q., Tang, Y., Nie, Z., & Hu, C. (2015). Xylem transport and gene expression play decisive roles in cadmium accumulation in shoots of two oilseed rape cultivars (*Brassica napus*). *Chemosphere, 119*, 1217–1223. https://doi.org/10.1016/j.chemosphere.2014.09.099
69. DalCorso, G., Farinati, S., & Furini, F. (2010). Regulatory networks of cadmium stress in plants. *Plant Signaling & Behavior, 5*(6), 663–667. https://doi.org/10.4161/psb.5.6.11425
70. Ghosh, R., & Roy, S. (2019). Cadmium toxicity in plants: Unveiling the physicochemical and molecular aspects. In *Cadmium tolerance in plants* (pp. 223–246). https://doi.org/10.1016/B978-0-12-815794-7.00008-4
71. Adrees, M., Ali, S., Rizwan, M., Zia-ur-Rehman, M., Ibrahim, M., Abbas, F., Farid, M., Qayyum, M. F., & Irshad, M. K. (2015). Mechanisms of silicon-mediated alleviation of heavy metal toxicity in plants: A review. *Ecotoxicology and Environmental Safety, 119*, 186–197. https://doi.org/10.1016/j.ecoenv.2015.05.011
72. Shahid, M., Niazi, N. K., Rinklebe, J., Bundschuh, J., Dumat, C., & Pinelli, E. (2020). Trace elements-induced phytohormesis: A critical review and mechanistic interpretation. *Critical Reviews in Environmental Science and Technology, 50*(19), 1984–2015. https://doi.org/10.1080/10643389.2019.1689061
73. Lal, N. (2010). Molecular mechanisms and genetic basis of heavy metal toxicity and tolerance in plants. In *Plant adaptation and phytoremediation*.https://doi.org/10.1007/978-90-481-9370-7
74. Shanmugaraj, B. M., Malla, A., & Ramalingam, S. (2018). Cadmium stress and toxicity in plants: An overview. In *Cadmium toxicity and tolerance in plants: From physiology to remediation*. Elsevier Inc. https://doi.org/10.1016/B978-0-12-814864-8.00001-2
75. Verma, V., Ravindran, P., & Kumar, P. P. (2016). Plant hormone-mediated regulation of stress responses. *BMC Plant Biology, 16*(1), 1–10. https://doi.org/10.1186/s12870-016-0771-y
76. Kumar, A., & Prasad, M. N. V. (2018). Plant-lead interactions: Transport, toxicity, tolerance, and detoxification mechanisms. *Ecotoxicology and Environmental Safety, 166*(September), 401–418. https://doi.org/10.1016/j.ecoenv.2018.09.113
77. He, M., He, C. Q., & Ding, N. Z. (2018). Abiotic stresses: General defenses of land plants and chances for engineering multistress tolerance. *Frontiers in Plant Science, 871*(December), 1–18. https://doi.org/10.3389/fpls.2018.01771

78. Vishwakarma, K., Upadhyay, N., Kumar, N., Yadav, G., Singh, J., Mishra, R. K., Kumar, V., Verma, R., Upadhyay, R. G., Pandey, M., & Sharma, S. (2017). Abscisic acid signaling and abiotic stress tolerance in plants: A review on current knowledge and future prospects. *Frontiers in Plant Science, 8*(February), 1–12. https://doi.org/10.3389/fpls.2017.00161
79. Lin, Y. F., & Aarts, M. G. M. (2012). The molecular mechanism of zinc and cadmium stress response in plants. *Cellular and Molecular Life Sciences, 69*(19), 3187–3206. https://doi.org/10.1007/s00018-012-1089-z
80. Liu, Y., Liu, L., Qi, J., Dang, P., & Xia, T. (2019). Cadmium activates ZmMPK3-1 and ZmMPK6-1 via induction of reactive oxygen species in maize roots. *Biochemical and Biophysical Research Communications, 516*(3), 747–752. https://doi.org/10.1016/j.bbrc.2019.06.116
81. Berni, R., Luyckx, M., Xu, X., Legay, S., Sergeant, K., Hausman, J. F., Lutts, S., Cai, G., & Guerriero, G. (2019). Reactive oxygen species and heavy metal stress in plants: Impact on the cell wall and secondary metabolism. *Environmental and Experimental Botany, 161*(October 2018), 98–106. https://doi.org/10.1016/j.envexpbot.2018.10.017
82. Jagodzik, P., Tajdel-Zielinska, M., Ciesla, A., Marczak, M., & Ludwikow, A. (2018). Mitogen-activated protein kinase cascades in plant hormone signaling. *Frontiers in Plant Science, 9*(October), 1–26. https://doi.org/10.3389/fpls.2018.01387
83. Ahmad, J., Ali, A. A., Baig, M. A., Iqbal, M., Haq, I., & Irfan Qureshi, M. (2018). Role of phytochelatins in cadmium stress tolerance in plants. In *Cadmium toxicity and tolerance in plants: From physiology to remediation*. Elsevier Inc. https://doi.org/10.1016/B978-0-12-814864-8.00008-5
84. Lux, A., Martinka, M., Vaculík, M., & White, P. J. (2011). Root responses to cadmium in the rhizosphere: A review. *Journal of Experimental Botany, 62*(1), 21–37. https://doi.org/10.1093/jxb/erq281
85. Pérez Chaca, M. V., Vigliocco, A., Reinoso, H., Molina, A., Abdala, G., Zirulnik, F., & Pedranzani, H. (2014). Effects of cadmium stress on growth, anatomy and hormone contents in Glycine max (L.) Merr. *Acta Physiologiae Plantarum, 36*(10), 2815–2826. https://doi.org/10.1007/s11738-014-1656-z
86. Asgher, M., Khan, M. I. R., Anjum, N. A., & Khan, N. A. (2015). Minimising toxicity of cadmium in plants—Role of plant growth regulators. *Protoplasma, 252*(2), 399–413. https://doi.org/10.1007/s00709-014-0710-4
87. Chen, J., Yan, Z., & Li, X. (2014). Effect of methyl jasmonate on cadmium uptake and antioxidative capacity in *Kandelia obovata* seedlings under cadmium stress. *Ecotoxicology and Environmental Safety, 104*(1), 349–356. https://doi.org/10.1016/j.ecoenv.2014.01.022
88. Bulak, P., Walkiewicz, A., & Ska, M. B. Ň. (2011). Plant growth regulators-assisted phytoextraction. *X*, 1–8.
89. Ito, S., Yamagami, D., Umehara, M., Hanada, A., Yoshida, S., Sasaki, Y., Yajima, S., Kyozuka, J., Ueguchi-Tanaka, M., Matsuoka, M., Shirasu, K., Yamaguchi, S., & Asami, T. (2017). Regulation of strigolactone biosynthesis by gibberellin signaling. *Plant Physiology, 174*(2). https://doi.org/10.1104/pp.17.00301
90. Qin, S., Liu, H., Nie, Z., Rengel, Z., Gao, W., Li, C., & Zhao, P. (2020). Toxicity of cadmium and its competition with mineral nutrients for uptake by plants: A review. *Pedosphere, 30*(2), 168–180. https://doi.org/10.1016/S1002-0160(20)60002-9
91. Munzuroglu, O., & Zengin, F. K. (2006). Effect of cadmium on germination, coleoptile and root growth of barley seeds in the presence of gibberellic acid and kinetin. *Journal of Environmental Biology, 27*(4), 671–677.
92. Gangwar, S., Singh, V. P., Tripathi, D. K., Chauhan, D. K., Prasad, S. M., & Maurya, J. N. (2014). Plant responses to metal stress: The emerging role of plant growth hormones in toxicity alleviation. In *Emerging technologies and management of crop stress tolerance* (Second Edi, Vol. 2). Elsevier Inc. https://doi.org/10.1016/B978-0-12-800875-1.00010-7
93. Yang, L. P., Zhu, J., Wang, P., Zeng, J., Tan, R., Yang, Y. Z., & Liu, Z. M. (2018). Effect of Cd on growth, physiological response, Cd subcellular distribution and chemical forms of *Koelreuteria paniculata*. *Ecotoxicology and Environmental Safety, 160*(May), 10–18. https://doi.org/10.1016/j.ecoenv.2018.05.026

94. Bhuyan, M. H. M. B., Parvin, K., Mohsin, S. M., Al Mahmud, J., Hasanuzzaman, M., & Fujita, M. (2020). Modulation of cadmium tolerance in rice: Insight into vanillic acid-induced upregulation of antioxidant defense and glyoxalase systems. *Plants, 9*(2), 1–21. https://doi.org/10.3390/plants9020188
95. Zhu, Q., Zhang, J., Yu, H., Li, L., Chen, X., Jiang, M., & Tan, M. (2019). Maize Cd-tolerant ZmVTE4 encoding γ-tocopherol-methyl-transferase alleviated Cd-toxicity through its product α-tocopherol. *Environmental and Experimental Botany, 158*, 171–179. https://doi.org/10.1016/j.envexpbot.2018.11.019
96. Rady, M. M., Elrys, A. S., Abo El-Maati, M. F., & Desoky, E. S. M. (2019). Interplaying roles of silicon and proline effectively improve salt and cadmium stress tolerance in Phaseolus vulgaris plant. *Plant Physiology and Biochemistry, 139*(April), 558–568. https://doi.org/10.1016/j.plaphy.2019.04.025
97. Kishor, P. B. K., Hima Kumari, P., Sunita, M. S. L., & Sreenivasulu, N. (2015). Role of proline in cell wall synthesis and plant development and its implications in plant ontogeny. *Frontiers in Plant Science, 6*(JULY), 1–17. https://doi.org/10.3389/fpls.2015.00544
98. Hatamian, M., Nejad, A. R., Kafi, M., Souri, M. K., & Shahbazi, K. (2020). Nitrate improves hackberry seedling growth under cadmium application. *Heliyon, 6*(1). https://doi.org/10.1016/j.heliyon.2020.e03247
99. Amari, T., Ghnaya, T., & Abdelly, C. (2017). Nickel, cadmium and lead phytotoxicity and potential of halophytic plants in heavy metal extraction. *South African Journal of Botany, 111*, 99–110. https://doi.org/10.1016/j.sajb.2017.03.011
100. Semida, W. M., Hemida, K. A., & Rady, M. M. (2018). Sequenced ascorbate-proline-glutathione seed treatment elevates cadmium tolerance in cucumber transplants. *Ecotoxicology and Environmental Safety, 154*(December 2017), 171–179. https://doi.org/10.1016/j.ecoenv.2018.02.036
101. Mishra, J., Singh, R., & Arora, N. K. (2017). Alleviation of heavy metal stress in plants and remediation of soil by rhizosphere microorganisms. *Frontiers in Microbiology, 8*(SEP), 1–7. https://doi.org/10.3389/fmicb.2017.01706
102. Pires, C., Franco, A. R., Pereira, S. I. A., Henriques, I., Correia, A., Magan, N., & Castro, P. M. L. (2017). Metal(loid)-contaminated soils as a source of culturable heterotrophic aerobic bacteria for remediation applications. *Geomicrobiology Journal, 34*(9), 760–768. https://doi.org/10.1080/01490451.2016.1261968
103. Seneviratne, M., Seneviratne, G., Madawala, H., & Vithanage, M. (2017). *Role of rhizospheric microbes in heavy metal uptake by plants*. Springer International Publishing AG. https://doi.org/10.1007/978-3-319-49727-3
104. Shahabivand, S., Parvaneh, A., & Aliloo, A. A. (2020). Different response of Alyssum montanum and Helianthus annuus to cadmium bioaccumulation mediated by the endophyte fungus Serendipita indica. *Acta Ecologica Sinica, 40*(4), 315–322. https://doi.org/10.1016/J.CHNAES.2019.09.002
105. Ren, A., Li, C., & Gao, Y. (2011). Endophytic fungus improves growth and metal uptake of *Lolium arundinaceum* Darbyshire ex. Schreb. *International Journal of Phytoremediation, 13*(3), 233–243. https://doi.org/10.1080/15226511003671387
106. Jiang, J., Pan, C., Xiao, A., Yang, X., & Zhang, G. (2017). Isolation, identification, and environmental adaptability of heavy-metal-resistant bacteria from ramie rhizosphere soil around mine refinery. *3 Biotech, 7*(1), 3–8. https://doi.org/10.1007/s13205-017-0603-2
107. Hussain, A., Amna, Kamran, M. A., Javed, M. T., Hayat, K., Farooq, M. A., Ali, N., Ali, M., Manghwar, H., Jan, F., & Chaudhary, H. J. (2019). Individual and combinatorial application of Kocuria rhizophila and citric acid on phytoextraction of multi-metal contaminated soils by Glycine max L. *Environmental and Experimental Botany, 159*(October 2018), 23–33. https://doi.org/10.1016/j.envexpbot.2018.12.006
108. Thijs, S., Sillen, W., Weyens, N., & Vangronsveld, J. (2017). Phytoremediation: State-of-the-art and a key role for the plant microbiome in future trends and research prospects. *International Journal of Phytoremediation, 19*(1), 23–38. https://doi.org/10.1080/15226514.2016.1216076

109. Fu, H., Yu, H., Li, T., & Zhang, X. (2018). Influence of cadmium stress on root exudates of high cadmium accumulating rice line (*Oryza sativa* L.). *Ecotoxicology and Environmental Safety, 150*(September 2017), 168–175. https://doi.org/10.1016/j.ecoenv.2017.12.014
110. Tameling, W. I. L., & Takken, F. L. W. (2008). Resistance proteins: Scouts of the plant innate immune system. *European Journal of Plant Pathology, 121*(3), 243–255. https://doi.org/10.1007/s10658-007-9187-8
111. Nabi, R. B. S., Tayade, R., Hussain, A., Kulkarni, K. P., Imran, Q. M., Mun, B. G., & Yun, B. W. (2019). Nitric oxide regulates plant responses to drought, salinity, and heavy metal stress. *Environmental and Experimental Botany, 161*(January), 120–133. https://doi.org/10.1016/j.envexpbot.2019.02.003
112. Dalcorso, G., Farinati, S., Furini, A., Dalcorso, G., Farinati, S., & Furini, A. (2016). Regulatory networks of cadmium stress in plants. *Plant Signaling & Behavior, 2324*(October), 663–667.
113. Montiel-Rozas, M. M., Madejón, E., & Madejón, P. (2016). Effect of heavy metals and organic matter on root exudates (low molecular weight organic acids) of herbaceous species: An assessment in sand and soil conditions under different levels of contamination. *Environmental Pollution, 216*, 273–281. https://doi.org/10.1016/j.envpol.2016.05.080
114. Luo, Q., Sun, L., Hu, X., & Zhou, R. (2014). The variation of root exudates from the hyperaccumulator Sedum alfredii under cadmium stress: Metabonomics analysis. *PLoS ONE, 9*(12), 1–16. https://doi.org/10.1371/journal.pone.0115581
115. Yu, R., Li, D., Du, X., Xia, S., Liu, C., & Shi, G. (2017). Comparative transcriptome analysis reveals key cadmium transport-related genes in roots of two pak choi (*Brassica rapa* L. ssp. chinensis) cultivars. *BMC Genomics, 18*(1), 1–14. https://doi.org/10.1186/s12864-017-3973-2
116. Zhang, X. D., Meng, J. G., Zhao, K. X., Chen, X., & Yang, Z. M. (2018). Annotation and characterization of Cd-responsive metal transporter genes in rapeseed (*Brassica napus*). *BioMetals, 31*(1), 107–121. https://doi.org/10.1007/s10534-017-0072-4
117. Hong, C., Cheng, D., Zhang, G., Zhu, D., Chen, Y., & Tan, M. (2017). The role of ZmWRKY4 in regulating maize antioxidant defense under cadmium stress. *Biochemical and Biophysical Research Communications, 482*(4), 1504–1510. https://doi.org/10.1016/j.bbrc.2016.12.064

Agronomic Performance of Mung Bean as Affected by Basal Zinc Supply and Cadmium Contamination

Md Harunur Rashid, Mohammad Mahmudur Rahman, and Ravi Naidu

Abstract Zinc (Zn) deficiency and cadmium (Cd) toxicity in soils are both responsible for poor agronomic performance of crops. Mung bean was grown in a greenhouse under Cd contamination. Zn was applied to reduce the adverse effect of Cd on growth and yield of Mung beans. Zn and Cd rates (mg kg^{-1}) were 0, 5, 10, 20 and 0, 2, 4, respectively. Basal Zn supply boosted about 26% of yield and replenished 30% yield loss of Mung beans due to Cd contamination. Approximately 35% decline of grain yield was found due to Cd contamination in the soils as compared to Cd control. The decrease in grain yield per plant under Cd toxicity conditions was the result of reduced number of seeds per pod, 100 seed weight and number of seeds per plant. Improved grain yield was due to a significant increase in the number of pods, number of grains per pod and 100 grain weight owing to enhanced chlorophyll pigments in Mung bean leaves. Among the basal Zn application @ 5 mg Zn kg^{-1} soil was the best viable option to obtain a larger yield of Mung beans for Cd contaminated soils.

Keywords Mung beans · Zinc · Cadmium · Dermosol · Tenosol · Agronomic biofortification

M. H. Rashid (✉) · M. M. Rahman · R. Naidu
Global Centre for Environmental Remediation (GCER), College of Engineering, Science and Environment, The University of Newcastle, Callaghan 2308, Australia
e-mail: mdharunur.rashid@uon.edu.au

Cooperative Research Centre for Contamination Assessment and Remediation of the Environment (CRC CARE), ATC Building, The University of Newcastle, Callaghan, Newcastle 2308, Australia

M. H. Rashid
Bangladesh Agricultural Research Institute (BARI), Gazipur 1701, Bangladesh

M. M. Rahman
Department of General Educational Development, Faculty of Science and Information Technology, Daffodil International University, Ashulia, Savar, Dhaka 1207, Bangladesh

A. K. Jha and N. Kumar (eds.), *Cadmium Toxicity in Water*, Springer Water,
https://doi.org/10.1007/978-3-031-54005-9_14

1 Introduction

Heavy metal contamination is a great concern worldwide [11, 12]. Cadmium (Cd) is one of the major toxic elements due to its pervasiveness in the environment. It translocates easily into tissues [7] and declines biomass yield of crops [16] through leaf rolling, chlorosis [25] and diminishing of gas exchange rate [23]. It produced excessive amount of reactive oxygen species (ROS), free radicals [4] and lower of antioxidant enzyme activities [1] resulted in alteration of lipid composition. It hampers photochemical and carboxylation reactions of photosynthesis, affects chloroplast metabolism [10, 23], stomatal conductance, transpiration etc. which lead to reduced photosynthesis [5].

Zinc (Zn) is an essential element for plants and humans. Being a regulatory co-factor and structural body in proteins and enzymes it is essential for many biochemical pathways [2, 6]. Having similarity with Cd, it reduces the harmful effects of Cd. It sustains plants' agronomic performance through enhancement of photosynthetic pigments, other growth hormones and other yield contributing traits. Mung bean is an important pulse crop and sensitive to Cd stress. Soil parameters also affect mung bean yield and quality to a great extent. Information on the effect of Zn supply on growth and yield performance of Mung bean under Cd contamination in different soil types is meagre. This investigation was carried out to ascertain the influence of basal Zn supply on photosynthetic pigments, yield and yield attributes of Mung beans under Cd stress.

2 Materials and Methods

2.1 Experimental Details

Jade-AU, a variety of Mung beans of Australian Mung bean Association (AMA) was used for this study. Mung beans was grown in a greenhouse (25/20 °C day/night temperature, 75% relative humidity, 16/8 h photoperiod). Plastic pots (diameter: 24 cm, height: 30 cm) were filled with 4 kg of soils. Tenosol and Dermosol were collected from Taree, New South Wales (32 87′ 58.3″ S, 151 64′ 00.6″ E) and Newcastle (32 52′ 34.7″ S, 151 38′ 15.8″ E), Australia. Tenosol was sand predominating while Dermosol with silt (Table 1). Zinc sulfate hepta hydrate and cadmium nitrate were used as Zn and Cd sources, respectively. The soils were previously spiked with Cd @ 2 and 4 mg kg^{-1} and aged for three months. To explore the effect, Zinc sulfate @ 0, 5, 10 and 20 mg Zn kg^{-1} was used as basal following a completely randomized design with three replicates.

Table 1 Physico-chemical properties of soils used for this study

Soil	pH (H_2O)	EC $\mu s\ cm^{-1}$	% C	% N	% S	Texture	% Sand	% Silt	% Clay	Cd (mg kg^{-1})	Zn (mg kg^{-1})	DTPA Zn (mg kg^{-1})	As (mg kg^{-1})	Pb (mg kg^{-1})	Cr (mg kg^{-1})
Dermosol	7.7	161	2.7	0.21	0.055	Silty loam	33.8	55	11.2	0.01	5.9	0.5	1.6	5.2	9.1
Tenosol	6.4	190	1.4	0.22	0.009	Sandy loam	63.8	23.8	12.5	0.05	37	5	6.2	8.8	6.2

2.2 Crop Husbandry

The soils of pots were prepared with hand drawn tillage equipment "khurpa", fertilized with essential plant nutrients along with Zn as per treatment combinations, and irrigated to reach the desirable moisture level for seed sowing. Fertilizers were applied at the rate of 23, 58 and 30 kg ha^{-1} of nitrogen, phosphorus, and potassium, respectively, using triple super phosphate, urea, and sulphate of potash as sources. Initially ten seeds were sown in each pot followed by a thinning to maintain 2 seedlings in each pot. All pots were uniformly irrigated with ultrapure water (ELGA PURELAB System, USA) to maintain approximately 70% of soil water holding capacity. Weeds were manually uprooted; pots were randomly replaced every three days to make sure the environmental conditions for each pot.

2.3 Yield and Yield Attributes

Plants were harvested at maturity to record the yield and yield attributes. The shoot length and pod length were taken using a scale and averaged. The plants were sun-dried, and the biological yield was calculated. The pods were harvested manually and calculated the grain yield per plant at 10% moisture content. Grains per plant was counted and weighed from which 100-grain weight was subsequently estimated. Harvest index was measured based on grain yield and the biological yield.

2.4 Photosynthetic Pigments

Chlorophyll a, b and total chlorophyll were measured with standard method [3] using the following formula:

$$\text{Chlorophyll a}(\text{mg g}^{-1}, \text{fw.}) = (12.7 \times A_{663} - 2.69 \times A_{645}) \times X/1000 \times n \quad \text{(i)}$$

$$\text{Chlorophyll b}(\text{mg g}^{-1}, \text{fw.}) = (22.9 \times A_{645} - 4.68 \times A_{663}) \times X/1000 \times n \quad \text{(ii)}$$

$$\text{Total Chlorophyll }(\text{mg g}^{-1}, \text{fw.}) = (20.2 \times A_{645} + 8.02 \times A_{663}) \times X/1000 \times n \quad \text{(iii)}$$

2.5 *Statistical Analysis*

Experimental data were analysed using JMP Pro 14.2.0 software [14]. Three-way full factorial analysis of variance (ANOVA) was applied to test the significance of the different treatments where means were compared using Student's t test at $p < 0.01$ or $p < 0.05$.

3 Results

3.1 *Agronomic Parameters*

3.1.1 Leaves Per Plant

Number of leaves per plants did not significantly differ due to added Zn and Cd contamination. However, mean effect of Zn showed a positive impact on plant growth for both the soils. Number of leaves was slightly decreased for the interactions between Zn and Cd as compared to the mean effect of Zn for Tenosol which was opposite for Dermosol (Fig. 1a).

3.1.2 Shoot Length

Length of shoot was significantly influenced by added Zn ($p < 0.05$) and interaction between soil and Cd ($p < 0.01$). Reduced shoot length was observed because of Cd for both the soils, with Tenosol predominating (24% less) as compared to the respective controls. Presumably, basal Zn supply enhanced the shoot growth to about 12 and 7% higher than no Zn supply (control) for Dermosol and Tenosol, respectively. Combined effect of Zn and Cd increased shoot length to some extent which indicates that Zn application recovered the loss of growth impacted by Cd contamination (Fig. 1b).

3.1.3 Days to First Flowering

Number of days to first flowering was significantly ($p < 0.05$) affected due to Cd contamination and its interaction with soil. For Dermosol, flower initiation significantly delayed due to added Zn and Cd to soil as compared to control. Late flowering was occurred for Dermosol than Tenosol. However, no significant effect of Zn and Cd regarding emergence of flowering for Tenosol was observed (Fig. 1c).

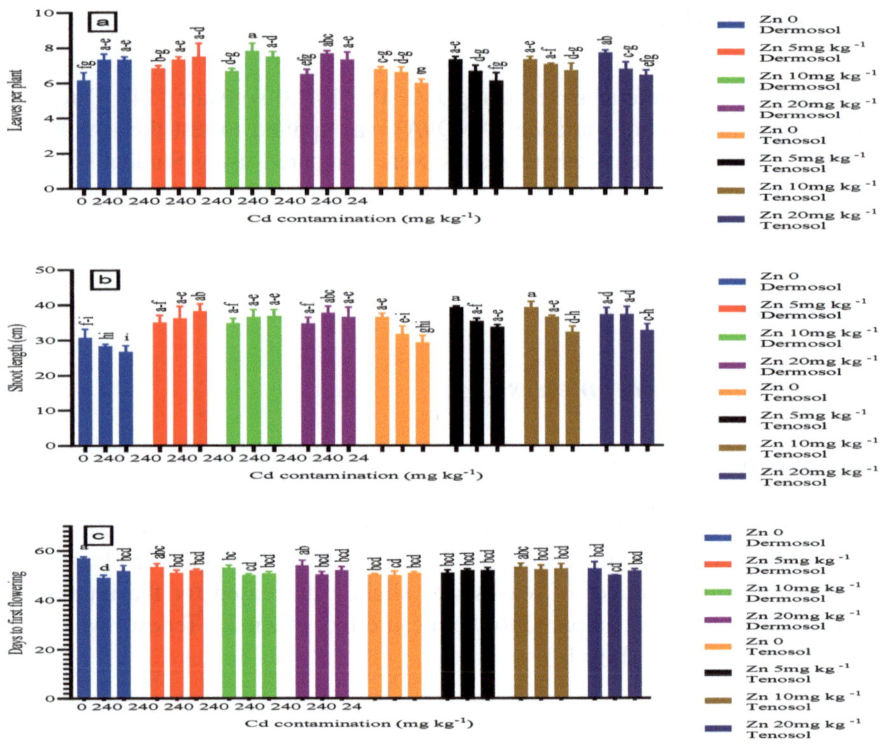

Fig. 1 **a–c** Agronomic parameters as affected by soil types, added Zn and Cd contamination

3.2 *Pod Parameters*

3.2.1 Pod Length

Interaction between soil types and Cd significantly ($p < 0.01$) influenced the length (mm) of pods in Mung beans. At control, Tenosol produced longer pods (90.5 mm) than Dermosol (54.3 mm). To investigate the effect of Zn and Cd on Dermosol it was clearly observed that sole Zn, Cd, and their interactions produced longer pods as compared to control. Length of pods gradually increased by basal Zn supply to soils up to 20 mg kg^{-1} (67.2 mm) for Dermosol and up to 5 mg kg^{-1} (92.5 mm) for Tenosol as compared to the respective controls. On the other hand, added Cd to the soils demonstrated contrast scenario for pod length where upward trend was observed for Dermosol and downward trend for Tenosol. Interaction effect of Zn and Cd caused a steady increase of pod length for Tenosol but not for Dermosol (Fig. 2a).

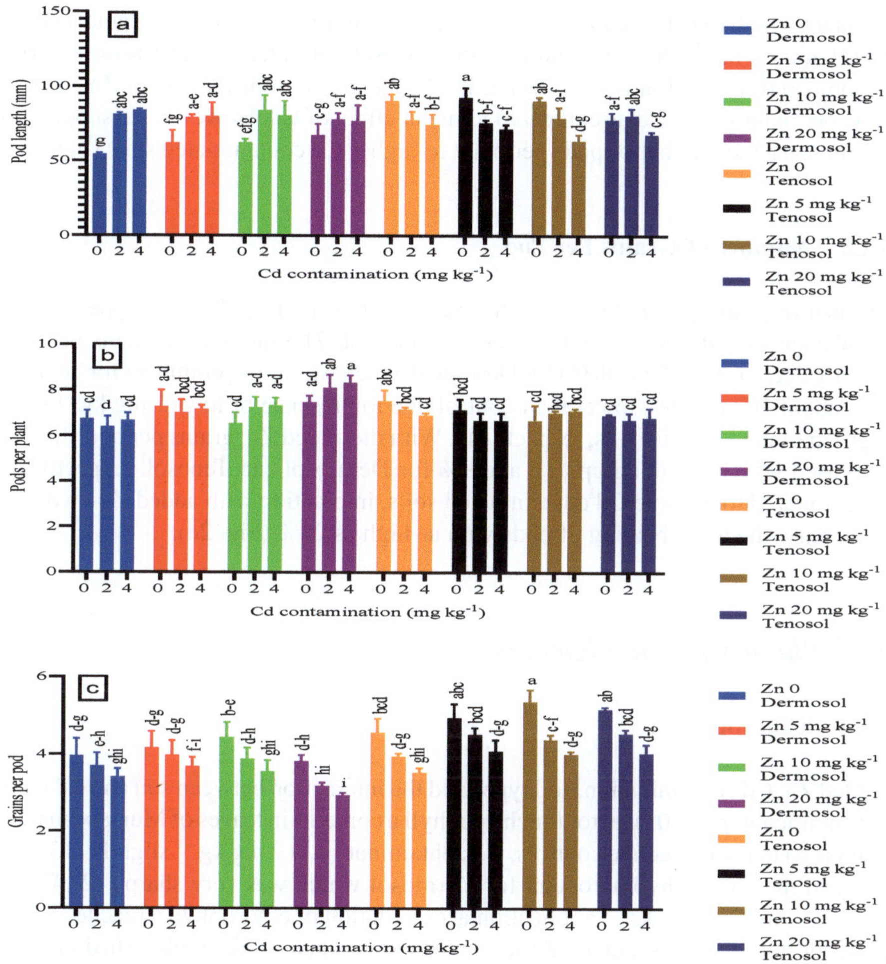

Fig. 2 a–c Pod parameters as affected by soil types, added Zn and Cd contamination

3.2.2 Pods Per Plant

Added Zn, Cd contamination and soil types were the significant factors ($p < 0.01$) for the number of pods per plant. In Dermosol the presence of low (2 mg Cd kg^{-1}) and high (4 mg Cd kg^{-1}) Cd decreased number of pods per plant by 8% and 24%, respectively as compared to control. Basal Zn supply to Dermosol gradually increased pod number and doubled (14.03) at 10 mg Zn kg^{-1}. As compared to Cd control, number of pods increased in case of its interactions with Zn which indicates that application of Zn enhanced crop growth affected by Cd stress. Cd also decreased the number of pods than Cd control, although no significant difference for the two levels of Cd contamination was observed. Sole application of Zn gradually increased pods per plant which reached to the peak (11.67) at 10 mg Zn kg^{-1}. With the same level of Zn (10 mg Zn ha^{-1}) the higher number of pods were produced for Dermosol (14.03) as compared for the Tenosol (11.67). Like Dermosol, combined effect of Zn and Cd, pods per plant increased as compared to mean effect of Cd. However, Zn supply did not increase the number of pods per plant to higher Cd contaminated soils (Fig. 2b).

3.2.3 Number of Grains Per Pod

Number of grains per pod was significantly ($p < 0.01$) affected by soil types, added Zn, Cd stress and interaction between soil and Cd. The number of grains per pod slightly decreased (3%) with Cd for Dermosol which was more prominent for Tenosol e.g., about 16 and 38% lower than control due to presence of low (2 mg kg^{-1}) and high Cd (4 mg kg^{-1}) in soils, respectively. With the added Zn, grains per pod slightly increased with amounting up to 6 and 9% for Dermosol and Tenosol, respectively. As compared to the sole Cd contaminated soils, interaction with added Zn favoured to increase the grain bearing of pods with as high as 24% (Fig. 2c).

3.3 Photosynthetic Pigments

3.3.1 Chlorophyll a

Added Zn, Cd contamination, soil types and the interaction between soil and Cd were the significant ($p < 0.01$) factor for chlorophyll a contents in leaves of Mung beans. As compared to the respective controls, Cd contamination at 2 mg kg^{-1} slightly (13.2%) reduced the chlorophyll a content for Dermosol which was very sharp (127%) for Tenosol. At 10 mg Zn kg^{-1} application accumulation of chlorophyll a reached to the peak for both the soils. Added Zn to the Cd contaminated soils replenished the loss of chlorophyll a content which was more prominent for Tenosol (Fig. 3a).

3.3.2 Chlorophyll b

Concentration of chlorophyll b was differed ($p < 0.01$) with soil types, added Zn, Cd contamination and interaction between soil and Cd. While Cd stress at lower (2.0 mg kg^{-1}) and higher levels (4.0 mg kg^{-1}) decreased chlorophyll b by 12% (2.90 mg g^{-1}, fw.) and 24.9% (2.48 mg g^{-1}, fw.) for Dermosol, the same toxicity reduced about 57% (1.53 mg g^{-1}, fw.) and 72% (2.54 mg g^{-1}, fw.), for Tenosol, respectively as compared to the respective controls. Added Zn upscaled the chlorophyll b content in both the soils, however, the effect was identical among the Zn levels. As compared to the elevated Zn applications to the soils, its interaction with Cd declined the content, with more prominent for the Tenosol than the Dermosol (Fig. 3b).

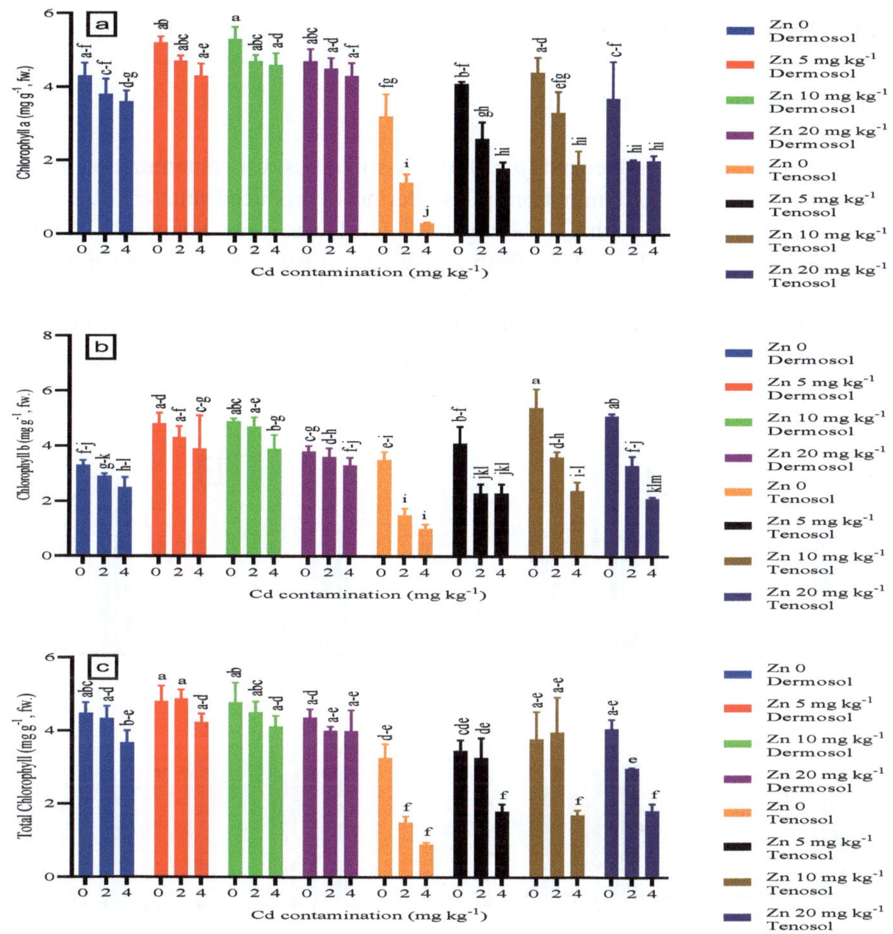

Fig. 3 **a–c** Photosynthetic pigments as affected by soil types, added Zn and Cd contamination

3.3.3 Total Chlorophyll

Concentration of total chlorophyll was significantly differed ($p < 0.01$) with soil types, added Zn, Cd stress. Prevalence of 2 mg Cd kg^{-1} in soil, total chlorophyll contents decreased by 3.13% (4.33 mg g^{-1}, fw.) and 54% (1.49 mg g^{-1}, fw.) for Dermosol and Tenosol, respectively. Further Cd contamination caused about 18% and 73% decrease in total chlorophyll content for Dermosol and Tenosol, respectively. The increment of total chlorophyll due to sole Zn supply was significant for Dermosol and non-significant for Tenosol. Zn supply to both the soils replenished the loss of total chlorophyll content caused by Cd contamination. Irrespective of treatments and control, Mung bean leaves in Dermosol contained significantly higher total chlorophyll contents (Fig. 3c).

3.4 *Yield Parameters*

3.4.1 100 Grain Weight

100 grain weight was influenced ($p < 0.01$) by soil types, Cd contamination and by their interactions. Cd contamination spiked Dermosol produced smaller sized grains for Cd contamination, with Cd@4 mg kg^{-1} level was more noticeable (5.99 g) as compared to control (7.43 g). With Zn supply to soils, grain weight gradually increased to reach to the peak (8.21 g) from control (7.43 g). However, added Zn to Cd contaminated soils produced slightly bigger grains as compared to that of sole Cd treatments. 100 grain weight for the Tenosol follows the same trend as of Dermosol (Fig. 4a).

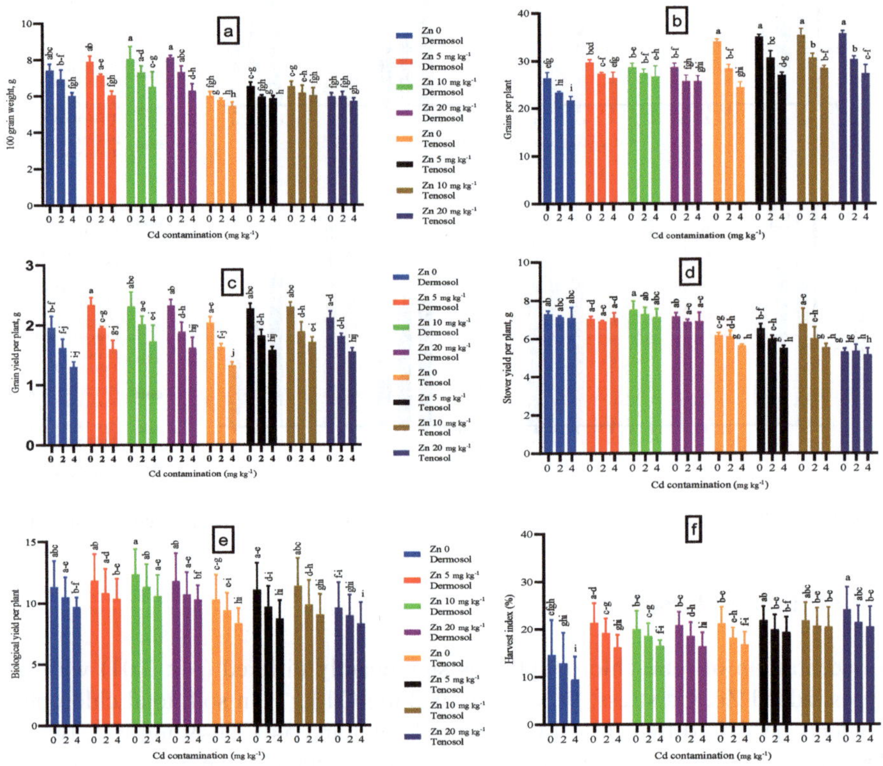

Fig. 4 **a–f** Yield parameters as affected by soil types, added Zn and Cd contamination

3.4.2 Number of Grains Per Plant

Number of grains per plant was significantly ($p < 0.01$) influenced by soil types, added Zn, Cd contamination and interaction between soil and Cd. Up to 18 and 28% decline of grain per plant was observed due to Cd toxicity in Dermosol and Tenosol, respectively as compared to control (26.3). Mean effect of Zn increased grains per plant gradually and reached to the peak at 5 mg Zn kg^{-1} application for Dermosol (29.7) and at 20 mg Zn kg^{-1} for Tenosol (35.7). However, as compared to the individual Zn application, combined presence of Zn and Cd reduced the number of grains per plant, might be due to the detrimental effect of Cd in soils. Combined application of Zn and Cd to Dermosol and Tenosol brought a significant benefit by increasing grain bearing as compared to the sole Cd contaminated soils. Up to 23 and 17% increased grains per plant was observed for Dermosol and Tenosol, respectively due to the interaction between 10 mg Zn kg^{-1} and 4 mg Cd kg^{-1} (Fig. 4b).

3.4.3 Grain Yield Per Plant

Grain yield (g) per plant was affected ($p < 0.01$) by added Zn, Cd contamination. However, the effect of interactions was not significant. As compared to soil, grain yield declined up to 35% which were about 1.30 g and 1.32 g per plant for Dermosol and Tenosol, respectively due to presence of Cd. The mean effect of Zn showed its beneficial effect to gain higher yield for both the soils. Up to 19 and 13% increased yield were observed for Dermosol and Tenosol, respectively. Prevalence of Cd in soils retarded seed yield considerably as compared to control which were compensated to some extent when Zn was applied. Application of 10 mg Zn kg^{-1} to 2 mg Cd kg^{-1} treated soil of Dermosol and Tenosol boosted the grain yield by about 24 and 16%, respectively than the mean effect of Cd. The interaction between Zn and higher Cd contamination, the increase of yield was about 30% for both the soils than the mean effect of higher Cd treatment (4 mg kg^{-1}) soil (Fig. 4c).

3.4.4 Stover Yield Per Plant

Stover yield of Mung beans was significantly affected by soil ($p < 0.01$), Zn ($p < 0.05$) and Cd contamination ($p < 0.05$). Due to presence of Cd contamination, stover yield significantly differed between two soils. As compared to control application of Zn up to 10 mg kg^{-1} increased the stover yield which was more prominent for Dermosol (7.52 g) than Tenosol (6.75 g). As compared to sole Zn application, combinedly Zn and Cd in the soils exerted less effect (Fig. 4d).

3.5 Biological Yield Per Plant

Biological yield (g) per plant was significantly ($p < 0.01$) affected by soil types, added Zn, Cd contamination although the interactions were not significant. While the main effect of Cd was the significant factor to decline the biological yield (about 11%) for Dermosol, more reduction (up to 18%) was observed for the Tenosol. Presumably, sole application of Zn increased the biological yield of Mung beans which were about 6% and 9% higher as compared to control for Dermosol and Tenosol, respectively. Zn and Cd combinedly exerted a bit positive effect for Dermosol which was inconsistent for the Tenosol (Fig. 4e).

3.6 Harvest Index

Harvest index of Mung beans was influenced ($p < 0.01$) by soil, added Zn and Cd contamination. About 13 and 26% less harvest index were observed for the presence of lower (2 mg kg^{-1}) and higher (4 mg kg^{-1}) contaminations of Cd for Dermosol which were about 15 and 24% for Tenosol, respectively. Although the mean effect of Zn on harvest index for both the soils were identical, the effect was more prominant for Dermosol than that of for Tenosol. The detrimental effect due to Cd contamination was compensated by the added Zn for both the soils. The harvest index was about 24% higher (19.4) in the interaction between 10 mg Zn kg^{-1} and 4 mg Cd kg^{-1} as compared to the mean effect of Cd for Dermosol. On the other hand, the increment was 25% higher (23.6) for the same levels of Zn and Cd for Tenosol (Fig. 4f).

4 Discussion

The experiment was carried out with a view to ascertain the effect of basal Zn supply to replenish the yield loss due to Cd contamination to soils. Added Zn boosted the agronomic performance of Mung beans. Shoot length increased up to 12% due to added Zn to soils. Plant height in green gram was increased with Zn at 15 kg ha^{-1}, revealed in a study by Usman et al. [22]. This might be for added Zn that produces auxin responsible for enhanced cell growth, improved plant height [9]. Number of grains in each pod was increased by 12% and 18% due to added Zn as compared to control for Dermosol and Tenosol, respectively. Nadergoli et al. [18] observed higher grains per pod and pods bearing branches where $ZnSO_4$ @ 20 kg ha^{-1} was applied in soil. Added Zn might be behind it, as it accelerates stamens and pollens which enhanced production of more sympodial branches. Added Zn increased 100-seed weight (up to 9%), seeds per plant (up to 13%), seed yield (up to 18%) as compared to control. Increased grain weight, grains per pod were observed in chickpea [15, 19]. Verma and Mishra [24] observed improved grain weight in Mung beans by supplying Zn and boron fertilizer together. Results revealed that Zn levels (5 mg Zn kg^{-1}) as basal application improved the biological yield and grain yield of Mung beans. Greater biological yield was observed in Maize due to basal Zn supply [20]. Khorgamy and Farnia [15] found the positive effect of Zn application on the harvest index which supports this study. Our findings are also in agreement with Montenegro et al. [17] who reported that among Zn, molybdenum (Mo) and boron (B) applications, added Zn produced the highest harvest index of chickpeas.

This study revealed that up to 35% losses in grain yield and up to 18% in biological yield for Cd contamination in soils. It is predictable that under Cd-stressed condition, decline in growth performance and photosynthesis happened [5, 16] through demonstrating chlorosis, necrosis [8]. The reduction of yield is highly related to the reduced photosynthetic activity in Mung bean leaves due to Cd stress (Fig. 2a–c) as photosynthetic systems in plants are specific targets of Cd [4, 25]. Chlorophylls and carotenoids favoured light harvesting and chlorophyll-to-carotenoid ratio is important for optimum light energy capture in photosynthesis [26].

Soil pH influences nutrient availability and thereby growth performance of plants [13]. Here the Dermosol is characterised with high pH (pH 7.7) as compared to the Tenosol (pH 6.4). Although seed yield per plant of the soils were identical, Dermosol produced higher stover yield which is supported by a recent study in rice [21].

5 Conclusions

About 30% grain yield was declined due to Cd contamination. Basal Zn supply reduced the negative effects of Cd to Mung beans by enhancing production of photosynthetic pigments, increasing pods per plant, grains per plant and seed size. Among the Zn treatments, Zn @ 5 mg kg^{-1} was found to be effective, to enhance mung bean growth and yield (up to 26% increase) under Cd stress. However, investigations at the field level with different crops, soil types, and other Zn sources are warranted.

6 Data Availability Statement

The data that support this research will be shared upon reasonable request to the corresponding authors.

Acknowledgements The first author is supported by the UNIPRS, and UNRS Central Scholarship and RTP allowances from The University of Newcastle, Australia. MHR duly acknowledges the BARI, Bangladesh, for the study leave. In addition, the authors are thankful to GCER Laboratory for providing laboratory and instrumental supports to conduct this research.

Author Contribution MHR has conceived the concept and written original draft. MMR and RN were involved in writing and editing of the manuscript.

Funding The research was partly funded by the International Zinc Association and CRC-CARE (G2000677).

Institutional Review Board Statement: Not applicable.

Informed Consent Statement: Not applicable.

Conflict of Interest The authors declare no competing interests.

References

1. Agrawal, V., & Sharma, K. (2006). Phytotoxic effects of Cu, Zn, Cd and Pb on in vitro regeneration and concomitant protein changes in *Holarrhena antidysenterica*. *Biologia Plantarum, 50*(2), 307–310.
2. Alloway, B. (2009). Soil factors associated with zinc deficiency in crops and humans. *Environmental Geochemistry and Health, 31*(5), 537–548.
3. Arnon, D. I. (1949). Copper enzymes in isolated chloroplasts. Polyphenoloxidase in Beta vulgaris. *Plant physiology, 24*(1), 1.
4. Balakhnina, T., Kosobryukhov, A., Ivanov, A., & Kreslavskii, V. (2005). The effect of cadmium on CO_2 exchange, variable fluorescence of chlorophyll, and the level of antioxidant enzymes in pea leaves. *Russian Journal of Plant Physiology, 52*(1), 15–20.
5. Bindhu, S., & Bera, A. (2001). Impact of cadmium toxicity on leaf area, stomatal frequency, stomatal index and pigment content in mungbean seedlings. *Journal of Environmental Biology, 22*(4), 307–309.
6. Cakmak, I., McLaughlin, M. J., & White, P. (2017). *Zinc for better crop production and human health*. Springer.
7. Carter, M. (2006). *Mineral nutrition of plants: Principles and perspectives, Emanuel Epstein, Arnold J. Bloom, Sinauer Associates, Sunderland, MA, USA (2005),(399 pp.) ISBN 0 87893 172 4*. Elsevier.
8. Dalla Vecchia, F., La Rocca, N., Moro, I., De Faveri, S., Andreoli, C., & Rascio, N. (2005). Morphogenetic, ultrastructural and physiological damages suffered by submerged leaves of *Elodea canadensis* exposed to cadmium. *Plant Science, 168*(2), 329–338.
9. Dashadi, M., Hossein, A., Radjabi, R., & Babainejad, T. (2013). Investigation of effect different rates phosphorus and zinc fertilizers on two cultivars lentil gachsaran and (flip92-121) in irrigation complement condition. *International Journal of Agriculture and Crop Sciences (IJACS), 5*(1), 1–5.
10. Ekmekçi, Y., Tanyolac, D., & Ayhan, B. (2008). Effects of cadmium on antioxidant enzyme and photosynthetic activities in leaves of two maize cultivars. *Journal of Plant Physiology, 165*(6), 600–611.
11. Helal, M., Baibagyshew, E., & Saber, S. (1998). Uptake of Cd and Ni by spinach, *Spinacea oleracea* (L.) from polluted soil under field conditions as affected by salt water irrigation. *Agronomie, 18*(7), 443–448.
12. Jamali, M., Kazi, T., Arain, M., Afridi, H., Jalbani, N., & Memon, A. (2007). Heavy metal contents of vegetables grown in soil, irrigated with mixtures of wastewater and sewage sludge in Pakistan, using ultrasonic-assisted pseudo-digestion. *Journal of Agronomy and Crop Science, 193*(3), 218–228.
13. Jaskulska, I., Jaskulski, D., & Kobierski, M. (2014). Effect of liming on the change of some agrochemical soil properties in a long-term fertilization experiment. *Plant, Soil and Environment, 60*(4), 146–150.
14. JMP (2019) *JMP®, Version 14.2.0* (pp. 1989–2019). SAS Institute Inc.
15. Khorgamy, A., & Farnia, A. (2009). *Effect of phosphorus and zinc fertilisation on yield and yield components of chick pea cultivars*. African Crop Science Society.

16. Mediouni, C., Benzarti, O., Tray, B., Ghorbel, M. H., & Jemal, F. (2006). Cadmium and copper toxicity for tomato seedlings. *Agronomy for Sustainable Development, 26*(4), 227–232.
17. Montenegro, J. B. V., Fidalgo, J. A. B., & Gabella, V. M. (2010). Response of chickpea (*Cicer arietinum* L.) yield to zinc, boron and molybdenum application under pot conditions. *Spanish Journal of Agricultural Research, 3*, 797–807.
18. Nadergoli, M., Yarnia, M., & Khoei, F. R. (2011). Effect of zinc and manganese and their application method on yield and yield components of common bean (*Phaseolus vulgaris* L. CV. Khomein). *Middle East Journal of Scientific Research, 8*(5), 859–865.
19. Rathi, B. K., Jain, A. K., Kumar, S., & Panwar, J. (2009). Response of Rhizobium inoculation with sulphur and micronutrients on yield and yield attributes of blackgram [*Vigna mungo* (L.) Hepper]. *Legume Research-An International Journal, 32*(1), 62–64.
20. Shahab, Q., Afzal, M., Hussain, B., Abbas, N., Hussain, S. W., Zehra, Q., Hussain, A., Hussain, Z., Ali, A., & Abbas, Y. (2016). Effect of different methods of zinc application on maize (*Zea mays* L.). *International Journal of Agronomy and Agricultural Research, 9*(3), 66–75.
21. Siddique, A. B., Rahman, M. M., Islam, M. R., Mondal, D., & Naidu, R. (2021). Response of iron and cadmium on yield and yield components of rice and translocation in grain: Health risk estimation. *Frontiers in Environmental Science, 9*, 716770.
22. Usman, M., Tahir, M., & Majeed, M. A. (2014). Effect of zinc sulphate as soil application and seed treatment on green gram (*Vigna radiata* L.). *Pakistan Journal of Life and Social Sciences, 12*(2), 87–91.
23. Vassilev, A., Perez-Sanz, A., Semane, B., Carleer, R., & Vangronsveld, J. (2005). Cadmium accumulation and tolerance of two Salix genotypes hydroponically grown in presence of cadmium. *Journal of Plant Nutrition, 28*(12), 2159–2177.
24. Verma, R., & Mishra, P. (1999). Effect of doses and methods of boron application on growth and yield of mungbean. *Indian Journal of Pulses Research, 12*(1), 115–118.
25. Wahid, A., Ghani, A., Ali, I., & Ashraf, M. (2007). Effects of cadmium on carbon and nitrogen assimilation in shoots of mungbean [*Vigna radiata* (L.) Wilczek] seedlings. *Journal of Agronomy and Crop Science, 193*(5), 357–365.
26. Wahid, A., & Ghazanfar, A. (2006). Possible involvement of some secondary metabolites in salt tolerance of sugarcane. *Journal of Plant Physiology, 163*(7), 723–730.

MIX
Papier aus verantwortungsvollen Quellen
Paper from responsible sources
FSC® C105338

FSC
www.fsc.org

If you have any concerns about our products,
you can contact us on
ProductSafety@springernature.com

In case Publisher is established outside the EU,
the EU authorized representative is:
Springer Nature Customer Service Center GmbH
Europaplatz 3, 69115 Heidelberg, Germany

Printed by Libri Plureos GmbH
in Hamburg, Germany